WATER CHEMISTRY

WATER CHEMISTRY

VERNON L. SNOEYINK
University of Illinois, Urbana

DAVID JENKINS
University of California, Berkeley

JOHN WILEY & SONS, New York · Chichester · Brisbane · Toronto

Library of Congress Cataloging in Publication Data:

Snoeyink, Vernon L.
  Water Chemistry

    Includes bibliographies and index.
    1. Water chemistry.   I. Jenkins, David, 1935–
joint author.   II. Title.

QD169.W3S66   546'.22'024628   79-21331
ISBN 0-471-05196-9

Printed in the United States of America

10 9 8 7 6 5 4 3 2 1

To Jeannie, Todd, and Craig
Joan, Daniel, and Sarah

PREFACE

This book is a first-level text on water chemistry that places special emphasis on the chemistry of natural and polluted waters and on the applied chemistry of water and wastewater treatment. It provides a comprehensive coverage of the dilute aqueous solution chemistry of acid-base reactions, complex formation, precipitation and dissolution reactions, and oxidation-reduction reactions. Although it is written primarily for students and those individuals who are currently working in environmental or sanitary engineering, much of its contents should also prove to be of value to chemists, biologists, ecologists, and geochemists. We assume that the readers will have a good background in general chemistry; additional knowledge in the fields of analytical and physical chemistry is useful but not essential. The material in the text has been satisfactorily taught to advanced undergraduates and first-level graduate students in environmental engineering with one year of elementary chemistry as a background.

This book provides an integrated coverage of chemical kinetic and equilibrium principles and applies them to water chemistry in its broadest sense, that is, to natural and polluted water and to water and wastewater treatment processes. Special attention is paid to the effects of temperature and ionic strength on reactions. In the first introductory chapter, the characteristics of the various types of water of concern are presented along with a discussion of the concentration units unique to water chemistry. Chapters 2 and 3 provide the background in chemical kinetics and thermodynamics necessary for the subsequent coverage of water chemistry. If the text is to be used for teaching students with background in physical chemistry, Chapters 2 and 3 may be read for review purposes only. Chapters 4 to 6, and 7 deal, respectively, with acid-base, complexation, precipitation-dissolution, and oxidation-reduction chemistry. In each of these chapters we first present the necessary principles for understanding the systems to be encountered and then the necessary tools to solve problems; following this, we present one or more applications of the topic in the water or wastewater areas. The graphical solution of equilibrium problems is stressed throughout the book. The acid-base chemistry chapter concludes with a detailed discussion of the carbonate system. The complexation chapter deals at length with the organic and inorganic complexes of metal ions found in natural waters. Phosphate chemistry and heterogenous calcium carbonate equilibria and kinetics are presented as applications of the principles of precipitation-dissolution presented in Chapter 6. In Chapter 7, the topics of corrosion, iron chemistry, chlorine chemistry, biologically important redox reactions, and principles of electrochemical measurements are presented to illustrate the importance and practical application of the principles of oxidation-reduction reactions.

Numerous example problems are presented throughout the discussion as aids to the understanding of the subject matter. Many problems are presented at the end of the chapters. The answers to these problems are given in an appendix so that readers can test their knowledge. References to more detailed treatments of selected topics are given at the end of each chapter.

The *Water Chemistry Laboratory Manual* by D. Jenkins, V. L. Snoeyink, J. F. Ferguson, and J. O. Leckie, 3rd ed., John Wiley & Sons, Inc., New York, 1980, initially published by the Association of Environmental Engineering Professors, contains many experiments that are based on the principles presented in this book. The laboratory manual is written for use in the same type of course for which this text is written. The nomenclature and format of these two texts are designed to make them companion volumes for an integrated lecture-laboratory course.

Our thanks are due to the students and many other persons who aided in the preparation of this book: to J. F. Fitzpatrick, R. A. Minear, and R. R. Trussell, who were selected by the publishers to review the text; to the University of California at Berkeley and the University of Illinois at Urbana-Champaign for their generous leave policies that enabled us to devote our time to writing this book; and to Miss Cathy Cassells, Mrs. Ruth Worner, Mrs. Flora Orsi, and especially to Mrs. Virginia Ragle for their typing services in the preparation of the manuscript.

Vernon L. Snoeyink
Urbana, Illinois

David Jenkins
Berkeley, California

CONTENTS

WATER CHEMISTRY

CHAPTER 1

INTRODUCTION

1.1. PROPERTIES OF WATER

Although the title of this book is *Water Chemistry*, the contents deals only slightly with the chemistry of H_2O. With a few exceptions it is only in this introductory chapter that we feel compelled to even talk about the properties of H_2O. In common with the usage adopted by the sanitary/environmental engineering profession and other branches of the scientific community, we use the word *water* to mean the dilute aqueous solution/suspension of inorganic and organic compounds that constitutes various types of aquatic systems. This book is about the interactions between these compounds. *The American Heritage Dictionary of the English Language*[1] concisely draws the distinction we want to make in its first three definitions of the word *water*.

"*Water* (wo'tar, wot'ar) *n.* *1.* A clear colorless, nearly odorless and tasteless liquid, H_2O, essential for most plant and animal life and the most widely used of all solvents. Melting point 0°C (32°F), boiling point 100°C (212°F), specific gravity (4°C) 1.0000, weight per gallon (15°C) 8.337 pounds. 2. Any of various forms of water such as rain. *3.* Any body of water such as a sea, lake, river, or stream."

This book is largely directed toward a consideration of the types of waters exemplified by definitions 2 and 3. However, it would be remiss if we did not consider the important properties of water that make it such a dominant aspect of our environment.

Most water molecules have a molecular weight of 18. However, since hydrogen and oxygen each have 3 isotopes, there exist 18 possible molecular weights for water. In the water molecule both hydrogen atoms are located on the same side of the oxygen atom; their bonds with the oxygen atom are 105° apart. The hydrogen atoms carry a positive charge while the oxygen atom is negatively charged. Because of this distribution of charge, H_2O is a strongly dipolar molecule. The water molecule dipoles

[1] *The American Heritage Dictionary of the English Language*, William Morris, ed. American Heritage and Houghton, 1969, p. 1447.

attract each other and form aggregates through bonds that are known as "hydrogen bonds":

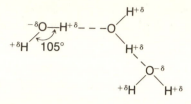

It is thought that these aggregates in water at room temperature can reach sizes of up to about 100 H_2O molecules.[2]

The hydrogen bonding in water is responsible for many of the unusual properties possessed by this substance. Water is the dihydride of oxygen. If we compare it with the dihydrides of the elements in the same family of the periodic table as oxygen, that is, hydrogen sulfide, H_2S; hydrogen selenide, H_2Se; and hydrogen telluride, H_2Te, we find that many of its physical properties are anomalous. At atmospheric pressure and room temperature (25°C) the heavier molecules, H_2S (molecular weight 34), H_2Se (molecular weight 81), and H_2Te (molecular weight 130), are all gases. Water is a liquid that becomes a gas only when the temperature is increased to 100°C and above. It is far denser than its related species at any given temperature; the maximum density is at 4°C. Its surface tension and dielectric constant are much higher than would be predicted from the properties of the other dihydrides. Its freezing point is lower than would be expected, and it freezes to form ice, an open-structured substance that is less dense than the liquid water from which it forms. All of these properties (and many more) are caused by the hydrogen bonding between H_2O molecules. The last property—that in which H_2O forms a less dense solid than the liquid from which it forms—has far-reaching ramifications. If solid H_2O were denser than liquid H_2O, ice would form at the bottom of natural bodies of water rather than at the top. Lakes would freeze from the bottom upward and, consequently, life in its present form in aquatic systems would not exist because natural bodies of water would freeze solid whenever the temperature fell below the freezing point of water.

The polarity of water is an important factor in determining its solvent properties. The minerals that make up the earth's crust are largely inorganic solids in which positively charged and negatively charged ions exist in a lattice structure, attracted to each other by electrostatic bonds. Water, with its dipolar character, has the power to surround a positively charged ion with the negatively charged part of its molecule (or conversely surround the negatively charged crystal ion with the positively charged part of its molecule), thereby isolating the ion from its surrounding ions

[2] G. Nemethy and G. H. Scheraga, "Structure of Water and Hydrophobic Bonding in Proteins. I. A Model for the Thermodynamic Properties of Liquid Water," *J. Phys. Chem.*, 36: 3382 (1962).

and neutralizing the forces of attraction that maintain the integrity of the crystal structure. The ion, surrounded (or hydrated) by water molecules, can then leave the crystal lattice and move out into solution—it becomes a dissolved ion.

In the last analysis the solvent properties of water make necessary the several dictionary definitions of the word *water*. Water dissolves some (small or large) amount of virtually every solid or gas with which it comes in contact. In the global cycle of water on the earth (the hydrologic cycle), water contacts the gases in the atmosphere (including air pollutants and volcanic emissions) and the minerals in the top few kilometers of the earth's crust. On a smaller scale water circulates in man-made systems (conduits and pipes made from synthetic minerals, such as concrete, and refined metals such as iron and copper). The solvent powers of water are exerted in these systems leading to such general phenomena as corrosion and scaling.

1.2. COMPOSITION OF SEVERAL TYPES OF WATER

The general composition of the various types of water in the hydrosphere can best be discussed within the framework of the hydrologic cycle. Of the total amount of water on the earth the oceans account for the vast majority: 97.13 percent. The polar ice caps and glaciers contain 2.24 percent; groundwater accounts for 0.61 percent; and the rivers, lakes, and streams contain only 0.02 percent of the total.

The ocean is an approximately 1.1 M solution of anions plus cations; its average composition is given in Table 1-1. In this table the category

TABLE 1-1   Major Constituents of Seawater

Constituent	mg/kg (ppm)
Sodium (Na^+)	10,500
Magnesium (Mg^{2+})	1,350
Calcium (Ca^{2+})	400
Potassium (K^+)	380
Chloride (Cl^-)	19,000
Sulfate (SO_4^{2-})	2,700
Bicarbonate (HCO_3^-)	142
Bromide (Br^-)	65
Other solids	34
Total dissolved solids	34,500
Water (balance)	965,517

Source: E. D. Goldberg, "Chemistry—The Oceans as a Chemical System" in H. M. Hill, Composition of Sea Water, Comparative and Descriptive Oceanography, Vol. 2 of The Sea. Wiley-Interscience, New York, 1963, pp. 3–25. Reprinted by permission of John Wiley & Sons, Inc.

of "other solids" incorporates a wide variety of species that includes just about every element present in the earth's crust. In this category average concentrations range from moderate (Sr (8 mg/liter), SiO_2 (6.4 mg/liter), B (4.6 mg/liter), and F (1.3 mg/liter)) to small (N (0.5 mg/liter), Li (0.17 mg/ liter), P (0.07 mg/liter), and I (0.06 mg/liter)) to minute (Cd (0.0001 mg/liter), Cr (0.00005 mg/liter), and Hg (0.00003 mg/liter)) to miniscule (Pa (2×10^{-9} mg/liter)and Ra (1×10^{-10} mg/liter)). The dissolved salt content of sea water (total dissolved solids of approximately 34,500 mg/liter) is sufficient to raise its specific gravity at 20°C to a value of 1.0243 g/cc, which is significantly greater than pure water.

From the oceans, water evaporates and then is transported over land masses, where it can be deposited as one or another form of precipitation (rain, snow, hail, etc.). During its passage from the ocean to the land surface, water passes through the earth's lower atmosphere. Because of this, the water has a chance to equilibrate with the gases in the atmosphere. The mean composition of the lower atmosphere is shown in Table 1-2. It should be realized that the mean composition of the earth's atmosphere is subject to considerable variation, especially in the levels of some of the minor constituents such as CO_2, CO, SO_2, NO_x, and so

TABLE 1-2   Mean Composition of the Atmosphere

Gas	Percentage by Volume	Partial Pressure (atm)
N_2	78.1	0.781
O_2	20.9	0.209
Ar	0.93	0.0093
H_2O	0.1–2.8	0.028
CO_2	0.03	0.0003
Ne	1.8×10^{-3}	1.8×10^{-5}
He	5.2×10^{-4}	5.2×10^{-6}
CH_4	1.5×10^{-4}	1.5×10^{-6}
Kr	1.1×10^{-4}	1.1×10^{-6}
CO	$(0.06–1) \times 10^{-4}$	$(0.6–1) \times 10^{-6}$
SO_2	1×10^{-4}	1×10^{-6}
N_2O	5×10^{-5}	5×10^{-7}
H_2	5×10^{-5}	5×10^{-7}
O_3	$(0.1–1.0) \times 10^{-5}$	$(0.1–1.0) \times 10^{-7}$
Xe	8.7×10^{-6}	8.7×10^{-8}
NO_2	$(0.05–2) \times 10^{-5}$	$(0.05–2) \times 10^{-8}$
Rn	6×10^{-18}	6×10^{-20}

Source: B. A. Mirtov, "Gaseous composition of the atmosphere and its analysis." Akad. Nauk. SSSR, Inst. Prikl. Geofiz Moskva (translated by the Israel Program for Scientific Translations, published in Washington, U.S. Dept. of Commerce, Office of Technical Services, 1961, 209 pp.).

forth, which are products of combustion processes and are associated with the air pollution that accompanies urban-industrial communities.

The major atmospheric constituents, N_2 and O_2, are both sparingly soluble in water (17.5 and 39.3 mg/kg, respectively, at 25°C)[3] but some of the minor constituents, for example, CO_2 and SO_2, are very soluble (1450 and 94,100 mg/kg, respectively, at 25°C). The composition of precipitation is quite variable and greatly influenced by atmospheric contaminants. Precipitation near the oceans contains more SO_4^{2-}, Cl^-, Na^+, and Mg^{2+} than precipitation that falls in the interior of a great land mass. Rain and snow that are generated from an atmosphere containing high concentrations of combustion-generated SO_2 can be very acidic. Even so, the figures in Table 1-3 show that rainwater is indeed an extremely dilute solution

TABLE 1-3   Composition of Rain and Snow (in mg/liter)

Constituent	1	2	3	4	5
SiO_2	0.0		1.2	0.3	
Al (III)	0.01				
Ca^{2+}	0.0	0.65	1.2	0.8	3.3
Mg^{2+}	0.2	0.14	0.7	1.2	0.36
Na^+	0.6	0.56	0.0	9.4	0.97
K^+	0.6	0.11	0.0	0.0	0.23
NH_4^+	0.0				0.42
HCO_3^-	3		7	4	0.0
SO_4^{2-}	1.6	2.18	0.7	7.6	6.1
Cl^-	0.2	0.57	0.8	17	2.0
NO_2^-	0.02		0.0	0.02	
NO_3^-	0.1	0.62	0.2	0.0	2.2
Total dissolved solids	4.8		8.2	38	
pH	5.6		6.4	5.5	4.4

1. Snow, Spooner Summit, U.S. Highway 50, Nevada (east of Lake Tahoe) altitude 7100 ft, Nov. 20, 1958. J. H. Feth, S. M. Rogers, and C. E. Roberson, Chemical Composition of Snow in the Northern Sierra Nevada and Other Areas, U.S. Geological Survey Water Supply Paper 1535J, 1964, 39 pp.
2. Average composition of rain from August 1962 to July 1963 at 27 points in North Carolina and Virginia. A. W. Gambell and D. W. Fisher, Chemical Composition of Rainfall, Eastern N. Carolina and Southeastern Virginia: U.S. Geological Survey Water Supply Paper 1535K, 1964, 41 pp.
3 and 4.   Rain, Menlo Park, Calif., 7:00 P.M. Jan. 9 to 8:00 A.M. Jan. 10, 1958. Whitehead, and J. H. Feth, Chemical Composition of Rain, Dry Fallout, and Bulk Precipitation at Menlo Park, Calif., 1957–1959, J. Geophys. Res., 69:3319–3333 (1964).
5. Station 526U, Belgium, European Atmospheric Chemistry Network. Average of 180 samples. L. Granat, On the Relation Between pH and the Chemical Composition in Atmospheric Precipitation, Tellus, 24, 550–556 (1972).

[3] The solubility of gas is expressed as the weight of gas (mg) dissolved in 1 kg of water at a total pressure (partial pressure of the indicated gas plus the vapor pressure of water at 25°C) of 760 mm Hg (760 torr or 1 atm).

of dissolved salts, indicating the remarkable efficiency of the distillation process that results in the production of "fresh water" from seawater.

Note that the pH of "unpolluted" precipitation (5.5 to 6.5) is significantly lower than that of the seawater from which it was generated. This is a reflection of the equilibration that has taken place with atmospheric CO_2. Rainwater is poorly buffered so that it cannot maintain its neutral pH in the presence of even these small amounts of acid-producing gases.

In precipitation, water falls onto the land surface and comes in contact with rocks, sediments, and soils and the animal and plant inhabitants of the land surface. Chemical reactions take place that further modify the composition of the water. The reactions can be viewed in general terms as a giant global acid-base titration in which the acids of the rainwater (CO_2, SO_2, and NO_x) titrate the bases of the rocks. Since the composition of the land surface and the extent and nature of biological activity (including man's activity) vary from one place to another, we can expect that waters of a variety of compositions will result from these reactions. The time and intimacy of contact between water and rocks also influences the composition of the solution.

Surface waters that originate in basins where the major rocks are granite contain very small amounts of dissolved minerals, not more than approximately 30 mg/liter. These are an important group of waters; for example, the water supplies of New York City (from the Catskill Mountains), San Francisco and Oakland (from the Sierra Nevada Mountains), Seattle (from the Cascade Mountains), and many rivers and lakes in New England are of this type. They are illustrated by Type A, in Table 1-4,

TABLE 1-4   Typical Analyses of Surface and Groundwaters in the United States

Constituent, mg/liter	A	B	C
SiO_2	9.5	1.2	10
Fe(III)	0.07	0.02	0.09
Ca^{2+}	4.0	36	92
Mg^{2+}	1.1	8.1	34
Na^+	2.6	6.5	8.2
K^+	0.6	1.2	1.4
HCO_3^-	18.3	119	339
SO_4^{2-}	1.6	22	84
Cl^-	2.0	13	9.6
NO_3^-	0.41	0.1	13
Total dissolved solids	34	165	434
Total hardness as $CaCO_3$	14.6	123	369

A.   Pardee Reservoir, East Bay Municipal Utility District, Oakland, Calif. Average data for 1976.
B.   Niagara River, Niagara Falls, N.Y.
C.   Well Water, Dayton, Ohio.

which is the yearly average composition of the outlet of Pardee Reservoir of the East Bay Municipal Utility District, the water supply for much of the East Bay Area of the San Francisco Bay region.

The second general type of surface water (Table 1-4, Type B) originates from basins other than the granite basins described above. This water, typified by that of the Great Lakes (except Lake Superior) is of intermediate hardness (calcium plus magnesium), alkalinity (HCO_3^-), and total mineral content. This type of water is widely used as public water supplies for cities such as Chicago, Cleveland, Buffalo, Niagara Falls, Detroit, Milwaukee, and many smaller places along the lakes and rivers of the St. Lawrence River Basin.

Groundwaters generally have higher dissolved mineral concentrations than surface waters. This is because of the intimate contact between the CO_2-bearing water and rocks and soils in the ground and the length of time for dissolution. Additionally, CO_2 may be added to the water in the soil by the activities of soil microorganisms. Example C in Table 1-4 is a groundwater from 30 to 60-ft-deep wells used for the public water supply of Dayton, Ohio. Waters such as this are widely distributed throughout the heartland of the United States. They usually require softening to make them acceptable for general domestic use and most industrial uses.

In an earth unpopulated by living things the natural waters formed by water-air-earth interactions would now flow back to the ocean carrying their various dissolved constituents and suspended clay and silt particles. For the moment let us assume that we have a sterile earth. On the land surface the acids of the atmosphere have attacked the bases of the rocks to produce a water containing dissolved minerals. The dissolution of these minerals from the rocks causes them to degrade and eventually to form clay minerals. Cations such as K^+, Na^+, Ca^{2+}, and Mg^{2+} are leached out of the rocks along with silica (SiO_2) leaving behind a clay mineral, for example, montmorillonite, illite, or kaolinite, that is less rich in silica than the parent rock. These particles of clay minerals can be washed down the rivers along with the dissolved salts into the ocean. It has been proposed by Mackenzie and Garrels[4] that in the oceans the reverse of the reaction that weathered the rocks takes place. Cations, bicarbonate ion, and dissolved silica and clay minerals react to generate new rocks, which are laid down at the bottom of the ocean. Although the detailed arguments in support of this view are beyond the scope of this book, it is believed that the constancy of the concentrations of SiO_2, K^+, and HCO_3^- in the ocean is attributable to this reaction. It is thought that this reaction controls the pH of the ocean at about pH 8.0. The reaction also helps to regulate the CO_2 content of the earth's atmosphere. Siever has summarized this global chemical scheme in the form of a series of chemical engineering processes (Fig. 1-1).

[4] F. T. Mackenzie and R. M. Garrels, "Chemical Mass Balance Between Rivers and Oceans," Am. J. Sci., 264: 507–525 (1966).

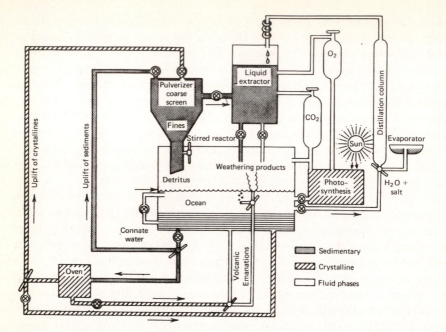

Fig. 1-1. Chemical engineering analogy to the surface geochemical cycle. From R. Siever, "Sedimentological Consequences of Steady-State Ocean-Atmosphere," *Sedimentology, 11,* 5–29 (1968). Reprinted by permission of Blackwell Scientific Publications Limited.

The interlocking nature of these reactions on a global scale has some important consequences for the composition of natural waters. For example, the concentrations of the major dissolved constituents (Ca^{2+}, Mg^{2+}, Na^+, Cl^-, SO_4^{2-}, HCO_3^-, $SiO_{2(aq)}$) vary over a rather narrow range because of the buffering action of these reactions. (See Fig. 1-2.) For the same reason, the composition of the ocean is fairly constant with respect to these major dissolved components.

Now let us inhabit the earth with plants and animals. Their activities are important in altering the composition of natural waters at all stages of the hydrologic cycle. We have already seen that man's industrial activities can increase the acidity of the atmosphere and the water that comes in contact with it through precipitation. This increased activity can dissolve more minerals than would be possible with precipitation from unpolluted air. It has resulted in depressed pH values and increased mineral contents in the poorly buffered lakes in the Scandinavian countries (polluted air from the industrial areas of northern Europe drifts north to Scandinavia) and in New England and Eastern Canada (polluted air from industrial sections of the United States drifts into New England and Canada).

Man's activities affect water quality in a variety of other ways including the discharge of municipal, industrial, and agricultural wastes into the surface- and groundwaters. Domestic wastes contribute increases in the mineral and organic matter content of natural waters. Typically, a single municipal use of a water will contribute about 300 mg/liter of total dissolved minerals to a water. The individual components that make up these dissolved minerals are not present in the same ratio that one would typically expect from the dissolution of rocks by water. Thus Table 1-5 shows that increases in Cl^- and Na^+, NH_4^+ and NO_3^- are disproportionately greater than increases in Ca^{2+}, Mg^{2+}, and SO_4^{2-}. The contribution of these dissolved salts by municipal reuse has an important influence on the degree to which wastewater can be recycled for reuse.

Significant contributions of dissolved salts also arise from irrigated agriculture. Again the quantities and concentrations of the specific types of dissolved minerals are in different ratios than would be expected for the dissolution of the minerals of rocks. An excellent example of the striking effect that water returned from agricultural irrigation can have on water quality is given in Table 1-6, which shows the water quality in the Upper Colorado River Basin as one progresses from the headwaters in western Colorado downstream through Utah to Arizona. The high salinity (TDS) of this river has led to both interstate and international disputes (between Mexico and the United States), since the upstream salt loads produced by irrigated agriculture and natural runoff make the water unusable for many purposes in the downstream regions.

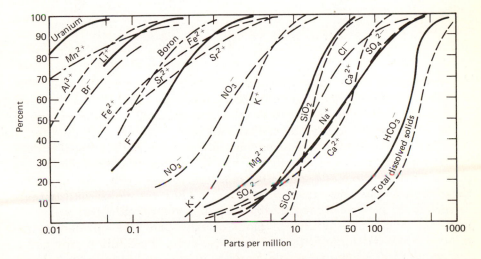

Fig. 1-2. Cumulative curves showing the frequency distribution of various constituents in terrestrial water. The data are mostly from the United States from various sources. From S. N. Davies and R. C. M. DeWiest, *Hydrogeology*, John Wiley, New York, 1966. Reprinted by permission from John Wiley & Sons, Inc.

TABLE 1-5   Increment Added by Municipal Use of Water,
Based on 22 U.S. Cities
(Tap Water to Secondary Effluent)

Constituent	Added Increment, mg/liter		
	Average	Minimum	Maximum
Cations			
Na^+	66	8	101
K^+	10	7	15
NH_4^+	15	0	36
Ca^{2+}	18	1	50
Mg^{2+}	6	Trace	15
Anions			
Cl^-	74	6	200
NO_3^-	10	−4.7	25.9
NO_2^-	1	0.1	2
HCO_3^-	100	−44	265
SO_4^{2-}	28	12	57
PO_4^{3-} (total)	24	7	50
PO_4^{3-} (ortho)	25	7.5	40
Others			
SiO_2	15	9	22
Hardness (as $CaCO_3$)	79	6	261
Alkalinity (as $CaCO_3$)	81	−36	217
Total solids	320	128	541
pH (in pH units)	−0.6	−1.7	0

Source. J. H. Neal, "Advanced Waste Treatment by Distillation,"
AWTR-7, U.S. Public Health Service Report 999-WP-9, 1964.

Wastewaters from industry, and domestic and agricultural sources
contain a wide variety of organic compounds (see Tables 1-7 and 1-8).
Typical domestic wastewater contains about 100 to 300 mg/liter of organic
carbon, 10 to 30 mg/liter of organic nitrogen, and 1 to 2 mg/liter of
phosphorus bound to organic compounds. By comparison natural surface
water concentrations of organic carbon, nitrogen, and phosphorus are
approximately 1 to 5, <1, and <0.5 mg/liter, respectively. The impact of
these organic materials from wastewater on natural water quality takes
a variety of forms. Biodegradable organic compounds can either be
removed in a waste treatment plant or they can be degraded by the
natural water flora and fauna. In the latter case, the dilution provided to
the wastewater by the receiving water is important because the concentrations of biodegradable organic matter in undiluted wastewater require
many times more oxygen for their aerobic degradation than is present in
water saturated with dissolved oxygen in contact with the earth's atmosphere. If sufficient dilution is not provided, the degradation of this organic

TABLE 1-6 Quality of Water in Upper Colorado River Basin

Station Location	Average Flow, cfs	Ca^{2+}	Mg^{2+}	Na^+	K^+	HCO_3^-	SO_4^{2-}	Cl^-	TDS	Boron
Hot Sulfur Springs, Col.	244	16	3.1	5.9	1.6	72	6.0	1.3	91	0
Near Dorsero, Col.	2399	48	9.8	40	2.2	120	77	54	334	0
Near Camero, Col.	4138	56	12	62	3.4	142	93	84	400	0.04
Near Asco, Utah	7639	70	25	79	3.6	178	281	11	660	0.08
Hite, Utah	14167	73	26	79	3.5	182	230	53	658	0.12
Lees Ferry, Ariz.	17550	76	25	74	4.0	184	235	48	580	0.11

Source. Data largely from W. V. Iorns, C. H. Hembree, and G. L. Oakland, "Water Resources of the Upper Colorado River Basin," U.S. Geological Survey Professional Paper 441, Washington, D.C., 1965.

TABLE 1-7   Composition of Stevenage, England, Raw Sewage

Constituent	Milligrams as Carbon per Liter
Fatty acids	71.0
Fatty acid esters	28.2
Proteins	31.0
Amino acids	5.0
Carbohydrates	55.0
Soluble acids	21.0
Amides	1.5
Anionic surface-active agents	14.0
Creatinine	3.5
Amino sugars	1.8
Muramic acids	0.2
Total identified	232
Total organic carbon	311
Proportion identified	75 percent

Source. H. A. Painter, M. Viney, and A. Bywaters, "Composition of Sewage and Sewage Effluents," J. Inst. Sewage Purif., 4:302 (1961). Copyright © Institute of Water Pollution Control, reprinted by permission.

matter can consume all the oxygen present and convert the water from an oxic state to an anoxic condition. This change can have far-reaching consequences for the chemistry of many species (see Chapter 7).

The presence of nonbiodegradable (persistent) organics in wastewaters affects water in ways that depend on the nature of the materials. For example, the products of microbial degradation (humic materials) can give the water a brown color and can modify metal ion behavior (see Chapter 5). Toxic materials (both inorganic and organic) can modify or eliminate biological processes occurring in natural waters. Organics of industrial origin may cause taste and odor, or they may have undesirable health effects if they are not removed during drinking water purification. Organics of all types may react with water treatment chemicals, and new organics that have undesirable effects may be produced.

Natural waters themselves are never free of organisms. Even water treated by disinfection for public supply contains microorganisms. The importance of organisms in natural water relates to the types of chemical transformations that they catalyze. We can divide these activities into two types: those that yield usable energy for the organism (catabolism) and those in which the organism expends energy (anabolism). Because the individual anabolic and catabolic reactions executed by microorganisms are too numerous to discuss individually, we will give a few examples to illustrate how they influence natural water quality.

TABLE 1-8   Typical Organic Compounds of Industrial Origin Found in Lower Mississippi River Water[a]

acetophenone	isopropylbenzene
a-camphanone	o-methoxy phenol
dicyclopentadiene	nitrobenzene
1,2-dimethoxybenzene	toluene
2,3-dimethylnaphthalene	o-cresol
2,3-dimethyloctane	phenyl cyclohexane
n-dodecane	1,3,5-trichlorophenol
4-ethyl pyridine	vinyl benzene
ethylbenzene	xylene

[a] Source. "Industrial Pollution of the Lower Mississippi River in Louisiana," Report of the U.S. Environmental Protection Agency, Region VI, Dallas, Tex., April 1972.

Photosynthesis is the reaction in which organisms utilize radiant energy (sunlight) to drive an oxidation-reduction reaction in which carbon dioxide is reduced to organic matter by water with the production of oxygen. This reaction is fundamentally important in natural systems because its occurrence dictates that the aquatic system cannot be treated as a closed system. Photosynthesis uses an external (i.e., nonchemical) source of energy to drive a reaction that is energetically impossible in a closed system. The occurrence of photosynthesis in natural systems also means that we may not treat them as equilibrium systems; we must find the *steady state* that results from the superimposition of the photosynthetic (nonequilibrium reaction) upon the chemical equilibria. Photosynthesis is ultimately responsible for most of the organic matter present on the earth and in natural waters; like many other biological reactions it affects water quality. Because photosynthetic organisms consume inorganic carbon (CO_2, HCO_3^-, and CO_3^{2-}) from solution, they reduce the alkalinity and increase the pH of waters in which they grow. Substances whose solubility is inversely related to pH (e.g., $Mg(OH)_{2(s)}$ and $CaCO_{3(s)}$) may precipitate from a photosynthesizing natural water. Photosynthesis produces oxygen so that we would expect oxic conditions in a water containing actively photosynthesizing organisms.

Other energy-producing reactions of organisms involve the reduction of oxygen to water, the reduction of nitrate to ammonia and nitrogen gas, the reduction of sulfate to sulfide, and the reduction of carbon dioxide to methane. All of these reactions can exert a profound effect on water quality especially when it is realized that the affected chemical species also engage in many other chemical reactions. For example, the sulfide ion forms precipitates with many heavy metals. The microbial reduction of sulfate to sulfide could be accompanied by a reduction in the dissolved heavy metal content in a natural water.

The anabolic activities of aquatic organisms also affect water quality.

Photosynthetic aquatic organisms withdraw inorganic carbon from solution to provide building blocks for the organic compounds they synthesize. Nonphotosynthetic organisms build their cell material from
organic compounds, thereby modifying natural water organics. Since the
construction of cell material requires the use of elements such as nitrogen,
phosphorus, and potassium beside carbon, hydrogen, and oxygen, the
growth of organisms in natural waters will modify the concentrations of
these so-called nutrient elements. The growth of algae in a productive
natural water can nearly completely deplete the nitrogen and phosphorus
concentrations of the water; in fact, the situation often exists where the
amount and rate of growth of aquatic photosynthetic organisms is
controlled by the concentrations of nutrients such as nitrogen (NH_4^+, NO_3^-)
and phosphorus (PO_4^{3-}).

In summary, the chemical reactions of atmosphere and the solids in
the rocks of the land surface form dilute mineral solutions that we call
surface water and groundwater. These waters and the degraded minerals
that result from their production flow into the ocean where a reverse
reaction to the weathering that produced them occurs. An ocean of the
present composition results and new rocks are produced. Man's activities
and the activities of organisms growing in and around natural waters
modify both the inorganic and organic composition of all of these
solutions.

1.3.   METHODS OF EXPRESSING CONCENTRATION

Several different methods of expressing concentration are commonly
used in various branches of water chemistry and it is appropriate to
present these here. This discussion will also provide definitions and
methods of calculation that will be useful in the later chapters of this
book.

1.3.1.   Mass Concentration

There are two basic ways to express the mass concentration of dissolved
species (solutes) in solution. The first is to state concentration in units of
mass of solute in a unit volume of solution—the so-called w/v (weight/
volume) basis. The second is a w/w basis, that is, weight of solute in a
given weight of solution.[5] Both of these methods of expressing concentration are widely used in water chemistry. For example, the units mg/liter
and ppm (parts per million) are, respectively, the w/v and w/w units most
often used to express the concentration of various materials in waters
and wastewaters. They can be interconverted if the density of the solution
is known.

---

[5] *Note:* For gaseous constituents in solution or for the various components of a
    mixture of gases, v/v, or the volume of a constituent at standard temperature
    and pressure in a unit volume of solution, is also in common usage.

If the density of the solution is unity, these two methods of concentration expression are identical, that is,

$$\text{mg/liter} = \frac{\text{mass of substance (mg)}}{\text{volume of solution (liter)}}$$

$$\text{ppm} = \frac{\text{mass of substance (mg)}}{\text{mass of solution (kg)}}$$

$$\text{density of solution, } \rho = \frac{\text{mass of solution (kg)}}{\text{volume of solution (liter)}}$$

$$\therefore \text{Concentration in ppm} \left(\frac{\text{mg}}{\text{kg}}\right) = \text{concentration in} \left(\frac{\text{mg}}{\text{liter}}\right) \times \frac{1}{\rho}\left(\frac{\text{liter}}{\text{kg}}\right)$$

If $\rho = 1$ kg/liter,

$$\text{Concentration in ppm} \left(\frac{\text{mg}}{\text{kg}}\right) = \text{concentration in} \frac{\text{mg}}{\text{liter}}$$

For wastewaters and most natural waters (with the exception of seawater and brines), we can make the assumption that the density of the solution is unity so that mg/liter and ppm are equivalent. Indeed, these concentration units are used interchangeably by practitioners in the field. In recent years, however, there has been a tendency to use the mg/liter unit rather than ppm. Possibly, this is because the mg/liter method represents the way in which most constituents are analyzed, that is, a mass of a material in a known volume of solution is analyzed rather than a mass of solute in a known mass of solution. However, for the analysis and expression of concentration of the constituents of sludges and sediments, the mg/kg (ppm) unit is usually used because, in this type of analysis, samples are usually weighed out rather than measured out volumetrically.

In both types of concentration expression the mass component of the concentration term can take a variety of forms. The simplest form is when specific components of a solution are expressed in terms of the specific components themselves, for example, if we were to analyze a solution of ammonium nitrate (NH_4NO_3) for ammonium ion (NH_4^+) and for nitrate ion (NO_3^-) and find that 100 ml of solution contained 34 mg NH_4^+ and 124 mg NO_3^-, we could compute that the solution contains:

$$\frac{36 \text{ mg } NH_4^+}{100 \text{ ml}} \times \frac{1000 \text{ ml}}{\text{liter}} = 360 \text{ mg } NH_4^+/\text{liter}$$

and

$$\frac{124 \text{ mg } NO_3^-}{100 \text{ ml}} \times \frac{1000 \text{ ml}}{\text{liter}} = 1240 \text{ mg } NO_3^-/\text{liter}$$

Each component of the solution is expressed as the form in which it appears in solution. We can express the components as constituents other than those in which they appear in solution. For example, if we were to express the constituents of the ammonium nitrate solution in terms of the amount of nitrogen they contained, we would find

$$\text{mg NH}_4^+\text{-N/liter} = 360 \text{ mg NH}_4^+/\text{liter} \times \frac{14 \text{ mg N}}{18 \text{ mgNH}_4^+}$$

$$= 280 \text{ mg NH}_4^+\text{-N/liter}$$

$$\text{mg NO}_3^-\text{-N/liter} = 1240 \text{ mg NO}_3^-/\text{liter} \times \frac{14 \text{ mg N}}{62 \text{ mg NO}_3^-}$$

$$= 280 \text{ mg NO}_3^-\text{-N/liter}$$

This particular method of expressing concentration in terms of a common constituent is useful when a variety of different components all containing the common constituent is present in a water and it is desired to find the total concentration of that constituent. It is used widely in water chemistry, especially for nitrogen, phosphorus, and carbon. This method is used frequently by oceanographers to express the concentration of micronutrients in seawater. The units μg-atoms/liter are often found in oceanographic literature. For example 1.0 μg-atom per liter of phosphate-phosphorus is 31 μg PO_4-P/liter, where 31 is the atomic weight of phosphorus. It is important to indicate clearly which method of concentration expression is being used or considerable errors can result. For example, the EPA Drinking Water Standards recommend an upper limit on nitrate for public water supplies of 10 mg/liter. In this instance the nitrate is expressed as N. If we had read this incorrectly and assumed that the standard was stated in terms of mg NO_3^-/liter, we would be led to believe that

$$10 \text{ mg NO}_3^-/\text{liter} \times \frac{14 \text{ mg NO}_3^-\text{-N}}{62 \text{ mg NO}_3^-} = 2.25 \text{ mg NO}_3^-\text{-N/liter}$$

instead of the 10 mg NO_3^--N/liter.

It is also important to remember that the expression of various constituents in terms of a common constituent is merely a convenience and has nothing to do with the nature of the constituent in a specific water. Often people speak of the amount of phosphorus in water or wastewater. There is no phosphorus in water or wastewater. If there were, we would be in trouble because the water would be unusable for most purposes. But there are various types of phosphates, which we analytically express as phosphorus.

1.3.2. Molar Concentration and Activity

The previous discussion has centered around the topic of expressing the mass of constituents either as "themselves" or as some common constituent for the sake of convenience. More common to chemistry in general is the use of molarity, the moles of solute per volume of solution in liters, and molality, the moles of solute per mass of solvent in kilograms; we assume that the reader is familiar with these.

Equilibrium constants, with which we deal in great detail later in this text, are based on molar concentrations. This is because the equilibrium constant of a reaction has as its basis the law of mass action in which quantities of reacting species are expressed in terms of moles. In the very dilute aqueous systems such as fresh natural waters we can use concentrations in terms of moles/liter directly in equilibrium expressions. In more concentrated solutions it is often necessary to account for the fact that a substance behaves in chemical reactions as though its concentration were somewhat less than actual. The ratio between the apparent, or active concentration, of the substance and the analytical, or actual, concentration of the substance is called the activity coefficient (γ). The active concentration—that concentration we use in many thermodynamic equations—is called the activity of the substance (see Chapter 3).

1.3.3. Equivalents and Normal Concentration

The expression of solute concentration as equivalents/liter, that is, normal concentration, is based on a definition that is related to the type of reaction in which the solution constituents are involved. The advantage in using normality is that when two substances react to produce other substances, the number of equivalents of each reacting species is equal to the number of equivalents of product. Knowing this, many problems are easier to solve. Expression of solute concentrations in terms of equivalents/liter or normal concentrations requires a knowledge of the nature of the reaction being considered. Cases may arise where a single substance has two different equivalent weights because of its involvement in two different types of reaction.

A one normal solution of a substance is a solution that contains one equivalent weight of a substance per liter of solution. Three methods of definition of equivalent weight are commonly encountered in water and wastewater chemistry. The methods are those based on (1) the charge of an ion, (2) the number of protons of hydroxyl ions transferred in an acid-base reaction, and (3) the number of electrons transferred in an oxidation-reduction reaction.

1. *Equivalent Weight Based on Ion Charge.*   The equivalent weight is defined as

$$\text{Equivalent weight} = \frac{\text{molecular weight}}{\text{ion charge}}$$

and the number of equivalents per liter, the normality, is

$$\text{Normality} = \frac{\text{mass of substance per liter}}{\text{equivalent weight}}$$

We use these definitions of equivalents and normality for determining the number of equivalents/liter of charge units, and for determining the number of equivalents per liter of species that participate in precipitation-dissolution reactions. The equivalents/liter of charge units (or the identical quantity, the "moles of charge units/liter") are useful in checking the accuracy of water analysis. To satisfy the law of electroneutrality, the total number of equivalents/liter of positively charged ions (i.e., the total amount of positive charge) must equal the total number of equivalents/liter of negatively charged ions (i.e., the total amount of negative charge). We illustrate the use of this concept in Chapter 4.

Example 1-1

Find the normality of the following solutions:

✓ 1.   60 mg CO_3^{2-}/liter, given that CO_3^{2-} participates in the precipitation reaction,

$$Ca^{2+} + CO_3^{2-} \rightarrow CaCO_{3(s)}$$

✓ 2.   155 mg $Ca_3(PO_4)_2$/liter given that $Ca_3(PO_4)_2$ participates in the dissolution reaction,

$$Ca_3(PO_4)_2 \rightarrow 3Ca^{2+} + 2PO_4^{3-}$$

Solution

1.   The molecular weight of CO_3^{2-} is 60.

$$\text{Gram equivalent weight} = \frac{\text{gram molecular weight}}{\text{ion charge}} = \frac{60 \text{ g/mole}}{2 \text{ eq/mole}} = 30 \text{ g/eq}$$

$$= 30 \text{ mg/meq}$$

$$\text{Normality} = \frac{60 \text{ mg/liter}}{30 \text{ mg/meq}} = 2 \text{ meq/liter}$$

2.   The molecular weight of $Ca_3(PO_4)_2$ is 310. Because each $Ca_3(PO_4)_2$ forms six positive and six negative charges,

$$\text{Gram equivalent weight} = \frac{310 \text{ g/mole}}{6 \text{ eq/mole}} = 51.67 \text{ g/eq} = 51.67 \text{ mg/meq}$$

$$\text{Normality} = \frac{155 \text{ mg/liter}}{51.67 \text{ mg/meq}} = 3 \text{ meq/liter}$$

---

2. Equivalent Weight Based on Acid-Base Reactions. The equivalent weight of a substance in acid-base reactions is defined as "the weight of a substance that will either replace one H^+ (hydrogen ion) (proton) in an acid, provide one H^+ for reaction, or react with one H^+ to form an acid." Another way of stating the same definition is "that weight of a substance that will either replace one OH^- (hydroxyl ion) from a base, provide one

OH^- for a reaction, or react with one OH^- to form a base." Generalizing, we can say

$$\text{Equivalent weight} = \frac{\text{molecular weight}}{n}$$

where n is the number of protons or hydroxyl ions that react.

Example 1-2

Find the normality of the following solutions:

1. 36.5 mg HCl/liter, with respect to the reaction

$$HCl + NaOH \rightleftharpoons NaCl + H_2O$$

2. 49 mg H_3PO_4/liter, with respect to the reaction

$$H_3PO_4 \rightleftharpoons 2H^+ + HPO_4^{2-}$$

3. 45 mg CO_3^{2-}/liter, with respect to the reaction

$$CO_3^{2-} + H_2O \rightleftharpoons HCO_3^- + OH^-$$

4. 45 mg CO_3^{2-}/liter, with respect to the reaction

$$CO_3^{2-} + 2H^+ \rightleftharpoons 2H_2CO_3$$

Solution

1. One H^+ reacts per HCl. Therefore, we find

$$\text{Gram equivalent weight} = \frac{\text{gram molecular weight}}{1\ \text{eq/mole}} = \frac{36.5\ \text{g/mole}}{1\ \text{eq/mole}}$$

$$= 36.5\ \text{g/eq} = 36.5\ \text{mg/meq}$$

$$\text{Normality} = \frac{36.5\ \text{mg/liter}}{36.5\ \text{mg/meq}} = 1\ \text{meq/liter}$$

2. $2H^+$ react per H_3PO_4. Therefore,

$$\text{Gram equivalent weight} = \frac{\text{gram molecular weight}}{2\ \text{eq/mole}} = \frac{98\ \text{g/mole}}{2\ \text{eq/mole}}$$

$$= 49\ \text{g/eq} = 49\ \text{mg/meq}$$

$$\text{Normality} = \frac{49\ \text{mg/liter}}{49\ \text{mg/meq}} = 1\ \text{meq/liter}$$

3. One OH^- results from this reaction. Thus

$$\text{Gram equivalent weight} = \frac{\text{gram molecular weight}}{1\ \text{eq/mole}}$$

$$= 60\ \text{g/eq} = 60\ \text{mg/meq}$$

$$\text{Normality} = \frac{45\ \text{mg/liter}}{60\ \text{mg/meq}} = 0.75\ \text{meq/liter}$$

4.   Two H^+ react with each CO_3^{2-}. Thus

$$\text{Gram equivalent} = \frac{\text{gram molecular weight}}{2\,\text{eq/mole}} = \frac{60\,\text{g}}{2\,\text{eq/mole}}$$

$$= 30\,\text{g/eq} = 30\,\text{mg/meq}$$

$$\text{Normality} = \frac{45\,\text{mg/liter}}{30\,\text{mg/meq}} = 1.5\,\text{meq/liter}$$

An acid or a base can have more than one equivalent weight. For example, in Example 1-2 the ion CO_3^{2-} had an equivalent weight of 30 for one reaction and 60 for another, and the normality of two CO_3^{2-} solutions, each with the same mass concentration of CO_3^{2-}, differed by a factor of 2. In much of the water chemistry literature, CO_3^{2-}, and correspondingly $CaCO_3$, is treated as though it has only one equivalent weight, 50, a value that is of one-half the molecular weight. This can be a source of confusion, but it can be dealt with if one is aware of it.

Another type of reaction that may cause some confusion is the combined reaction. For example, the reaction

$$Ca^{2+} + HCO_3^- \rightleftharpoons CaCO_{3(s)} + H^+ \tag{1-1}$$

is the sum of the acid-base reaction,

$$HCO_3^- \rightarrow H^+ + CO_3^{2-} \tag{1-2}$$

and the precipitation reaction,

$$Ca^{2+} + CO_3^{2-} \rightarrow CaCO_{3(s)} \tag{1-3}$$

Since Ca^{2+} participates only in the precipitation reaction, it has 2 equivalents/mole for reactions 1-1 and 1-3. However, HCO_3^- has 1 equivalent/mole for the acid-base reaction, Eq. 1-2, and 2 equivalents/mole for the precipitation step in Eq. 1-1. In Eq. 1-1 it is the CO_3^{2-} within each HCO_3^- that reacts to form the precipitate, and each CO_3^{2-} has a charge of 2.

3. Equivalent Weight for Oxidation-Reduction Reactions.   For oxidation-reduction reactions, the equivalent weight is defined as "the mass of substance per mole of electrons transferred." One mole of electrons is one equivalent of electrons. Thus, in the half reaction for the reduction of oxygen to water,

$$O_2 + 4H^+ + 4e^- \rightarrow 2H_2O$$

the equivalent weight of oxygen is

$$\text{Gram equivalent weight} = \frac{\text{gram molecular weight}}{4\,\text{eq/mole}} = \frac{32\,\text{g/mole}}{4\,\text{eq/mole}} = 8\,\text{g/eq}$$

Similarly, for an overall oxidation-reduction reaction, such as the one in which ferrous ion, Fe^{2+}, is oxidized to ferric ion, Fe^{3+}, by oxygen,

$$4Fe^{2+} + O_2 + 4H^+ \rightarrow 4Fe^{3+} + 2H_2O$$

we need to determine the number of electrons transferred per mole of species. We do this by separating the overall reaction into half-reactions,

$$O_2 + 4H^+ + 4e^- \rightarrow 2H_2O$$

and

$$4Fe^{2+} \rightarrow 4Fe^{3+} + 4e^-$$

or

$$Fe^{2+} \rightarrow Fe^{3+} + e^-$$

One mole of Fe^{2+} or Fe^{3+} reacts or is produced per mole of electrons so that the equivalent weights of each of these species are the same as the molecular weights. Note that if Fe^{2+} and Fe^{3+} were to participate in a precipitation-dissolution reaction, the equivalent weights would be one half and one third of the molecular weights, respectively.

1.3.4. Mass/Volume Concentrations as O_2

It is common in water chemistry to express the results of oxidation-reduction reactions [such as those that take place in the biochemical oxygen demand (BOD) or chemical oxygen demand (COD) tests] in terms of oxygen much the same way that we earlier expressed ammonia and nitrate in terms of nitrogen. To do this we assume that in all cases, each mole of oxygen can accept 4 moles of electrons as in Section 1.3.4. (There are other reactions for O_2 in which the number of electrons transferred is other than 4, but they will not take place under the conditions of interest in the BOD or COD tests.) Oxygen thus has an equivalent weight of $\frac{32}{4}$, or 8.

In the BOD test, let us assume that 0.25×10^{-3} moles of O_2 react with the organic matter in 1 liter of water. We then say that the concentration of organic matter is

$$0.25 \times 10^{-3} \frac{\text{moles}}{\text{liter}} \times \frac{4 \text{ eq}}{\text{mole}} \times 8000 \frac{\text{mg}}{\text{eq}} = 8 \text{ mg/liter as } O_2$$

In the COD test, the oxidizing agent dichromate, $Cr_2O_7^{2-}$, is used in place of oxygen. During the test the number of equivalents/liter of $Cr_2O_7^{2-}$ used to oxidize organic matter is determined; the assumption is then made that if O_2 had been used in place of $Cr_2O_7^{2-}$, the same number of eq/liter of it would react. Thus the "number of eq/liter $\times$ 8000 mg O_2/eq" yields the concentration of organic matter in "mg/liter as O_2."

Similar computations can be used to express the concentration of available chlorine in various compounds (see Chapter 7).

1.3.5. Equivalents per Million, epm

The term *equivalents per million*, epm, is the w/w version of meq/liter. It is a term that was once commonly used but rarely is found in current literature; it commonly appears in older literature, especially in geochemistry. Since

$$1 \text{ meq/liter} = 10^{-3} \text{ eq/liter}$$

and

$$10^{-3} \text{ eq/liter} = 10^{-6} \text{ eq/ml} = 1 \text{ eq/10}^6 \text{ ml}$$

When solution density = 1 g/ml, 10^6 ml = 10^6 g and

$$1 \text{ epm} = 1 \text{ meq/liter}$$

1.3.6. Mass Concentration as $CaCO_3$

A very widespread system for expressing hardness (calcium and magnesium) and alkalinity (HCO_3^-, CO_3^{2-} and OH^-) concentrations in water and wastewater engineering and water chemistry is the calcium carbonate system. This method of expressing concentration has the identical advantages (and disadvantages) to the "equivalents" method; it probably arose because early sanitary engineers were attracted by the prospect of normalizing concentrations to $CaCO_3$, a substance commonly used and referred to in water chemistry. In this method of expression, the concentration of a substance as "mg/liter as $CaCO_3$" is determined by the equation

Number of equivalents of substance per liter

$$\times \frac{50 \text{ g } CaCO_3}{\text{equivalent of } CaCO_3} \quad (1\text{-}4)$$

The equivalent weight of $CaCO_3$ in this system is defined on the basis of "charge" or "acid-base reactions" (i.e., cases 1 and 2 in Section 1.3.3.).

For example, for hardness,

$$CaCO_{3(s)} \rightarrow Ca^{2+} + CO_3^{2-} \quad (1\text{-}5)$$

Each mole of $CaCO_3$ yields 1 mole, or 2 equivalents of Ca^{2+}. Therefore, with respect to Ca^{2+} in precipitation or dissolution reactions,

$$\text{Equivalent weight of } CaCO_3 = \frac{100 \text{ g/mole}}{2 \text{ eq/mole}} = 50 \text{ g/eq}$$

Because Mg^{2+} is also divalent and its properties as a hardness ion are based on charge just as are those of Ca^{2+}, we treat Mg^{2+} the same as Ca^{2+}.

For the total alkalinity reaction we are concerned with the CO_3^{2-} portion of $CaCO_3$.

$$CaCO_3 + 2H^+ \rightarrow H_2CO_3 + Ca^{2+} \quad (1\text{-}6)$$

and, since $2H^+$ react per mole of $CaCO_3$, for this reaction $CaCO_3$ once again has an equivalent weight of 50. Other bases such as OH^- and HCO_3^- are part of alkalinity and react as follows,

$$OH^- + H^+ \rightarrow H_2O \tag{1-7}$$

$$HCO_3^- + H^+ \rightarrow H_2CO_3^* \tag{1-8}$$

Each mole of these bases reacts with 1 mole of H^+; therefore, each mole contains one equivalent of capacity for reacting with H^+ and we convert the moles/liter (= eq/liter) of these species to "mg/liter as $CaCO_3$" again as shown in Eq. 1-4.

Calcium carbonate does not always have an equivalent weight of 50 g/mole, however. In the following reaction,

$$CaCO_3 + H^+ \rightarrow Ca^{2+} + HCO_3^-$$

1 mole of $CaCO_3$ reacts with 1 mole of H^+. Thus, *for this reaction*, $CaCO_3$ has an equivalent weight of 100 g/eq. This point is overlooked in much of the water chemistry literature. For this reason the reader should always examine the basis used to express concentrations as "mg/liter as $CaCO_3$."

1.4. ADDITIONAL READING

Davies, S. N., and R. C. M. DeWiest, *Hydrogeology*. John Wiley, New York, 1966.

Garrels, R. M., and C. L. Christ, *Solutions, Minerals and Equilibria*. Harper, New York, 1965.

Holland, H.D., *The Chemistry of the Atmosphere and Oceans*. Wiley-Interscience, New York, 1978.

Stumm, W., and J. J. Morgan, *Aquatic Chemistry*. Wiley-Interscience, New York, 1970.

CHEMICAL KINETICS

2.1. INTRODUCTION

The topic of chemical kinetics deals with the rate and mechanism of chemical reactions. The subject is of considerable concern to the aquatic chemist because in many instances the rate at which a reaction proceeds toward equilibrium, rather than the equilibrium condition, determines the design and performance of treatment processes and the behavior of natural water systems.

In water and wastewater treatment, for example, the chemical oxidation of organic compounds by chlorine and ozone, the precipitation of calcium phosphates, calcium carbonate, and magnesium hydroxide, and the chemical oxidation of iron(II) and manganese(II), are reactions whose rates control the design and efficiency of treatment processes.[1] In natural waters, reaction rates control such diverse reactions as precipitation, dissolution, and oxidation-reduction and concentrations exist that are not those that would be predicted by equilibrium calculations. These none-quilibrium situations are particularly common for reactions involving oxidation-reduction and precipitation-dissolution. For example, by equilibrium calculations alone we can predict that it woud be impossible for sulfide to exist in water that contained dissolved oxygen. However, data collected in South San Francisco Bay in 1962 showed that sulfide could be detected in waters that contained up to 3 mg/liter dissolved oxygen.[2] This occurrence is possible because the reaction rate between oxygen and sulfide in dilute aqueous solution is not rapid.

The oxidation of Fe(II) to Fe(III) by molecular oxygen is a reaction that equilibrium calculations predict will go to completion in waters where dissolved oxygen is present. However, we find that solutions of Fe(II) salts at low pH values are stable indefinitely even when a stream of pure

---

[1] In this chapter we concern ourselves only with systems that are uniform in concentration throughout. In process design the degree and type of mixing are important in addition to the rate at which chemical species react in uniform systems.

[2] P. N. Storrs, R. E. Selleck, and E. A. Pearson, "A Comprehensive Study of San Francisco Bay, 1961–62, South San Francisco Bay Area, Suisun Bay—Lower San Joaquin River Area, San Pablo Bay Area," SERL Report No. 63-3, 1963, University of California, Berkeley, 2nd annual report.

oxygen gas is bubbled through them.[3] At pH < 4, less than 0.1 percent Fe(II) will oxidize each day. In Fe(II) solutions with pH values > 7 to 8 the rate of oxidation of Fe(II) to Fe(III) is so rapid that it is limited by the rate at which oxygen can be supplied by dissolution from the gas phase. In this example we see that pH exerts a profound effect on whether equilibrium conditions or kinetic considerations better describe the system.

If natural water phosphate levels were controlled by equilibrium of the water with the thermodynamically stable calcium phosphate solid— calcium hydroxyapatite, $Ca_5OH(PO_4)_3$—phosphate levels would be so low that we would not be concerned with phosphate as a nutrient for photosynthetic aquatic organisms. However, levels of phosphate exist in receiving waters that are far in excess of those predicted by equilibrium with hydroxyapatite because, not only is the rate of formation and dissolution of this thermodynamically predicted solid slow, but also calcium phosphate solids of higher solubility form and then transform very slowly into hydroxyapatite.

The information derived from chemical kinetics should help us understand why, when we mix hydrogen and oxygen gases together, there is no reaction; yet the presence of a little platinum or a spark in a mixture of these two gases will produce water with explosive violence. From our investigation of chemical kinetics we should understand why some oxidation-reduction reactions proceed very slowly in the absence of microorganisms while in their presence they occur rapidly. For example, thermodynamic or equilibrium predictions would lead one to conclude that ammonia would be oxidized to nitrate in oxygenated water. Yet an aerated sterile solution of ammonium chloride is stable indefinitely. However, if we introduce microorganisms of the genre *Nitrosomonas* and *Nitrobacter*, a rapid conversion of ammonium ion to nitrite and then to nitrate will take place.

In similar fashion for an unoxygenated water in equilibrium with the earth's atmosphere, which contains 78 percent nitrogen gas, we would predict that nitrate in water should be converted to nitrogen gas. This reaction, which is called denitrification, again only takes place in the presence of suitable microorganisms. In their absence sodium nitrate solutions are perfectly stable. In this chapter and in succeeding chapters we will examine the reasons for these observations in an effort to better understand the chemistry of aqueous systems.

2.2. COLLISIONS OF REACTING SPECIES

With a few exceptions, such as radioactive decay, collisions between reacting species are necessary for chemical reactions to occur. Although

[3] W. Stumm and J. J. Morgan, *Aquatic Chemistry*, Wiley-Interscience, 1970, pp. 534–538.

collisions bring the reacting species close enough together for reaction to occur, not all collisions successfully produce a chemical reaction.

Collisions between two species (a bimolecular collision) are a far more common occurrence than the simultaneous collision of three (trimolecular) or more species. For example, in air at ordinary laboratory conditions, where less than 0.1 percent of the gas volume is occupied by gas molecules, one molecule hits another (bimolecular collision) approximately 10^9 times per second. Three molecules collide simultaneously (trimolecular collision) at a rate of about 10^5 times per second.[4] We can reason that on the basis of the much greater frequency of bimolecular collisions, it is more likely that these are usually responsible for chemical reactions.

A similar line of reasoning can be followed for substances reacting in solution. The number of collisions between species other than the solvent in a solution is only slightly greater than in the gas phase.[5] Thus, for the same concentration and temperature, reactions in solution should occur at approximately the same rate as in the gas phase. However, reacting species, especially ionic species, frequently interact with the solvent, and this can significantly affect the rate at which collisions occur.

Differences between liquid phase reactions and gas phase reactions occur largely because of the so-called "cage effect." In a gas phase when two reacting species collide and subsequently rebound they can become separated by a great distance in a very short period of time. However, in solution the solvent molecules that surround the colliding species serve to trap them subsequent to collision by immediately colliding with the rebounding species. Indeed, in many instances, the reactants are deflected so that another collision between the reactants can occur. Because of this, successive collisions of the same molecules tend to occur in a solvent.

We can increase the number of collisions between reacting species by increasing the temperature. If the reacting medium is a compressible gas, we can also increase the number of collisions by increasing the pressure.

2.3.   ORIENTATION OF REACTANTS

Reactions between species that require no special orientation during collision to result in a reaction tend to be more rapid than reactions between species that must be properly aligned for a collision to result in a reaction. This so-called "orientation effect" partially explains why many collisions between reacting species are not successful in producing a reaction. For example, the spherical Ag^+ ion and the spherical Cl^- ion combine very rapidly to form the precipitate $AgCl(s)$ when present in a

[4] J. A. Campbell, *Why Do Chemical Reactions Occur?* Prentice-Hall, Englewood Cliffs, N.J., 1965.
[5] K. J. Laidler, *Chemical Kinetics*, 2nd ed. McGraw-Hill, New York, 1965.

supersaturated solution. A very large percentage of the collisions between these species is successful in producing reaction because each is spherical and no special orientation is required (Fig. 2-1).

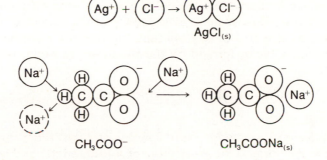

Fig. 2-1. Orientation effect.

The formation of solid sodium acetate[6] from supersaturated sodium acetate solutions is slow by comparison and the supersaturated solutions can exist for long periods of time without a precipitate forming. Many collisions between the spherical Na^+ ion and the linear CH_3COO^- (acetate) ion occur that do not result in reaction because the only collision that can result in reaction is between Na^+ and the oxygen end of CH_3COO^- (Fig. 2-1).[7] A collision between Na^+ and any other part of the acetate ion cannot form a sodium acetate precipitate.

2.4.   THE RATE LAW

We can show experimentally that for the general irreversible reaction

$$A + 2B + \ldots \rightarrow P + 2Q \ldots$$
$$\text{reactants} \quad \text{products}$$

we can write the rate law,

$$\frac{d[A]}{dt} = -k[A]^a[B]^b[P]^p[Q]^q \ldots \tag{2-1}$$

where

$$\frac{d[A]}{dt} = \text{time rate of change in molar concentration of species A,}$$

$$k = \text{reaction rate constant, and}$$

$$a, b, p, q, \ldots = \text{constants}$$

[6] However, sodium acetate will not precipitate unless the concentrations of Na^+ and CH_3COO^- each exceed approximately 15 moles per liter, concentrations far in excess of those encountered in dilute aqueous solution.

[7] J. A. Campbell, *Why Do Chemical Reactions Occur?* Prentice-Hall, Englewood Cliffs, N.J., 1965.

In this book, [ ] is used to signify concentration in moles/liter. We may use concentration units other than moles/liter in the rate law but in doing so we should use the same concentration unit for each species and realize that both the numerical value and units of the reaction rate constant will differ from those found when molecular concentrations are used.

Using our knowledge of the stoichiometry of the reaction, that is, the relative number of moles of species reacting and the relative number of moles of products being formed as the reaction proceeds, we can state that

$$\frac{d[A]}{dt} = \frac{1}{2}\frac{d[B]}{dt} \ldots = \frac{-d[P]}{dt} = \frac{-1}{2}\frac{d[Q]}{dt} \ldots \tag{2-2}$$

because 1 mole of A reacts for every 2 moles of B that react, and so forth, and 1 mole of P is formed for every mole of A that reacts, and so forth. We can determine the *reaction order* from the rate law. The *overall* reaction order is

$$a + b + p + q \ldots \tag{2-3}$$

while the order *with respect to* A is a, the order with respect to B is b, and so forth. If the reaction is irreversible, then $p, q, \ldots$, the exponents of the product concentration, are usually zero. For example, if

$$\frac{d[A]}{dt} = -k[A][B]^2$$

then we would say that the reaction was first order with respect to A, second order with respect to B, and third order overall. It is important to note that reaction order is generally not determined by the stoichiometry of the overall reaction. Laboratory experimentation is necessary to determine the order.

The following example illustrates several points that are important for a good understanding of the rate law.

Example 2-1

Ammonia, NH_3, is a common constituent of many natural waters and wastewaters. It reacts with the disinfectant hypochlorous acid, HOCl, in solution to form monochloroamine, NH_2Cl, as follows.

1. $NH_3 + HOCl \rightarrow NH_2Cl + H_2O$

The rate constant, k, was found by experiment to be 5.1×10^6 (liters/mole sec) at 25°C.[8] The rate law was determined to be

---

[8] J. C. Morris, chapter in *Principles and Applications of Water Chemistry*, John Wiley, New York, 1967.

2. $\dfrac{d[HOCl]}{dt} = -k[HOCl][NH_3]$

    (a) What is the overall reaction order?

    (b) What percent decrease in reaction rate occurs if the concentration of each reactant is reduced by 50 percent?

    (c) Determine the value of the rate constant when concentrations are expressed in units of mg/liter rather than moles/liter.

Solution

1. Equation 2-2 can be written more explicitly as

$$\frac{d[HOCl]}{dt} = -k[HOCl]^a[NH_3]^b$$

where $a = 1$ and $b = 1$.

The overall reaction order is $a + b = 2$. The experienced eye will immediately recognize a second-order reaction from the units of the rate constant which, for reactions with an overall order of 2, are always liter/mole sec or more generally (time^{-1})(concentration^{-1}). This derives from a rearrangement and dimensional analysis of Eq. 2-2 thus,

$$k = \frac{\dfrac{d[HOCl]}{dt}}{[HOCl]^1[NH_3]^1} = \frac{\text{mole/liter sec}}{(\text{mole/liter})(\text{mole/liter})} = \frac{\text{liter}}{\text{mole sec}}$$

2. Let the initial concentrations of NH_3 and $HOCl$ be x and y, respectively. The rate of reaction is then,

$$\frac{(d[HOCl])}{(dt)_{\text{original}}} = -kyx$$

When x and y are each reduced by 50 percent, the reaction rate becomes

$$\frac{(d[HOCl])}{(dt)_{\text{new}}} = -k\left(\frac{y}{2}\right)\left(\frac{x}{2}\right) = \frac{1}{4}kyx$$

The new reaction rate is 25 percent of the initial reaction rate.

3. When the concentrations of $HOCl$ and NH_3 are both expressed as mole/liter, the rate constant can be written,

    (a) $k = \dfrac{d[HOCl]/dt}{[NH_3][HOCl]} = \left(\dfrac{1}{[NH_3]}\right)\left(\dfrac{1}{t}\right)$

    Setting k' = rate constant when concentrations are expressed in mg/liter, we can state

    (b) $k' = \left(\dfrac{1}{\text{mg } NH_3/\text{liter}}\right)\left(\dfrac{1}{t}\right)$

    Since 1 mole NH_3/liter = 17,000 mg NH_3/liter,

    (c) mg NH_3/liter = $[NH_3] \times 17,000$

Combining (a), (b), and (c), we have

$$\frac{k'}{k} = \frac{\dfrac{1}{[NH_3] \times 17{,}000\,t}}{\dfrac{1}{[NH_3]\,t}} = \frac{1}{17{,}000}$$

Since $k = 5.1 \times 10^6$ liter/mole sec,

$$k' = \frac{5.1 \times 10^6}{1.7 \times 10^4} = 300 \text{ liter/mg sec}$$

Integrated forms of the rate law are very useful for analyzing rate data to determine reaction rate constants and reaction order. Let us first consider the irreversible reaction

$$A \rightarrow \text{products}$$

which has the rate law

$$\frac{d[A]}{dt} = -k[A]^n$$

To determine the behavior of [A] as a function of time, we must integrate the rate expression with respect to time. We will do this for several values of the reaction order, n. When $n = 0$, the reaction is zero order, and

$$\frac{d[A]}{dt} = -k[A]^0 = -k \tag{2-4}$$

Upon integrating, we obtain

$$[A] = [A]_0 - kt \tag{2-5}$$

where $[A]_0 =$ the concentration of A at $t = 0$, that is, the initial concentration of A. The *half-life*, $t_{1/2}$, or time for 50 percent of the initial concentration to react can be obtained from Eq. 2-5 by setting $[A] = 0.5\,[A]_0$ when $t = t_{1/2}$. Then

$$t_{1/2} = \frac{0.5[A]_0}{k}$$

When $n = 1$, the reaction is *first order*, both with respect to A and overall, and we can write,

$$\frac{d[A]}{dt} = -k[A] \tag{2-6}$$

Rearranging Eq. 2-6 and solving the integral,

$$\int_{[A]_0}^{[A]} \frac{d[A]}{[A]} = -\int_0^t k\,dt$$

we find

$$\ln [A] = \ln [A]_0 - kt \tag{2-7}$$

or

$$[A] = [A]_0 e^{-kt} \tag{2-8}$$

Examination of Eq. 2-7 suggests that the rate constant k may be determined experimentally from a plot of $\ln [A]$ versus t, which has a slope of $-k$. Also, from Eq. 2-8, when $[A] = 0.5 [A]_0$, we find the half-life to be

$$t_{1/2} = \frac{0.693}{k}$$

If the reaction is greater than first order, then we can write

$$\frac{d[A]}{dt} = -k[A]^n \tag{2-9}$$

and integrating, we obtain

$$\int_{[A]_0}^{[A]} \frac{d[A]}{([A])^n} = -\int_0^t k \, dt$$

or

$$\left(\frac{-1}{n-1}\right)\left(\frac{1}{[A]^{n-1}}\right) - \left(\frac{-1}{n-1}\right)\left(\frac{1}{[A]_0^{n-1}}\right) = -kt$$

or

$$\frac{1}{n-1}\left(\frac{1}{[A]_0^{n-1}} - \frac{1}{[A]^{n-1}}\right) = -kt \tag{2-10}$$

If $n = 2$, for example, the reaction is *second order*, both with respect to A and overall, and we can write

$$\frac{1}{[A]} = \frac{1}{[A]_0} + kt \tag{2-11}$$

and the half-life is

$$t_{1/2} = \frac{1}{k[A]_0}$$

For a second-order reaction involving one reactant, we would determine the rate constant k by plotting $1/[A]$ versus t to yield a straight line with a slope of k in accordance with Eq. 2-11.

Let us now proceed to the slightly more complex case of the irreversible reaction in which two reactants react to give products. We can write

$$A + B \rightarrow \text{products} \tag{2-12}$$

which has the rate law

$$\frac{d[A]}{dt} = \frac{d[B]}{dt} = -k[A]^a[B]^b \tag{2-13}$$

When $a = 1$ and $b = 1$, the reaction is second order overall and first order with respect to both A and B. There are two general situations we must examine. These are when $[A]_0 = [B]_0$, and when $[A]_0 \neq [B]_0$. For the first case,

$$\frac{d[A]}{dt} = -k[A][B] \tag{2-14}$$

From the reaction stoichiometry in Eq. 2-12, we can see that for each molecule of A which reacts, one molecule of B will react. The concentrations of A and B will therefore decrease at the same rate, and since they were initially equal this means they will always be equal. We can therefore substitute [A] for [B] to give

$$\frac{d[A]}{dt} = -k[A]^2 \tag{2-15}$$

which, when integrated, will yield the same expression as Eq. 2-11.

When $a = b = 1$ and $[A]_0 \simeq [B]_0$, it is useful to use the substitution

$$[A] = [A]_0 - X \tag{2-16}$$

$$[B] = [B]_0 - X \tag{2-17}$$

where X is the concentration in moles/liter of each species that has reacted. Differentiating Eq. 2-16, we obtain

$$\frac{dX}{dt} = -\frac{d[A]}{dt} \tag{2-18}$$

From Eq. 2-14,

$$\frac{d[A]}{dt} = -k[A][B]$$

or

$$\frac{dX}{dt} = \frac{-d[A]}{dt} = k[A][B] \tag{2-19}$$

and substituting for [A] and [B] from Eqs. 2-16 and 2-17 in Eq. 2-19, we obtain

$$\frac{dX}{dt} = k([A]_0 - X)([B]_0 - X) \tag{2-20}$$

Integrating, rearranging, and substituting for X from Eqs. 2-16 and 2-17, we obtain

$$\ln \frac{[B]}{[A]} = \ln \frac{[B]_0}{[A]_0} + ([B]_0 - [A]_0)\, kt \qquad (2\text{-}21)$$

A plot of experimental data in the form $\ln([B]/[A])$ versus t will yield a straight line with a slope of $([B]_0 - [A]_0)k$, and since $[B]_0$ and $[A]_0$ are known, k can be determined as $k = \text{slope}/([B]_0 - [A]_0)$. It is important to note that Eq. 2-21 is only valid for the reaction stoichiometry given in Eq. 2-12. If the stoichiometry is different, the integrated form of the rate law will also be different. For example, for the reaction

$$2A + B \rightarrow \text{products} \qquad (2\text{-}22)$$

For the initial condition of $[A]_0 = 2[B]_0$ and a rate law of

$$\frac{d[B]}{dt} = -k[A]^2[B] \qquad (2\text{-}23)$$

the integrated form of the rate expression is

$$\frac{1}{[B]^2} = \frac{1}{[B]_0^2} + 8kt \qquad (2\text{-}24)$$

To obtain k from experimental data for a reaction that follows this rate law, we must plot $1/[B]^2$ versus t and obtain the slope which is equal to $8k$. Then $k = \text{slope}/8$.

One other reaction order will be considered here, namely the *pseudo first-order* reaction. The reaction

$$A + B \rightarrow \text{products} \qquad (2\text{-}25)$$

which has the rate law

$$\frac{d[A]}{dt} = -k[A][B] \qquad (2\text{-}26)$$

can be treated as a first-order reaction if one of the reactants is present in such excess that its concentration is virtually unchanged during the course of the reaction. For example, if reactant B is present in large excess over reactant A, then

$$\frac{d[A]}{dt} = -k[A][B] = -k'[A] \qquad (2\text{-}27)$$

where $k' = k[B]$ and k' is a pseudo first-order reaction rate constant. This equation can be integrated to give

$$[A] = [A]_0 e^{-k't} \qquad (2\text{-}28)$$

When reactions taking place in dilute aqueous solution have water as one of the reactants, it is often possible to assume that the water is present in large excess. For example, the hydrolysis of one molecule of the disaccharide sucrose to yield two molecules of monosaccharides (one molecule each of glucose and fructose) can be written as

$$H_2O + C_{12}H_{22}O_{11} \rightarrow C_6H_{12}O_6 + C_6H_{12}O_6$$

$$\text{sucrose} \qquad \text{glucose} \qquad \text{fructose}$$

Experiments show this reaction to be second order overall, but because the H_2O concentration remains essentially constant in dilute aqueous solution it can be considered as a pseudo first-order reaction.

A common experimental technique used in the study of reaction rates is to make each of the reactants in turn present in large excess. The dependence of the reaction rate on the concentration of the other reactants can then be studied.

In each of the situations already discussed, we assumed that the reactions were irreversible, that is, once products formed, no reverse reaction to form the reactants occurred. Very few reactions are strictly irreversible; however, we can make use of the equations applicable to irreversible reactions by controlling the reaction conditions used to collect rate data so that the assumption of irreversibility can legitimately be made. One way of doing this is to collect rate data only during a very short time period following the mixing of reactants. Even if the reaction is reversible there will be little back reaction during this time period because the products that participate in the back reaction are present only in small quantities.

The following example illustrates a procedure for determining the reaction order and rate constant. The reader is referred to other texts (see Section 2-10, Additional Reading) for procedures on handling more complex problems and for a more detailed treatment of the topic of rate equations.

Example 2-2

Hydrogen peroxide, H_2O_2, is an oxidizing agent that often finds use in water purification processes. It readily decomposes to oxygen and water in the presence of a manganese dioxide catalyst (see Section 2-7). Given that the reaction is irreversible, we find that

$$2H_2O_2 \xrightarrow{\quad MnO_{2(s)} \quad} 2H_2O + O_2$$

Determine the rate constant and order of this reaction.

Solution

The data in Table 2-1 were obtained during an experiment to determine the rate constant and order of the reaction. First, let us test whether the reaction is first or second order. For a first-order reaction the plot of $\ln [H_2O_2]_t$ versus t should be straight line while for a second-order reaction the plot of $1/[H_2O_2]$ versus t should be linear. Figure 2-2 shows that $\ln [H_2O_2]$ versus t is linear and $1/[H_2O_2]$ versus t is curvilinear. Thus the reaction can be treated as first order with respect to H_2O_2.

TABLE 2-1

Time, minutes	$[H_2O_2]$ M	ln $[H_2O_2]$	$1/[H_2O_2]$
0	0.032	−3.44	31.25
10	0.023	−3.77	43.2
20	0.018	−4.02	55.6
30	0.013	−4.34	76.9
40	0.0099	−4.62	101
50	0.0071	−4.95	141

The rate constant can be determined from the slope of the plot of ln $[H_2O_2]$ versus t. From Eq. 2-27,

$$\text{slope} = \frac{\Delta \ln [A]}{\Delta t} = -k \frac{[-4.05 - (-3.45)]}{20 - 0} = -0.03/\text{min}$$

$$k = +0.03 \times \frac{1}{60}/\text{sec}$$

$$k = 5 \times 10^{-4}/\text{sec}$$

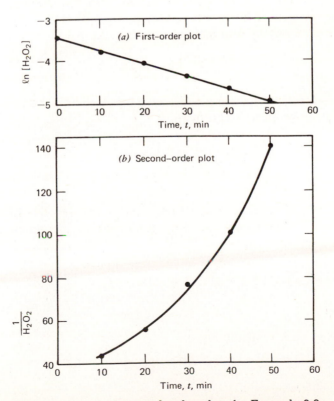

Fig. 2-2. First- and second-order plots for Example 2-2.

This trial-and-error approach to finding reaction order has definite limitations because (1) it is difficult to apply if the reaction order is not an integer, and (2) the scatter of data points owing to experimental error makes it difficult to judge whether the straight line fit for one assumed order is better than another. Most reaction orders are integers, however, and if the data are carefully collected and the experiment is properly designed, the procedure gives fairly reliable results. (The data used in this example were collected by one of us when he was an undergraduate. Thus, while the design of the experiment was most likely well done, the reader is left to judge for himself whether the data were carefully collected.)

2.5.   REACTION MECHANISM

We have noted previously that the order of a reaction cannot necessarily be determined from an examination of the stoichiometry of the reaction. Indeed we can define a particular type of reaction—the *elementary reaction*—which has the special property that its reaction order can be determined from its stoichiometry. Elementary reactions may be *monomolecular*, where a single species reacts and its concentration alone determines the rate of the reaction. Generally stated, if the reaction A → products is elementary and monomolecular, the rate law will be

$$\frac{d[A]}{dt} = -k[A]$$

Radioactive decay is an example of such a reaction as is the decomposition of the gas cyclopropane to propylene,

$$CH_2 - CH_2 \rightarrow CH_3 - CH = CH_2$$
$$\diagdown \diagup$$
$$CH_2$$

Elementary reactions that are bimolecular involve the interaction of two molecules which may be two molecules of the same species or molecules of different species. Thus

$$A + A \rightarrow products$$

or

$$A + B \rightarrow products$$

The rate laws are, respectively,

$$\frac{d[A]}{dt} = -k[A]^2$$

and

$$\frac{d[A]}{dt} = \frac{d[B]}{dt} = -k[A][B]$$

Examples of such reactions are the decomposition of nitrogen dioxide to nitric oxide and oxygen in the gas phase.

$$2NO_2 \rightarrow 2NO + O_2$$

$$\frac{d[NO_2]}{dt} = -k[NO_2]^2$$

and the formation of iodomethane from bromomethane.

$$\begin{array}{ccccccc}
CH_3Br & + & I^- & \rightarrow & CH_3I & + & Br^- \\
\text{bromomethane} & & \text{iodide} & & \text{iodomethane} & & \text{bromide}
\end{array}$$

$$\frac{d[CH_3Br]}{dt} = \frac{d[I^-]}{dt} = -k[CH_3Br][I^-]$$

More complex reactions than these are the rule rather than the exception. Overall reactions, when written stoichiometrically, may lead one to believe that they have simple mechanisms. However, experimentation may well reveal a complicated rate law that indicates a complex mechanism.

For example, contrast the following two reactions: The formation of hydrogen iodide from hydrogen gas and iodine gas is stoichiometrically represented as

$$H_{2(g)} + I_{2(g)} \rightarrow 2HI_{(g)}$$

The reaction is elementary, bimolecular and first order with respect to both $[I_{2(g)}]$ and $[H_{2(g)}]$. The rate law is

$$\frac{1}{2}\frac{d[HI]}{dt} = \frac{-d[I_2]}{dt} = \frac{-d[H_2]}{dt} = k[H_{2(g)}][I_{2(g)}]$$

The formation of hydrogen bromide from hydrogen gas and bromine gas can be stoichiometrically stated as

$$H_{2(g)} + Br_{2(g)} \rightarrow 2HBr_{(g)}$$

The rate law is found by experiment to be

$$\frac{1}{2}\frac{d[HBr]}{dt} = \frac{k[H_{2(g)}][Br_{2(g)}]^{1/2}}{1 + k'[HBr_{(g)}]/[Br_{2(g)}]}$$

The reaction is evidently not elementary. Such complex reactions as these, however, are composed of a series or a sequence of elementary reactions. We can determine the reaction order with respect to each reactant (and product) and the influence of the concentration of all species on the rate of the overall reaction if we know the sequence and rates of the elementary reactions that make up the overall reaction.

To illustrate these points further, let us examine the results from a biochemical experiment in which the rate of hydrolysis of sucrose by the enzyme saccharase is investigated. This enzyme, or biological catalyst (of which we shall speak in more detail later), catalyzes the reaction:

$$\underset{\text{sucrose}}{C_{12}H_{22}O_{11}} + H_2O \quad \overset{\text{saccharase}}{\longrightarrow} \quad \underset{\text{glucose}}{C_6H_{12}O_6} + \underset{\text{fructose}}{C_6H_{12}O_6}$$

The data of Kuhn[9] showed that the initial rate of hydrolysis of the substrate sucrose $(-d[S]/dt)$ was a function of the substrate concentration [S] when the saccharase enzyme concentration [E] was constant (Fig. 2-3). At low substrate concentrations the rate of hydrolysis appeared to be proportional to sucrose concentration, that is, $-d[S]/dt = k[S]$, while at higher sucrose concentrations the rate of hydrolysis approached a maximum rate that was apparently independent of sucrose concentration, that is, $-d[S]/dt$ = constant.

This rate behavior can be rationalized by assuming that the sucrose and enzyme react to form a complex which then may either revert to sucrose and enzyme or react to form the products, [P], glucose, and fructose. The reaction scheme proposed is

$$S + E \underset{k_2}{\overset{k_1}{\rightleftharpoons}} ES \underset{k_4}{\overset{k_3}{\rightleftharpoons}} P \ldots + E \tag{2-29}$$

We can reason that the rate of reaction should be related to the concentration of ES complex because, for the reaction to take place, ES must be formed. Our first step will be to determine ES in terms of the total amount of E-containing species and substrate concentration.

When the reaction depicted in Eq. 2-29 is at steady state, $d[ES]/dt = 0$, and

Rate of ES formation = rate of ES removal

Thus, because the reactions in Eq. 2-29 are elementary,

$$k_1[E][S] + k_4[E][P] = k_2[ES] + k_3[ES]$$

Dividing through by [E] and rearranging yields

$$\frac{[E]}{[ES]} = \frac{(k_2 + k_3)}{k_1[S] + k_4[P]}$$

Assuming $k_4[P] \ll k_1[S]$ and defining the total amount of enzyme as $[E]_t$ = [E] + [ES], we can derive the expression,

[9] R. Kuhn "Über Spezifität der Enzyme II. Saccharase und Raffinase wirkung des Invertins," *Zts. Physiol. Chem*, 125: 28 (1923), in "Enzymes" by J. B. S. Haldane, Longman, London, 1930.

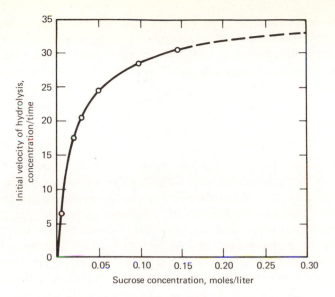

Fig. 2-3. Rate of sucrose hydrolysis by yeast saccharase as a function of substrate concentration. Adapted
from R. Kuhn in J. B. S. Haldane, "Enzymes," Long-
mans Greene and Co., London, 1930. Reprinted by
permission of the Longman Group Limited.

$$[ES] = \frac{[E]_t\,[S]}{[(k_2 + k_3)/k_1] + [S]}$$

If we now assume that the rate of reaction, V, is proportional to [ES],
then the maximum rate of reaction V_{max} will occur when all of the enzyme
is present as [ES], when $[ES] = [E]_t$. Further, if we set $K_m = (k_2 + k_3)/k_1$,
we obtain

$$V = \frac{V_{max}\,[S]}{K_m + [S]} \tag{2-30}$$

A plot of V versus [S] using this expression, the Michaelis-Menten
equation, gives a rectangular hyperbola—a curve of the form depicted in
Fig. 2-3. This lends support to the contention that our kinetic model is
descriptive of the reaction mechanism.

As we shall see later in this chapter, this rectangular hyperbolic
expression provides a useful model for microbial growth kinetics. At this
point we should examine the two constants that determine the shape of
the V versus [S] curve. The value of V_{max} is the maximum rate of reaction,
that is, the rate attained at high values of [S]. When V_{max} is reached,
further increases in [S] have no effect on reaction rate.

When $[S] \gg K_m$, analysis of Eq. 2-30 shows that $V = V_{max}$. The reaction
then is zero order with respect to [S], a property that agrees with one of

our initial experimental observations. The value of K_m is equal to the value of [S] when $V = V_{max}/2$. We call this value of the substrate concentration $[S]_{1/2}$ (see Fig. 2-3). This can be shown using Eq. 2-3. If $V = V_{max}/2$ and $[S] = [S]_{1/2}$, for example,

$$\frac{V_{max}}{2} = \frac{V_{max}\,[S]_{1/2}}{K_m + [S]_{1/2}}$$

or $K_m = [S]_{1/2}$

This constant, K_m, is variously called the half-velocity constant, the Michaelis-Menten constant, and is indicative of the strength of the bond between enzyme and substrate. The lower the value of K_m, the greater is the affinity between enzyme and substrate. Values of K_m for single substrate-enzyme reactions are generally between 10^{-2} and 10^{-5} M. The significance of this range of values is that it only requires 10^{-5} to 10^{-2} M of substrate to allow an enzyme to operate at half of its maximum rate.

In the substrate concentration range where $K_m \gg$ [S], Eq. 2-30 predicts that

$$V = \frac{V_{max}}{K_m}\,[S] = K'\,[S]$$

This is the expression for a reaction which is first order with respect to [S], an expression that fits the experimentally observed behavior in Fig. 2-3. Thus we see that the same reaction can be either zero or first order depending on the concentration of reactants.

2.6. EFFECT OF TEMPERATURE ON REACTION RATE

Experiments have shown that chemical reaction rates increase with increasing temperatures.[10] In many instances, the effect of temperature on reaction rate is related to its effect on the reaction rate constant. Arrhenius formulated the empirical rate law,

$$k = A e^{(Ea/RT)} \tag{2-31}$$

Equation 2-31 can be linearized as follows,

$$\ln k = \ln A - \frac{E_a}{RT} \tag{2-32}$$

---

[10] One exception is the effect of temperature on enzymatically catalyzed reactions. Rather than show a consistent increase with increasing temperature, these reactions have temperature optima. Above a certain temperature the structure of the enzyme becomes altered (denaturation) and its catalytic properties are reduced and eventually destroyed.

so that a plot of ln k versus $1/T$ should give a straight line with a slope of $-E_a/R$ and an intercept on the $1/T$ axis of ln $(A/(E_a/R))$,

where

A = the pre-exponential factor, or frequency factor, and is usually treated as a constant that is independent of temperature for a particular reaction,

E_a = the activation energy and is also treated as a constant for a particular reaction

R = the ideal gas constant

T = the temperature in °K

It is possible to relate the experimentally observed and empirically stated effect of temperature on reaction rate to a theoretical description of the effect of temperature on the energy level distribution of reacting species. To do this, we assume that the reaction

$$A + B \rightarrow \text{products}$$

proceeds through a high-energy, unstable intermediate known as a transition complex or an activated complex

$$A + B \rightarrow \text{activated complex}$$
$$\text{activated complex} \rightarrow \text{products}$$

We can illustrate the relative average energy levels of reactants, products, and activated complex by a diagram such as that presented in Fig. 2-4.

Figure 2-4 shows that the complete reaction proceeds, in this example, with a release of energy, ΔH. However, the intermediate formation of the activated complex requires the input of an average amount of energy E_a, the activation energy of the forward reaction. This energy, as well as ΔH, the heat of reaction, is released when the activated complex decomposes

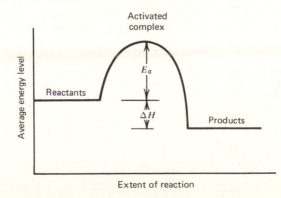

Fig. 2-4. Activation energy.

into the products of the reaction. For the reverse of this reaction to proceed, an energy input of $(\Delta H + E_a)$ is required to form the activated complex of which only E_a is released when A and B are reformed. The quantity $(\Delta H + E_a)$ is the activation energy of the reverse reaction.

Figure 2-4 is based on the average energy levels of reactants, products, and the activated complex. To understand how the idea of a high-energy activated complex is related to the effect of temperature on the reaction rate, we must examine the distribution of energy levels in a reactant at a given temperature (see Fig. 2-5).

The effect of temperature on the distribution of energy levels for a given species can be described by the Maxwell-Boltzmann theory, which relates the variation of the number of molecules with an energy *equal to or greater than* a given energy level to the absolute temperature. Thus

$$N = N_0 D \exp\left(\frac{-E}{RT}\right) \tag{2-33}$$

where

E = a stipulated energy level (see Fig. 2-5)
N_0 = total number of molecules
N = number of molecules with energy equal to or greater than E
R = gas constant
T = absolute temperature, °K
D = constant

From this equation and from Fig. 2-5 we can see that, for a given temperature, the value of N will decrease as the value of E is increased. Simply stated, as the stipulated energy level is raised, a smaller and smaller fraction of the total molecules are included in N.

Further examination of Eq. 2-33 reveals that for a given energy level, increasing the temperature should exponentially increase the population of molecules present at or above this energy level.

We can now postulate that only molecules with an energy level equal to or greater than E are capable of forming the high energy-level transition complex. Making this postulate draws our attention to the similarity

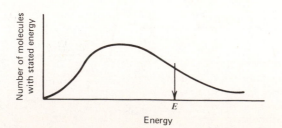

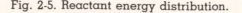

Fig. 2-5. Reactant energy distribution.

between Eqs. 2-31 and 2-33 and provides some theoretical basis for the empirically derived Arrhenius equation.[11]

The procedure for determining the values of E_a and A for a reaction is illustrated in the following example.

Example 2-3

The rate constant k for the reaction between hydrogen peroxide and potassium iodide to form iodine and water, $H_2O_2 + 2KI + 2H^+ \rightleftharpoons 2H_2O + I_2(aq) + 2K^+$, was found to vary as follows with temperature when the reactant concentrations were held constant at $[H_2O_2] = 5.56 \times 10^{-1} M$ and $[KI] = 1.2 \times 10^2 M$.

Temperature, °C	Rate Constant, k
44.5	1.66×10^{-3}
35.0	1.02×10^{-3}
25.7	6.63×10^{-4}
15.1	2.98×10^{-4}
4.5	1.17×10^{-4}

Determine the activation energy for this reaction and the value of the pre-exponential factor.

Solution

Plot the data according to Eq. 2-32.

$\ln k$	Temperature, °K (T)	$\frac{1}{T} \times 10^3$
−6.401	317.7	3.15
−6.888	308.2	3.25
−7.319	298.9	3.35
−8.118	288.3	3.47
−9.053	277.7	3.60

See Fig. 2-6. From this figure, slope $= \Delta(\ln k)/\Delta(1/T) = -E_a/R = -5750$ °K.
Since $R = 1.99$ cal/mole °K,

$$E_a = 11,400 \text{ cal/mole}$$
$$= 11.4 \text{ kcal/mole}$$

Using Eq. 2-32 and observing from Fig. 2-6 that when $\ln k = -7$, $1/T = 0.00327$ °K^{-1} yields

$$\ln A = -7 + \frac{11,400}{1.99 \times 305.8} = 11.73$$
$$A = 1.24 \times 10^5 \text{ liter/mole sec}$$

The units of A are the same as the units of k (see Eq. 2-31), since $e^{-E_a/RT}$ is dimensionless.

---

[11] The Maxwell-Boltzmann theory strictly applies to molecules in the gas phase, but the same concepts apply equally well to the distribution of the velocities among molecules in a solution. See J. A. Campbell, *Why Do Chemical Reactions Occur?* Prentice-Hall, Englewood Cliffs, N.J., 1965, for a discussion of this topic.

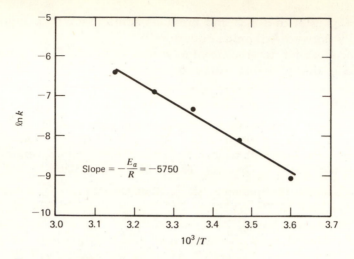

Fig. 2-6. Arrhenius plot for Example 2-3.

2.7. CATALYSIS

Catalysts are substances that increase the rate of reaction. They act by modifying the reaction pathway or the nature of the activated complex so that the reaction may proceed through an activated complex with a lower activation energy (Fig. 2-7).

For example, we can show that if the rate of a reaction (at 25°C) is increased by a factor of two upon the addition of a catalyst, the activation energy is reduced by 409 cal/mole. With no catalyst,

1.
$$\ln k_1 = \ln A - \frac{E_{a,1}}{RT}$$

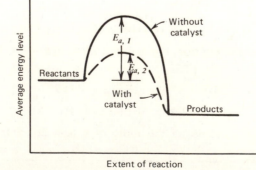

Fig. 2-7. Effect of a catalyst on activation energy.

With a catalyst,

2.
$$\ln k_2 = \ln (2k_1) = \ln A - \frac{E_{a,2}}{RT}$$

1–2.
$$\ln k_1 - \ln 2k_1 = \frac{E_{a,2}}{RT} - \frac{E_{a,1}}{RT}$$

$$RT \ln (\tfrac{1}{2}) = E_{a,2} - E_{a,1}$$

$$1.99 \times 298 \times \ln (\tfrac{1}{2}) = E_{a,2} - E_{a,1} = -409 \text{ cal/mole}$$

Thus E_a is decreased by 0.409 kcal/mole.

Although there is a change in E_a attributable to the presence of a catalyst the heat of the reaction, ΔH, is not altered. We can therefore deduce that a catalyst will only influence the rate of a reaction, not its extent. A catalyst may participate in the reaction by, for example, becoming part of the activated complex. Its concentration is not changed by the overall reaction, that is,

$$A + B + \text{catalyst} = \text{products} + \text{catalyst}$$

Catalysts may be generally classified as *homogeneous* when the catalyst is uniformly distributed on a molecular level throughout the reacting medium, or *heterogeneous* when the catalyst is present as a distinctly separate phase. Both types of chemical catalysts as well as the specific biological catalysts known as *enzymes* are important in aquatic chemistry.

Hydrogen ion (H^+) and hydroxyl ion (OH^-) are common catalysts in aquatic systems. The effect of their catalysis is manifested by changes in reaction rate that occur with changes in pH. For example, one of the major constituents of household synthetic detergents is the salt of condensed phosphoric acid such as pyrophosphoric acid, $H_4P_2O_7$, and tripolyphosphoric acid, $H_5P_3O_{10}$. These compounds react with water in a reaction known as hydrolysis to form orthophosphoric acid thus:

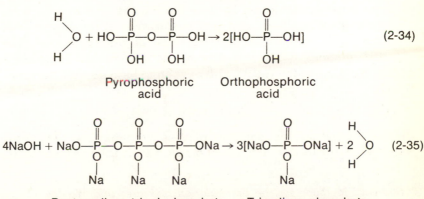

The condensed phosphates contribute approximately 50 to 60 percent of the total phosphate in domestic wastewater. They are used in the commercial synthetic detergent formulation to complex the hardness ions, Ca^{2+} and Mg^{2+}, and prevent their interaction with the surfactant. Ortho-phosphates do not perform this complexing function as well as the condensed phosphates, so it is important that the condensed phosphates are stable (i.e., do not undergo hydrolysis as in Eqs. 2-34 and 2-35) during the cleansing process. The hydrolysis of condensed phosphates is cata-lyzed by H^+. As Fig. 2-8 shows, the time for 5 percent hydrolysis of a pyrophosphate solution at 10°C is about 1 year at pH 4, many years at pH 7, and over a century at pH 10. In the typical washing machine environment of 65°C and pH 9, a 5 percent decomposition of pyrophosphate would take several days so that in the typical wash cycle of 10 to 15 min, hydrolysis will be insignificant.

The rate equation for the hydrolysis of pyrophosphate is $-d[P_2O_7^{4-}]/dt = k[P_2O_7^{4-}]$. The rate constant is highly dependent on $[H^+]$ and varies from 0.534/hr at pH = 0 to 0.0318 at pH = 1.1 to 0.00272/hr at pH 3.3. Most domestic wastewaters have pH values in the range 6.5 to 8 and temper-atures between 10 to 20°C. From Fig. 2-8 we would predict that pyro-phosphates would be quite stable under these conditions. However, we observe that much of the condensed phosphate in domestic wastewater has reverted to orthophosphate by the time that the wastewater reaches the treatment plant, usually a time period of significantly less than 1 day. Moreover, no condensed phosphates ever survive biological waste treat-ment processes—again a maximum period of about 1 day. The reason for these observations is again catalysis but in this instance catalysis by enzymes. Most microorganisms possess an enzyme (a specific biological catalyst) that will mediate the hydrolysis of inorganic condensed phos-phates. An example of such an enzyme is pyrophosphate phosphohydro-lase, which catalyzes the hydrolysis of inorganic pyrophosphate to 2 moles of orthophosphate.

The efficiency with which aeration devices transfer oxygen to water or to wastewater is measured by determining the rate at which the device increases the dissolved oxygen concentration of the liquid. Prior to the test, the liquid under investigation must be deoxygenated. The recom-mended method[12] for accomplishing deoxygenation is to add sodium sulfite, Na_2SO_3, which reacts with oxygen to form sodium sulfate, Na_2SO_4, thus:

$$2SO_3^{2-} + O_2 \rightleftharpoons 2SO_4^{2-}$$

thereby depleting the solution of dissolved oxygen. This reaction is extremely slow in the absence of a catalyst. It takes about 10 minutes to

---

[12] *Standard Methods for the Examination of Water and Wastewater.* American Public Health Assoc., 14th ed., 1975, p. 85.

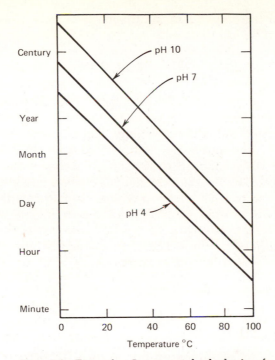

Fig. 2-8. Time for 5 percent hydrolysis of pyrophosphate (sodium salt) in a 1 percent (approximate) solution. Adapted from J. R. Van Wazer, *Phosphorus and Its Compounds*, Vol. 1, Interscience, New York, 1966, p. 454. Reprinted by permission of John Wiley & Sons, Inc.

reduce the concentration of dissolved oxygen from 10 to 7 mg/liter when catalysts are not present. The introduction of 0.01 mg/liter cobalt ion allows the complete removal of dissolved oxygen in 15 to 20 seconds.[13]

A further example of catalysis by metal ions is the effect of cupric ion, Cu^{2+}, on the oxidation of cyanide by ozone.[14] In wastes from industries such as metal plating and coal carbonization, cyanide can be oxidized to cyanate by ozone.

$$3CN + {}^-O_3 \rightarrow 3CNO^-$$

The rate of this reaction is more than doubled in the presence of Cu^{2+}.

[13] D. J. Pye, "Chemical Fixation of Oxygen," *J. Am. Water Works Association*, **39**: 1121–1127 (1947).

[14] K. K. Khandelwal, A.J. Barduhn, and C. S. Grove, Jr., "Kinetics of Ozonation of Cyanides," *Ozone Chemistry and Technology*, Advances in Chemistry Series 21, American Chemical Society, Washington, D.C., 1959, p. 78.

The rate law for the reaction in the presence of an excess and constant ozone concentration is

$$\frac{-d[CN^-]}{dt} = k[CN^-]^{1/3}$$

It shows no dependence on Cu^{2+} concentration above a certain minimum concentration indicating that its effect is purely catalytic. The fractional order of the reaction indicates a complex mechanism.

An example of heterogeneous catalysis related to the field of water chemistry is the use of cobaltic oxide (Co_3O_4) catalyst in the determination of organic carbon. The Beckman Model 900 carbon analyzer consists of a quartz tube heated to 950°C, containing asbestos that has been impregnated with cobalt nitrate. At 950°C the cobalt nitrate decomposes to cobaltic oxide. A sample of 20 μl is injected into a stream of oxygen that passes over the heated catalyst that enhances the oxidation of organic carbon to CO_2. The CO_2 concentration is determined by an infrared detector.

2.8.   EMPIRICAL RATE LAWS

Our discussion up to this point has dealt with the use of kinetics to describe the rates of chemically well-defined reactions and to explore the mechanism of reactions. Kinetic formulations can be used in what one might call the opposite sense—that is, to provide an empirical mathematical framework in which data from complex reactions can be analyzed. The objective here is a simplification of complex situations, not the discovery of exact mechanism from kinetic analysis. Two examples of this use of kinetic formulations that are relevant to aquatic systems will be given here. The first concerns treatment of data from the biochemical oxygen demand (BOD) test that is used variously to determine the strength of wastewater, the degree of organic pollution in receiving waters and the rate at which the organic matter in wastewater can be degraded by microorganisms under aerobic conditions [i.e., when dissolved oxygen (D.O.) is present].

In the BOD test, a diluted sample of wastewater is incubated for a period (5 days at 20°C in the standard BOD_5 test) and the amount of dissolved oxygen consumed in this time period is measured. The BOD_5 is then calculated as

$$BOD_5 = \frac{dissolved\ oxygen\ (mg/liter)}{dilution\ factor}$$

where dilution factor = volume sample/(volume of sample + volume dilution water).

The BOD_5 measurement determines only one point on the curve that relates dissolved oxygen uptake to time. The reaction monitored by measuring oxygen uptake has often been expressed as

$$\text{organics} + O_2 \rightarrow CO_2 + H_2O + \text{oxidized products} + \ldots$$

and the rate of this reaction in the presence of an excess of dissolved oxygen has often been stated to be first order with respect to organic matter. It must be recognized that this is a totally empirical statement because the nature of the degradable organic matter in wastewater is not well-defined and certainly the individual microbial degradation rates of all organic compounds in water are not known. Nevertheless, the overall time course of the consumption of oxygen for degradation of carbonaceous organic matter in the BOD test has many of the characteristics of a first-order reaction. Since the rate of BOD exertion (or organic matter degradation) is of importance in waste treatment plant design and receiving water management, we can usefully employ the empirical first-order relationship to formulate the "BOD curve" (see Fig. 2-9).

If L = the concentration of degradable organic matter at any time t (days), then for a first-order reaction we can write

$$\frac{dL}{dt} = -kL \tag{2-36}$$

Integrating, we obtain

$$L = L_0 e^{-kt} \tag{2-37}$$

where

L_0 = the original concentration of biodegradable organic matter
k = rate constant/day

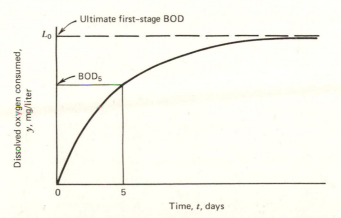

Fig. 2-9. "First-order" representation of first-stage carbonaceous BOD.

Because L cannot be measured directly, the equation must be modified to replace L with a parameter that can be measured as a function of time. We can achieve this modification by setting

$$y = L_0 - L \tag{2-38}$$

where y = the amount of material that has been degraded at any time, t.

The value of y can be assessed by determining dissolved oxygen consumption at any time, through measurements of dissolved oxygen concentration of the sample,

$$y = \text{D.O.}_{\text{initial}} - \text{D.O.}_{\cdot t} \tag{2-39}$$

Substituting Eq. 2-38 in Eq. 2-37, we obtain

$$L_0 - L = y = L_0 - L_0 e^{-kt}$$

or

$$y = L_0(1 - e^{-kt}) \tag{2-40}$$

which is the classical first-order empirical rate equation for BOD.

Various procedures exist for analyzing y versus t data to obtain the constants k and L_0. One of these, the Thomas slope method, is presented here.[15] This procedure involves developing a straight line equation that approximates the relationship of y and t.

If we expand the $(1 - e^{-kt})$ portion of Eq. 2-40, we obtain

$$1 - e^{-kt} = kt \left[1 - \frac{kt}{2} + \frac{(kt)^2}{6} - \frac{(kt)^3}{24} + \cdots \right] \tag{2-41}$$

The quantity $[kt(1 + kt/6)^{-3}]$ has a similar expansion,

$$kt \left(1 + \frac{kt}{6} \right)^{-3} = kt \left(1 - \frac{kt}{2} + \frac{(kt)^2}{6} - \frac{(kt)^3}{21.6} + \cdots \right) \tag{2-42}$$

Comparing these two expansions, we see that a minor difference appears only in the fourth term; on this basis we can say

$$L_0 (1 - e^{-kt}) \cong L_0 \left[kt \left(1 + \frac{kt}{6} \right)^{-3} \right] \tag{2-43}$$

or

$$y \cong L_0 \left[kt \left(1 + \frac{kt}{6} \right)^{-3} \right] \tag{2-44}$$

Rearranging Eq. 2-44, we obtain an equation of a straight line

$$\left(\frac{t}{y} \right)^{1/3} = (L_0 k)^{-1/3} + \left(\frac{k^{2/3}}{6L_0^{1/3}} \right) t \tag{2-45}$$

[15] H. A. Thomas, Jr., "Graphical Determination of BOD Curve Constants," *Water and Sewage Works*, 97: 123 (1950).

Letting

$$(L_0 k)^{-1/3} = a \qquad (2\text{-}46)$$

and

$$\frac{k^{2/3}}{6L_0^{1/3}} = b \qquad (2\text{-}47)$$

it can be shown that

$$k = \frac{6b}{a} \qquad (2\text{-}48)$$

and

$$L_0 = \frac{1}{ka^3} \qquad (2\text{-}49)$$

With experimental data relating y and time, we can plot $(t/y)^{1/3}$ versus time. However, experimental values of $y > 0.9L_0$ should not be used in fitting the straight line because Eq. 2-42 is significantly different from Eq. 2-41 in this range. Fitting the plot with a straight line and determining the intercept, a, and the slope, b, Eqs. 2-47 and 2-48 can be used to give k and L_0. The following example illustrates the procedure.

We must be careful to bear in mind as we proceed with these calculations that they are empirically based. It is neither wise nor proper to read any more significance into them than that; it is certainly improper to infer, as many have done, that we can deduce anything about the reaction mechanism from these calculations.

Example 2-4

Standard Methods[16] states that the rate constant (base 10) for the oxygen uptake of a mixture of glucose (150 mg/liter) and glutamic acid (150 mg/liter) should be between 0.16 to 0.19/day when proper seeding with microorganisms is practiced. The data in Table 2-2 were obtained at 20°C on a candidate seed material by a University of California graduate student. Does the seed satisfy the above specifications?

TABLE 2-2

Time, t, days	BOD, y, mg/liter	t/y	$(t/y)^{1/3}$
1	122	0.0082	0.202
2	117	0.017	0.25
3	184	0.016	0.25
4	193	0.021	0.276
5	203	0.025	0.292
6	205	0.029	0.307
7	207	0.034	0.324

[16] *Standard Methods for the Examination of Water and Wastewater*, 14th ed., American Public Health Association, 1975, p. 548.

Solution

From Fig. 2-10, a plot of $(t/y)^{1/3}$ versus t according to Eq. 2-45, we find

$$a = 0.182 \quad \text{and} \quad b = 0.0216$$

From Eq. 2-48,

$$k(\text{base } e) = 6b/a = 6 \times 0.0216/0.182 = 0.71/\text{day}$$

or $k(\text{base } 10) = 0.71/2.3 = 0.31$. From Eq. 2-49,

$$L_0 = \frac{1}{ka^3} = \frac{1}{0.71(0.182)^3} = 234 \text{ mg/liter}$$

Since k is much greater than 0.16 to 0.19/day, the seed is unacceptable.

The second widespread use of empirical kinetic expressions in the wastewater and water pollution control fields is in the kinetics of growth of mixed microbial cultures on single or multiple substrates. Typical examples are the growth of activated sludge solids on wastewater BOD

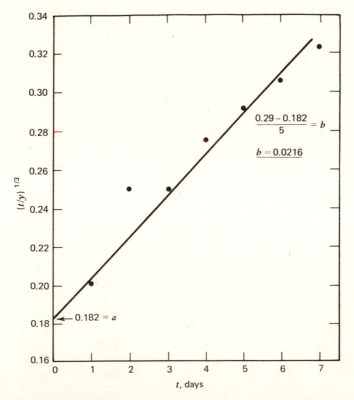

Fig. 2-10. Plot of $(t/y)^{1/3}$ versus t to determine k and L_0 for Example 2-4.

and the growth of algae on nutrients such as nitrate and phosphate. The French microbiologist Monod[17] derived a relationship between the growth rate of a pure culture of bacteria and the concentration of the substrate that limited the growth of the bacteria (the so-called growth limiting substrate).

$$M = \frac{M_{max}\,[S]}{K_s + [S]} \qquad (2\text{-}50)$$

where

M = growth rate = $(1/X)(dX/dt)$
M_{max} = maximum growth rate
X = microbial mass or number concentration
K_s = substrate concentration when $M = M_{max}/2$

The relationship, expressed in Eq. 2-50, bears a striking resemblance to the Michaelis–Menten expression developed previously for enzyme substrate interaction. Indeed, some workers believe that the analogy is not coincidental, stating that microorganisms are 'bags full of enzymes' so that it is not surprising that their growth rate is related to the behavior of the catalysts that mediate the many reactions that contribute to growth.

The growth-rate-substrate relationship is expressed empirically in applications such as biological wastewater treatment. In these processes we grow a mixed and variable population of microorganisms (e.g., activated sludge) on a mixed and variable substrate (e.g., domestic sewage). We have the same problem that arose in attempting to assess the kinetics of the BOD reaction. We do not have a definitive measure of growth-limiting substrate concentration. Neither do we have a well-defined measurement of microorganism concentration. Without going into the rationale for their selection, the parameter used for microorganisms is usually the mass of suspended solids or the mass of volatile suspended solids in the activated sludge suspension. For growth-limiting substrate we usually use BOD or COD concentration. These parameters are readily measured, but they are collective parameters that indicate rather than specify exactly what we should be measuring. As a consequence the empirical nature of the microbial growth expression is compounded. In spite of this, however, the expression has useful application in the design and prediction of performance of biological waste treatment plants. For example, the data in Fig. 2-11 show the relationship between the growth rate and substrate concentration, expressed as soluble COD, for an activated sludge unit operated on a protein/beef extract substrate. The growth rate is a first-order function of the degradable organic matter concentration. We can see that for the range of conditions studied, there

[17] J. Monod, "La Technique de Culture Continue, Théorie et Applications," *Ann. Inst. Pasteur, 79:* 390–410 (1950).

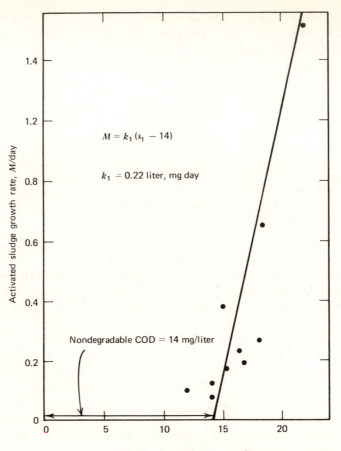

Fig. 2-11. First-order relationship between the activated sludge growth rate and the degradable effluent soluble COD concentration. From D. W. Eckhoff and D. Jenkins, "Activated Sludge Systems, Kinetics of the Steady and Transient States," SERL Rept. 67-12, University of California, Berkeley.

was a linear relationship between M and [S] indicating that we are in the region where $K_s \gg$ [S] and $M = k'$[S]. Such an observation is indeed commonplace for activated sludge systems treating wastewaters at the practically employed rates of treatment. The curve in Fig. 2-11 would allow us to state that if we wished to produce an effluent with a soluble COD of less than 20 mg/liter from the activated sludge treatment of this waste, we should design the treatment plant to accommodate an activated sludge growth rate of less than about 1.2/day.

2.9. PROBLEMS

1. The rate constant of a first-order reaction is 2.5×10^{-6}/sec, and the initial concentration is $0.1\ M$. What is the initial rate in moles/liter sec, moles/cc sec, and in moles/cc min?

2. A first-order reaction is 40 percent complete at the end of 50 minutes. What is the value of the rate constant in $\sec^{-1}$? In how many minutes will the reaction be 80 percent complete?

3. If the initial rate of a second-order reaction at 20°C is 5×10^{-7} moles/liter sec, and the initial concentrations of the two reacting substances are each $0.2\ M$, what is the k in liters/mole sec? If the activation energy is 20 kcal/mole, what is k at 30°C?

4. A reaction

$$2A \rightarrow 3P$$

is first order with respect to A. If the half-life of A (time for 50 percent to react) is t_1 seconds, and the reaction is irreversible,
(a) Write an expression for the reaction rate constant in terms of t_1.
(b) Using the expression for the reaction rate constant determined in part (a), develop an expression for the time required for 90 percent of A to react.

5. Evaluate the following data to determine whether the reaction

$$A \rightarrow products$$

is first or second order. Calculate the rate constant and be certain to give its units.

A (mM/liter)	Time (sec)
1.00	0
0.50	11
0.25	20
0.10	48
0.05	105

6. The forward rate law for the reaction

$$H^+ + OH^- \rightleftharpoons H_2O$$

is given as

$$\frac{d[H^+]}{dt} = \frac{d[OH^-]}{dt} = -k[H^+][OH^-]$$

The rate constant, k, at 20°C was found to be 1.3×10^{11} liters/mole sec. Assuming NaOH is rapidly added and mixed with HCl in aqueous solution such that the initial concentrations of H^+ and OH^- are $10^{-4}\ M$, how much time is required for one half of the acid and base to react? (Assume for this calculation that the reaction is irreversible.)

7. The temperature dependence of the reaction rate is frequently expressed quantitatively using parameters other than E_a. For example, the following expression for the reaction rate constant for the BOD test is often used:

$$k_{T_2} = k_{20}(\theta)^{(T_2-293)}$$

where T_2 is the temperature in °K and k_{T_2} is the rate constant at that temperature.

(a) Show that

$$\theta = \exp \frac{E_a}{RT_1T_2}$$

and thus that θ is a function of T_1 and T_2.

(b) Determine E_a if $\theta = 1.047$ and $T_2 = 293°K$.

(c) If $T_2 = 283°K$ and θ remains the same, what is the value of E_a?

8. In the field of biology the term Q_{10} where

$$Q_{10} = \frac{k_{(T+10)}}{k_T}$$

is frequently used. Although Q_{10} does vary with E_a and temperature, if E_a is approximately a constant for certain types of reactions, Q_{10} values can be used to good advantage. If the temperature is 25°C, what is E_a if $Q_{10} = 1.8$?

9. A recent study has shown that monochloramine, NH_2Cl, "decay" in wastewater is very slow, especially in comparison with free chlorine, $HOCl$ and OCl^-. The exposure of NH_2Cl to light was found to increase the rate of decay significantly. When light was completely excluded from the sample, 20 percent decay took place in 8 hours and the data conformed to the first-order rate law.

(a) Assuming (1) a treatment plant is discharging effluent containing 2 mg/liter NH_2Cl, as Cl_2; (2) a 1:10 dilution (one part effluent plus nine parts receiving water) is achieved with complete mixing; (3) the receiving water is not exposed to light; and (4) a level greater than 0.002 mg/liter NH_2Cl, as Cl_2, is deleterious to trout, how long after discharge will it take for the receiving water to become "acceptable" for trout?

(b) Assuming that each 12-hour period of no light is followed by a 12-hour period of light during which the applicable first-order reaction rate constant is 0.3/hr, how long will it take for the receiving water to become acceptable for trout?

10. The effluent from a secondary clarifier has a pH of 8.3, an ammonia (NH_3) concentration of 34 mg/liter, and is dosed with 10^{-3} M $HOCl$. The reaction is reversible but assume conditions are such that only the forward reaction must be considered.

$$NH_3 + HOCl \rightarrow NH_2Cl + H_2O$$

It has a rate law,

$$\frac{d[NH_3]}{dt} = -k[NH_3][HOCl]$$

where $k = 5.5 \times 10^6$ liters/mole sec at 15°C.

(a) Calculate the time for 90 percent of the HOCl to react.

(b) If $k = 1 \times 10^8$ liters/mole sec at 40°C, what is the activation energy?

11.   BOD data have been determined as follows:

t (days)	Y (mg/liter)
1	0.705
2	1.060
3	1.540
4	1.700
5	1.880
7	2.310
10	2.570
12	2.805

Find L_0 and k (both base e and base 10) assuming a first-order reaction.

2.10.   ADDITIONAL READING

Campbell, J.A., *Why Do Chemical Reactions Occur?* Prentice-Hall, Englewood Cliffs, N.J., 1965.

Frost, A. A., and R. G. Pearson, *Kinetics and Mechanisms.* John Wiley, New York, 1961.

Hammes, G. G., *Principles of Chemical Kinetics.* Academic Press, New York, 1978.

King, E. L., *How Chemical Reactions Occur.* W. A. Benjamin, Menlo Park, Calif., 1964.

Laidler, K. J., *Chemical Kinetics.* McGraw-Hill, New York, 1965.

Skinner, G. B., *Introduction to Chemical Kinetics.* Academic Press, New York, 1974.

CHAPTER
3
CHEMICAL
EQUILIBRIUM

3.1. INTRODUCTION

In this chapter we will describe methods that will allow us to answer the questions "Will this reaction go?" and, if so, "How far can it proceed?." These are extremely important questions to pose prior to setting about the investigation of any chemical or biochemical system. If a reaction is impossible, there is absolutely no sense whatsoever in attempting to determine its rate, its mechanism, or its use in a water or waste treatment process. If, for example, you were to be approached by a salesman trying to sell you a chemical that would prevent carbon dioxide and nitrogen from combining with chloride to form dangerous hydrogen cyanide and hypochlorous acid during treatment of a particular wastewater, you could, by using techniques similar to those developed in this chapter, tell him that the reaction was not possible with or without the chemical and send him on his way. And you could do this conclusively without conducting a single experiment!

The techniques we will develop are based on the branch of science known as thermodynamics. The theoretical aspects of thermodynamics are extremely precise and orderly; its mathematical basis is complex. We, however, are only interested in what thermodynamics can do for us as a tool in solving problems of chemical equilibrium. We are in a situation similar to the automobile driver using a road map. Not many drivers thoroughly understand the principles of geometry and plane trigonometry that were used to draw the map. However, most know how to read a map and in doing so could manage reasonably well to get from Urbana to Berkeley.

3.2. THE DYNAMIC NATURE OF CHEMICAL EQUILIBRIUM

Let us examine the hypothetical, elementary, reversible reaction taking place at constant temperature,

$$aA + bB \rightleftharpoons cC + dD \tag{3-1}$$

The *reactants* A and B combine to form the *products* C and D. In this reaction, a moles of A combine with b moles of B to form c moles of C and d moles of D. If we were to introduce *A* and *B* into a suitable reaction

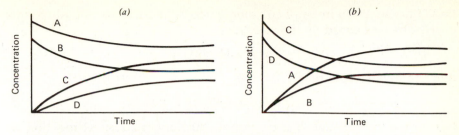

Fig. 3-1. Course of reaction between A, B, C, and D. (a) Initially only A and B are present; (b) initially only C and D are present.

vessel and at time intervals analyze the contents of the reaction vessel, we would obtain a concentration versus time profile such as in Fig. 3-1(a).

The concentrations of A and B decrease until they reach values that do not change with time, while the concentrations of C and D increase from zero to time-invariant values. If we were to add only the products of the reaction, C and D, to the reaction vessel under the same experimental conditions, we would observe a concentration-time change such as that depicted in Fig. 3-1(b). After a while A, B, C, and D concentrations all reach constant, time-invariant levels. If the ratio of the concentration of products to reactants, $[C]^c[D]^d/[A]^a[B]^b$ is the same as that attained in the previous experiment when A and B were initially present, we say that the reaction is at equilibrium and we call the concentrations of species present at that time equilibrium concentrations. The ratio is the so-called equilibrium constant, K.

$$\frac{[C]^c[D]^d}{[A]^a[B]^b} = K \qquad (3\text{-}2)$$

The unit of concentration as indicated by [ ] is mole/liter, but other concentration units also can be used.[1] From this discussion we learn that the equilibrium state can be approached from both directions.

When we investigate the rate at which the equilibrium condition is approached, we can deduce that the equilibrium condition is a dynamic one, not a static situation. The interaction of reactants and products does not cease when equilibrium is reached. The forward and reverse reactions proceed at such a rate that the ratio of concentrations of products to reactants (as described by the equilibrium constant, Eq. 3-2), remains constant. Another way of stating this is that a chemical reaction is at equilibrium if its forward rate of reaction, v_f, is equal to the rate of the reverse reaction, v_r. For example, we have seen previously in Chapter 2

---

[1] Strictly speaking, the equilibrium constant is defined in terms of activity, or active concentrations, as discussed later in this chapter. Concentration, such as the molar concentration, often can be used as a good approximation of activity for species in dilute aqueous solutions.

that if the reaction in Eq. 3-1 is elementary, the forward rate of reaction, v_f, can be expressed by the rate law,

$$v_f = k_1 [A]^a [B]^b \qquad (3\text{-}3)$$

and the reverse rate by v_r,

$$v_r = k_2 [C]^c [D]^d \qquad (3\text{-}4)$$

Initially, if only A and B are present, the forward rate of reaction will proceed at a finite rate while there will be no reverse reaction because no C and D are present. However, as soon as the reaction of A and B produces C and D, they will combine, and by the reverse reaction produce A and B. The reaction will proceed until the opposing reaction rates are equal, and

$$v_f = v_r \qquad (3\text{-}5)$$

Therefore,

$$k_1 [A]^a [B]^b = k_2 [C]^c [D]^d \qquad (3\text{-}6)$$

and

$$\frac{[C]^c [D]^d}{[A]^a [B]^b} = \frac{k_1}{k_2} = K \qquad (3\text{-}7)$$

The equilibrium constant is thus the ratio of the rate constants of the forward and the reverse reactions—a fact that underscores the dynamic nature of equilibrium.[2]

3.3.   THE THERMODYNAMIC BASIS OF CHEMICAL EQUILIBRIUM

A knowledge of the position of chemical equilibria is of the utmost importance to us in water chemistry. By knowing the position of chemical equilibria, we can determine whether it is possible for certain reactions between reactants at given concentrations to proceed. For example, we can provide answers to questions such as: Will calcium carbonate tend to precipitate or dissolve in this water? Can I possibly oxidize sulfide with nitrate? and so on. There are two general ways to answer questions like these. The first is to do an experiment and the second is to calculate the answer using previously determined equilibrium data. Although the first way may be more enjoyable to those who like puttering around in the laboratory, the second approach is far superior if time is of the essence.

To explore the various techniques that can be used to answer the

[2] If the chemical reaction is not elementary, derivation of K from rate constants is still possible but is more complex. See, e.g., T. S. Lee, in *Treatise on Analytical Chemistry*, I. M. Kolthoff, and P. J. Elving, eds., Part I, Vol. 1, Wiley-Interscience, New York, 1959.

question "Will this reaction go?" we must delve a little into the thermo-dynamic basis of equilibrium. Our treatment of thermodynamics will be brief; we will present only that which is useful and immediately applicable for our purpose. For a more comprehensive coverage of the topic, the student is referred to other texts (see Section 3-7).

Most of the reactions with which we are concerned in water chemistry take place in a closed system or can be analyzed as though they take place in a closed system. For thermodynamic purposes a closed system is one to which matter cannot be added or removed. Energy, however, may flow across its boundaries. Since, in addition to working with closed systems, we usually are interested in systems at constant temperature and pressure, we can make extensive use of the thermodynamic expression for free energy,

$$G = H - TS \qquad (3\text{-}8)$$

where

G = Gibbs free energy, kcal
T = absolute temperature, °K
S = entropy, kcal/°K
H = enthalpy, kcal

The enthalpy is the total energy content of an element or compound. The free energy is that part of the total energy that is available to perform "useful work," that is, other than pressure-volume work. The entropy is a sort of internal manifestation of energy that can be visualized in several ways; for example, entropy is often defined as describing the degree of order or organization in a system. Highly structured materials (e.g., well-formed crystals) have low entropy while randomly arranged systems (e.g., a gas) have high entropy. The product TS is that part of the total energy which is not available for useful work.

For closed systems at constant pressure and constant temperature, *the criterion for equilibrium is that the total free energy of the system (G_T) is a minimum.* For example, consider the reversible reaction previously examined in Eq. 3-1,

$$aA + bB \rightleftarrows cC + dD$$

If we were to add A and B to a reaction vessel and calculate the total free energy of the system as a function of extent of reaction as the reaction proceeded, we would find something like that depicted by the solid line at the left side of the diagram in Fig. 3-2. The total free energy, G_T, is the sum of the free energies of each of the reaction components. For example, if n_A, n_B, n_C and n_D represent the number of moles of A, B, C and D that are present and $\overline{G}_A$, $\overline{G}_B$, $\overline{G}_C$, and $\overline{G}_D$ represent the free energy/mole of each substance, then

$$G_T = n_A\overline{G}_A + n_B\overline{G}_B + n_C\overline{G}_C + n_D\overline{G}_D$$

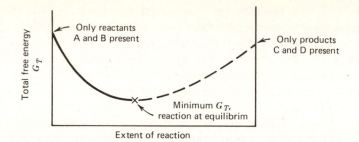

Fig. 3-2. Variation of Gibbs free energy for the chemical reaction aA + bB ⇌ cC + dD. Only reactants are present at the far left side of the diagram and only products at the far right side.

Conversely, if we were initially to add only C and D to the reaction vessel, calculation of the total free energy of the system as the reaction proceeded to form A and B would produce a curve of the form shown by the dashed line on the right-hand side of Fig. 3-2. In each case the reaction only proceeds spontaneously, or without any external help, as long as the value of G_T decreases. Because of this there exists, at a certain extent of reaction, a point where G_T is at a minimum. This point can be spontaneously reached from either the product or reactant side, and it is the equilibrium point of the system. Thus we may state that the equilibrium condition of a reaction is the point at which G_T is a minimum. Also, we may deduce that reaction in the direction that decreases G_T is spontaneous while reaction in the direction that increases G_T is not spontaneous or will not occur in a closed system.

It can be shown that as any reaction proceeds an incremental amount, the change in G_T is proportional to ΔG where

$$\Delta G = \left(\sum_i \nu_i \overline{G}_i \right)_{products} - \left(\sum_i \nu_i \overline{G}_i \right)_{reactants} \tag{3-9}$$

where ν_i is the stoichiometric coefficient (e.g., a, b, c, and d in Eq. 3-1) and $\overline{G}_i$ is the free energy per mole.

We can therefore state that if,

1. ΔG is < 0 (i.e., ΔG is negative and thus G_T decreases as the reaction proceeds); the reaction may proceed spontaneously as written.

2. ΔG is > 0 (i.e., ΔG is positive and thus G_T would increase if the reaction were to proceed); the reaction cannot proceed spontaneously as written. Conversely, it may proceed spontaneously in the opposite direction to which it is written.

3. $\Delta G = 0$ (i.e., G_T is at a minimum); the reaction is at equilibrium and will not proceed spontaneously in either direction.

Values of ΔG for a reaction provide us with a powerful tool to predict whether or not reactions are possible. To calculate ΔG for the general reaction given by Eq. 3-1, we use the relationship

$$\Delta G = \Delta G^\circ + RT \ln \frac{\{C\}^c \{D\}^d}{\{A\}^a \{B\}^b} \qquad (3\text{-}10)$$

where

$$\Delta G^\circ = \left(\sum_i \nu_i \overline{G}_i^{\,\circ} \right)_{products} - \left(\sum_i \nu_i \overline{G}_i^{\,\circ} \right)_{reactants} \qquad (3\text{-}11)$$

$\{\ \}$ = activity, or active concentration (discussed later in this section)

$\overline{G}_i^{\,\circ}$ = free energy/mole of species i at 25°C and 1 atmosphere pressure

Equation 3-10 is developed in more detail later. The first step in determining ΔG is to determine $\overline{G}_i^{\,\circ}$, the free energy per mole of each reactant and product at the standard state condition of 25°C and 1 atm total pressure. The term $\overline{G}_i^{\,\circ}$ is called the standard free energy per mole of species i.

At standard state *every element* is assigned, by convention, a free energy of zero per mole. Thus $H_{2(g)}$, $O_{2(g)}$, $C_{graphite(s)}$, and so forth, all are assigned free energy values of zero kcal/mole. Also, to establish a baseline for ionic substances in solution, H^+ at a concentration of 1 mole/liter in an ideal solution and at standard state conditions has been assigned a free energy of zero.

This convention is necessary because it is impossible to measure absolute values of free energy. However, we can measure changes in free energy. Thus, assigning a value of zero to elements at standard state allows us to measure the free energy change involved in forming compounds at standard state from their component elements at standard state. This free energy change is called the standard free energy of formation, $\Delta \overline{G}_f^{\,\circ}$. Table 3-1 presents a summary of $\Delta \overline{G}_f^{\,\circ}$ values for some substances commonly encountered in water chemistry; for example, we can determine that the formation of calcite ($CaCO_{3(s)}$) from carbon, calcium, and oxygen has the following free energy,

$$C_{(graphite)} + \tfrac{3}{2}O_{2(g)} + Ca_{(s)} \rightarrow CaCO_{3(s)}; \qquad \Delta \overline{G}_f^{\,\circ} = -269.78 \text{ kcal/mole}$$
$$\text{calcite}$$

The minus sign indicates that free energy is released or given off. We can also deduce that to break down 1 mole of calcite into 1 mole of Ca metal, 1 mole of graphite, and $1\tfrac{1}{2}$ moles of oxygen would require the input of $+269.78$ kcal of free energy per mole.

TABLE 3-1 Thermodynamic Constants for Species of Importance in Water Chemistry.[a]

Species	$\Delta \bar{H}_f^\circ$ kcal/mole	$\Delta \bar{G}_f^\circ$ kcal/mole
$Ca^{2+}_{(aq)}$	-129.77	-132.18
$CaCO_{3(s)}$, calcite	-288.45	-269.78
$CaO_{(s)}$	-151.9	-144.4
$C_{(s)}$, graphite	0	0
$CO_{2(g)}$	-94.05	-94.26
$CO_{2(aq)}$	-98.69	-92.31
$CH_{4(g)}$	-17.889	-12.140
$H_2CO^*_{3(aq)}$	-167.0	-149.00
$HCO^-_{3(aq)}$	-165.18	-140.31
$CO^{2-}_{3(aq)}$	-161.63	-126.22
CH_3COO^-, acetate	-116.84	-89.0
$H^+_{(aq)}$	0	0
$H_{2(g)}$	0	0
$Fe^{2+}_{(aq)}$	-21.0	-20.30
$Fe^{3+}_{(aq)}$	-11.4	-2.52
$Fe(OH)_{3(s)}$	-197.0	-166.0
$Mn^{2+}_{(aq)}$	-53.3	-54.4
$MnO_{2(s)}$	-124.2	-111.1
$Mg^{2+}_{(aq)}$	-110.41	-108.99
$Mg(OH)_{2(s)}$	-221.00	-199.27
$NO^-_{3(aq)}$	-49.372	-26.43
$NH_{3(g)}$	-11.04	-3.976
$NH_{3(aq)}$	-19.32	-6.37
$NH^+_{4(aq)}$	-31.74	-19.00
$HNO_{3(aq)}$	-49.372	-26.41
$O_{2(aq)}$	-3.9	3.93
$O_{2(g)}$	0	0
$OH^-_{(aq)}$	-54.957	-37.595
$H_2O_{(g)}$	-57.7979	-54.6357
$H_2O_{(\ell)}$	-68.3174	-56.690
$SO^{2-}_{4(aq)}$	-216.90	-177.34
$HS^-_{(aq)}$	-4.22	3.01
$H_2S_{(g)}$	-4.815	-7.892
$H_2S_{(aq)}$	-9.4	-6.54

Source. Condensed from the listing of R. M. Garrels and C. L. Christ, *Solutions, Minerals, and Equilibria*, Harper & Row, New York, 1965; and *Handbook of Chemistry and Physics*, Chemical Rubber Publishing Company, Cleveland, Ohio.

[a] For a hypothetical ideal state of unit molality, which is approximately equal to that of unit molarity.

Now let us use this information to determine the free energy change associated with the reaction in which calcite dissolves in acid, H^+, to form the calcium and bicarbonate ions.

$$CaCO_{3(s)} + H^+ \rightarrow Ca^{2+} + HCO_3^- \tag{3-12}$$

Even though it does not occur physically, we can visualize this reaction as taking place by $CaCO_{3(s)}$ and H^+ reacting to form their component elements; these elements then recombine to form Ca^{2+} and HCO_3^-. Thus

		$\Delta \overline{G}_f^\circ$
$CaCO_{3(s)}$	$\rightarrow Ca + C + \frac{3}{2}O_{2(g)}$	$+269.78$
H^+	$\rightarrow H^+$	0
Ca	$\rightarrow Ca^{2+}$	-132.18
$H^+ + C + \frac{3}{2}O_{2(g)}$	$\rightarrow HCO_3^-$	-140.31
$CaCO_{3(s)} + H^+$	$\rightarrow Ca^{2+} + HCO_3^-;$	$\Delta G^\circ = -2.71$

This computation allows us to see two things. First, with respect to the dissolving of $CaCO_{3(s)}$ in acid, if we react 1 mole of H^+ with 1 mole of calcite to produce 1 mole of Ca^{2+} and 1 mole of HCO_3^- at 25°C, free energy is released because ΔG° is negative. Second, and more generally, we can see it is possible to determine the overall standard free energy change for the reaction as written, ΔG°, from the free energies of formation of reactants and products by using Eq. 3-11 with $\Delta \overline{G}_{f,i}^\circ$ being used in place of $\overline{G}_i^\circ$. Thus

$$\Delta G^\circ = \left(\sum_i \nu_i \Delta \overline{G}_f^\circ \right)_{\text{products}} - \left(\sum_i \nu_i \Delta \overline{G}_f^\circ \right)_{\text{reactants}} \tag{3-13}$$

where ν_i is the stoichiometric coefficient of species i. Or for Eq. 3-12

$$\Delta G^\circ = \Delta \overline{G}_{f,HCO_3^-}^\circ + \Delta \overline{G}_{f,Ca^{2+}}^\circ - \Delta \overline{G}_{f,H^+}^\circ - \Delta \overline{G}_{f,CaCO_3}^\circ \tag{3-14}$$

Equation 3-13 has limited usefulness because it only refers to the situation when all reactants and products are at standard state. We are almost always interested in systems whose components are at other than standard state. We can relate the free energy of a single substance at a state other than standard to its free energy of formation at standard state by the expression

$$\overline{G}_i = \Delta \overline{G}_{f,i}^\circ + RT \ln \{i\} \tag{3-15}$$

where

$\overline{G}_i$ = the free energy per mole of substance i in a state other than standard measured relative to our established datum

$\{i\}$ = the active concentration, or activity, of species i

The value of the activity of a substance is dependent on the choice of standard state conditions—the conditions that result in unit activity—just as the value of $\Delta \overline{G}_f^\circ$ is dependent on the choice of standard state. Consistent with the choice of standard state conditions used to develop Table 3-1 and in most publications on aqueous chemistry, the following interpretation of activity is used throughout this book:

1. For ions and molecules in solution, $\{i\}$ is related to the molar concentration, $[i]$, by $\{i\} = \gamma_i[i]$ where γ_i = activity coefficient. As the solution becomes dilute (most cases of interest to us), γ_i approaches 1 and $\{i\}$ approaches $[i]$.

2. For the solvent in a solution, $\{i\} = \gamma_i X_i$ where X_i is the mole fraction. As the solution becomes more dilute, γ_i approaches 1. The activity generally is assumed to be 1 for the dilute solutions of concern to us.

3. For pure solids or liquids in equilibrium with a solution, $\{i\} = 1$.

4. For gases in equilibrium with a solution, $\{i\} = \gamma_i P_i$ where P_i is the partial pressure of the gas in atmospheres. As the total pressure decreases, γ_i approaches 1. When reactions take place at atmospheric pressure, the activity of a gas can be approximated closely by its partial pressure.

5. For mixtures of liquids, $\{i\} = X_i$ where X_i is the mole fraction.

Returning to our example of the dissolution of $CaCO_{3(s)}$ in acid, Eq. 3-12, and combining Eqs. 3-9 and 3-15, we can write

$$\Delta G = \Delta \overline{G}_{f,HCO_3^-}^\circ + RT \ln \{HCO_3^-\} + \Delta \overline{G}_{f,Ca^{2+}}^\circ + RT \ln \{Ca^{2+}\}$$
$$- \Delta \overline{G}_{f,H^+}^\circ - RT \ln \{H^+\} - \Delta \overline{G}_{f,CaCO_{3(s)}}^\circ - RT \ln \{CaCO_{3(s)}\}$$

Collecting terms, we obtain

$$\Delta G = \Delta \overline{G}_{f,HCO_3^-}^\circ + \Delta \overline{G}_{f,Ca^{2+}}^\circ - \Delta \overline{G}_{f,H^+}^\circ - \Delta \overline{G}_{f,CaCO_{3(s)}}^\circ$$
$$+ RT \ln \{HCO_3^-\} + RT \ln \{Ca^{2+}\} - RT \ln \{H^+\} - RT \ln \{CaCO_{3(s)}\}$$

Since, from Eq. 3-13,

$$\Delta G^\circ = \Delta \overline{G}_{f,HCO_3^-}^\circ + \Delta \overline{G}_{f,Ca^{2+}}^\circ - \Delta \overline{G}_{f,H^+}^\circ - \Delta \overline{G}_{f,CaCO_{3(s)}}^\circ$$

we can state

$$\Delta G = \Delta G^\circ + RT \ln \frac{\{HCO_3^-\}\{Ca^{2+}\}}{\{H^+\}\{CaCO_{3(s)}\}} \tag{3-16}$$

This is a specific form of Eq. 3-10 applied to the acid dissolution of

$CaCO_{3(s)}$ reaction. For dilute solutions, since the activity of pure solid is unity and assuming that the activity coefficients of the ionic component are equal to 1, we obtain

$$\Delta G = \Delta G^\circ + RT \ln \frac{[HCO_3^-][Ca^{2+}]}{[H^+]} \tag{3-17}$$

In examining Eqs. 3-16 and 3-17, and the general form of the equation, Eq. 3-10, we can see that the logarithmic term has a form that is reminiscent of the equilibrium constant for the reaction as written in Eq. 3-12. However, because the magnitude of the logarithmic term is not equal to the equilibrium constant except at equilibrium, we call this term the reaction quotient, Q, where

$$Q = \frac{\{HCO_3^-\}\{Ca^{2+}\}}{\{H^+\}\{CaCO_{3(s)}\}} \tag{3-18}$$

or in dilute solution

$$Q = \frac{[HCO_3^-][Ca^{2+}]}{[H^+]}$$

More generally, for the reaction

$$aA + bB \rightleftharpoons cC + dD$$

$$Q = \frac{\{C\}^c \{D\}^d}{\{A\}^a \{B\}^b} \tag{3-19}$$

or in dilute solution,

$$Q = \frac{[C]^c [D]^d}{[A]^a [B]^b}$$

Combining Eqs. 3-10 and 3-19 (or Eqs. 3-16 and 3-18), we can state,

$$\Delta G = \Delta G^\circ + RT \ln Q \tag{3-20}$$

When the value of Q is identical to that of the equilibrium constant K, that is, the system is at equilibrium and $\Delta G = 0$, we can then write

$$0 = \Delta G^\circ + RT \ln K \tag{3-21}$$

or

$$\Delta G^\circ = -RT \ln K \tag{3-22}$$

Substituting Eq. 3-22 into Eq. 3-20, we obtain

$$\Delta G = -RT \ln K + RT \ln Q = RT \ln \frac{Q}{K} \tag{3-23}$$

Equation 3-23 allows us to develop another set of criteria to determine whether reactions are possible or not because the ratio of Q/K will determine the sign of ΔG for a reaction.

1. If $Q/K > 1$, then ΔG is positive and the reaction is impossible as written.

2. If $Q/K = 1$, then $\Delta G = 0$ and the system is at equilibrium.

3. If $Q/K < 1$, then ΔG is negative and the reaction is spontaneous as written.

The use of the tables of free energy values and the determination of whether a reaction is at equilibrium is illustrated in the following examples.

Example 3-1

1. Determine the equilibrium constant for the reaction in which liquid H_2O dissociates to H^+ and OH^- at 25°C.

$$H_2O \rightleftharpoons H^+ + OH^-$$

2. Is this reaction proceeding as written when $[H^+] = 10^{-6} M$ and $[OH^-] = 5 \times 10^{-8} M$?

Solution

1. From Table 3-1

	ΔG_f°
$H_2O_{(\ell)}$	-56.69
H^+	0
OH^-	-37.60

From Eq. 3-13, $\Delta G^\circ = (1)(0) + (1)(-37.60) - (1)(-56.69) = +19.09$ kcal

$$\Delta G^\circ = 19.09 = -RT \ln K$$
$$\frac{-19.09}{1.987 \times 10^{-3} \times 298} = \ln K = -32.24$$

$$K = 9.96 \times 10^{-15} \ (\cong 1 \times 10^{-14})$$

Based on the selection of standard state conditions given previously in this section and neglecting ionic strength effects, K can be written as follows for dilute solutions,

$$K = [H^+][OH^-]$$

This equilibrium constant is usually given the special designation K_w.

2. From Eq. 3-20,

$$\Delta G = \Delta G^\circ + RT \ln [H^+][OH^-]$$
$$\Delta G = 19.09 + 2.3 \times 1.98 \times 10^{-3} \times 298 \times \log \ [(10^{-6})(5 \times 10^{-8})]$$
$$\Delta G = +1.03$$

Because $\Delta G > 0$, the reaction is not spontaneous as written and can proceed spontaneously only in the opposite direction, that is, H^+ and OH^- are combining to form H_2O molecules.

We can solve this problem alternatively using

$$\Delta G = RT \ln \frac{Q}{K}$$
$$Q = [H^+][OH^-] = (10^{-6})(5 \times 10^{-8}) = 5 \times 10^{-14}$$
$$K = 10^{-14}$$

Since $Q/K > 1$, ΔG must be positive. Therefore, the reaction as written is not possible.

Example 3-2

Find the equilibrium constant for the reaction at 25°C,

$$2Fe^{2+} + \tfrac{1}{2}O_{2(g)} + 5H_2O \rightleftharpoons 2Fe(OH)_{3(s)} + 4H^+$$

in which ferrous iron is oxidized by molecular oxygen to ferric hydroxide in aqueous solution. Atmospheric oxygen is in equilibrium with the dissolved oxygen.

Give the form of the equilibrium constant making the assumption that activity coefficients have a value of 1.

Solution

From Table 3-1,

	$\Delta \overline{G}_f°$ (kcal/mole)
Fe^{2+}	-20.3
$O_{2(g)}$	0
$H_2O_{(\ell)}$	-56.7
$Fe(OH)_{3(s)}$	-166.0
H^+	0

$$\Delta G° = 4 \times 0 + 2(-166.0) - 2(-20.3) - \tfrac{1}{2}(0) - 5(-56.7)$$
$$\Delta G° = -7.9 = -RT \ln K$$
$$RT \ln K = 2.3 \, RT \log K = 1.364 \log K = 7.9$$
$$K = 6.2 \times 10^5 = 10^{5.8}$$
$$K = \frac{[H^+]^4 [Fe(OH)_{3(s)}]^2}{(X_{H_2O})^5 (P_{O_2})^{1/2} [Fe^{2+}]^2} = \frac{[H^+]^4}{(P_{O_2})^{1/2} [Fe^{2+}]^2}$$

It is important to note that we may only use gaseous oxygen as a reactant in this equation if it is in true equilibrium with the solution, that is, if the equilibrium

$$O_{2(g)} \rightleftharpoons O_{2(aq)}$$

is satisfied. Otherwise we should use $O_{2(aq)}$ as a reactant and, correspondingly, the value of $\Delta \overline{G}_f°$ for $O_{2(aq)}$.

We now have sufficient information to provide answers to the question posed earlier in the chapter about sulfide stability in the presence of nitrate.

Example 3-3

Is it possible at 25°C to oxidize sulfide in natural waters with nitrate? Typical concentrations that exist are 10^{-4} M of reacting species and pH 8 ([H$^+$] = 10^{-8} M). Assume that ionic strength effects are negligible (these effects will be discussed later in this chapter). The reaction is

$$H^+ + NO_3^- + HS^- + H_2O \rightleftharpoons SO_4^{2-} + NH_4^+$$
$$\text{nitrate \ bisulfide} \qquad \text{sulfate \ ammonium}$$

We must find out whether the reaction is spontaneous as written.
From Table 3-1,

	$\overline{\Delta G_f}^{\,\circ}$
$H_2O_{(\ell)}$	-56.69
HS^-	$+3.01$
NO_3^-	-26.41
H^+	0
SO_4^{2-}	-177.34
NH_4^+	-19

From Eq. 3-13,

$$\Delta G^\circ = (-19 - 177.34) - (-26.41 + 3.01 - 56.69)$$
$$= -116.25$$
$$\Delta G = \Delta G^\circ + RT \ln Q$$
$$Q = \frac{[NH_4^+][SO_4^{2-}]}{[NO_3^-][HS^-][H^+]}$$

and since [NH$_4^+$] = 10^{-4}, [SO$_4^{2-}$] = 10^{-4}, [NO$_3^-$] = 10^{-4}, [HS$^-$] = 10^{-4}, and [H$^+$] = 10^{-8},

$$Q = \frac{(10^{-4})(10^{-4})}{(10^{-4})(10^{-4})(10^{-8})} = 10^8$$

Therefore,

$$\Delta G = -116.25 + 1.987 \times 10^{-3} \times 298 \times 2.3 \log 10^{\,8}$$
$$= -116.25 + 10.9 = -105.35$$

Because ΔG is negative, the reaction will proceed spontaneously as written at the reactant concentrations indicated.

An important property of standard free energy changes, ΔG°, is that they are additive. For example, if in two reactions the product of one reaction serves as a reactant in the other, the ΔG° value for the combined reactions is the sum of the ΔG° values of the two reactions. In our previous example of the dissolution of $CaCO_3$ in acid, we wrote the equation $CaCO_{3(s)} + H^+ \rightleftharpoons Ca^{2+} + HCO_3^-$. In the strict chemical sense this overall reaction is a composite of

	ΔG° (kcal)
$CaCO_{3(s)} \rightleftharpoons Ca^{2+} + CO_3^{2-}$	$+11.38$

and

$CO_3^{2-} + H^+ \rightleftharpoons HCO_3^-$	-14.09

with a standard free energy change of $\Delta G° = 11.38 - 14.09 = -2.71$ kcal. Whether the composite reaction is spontaneous depends upon the value of ΔG for that reaction, a function of $\Delta G°$ and Q. If ΔG for the composite reaction is negative (positive), it is not necessary that each of the individual reactions have a negative (positive) value of ΔG. Note further that it is the value of ΔG, not $\Delta G°$, which controls the direction of a reaction.

3.4. ENTHALPY AND THE TEMPERATURE DEPENDENCE OF THE EQUILIBRIUM CONSTANT

The enthalpy change of a chemical reaction (ΔH) is the amount of heat that is released or taken up during the course of the reaction. If ΔH is negative, heat is evolved and the reaction is called exothermic; if ΔH is positive, heat is taken up and the reaction is called endothermic.

The term $\Delta H°$ is of most interest to us. For a reversible reaction in a closed system it is related to $\Delta G°$ by

$$\Delta H° = \Delta G° + T\Delta S° \tag{3-24}$$

Similarly to $\Delta G°$, $\Delta H°$ for a reaction can be calculated from

$$\Delta H° = \left(\sum_i \nu_i \overline{H}_i°\right)_{\text{products}} - \left(\sum_i \nu_i \overline{H}_i°\right)_{\text{reactants}} \tag{3-25}$$

where

ν_i = the stoichiometric coefficient
$\overline{H}_i°$ = the enthalpy of species i in kcal/mole at standard conditions of 25°C and 1 atm pressure

For the general reaction, $aA + bB \rightleftharpoons cC + dD$, $\Delta H°$ is equal to the amount of heat taken up or released when a moles of A and b moles of B, each in their standard states, are completely converted to c moles of C and d moles of D, each in their standard states.

Just as we could not determine the absolute value of $\overline{G}_i°$, we also cannot measure $\overline{H}_i°$. As with $\overline{G}_i°$ we circumvent this problem by assigning $\overline{H}_i°$ a value of zero to all elements in their most stable form at 25°C and 1 atm pressure. In aqueous solution 1 mole/liter of the hydrogen ion, H^+, in ideal solution ($\gamma = 1$) also is assigned an $\overline{H}°$ value of zero. We can determine values of enthalpy of species based on these assignments and call these the enthalpy of formation, $\Delta \overline{H}_f°$. Similarly to the computations for $\Delta \overline{G}_f°$ values, we can compute the $\Delta \overline{H}_f°$ values of various compounds from the assigned $\overline{H}°$ values of their component elements. A selection of these $\Delta \overline{H}_f°$ values is given in Table 3-1.

Standard enthalpy change values, $\Delta H°$, for reactions are most commonly used in water chemistry to determine the effect of temperature on the position of equilibrium. A useful expression in this regard is due to Van't Hoff, which states that

$$\frac{d \ln K}{dT} = \frac{\Delta H°}{RT^2} \tag{3-26}$$

If we assume that, over a limited temperature range, $\Delta H°$ is not a function of temperature, integration of Eq. 3-26 yields

$$\ln \frac{K_1}{K_2} = \frac{\Delta H°}{R} \left(\frac{1}{T_2} - \frac{1}{T_1} \right) \tag{3-27}$$

or

$$\ln K = -\frac{\Delta H°}{RT} + \text{constant} \tag{3-28}$$

or

$$K = C' \exp \left(\frac{-\Delta H°}{RT} \right) \tag{3-29}$$

where C' is a constant. Application of the Van't Hoff relationship is illustrated in the following example.

Example 3-4

A municipal water supply enters a residence at 15°C and is heated to 60°C in the home water heater. If the water is just saturated with respect to $CaCO_{3(s)}$ at 25°C, what will be the condition of the water (i.e., oversaturated or undersaturated) with respect to $CaCO_{3(s)}$ (1) as it enters the residence and (2) as it leaves the water heater?

Solution

Let us tackle this problem using the equation

$$CaCO_{3(s)} + H^+ \rightleftharpoons HCO_3^- + Ca^{2+}$$

as the basis for our solution. We are told that at 25°C the concentrations of Ca^{2+}, HCO_3^-, and H^+ are such that the system is at equilibrium. First we will calculate the value of the equilibrium constant using the $\Delta G_f°$ value for this reaction of -2.71 kcal, which was calculated in Section 3-3, and then determine the value of the equilibrium constant at 15°C and at 60°C using $\Delta H_f°$ values from Table 3-1.

At 25°C,

$$\Delta \overline{H}_f{}^\circ$$
$$\text{kcal/mole}$$

$CaCO_{3(s)}$	-288.45
H^+	0
$Ca^{2+}_{(aq)}$	-129.77
$HCO^-_{3(aq)}$	-165.18

Let K at 25°C $= K_1$, then

$$-RT \ln K_1 = -2.71$$
$$\log K_1 = 1.99$$
$$K_1 = 10^{+1.99}$$

Calculate ΔH° from Eq. 3-25,

$$\Delta H^\circ = \sum_i (\nu_i \Delta \overline{H}^\circ_{f,i})_{products} - \sum_i (\nu_i \Delta \overline{H}^\circ_{f,i})_{reactants}$$

$$\Delta H^\circ = -129.77 - 165.18 - (-288.45 - 0)$$

$$\Delta H^\circ = -6.5\,kcal$$

Let $K_2 = K$ at 15°C (288°K), (T_2); and $K_3 = K$ at 60°C (333°K), (T_3). From Eq. 3-27,

$$\ln K_1 - \ln K_2 = \frac{\Delta H^\circ}{R} \left(\frac{1}{T_2} - \frac{1}{T_1} \right)$$

$$-\ln K_2 + \ln 10^{+1.99} = \frac{-6.5}{1.98 \times 10^{-3}} \left(\frac{1}{288} - \frac{1}{298} \right)$$

$$K_2 = 10^{+2.15}$$

Similarly, from Eq. 3-27,

$$\ln K_1 - \ln K_3 = \frac{\Delta H^\circ}{R} \left(\frac{1}{T_3} - \frac{1}{T_1} \right)$$

$$-\ln K_3 + \ln 10^{+1.987} = \frac{-6.5}{1.98 \times 10^{-3}} \left(\frac{1}{333} - \frac{1}{298} \right)$$

$$K_3 = 10^{+1.48}$$

At 15°C the equilibrium constant is greater, and at 60°C it is less, than that at 25°C. Thus the term

$$Q = \frac{[Ca^{2+}][HCO_3^-]}{[H^+]}$$

is greater than K at 60°C and less than K at 15°C. The water had a value of Q that met equilibrium conditions at 25°C and the value of this quantity will not change as temperature decreases. Therefore, if we may assume that $[H^+]$ remains constant, the water at 15°C contains less Ca^{2+} and HCO_3^- than allowed by equilibrium, that is, it is undersaturated. By similar reasoning, at 60°C the water is oversaturated with respect to $CaCO_{3(s)}$ and $CaCO_{3(s)}$ will be deposited within the heater. Note that these conclusions have some bearing on the life of home water heaters.

We see from the example that the equilibrium constant for $CaCO_{3(s)}$ + $H^+ \rightleftharpoons Ca^{2+} + HCO_3^-$ at 15°C is approximately 1.5 times larger than at 25°C; at 60°C it is less than $\frac{1}{3}$ as large as at 25°C. One could have sensed that K would decrease as the temperature increases from the value of $\Delta H°$, which is negative. Since heat is produced as the reaction goes to the right, raising the temperature tends to drive the reaction to the left, thus decreasing the equilibrium constant. If we take heat away from the reaction by lowering the temperature, the equilibrium is shifted to the right and the equilibrium constant increases. This is an example of the general situation that for exothermic reactions, an increase in temperature will shift the equilibrium in the direction of less complete reaction; for endothermic reactions, an increase in temperature will shift the equilibrium in the direction of more complete reaction.

The $\Delta H°$ values for a series of consecutive reactions can be added, just as were $\Delta G°$ values, to yield an overall $\Delta H°$ value for the overall reaction.

Example 3-5

Find the standard enthalpy of formation, $\Delta H°$ for $H_2SO_{4(l)}$ given the following reactions:

		$\Delta H°$ (kcal)
1.	$S_{(s)} + O_{2(g)} \rightleftharpoons SO_{2(g)}$	−70.96
2.	$SO_{2(g)} + \frac{1}{2}O_2 \rightleftharpoons SO_{3(g)}$	−23.5
3.	$SO_{3(g)} + H_2O_{(l)} \rightleftharpoons H_2SO_{4(l)}$	−31.14
4.	$H_{2(g)} + \frac{1}{2}O_{2(g)} \rightleftharpoons H_2O_{(l)}$	−68.32

Solution

Adding reactions (1) through (4) and the $\Delta H°$ values, we obtain

$$S_{(s)} + 2O_{2(g)} + H_{2(g)} \rightleftharpoons H_2SO_{4(l)}$$

where $\Delta H° = -70.96 - 23.5 - 31.4 - 68.32 = -193.92$ kcal. Applying Eq. 3-25, we obtain

$$\Delta \overline{H}_f° \text{ for } H_2SO_{4(l)} = -193.92 \text{ kcal}$$

3.5. NONIDEAL BEHAVIOR OF IONS AND MOLECULES IN SOLUTION

In very dilute aqueous solutions, ions behave independently of one another and it is valid to assume that activity coefficients of ions have values of unity. However, as the concentration of ions in solution increases, electrostatic interactions between the ions also increase and the activity of ions becomes somewhat less than their measured or analytical concentration. Thus, in the equation,

$$\{i\} = \gamma_i [i] \tag{3-30}$$

the activity coefficient, γ_i, becomes less than 1 and chemical equilibrium is affected. For example, for the general reaction,

$$aA + bB \rightleftharpoons cC + dD$$

it becomes necessary to write the equilibrium constant and reaction quotient in the form

$$Q \text{ or } K = \frac{\{C\}^c\{D\}^d}{\{A\}^a\{B\}^b} = \frac{(\gamma_C[C])^c(\gamma_D[D])^d}{(\gamma_A[A])^a(\gamma_B[B])^b} \tag{3-31}$$

To calculate activity coefficients of ions in aqueous solution, we must employ the quantity ionic strength, μ, which was devised by Lewis and Randall[3] to describe the intensity of the electric field in a solution:

$$\mu = \tfrac{1}{2}\sum_i (C_i Z_i^2) \tag{3-32}$$

where

C_i = concentration of ionic species, i
Z_i = charge of species i

Application of this equation is shown in the following two examples.

Example 3-6

Compute the ionic strength of a solution containing the following concentration of ions:

$$[Ca^{2+}] = 10^{-4}M, \quad [CO_3^{2-}] = 10^{-5}M, \quad [HCO_3^-] = 10^{-3}M$$
$$[SO_4^{2-}] = 10^{-4}M, \quad [Na^+] = 1.02 \times 10^{-3}M$$

Solution

$$\mu = \tfrac{1}{2}([10^{-4} \times 2^2] + [10^{-5} \times 2^2] + [10^{-3} \times 1^2] + [10^{-4} \times 2^2] + [1.02 \times 10^{-3} \times 1^2])$$
$$\mu = 1.43 \times 10^{-3}$$

Example 3-7

Which of the following brines has the greater ionic strength?

1. Brine a: 5800 mg/liter NaCl.

2. Brine b: 3100 mg/liter MgSO₄.

Solution

The atomic weights are Na, 23; Cl, 35.5; Mg, 24; S, 32; and O, 16.
For brine a,

$$[Na^+] = \frac{5800\ \text{mg/liter}}{58,500\ \text{mg/mole}} = 0.099\ M$$

$$[Cl^-] = \frac{5800\ \text{mg/liter}}{58,500\ \text{mg/mole}} = 0.099\ M$$

$$\mu = \tfrac{1}{2}(0.099 \times 1^2 + 0.099 \times 1^2) = 0.099 \approx 0.1$$

[3] G. N. Lewis and M. Randall, *J. Am. Chem. Soc.*, 43: 1111 (1921).

For brine b,

$$[Mg^{2+}] = \frac{3100 \text{ mg/liter}}{120,000 \text{ mg/mole}} = 0.026\,M$$

$$[SO_4^{2-}] = \frac{3100 \text{ mg/liter}}{120,000 \text{ mg/mole}} = 0.026\,M$$

$$\mu = \tfrac{1}{2}(0.026 \times 2^2 + 0.026 \times 2^2) = 0.104 \approx 0.1$$

They both have virtually the same ionic strength.

Rather than work through laborious calculations of ionic strength for more complex solutions than those given in the examples, it is usually precise enough for most purposes in water chemistry to use an approximation of ionic strength derived from a correlation with specific conductivity or TDS (total dissolved solids). The conductivity of a solution is a measure of the ability of a solution to conduct a current. It is a property attributable to the ions in solution. Electrical current is transported through solutions via the movement of ions, and conductivity increases as ion concentration increases. Figure 3-3 is a presentation of data given by Lind[4] on the conductivity versus ionic strength of various surface and groundwaters. If the conductivity of a water sample is known, this figure can be used to provide a crude estimate of ionic strength.

Langelier presented the approximation,[5]

$$\mu = 2.5 \times 10^{-5} \times \text{TDS} \tag{3-33}$$

for several waters he examined. In this expression TDS is the total dissolved solids in mg/liter. Russell[6] derived the following correlation between ionic strength and conductivity for 13 waters of widely varying composition,

$$\mu = 1.6 \times 10^{-5} \times \text{conductivity } (\mu\text{mho})$$

This is similar to the expression that can be derived from the data in Fig. 3-3.

Ionic strength appears in each of the various expressions used to calculate activity coefficients in aqueous solutions. The DeBye–Hückel theory of interaction of ions in aqueous solution incorporates both the electrostatic interactions between ions and the thermal motion of the ions. The basic equation, called the DeBye–Hückel limiting law, was

---

[4] C. J. Lind, U.S. Geological Survey Professional Paper 700 D, 1970, pp. D272–D280.
[5] W. F. Langelier, "The Analytical Control of Anti-Corrosion Water Treatment," J. Am. Water Works Assoc., 28: 1500 (1936).
[6] L. L. Russell, "Chemical Aspects of Groundwater Recharge with Wastewaters," Ph.D. Thesis, University of California, Berkeley, Dec. 1976.

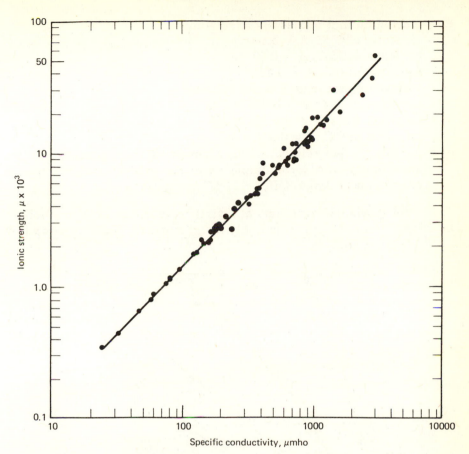

Fig. 3-3. Specific conductance as a means of estimating ionic strength. After C. J. Lind, U.S. Geological Survey Professional Paper 700D, pp. D272–D280, 1970.

developed for ionic strengths of less than approximately 5×10^{-3} and can be stated as

$$-\log \gamma_i = 0.5 Z_i^2 \, \mu^{1/2} \tag{3-34}$$

An alternative equation, the extended DeBye–Hückel approximation of the DeBye–Hückel limiting law, which is applicable for ionic strength of less than approximately 0.1, is

$$-\log \gamma_i = \frac{A Z_i^2 \, \mu^{1/2}}{1 + B a_i \mu^{1/2}} \tag{3-35}$$

where

A = a constant that relates to the solvent:
  for water at 25°C, $A = 0.509$
  for water at 15°C, $A = 0.50$
  for water at 0°C, $A = 0.488$
B = a constant that relates to the solvent:
  for water at 25°C, $B = 0.328 \times 10^8$
  for water at 15°C, $B = 0.326 \times 10^8$
  for water at 0°C, $B = 0.324 \times 10^8$
a_i = a constant that relates to the diameter of the hydrated ion: for monovalent ions, for example, with the exception of H^+, this is usually about 3 to 4×10^{-8}

Figure 3-4 shows the variations of activity coefficients as a function of ionic strength for some ions commonly found in water. These curves were calculated using Eq. 3-35.

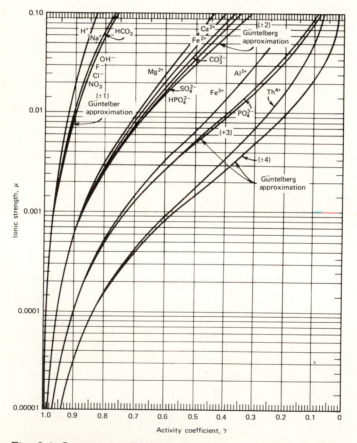

Fig. 3-4. Activity coefficients of aqueous ions based on the extended DeBye–Hückel equation (Eq. 3-35) and the Güntelberg approximation (Eq. 3-36).

For monovalent ions at 15°C, Eq. 3-35 becomes

$$-\log \gamma_i = \frac{0.5Z_i^2\mu^{1/2}}{1 + (0.326 \times 10^8 \times 3 \times 10^{-8})\mu^{1/2}} = \frac{0.5Z_i^2\mu^{1/2}}{1 + \mu^{1/2}} \qquad (3\text{-}36)$$

Equation 3-36 is also called the Güntelberg approximation of the DeBye–Hückel theory and is also commonly used to calculate γ for ions of various charge at temperatures other than 15°C. It is also plotted in Fig. 3-4. There have been other extensions of the theory to make it apply at higher ionic strengths. There is, however, no satisfactory theory that provides a good estimate of the activity coefficient for ionic strengths of greater than about 0.5.

Because anions cannot be added to a solution without an equivalent number of cations (and vice versa), it is impossible to determine experimentally the activity coefficient of a single ion. Therefore, Eqs. 3-34, 3-35, and 3-36 cannot be verified directly. However, it is possible to define, and measure experimentally, a mean activity coefficient, $\gamma_\pm$, as,

$$\gamma_\pm = (\gamma_+\gamma_-)^{1/2} \qquad (3\text{-}37)$$

The DeBye–Hückel and Güntelberg relationships can be extended to the mean activity coefficient thus:

$$-\log \gamma_\pm = 0.5 \, |Z_+Z_-| \, \mu^{1/2} \qquad (3\text{-}38)$$

$$-\log \gamma_\pm = \frac{0.5 \, |Z_+Z_-| \, \mu^{1/2}}{1 + \mu^{1/2}} \qquad (3\text{-}39)$$

where Eq. 3-38 is the DeBye–Hückel limiting law, Eq. 3-39 is the Güntelberg approximation, and

Z_+ = charge of the positive ion
Z_- = charge of the negative ion
$\gamma_\pm$ = mean of the two activity coefficients

Example 3-8

Find the ratio, at equilibrium, of the molar concentrations of carbonate to bicarbonate at 25°C in a solution with an ionic strength of 10^{-3}. Also find the pH. The equilibrium constant, $K_{a,2}$, for the reaction

$$HCO_3^- \rightleftharpoons H^+ + CO_3^{2-}$$

is $10^{-10.3}$. $[H^+] = 10^{-10}$ moles/liter.

Solution

$$K = 10^{-10.3} = \frac{\{H^+\}\{CO_3^{2-}\}}{\{HCO_3^-\}} = \frac{\gamma_{H^+}[H^+]\gamma_{CO_3^{2-}}[CO_3^{2-}]}{\gamma_{HCO_3^-}[HCO_3^-]}$$

Using the DeBye–Hückel limiting law,

$$-\log \gamma_{H^+} = -\log \gamma_{HCO_3^-} = 0.5 \times 1^2(10^{-3})^{1/2} = 0.0158$$
$$\gamma_{H^+} = \gamma_{HCO_3^-} = 0.96$$
$$-\log \gamma_{CO_3^{2-}} = 0.5 \times 2^2 \times (10^{-3})^{1/2}$$
$$\gamma_{CO_3^{2-}} = 0.86$$

$$10^{-10.3} = \frac{(0.96)(10^{-10})(0.86)[CO_3^{2-}]}{(0.96)[HCO_3^-]}$$

$$\frac{[CO_3^{2-}]}{[HCO_3^-]} = 0.58$$

If we had not taken activity coefficients into account the ratio $[CO_3^{2-}]/[HCO_3^-]$ would have been

$$10^{-10.3} = \frac{10^{-10}[CO_3^{2-}]}{[HCO_3^-]}$$
$$\frac{[CO_3^{2-}]}{[HCO_3^-]} = \frac{10^{-10.3}}{10^{-10}} = 0.501$$

Also

$$pH = -\log \{H^+\} = -\log(\gamma_{H^+}[H^+])$$
$$= -\log(0.96 \times 10^{-10}) = 10.02$$

The theory for predicting the activity coefficients of nonelectrolytes in aqueous solution is not as well developed as for electrolytes. An empirical equation of the form

$$\log \gamma = k_s \mu \tag{3-40}$$

is generally used to relate activity to ionic strength. The salting-out coefficient k_s must be experimentally determined. Because values of k_s generally fall in the range from 0.01 to 0.15, nonionic or molecular solutes have activity coefficients of approximately 1 for ionic strengths of less than 0.1.

For example, if $k_s = 0.132$ for oxygen in NaCl solution[7] and

$$\log \gamma = 0.132 \times 0.05 = 0.0066$$

then

$$\gamma \cong 1.02$$

As the ionic strength increases, for example, in seawater ionic strength is approximately 0.7, Eq. 3-40 predicts that the activity coefficient of oxygen will be

$$\log \gamma = 0.132 \times 0.7 = 0.0924$$
$$\gamma = 1.24$$

[7] M. Randall and C. F. Failey, Chem. Rev., 4: 285 (1927).

The activity coefficient is greater than one. We can visualize that this is indeed possible if we think of the condition of the water in seawater. Because there is so much salt present, there will be a significant quantity of the H_2O present as water of hydration associated with the ions present in the seawater. Oxygen dissolves less well in water which is bonded to ions so that a given total volume of water will dissolve a smaller concentration of oxygen at equilibrium if it contains much salt. If the oxygen dissolved in the seawater is in equilibrium with the atmosphere, the activity of oxygen in the water is controlled only by the partial pressure of oxygen in the atmosphere. It is not a function of salt content of the water. Therefore, if the dissolved oxygen activity is constant while the dissolved oxygen concentration decreases as the salt concentration increases, we see from the equation, $\{i\} = \gamma_i[i]$, that γ must be greater than 1. This phenomenon is illustrated by Example 3-9. The effect of decreasing the solubility of molecular species, such as dissolved oxygen, by increasing salt concentration is known as the "salting-out effect."

Example 3-9

Given that the equilibrium constant, Henry's constant, K_H, for the reaction

$$O_{2(g)} \rightleftharpoons O_{2(aq)}$$

is 1.29×10^{-3} at 25°C and assuming a salting-out coefficient, k_s, of 0.132, determine the molar dissolved oxygen activity and concentration at 25°C for (1) distilled water, (2) the Sacramento River water (specific conductivity = 450 μmho), and for (3) Pacific Ocean water (ionic strength = 0.7).

Solution

Dissolved oxygen in distilled water:

Since $\mu = 0$, $\{O_{2(aq)}\} = [O_{2(aq)}]$. Knowing P_{O_2} we can calculate $[O_{2(aq)}]$ from the equilibrium constant,

$$K = \frac{\{O_{2(aq)}\}}{P_{O_2}} = 1.29 \times 10^{-3}$$

In dry air the volume fraction of oxygen is 0.21. At 25°C the vapor pressure of water is 23.8 mm Hg. Given that X_{O_2} is the volume fraction of O_2 in dry air,

$$P_{O_2} = \left(\frac{\text{atmospheric pressure (mmHg)} - \text{vapor pressure (mmHg)}}{760} \right) X_{O_2}$$

$$= \left(\frac{760 - 23.8}{760} \right) (0.21) = 0.203$$

$$[O_{2(aq)}] = \{O_{2(aq)}\} = 1.29 \times 10^{-3} \times 0.203$$

$$= 2.62 \times 10^{-4} M \text{ (or 8.4 mg/liter)}$$

The ionic strength estimate for Sacramento River water:

From Fig. 3-3 for a specific conductance of 460 μmho, $\mu = 0.007$.

Let us assume that the partial pressure of oxygen is the same over the Pacific Ocean as it is over the Sacramento River (a reasonable assumption). Because of this the dissolved oxygen activity will be the same in both solutions.

For Sacramento River water,

$$\mu = 0.007 \quad \text{and} \quad k_s = 0.132$$
$$\log \gamma = 0.132 \times 0.007$$
$$\gamma = 1.002$$

For Pacific Ocean water,

$$\mu = 0.7 \quad \text{and} \quad k_s = 0.132$$
$$\log \gamma = 0.7 \times 0.132$$
$$\gamma = 1.24$$

Now since

$$K = \frac{\{O_{2(aq)}\}}{P_{O_2}} = \frac{\gamma[O_{2(aq)}]}{P_{O_2}}$$

For Sacramento River water,

$$[O_{2(aq)}] = \frac{2.62 \times 10^{-4}}{1.002} = 2.62 \times 10^{-4}\,M$$
$$= 8.4\,\text{mg/liter}$$

For Pacific Ocean water,

$$[O_{2(aq)}] = \frac{2.62 \times 10^{-4}}{1.24} = 2.11 \times 10^{-4}\,M$$
$$= 6.75\,\text{mg/liter}$$

For both waters the activity of oxygen is the same at $2.62 \times 10^{-4}\,M$ or 8.4 mg/liter. It is significant to note that a dissolved oxygen test conducted on the two waters using the wet chemical method (Winkler method) will produce a result of 8.4 mg/liter for the river water and 6.75 mg/liter for the ocean water. The same analysis conducted using a membrane-covered, specific oxygen electrode will produce an identical reading in both solutions because it responds to activity rather than concentration. We might ask how a fish feels with respect to the dissolved oxygen content of these two waters. Does it care about dissolved oxygen activity or dissolved oxygen concentration? We leave the reader to ponder this question.

3.6.   PROBLEMS

---

1.   The heat of combusion of a substance is defined as the enthalpy change that occurs when 1 mole of a substance reacts with elemental oxygen to form liquid water and gaseous CO_2. Determine the heat available from the combustion of methane, $CH_{4(g)}$ at 25°C and a constant pressure. Express the answer in terms of kcal/mole and kcal/g.

2.   Calculate the standard enthalpy of formation, $\Delta \overline{H}_f^\circ$, for $CO_{2-(g)}$ using the following information:

$$C_{(s)} + O_{2(g)} \rightarrow CO_{2(g)} \qquad \Delta H^\circ = -94.0\,\text{kcal}$$
$$CO + \tfrac{1}{2}O_{2(g)} \rightarrow CO_2 \qquad \Delta H^\circ = -67.6\,\text{kcal}$$

3. An average man produces about 2500 kcal of heat a day through metabolic activity. If a man had a mass of 70 kg with the heat capacity of water (1 cal/$g \cdot °C$), what would be the temperature rise in a day if no heat were lost? Man is actually an open system, and the main mechanism of heat loss is the evaporation of water. How much water would he need to evaporate in a day to maintain constant temperature? ($\Delta H°$ for vaporization at 37°C is 575 cal/g.)

4. An important reaction in muscular activity is the oxidation of lactic to pyruvic acid. Calculate $\Delta H°$ for the reaction given that the $\Delta H°$ of combustion (see Problem 1) is -279 kcal/mole for pyruvic acid and -326 kcal/mole for lactic acid (lactic acid = $CH_3CHOHCOOH$; pyruvic acid = $CH_3COCOOH$).

5. Calculate the amount of energy *available* (that is, $\Delta G°$) for all maintenance and synthesis of new bacterial cells when 1 mole of acetate ion undergoes aerobic oxidation by bacteria. The reaction can be represented as follows:

$$CH_3COO^- + 2O_{2(aq)} \rightarrow HCO_3^- + CO_{2(aq)} + H_2O_{(\ell)}$$

The bacteria function as a catalyst in the reaction but do not alter the amount of energy that can be obtained from the reaction.

Note: Not all of the available energy is used for synthesis and maintenance. There is a certain amount of inefficiency involved.

6. Ammonia, NH_3, is a base and will readily accept a proton in accordance with the following reaction

$$NH_{3(aq)} + H_2O \rightleftharpoons NH_4^+ + OH^-$$

(a) Calculate the equilibrium constant, K, for this reaction at 25°C.
(b) If, at some time, pH = 9.0, $[NH_3] = 10^{-5}$ M, and $[NH_4^+] = 10^{-6}$ M, is the reaction at equilibrium? If not, in which direction is the reaction going?

7. Assuming that the reaction

$$H^+ + NO_3^- \rightleftharpoons HNO_{3(aq)}$$

is at equilibrium in aqueous solution, what percent of C_{T,NO_3} (where $C_{T,NO_3} = ([NO_3^-] + [HNO_3])$) is present as HNO_3 at pH = 1? (Use standard free energy tables to make your calculations and assume that all activity coefficients have a value of 1.0.)

8. $Mg(OH)_{2(s)}$ is precipitated according to the following reaction

$$Mg^{2+} + 2OH^- \rightleftharpoons Mg(OH)_{2(s)}$$

How much Mg^{2+}, moles/liter, is present in solution at equilibrium when the pH is 10.0? (Neglect ionic strength effects.)

9. (a) Find the equilibrium constant, K_w, at 25 and 40°C for the following reaction,

$$H_2O \rightleftharpoons H^+ + OH^-$$

(b) Is the reaction exothermic or endothermic?

10. Calculate
    (a) Henry's constant, K_H, for H_2S. Note that $K_H = K$ for the reaction,

    $$H_2S_{(g)} \rightleftharpoons H_2S_{(aq)}$$

    (b) The partial pressure of $H_2S_{(g)}$ overlying the water if the total soluble sulfide concentration, $C_{T,S} = [H_2S_{(aq)}] + [HS^-] + [S^{2-}]$, is $1 \times 10^{-3}\,M$ and the solution pH is 8.5. Assume that the gas is in equilibrium with the water. ($K_{a,1} = 10^{-7}$ and $K_{a,2} = 10^{-14}$ for H_2S where $K_{a,1}$ is the equilibrium constant for $H_2S_{(aq)} \rightleftharpoons HS^- + H^+$ and $K_{a,2}$ is the equilibrium constant for $HS^- \rightleftharpoons S^{2-} + H^+$.)

11. The reaction of divalent manganese with O_2 in aqueous solutions is given as follows:

    $$Mn^{2+} + \tfrac{1}{2}O_{2(aq)} + H_2O \rightleftharpoons MnO_{2(s)} + 2H^+$$

    Delfino and Lee (*Environ. Sci. and Technol.*, December 1968) found that a lake water sample devoid of oxygen, pH = 8.5, originally contained 0.6 mg/liter of Mn^{2+}. The sample was aerated (atmospheric conditions) and after 10 days of saturation with atmospheric O_2 the Mn^{2+} concentration was 0.4 mg/liter.
    (a) Assuming that the pH remains constant during aeration, will the precipitate continue to form after the measurement on the tenth day?
    (b) What should the Mn^{2+} concentration be at equilibrium?

12. The following quantities of salts were added to a volume of water to make 1 liter of solution:

    $$1 \times 10^{-2}\,\text{moles NaCl}$$
    $$2 \times 10^{-2}\,\text{moles CaCl}_2$$
    $$2 \times 10^{-2}\,\text{moles BaCl}_2$$

    (a) What is the ionic strength of the solution?
    (b) A small amount of phosphate salt is added to the same solution with negligible change in ionic strength. Given $K = 10^{-7.2}$ for the reaction

    $$H_2PO_4^- \rightleftharpoons H^+ + HPO_4^{2-}$$

    Calculate $[H^+][HPO_4^{2-}]/[H_2PO_4^-]$, called ^cK, using the Güntelberg approximation of the DeBye–Hückel law.
    (c) Calculate the "salting-out" coefficient, k_s, for a nonelectrolyte in the *same solution* if its activity is $10^{-3}\,M$ and its concentration is $9.5 \times 10^{-4}\,M$.

13. Permanganate, MnO_4^-, decomposes in a solution that is in equilibrium with the atmosphere in accordance with the following reaction,

    $$4MnO_4^- + 4H^+ \rightleftharpoons 4MnO_{2(s)} + 2H_2O + 3O_{2(g)}$$

    The equilibrium constant for this reaction is 10^{+68}.
    (a) If $[MnO_4^-] = 10^{-10}\,M$ and pH = 7, is the reaction at equilibrium or proceeding to the right or the left?
    (b) If the reaction is not at equilibrium for the conditions in part (a), calculate the pH at which the reaction will be at equilibrium if $[MnO_4^-] = 10^{-10}\,M$.

14. At 25°C, aqueous solutions of $CO_{2(aq)}$, $NaHCO_3$, and HCl are mixed instantaneously so that the concentrations of $CO_{2(aq)}$, HCO_3^-, and H^+ are each 10^{-5} M initially. In what direction does the following reaction proceed initially?

$$CO_{2(aq)} + H_2O \rightleftharpoons HCO_3^- + H^+; \quad K = 10^{-6.3}$$

15. In order to cut costs at the lime softening plant, the operator decides to recalcinate the calcium carbonate sludge according to the reaction

$$CaCO_{3(s)} \rightleftharpoons CaO_{(s)} + CO_{2(g)}$$

The decision is made to store the sludge in a container open to the atmosphere prior to recalcination. Knowing that $P_{CO_2} = 10^{-3.5}$ atm, will any of the $CaCO_{3(s)}$ decompose according to the given reaction?

16. The reaction

$$O_{2(g)} \rightleftharpoons O_{2(aq)}$$

has an equilibrium constant, K ($= K_H$, Henry's constant) $= 1.29 \times 10^{-3}$ at 25°C and $\Delta H° = -3.9$ kcal.
(a) Calculate the free energy of formation and the enthalpy of formation for the reactant and the product and compare to the values listed in Table 3-1.
(b) Calculate the equilibrium constant at 50°C.
(c) Given that the vapor pressure is as follows:

Temperature, °C	Vapor Pressure, mm Hg
25	23.8
50	92.5
100	760

and that dry air is 21 percent O_2 by volume, calculate the equilibrium concentration of O_2 in water at 25, 50, and 100°, for atmospheric pressure.
(d) Given that 9.5 mg O_2/liter is found in solution at 25°C, is the reaction at equilibrium? Why?

3.7. ADDITIONAL READING

Lewis, G. N., and M. Randall, *Thermodynamics*, revised by K. S. Pitzer and L. Brewer, 2nd ed., McGraw-Hill, 1961.

Moore, W. J., *Physical Chemistry*, 4th ed., Prentice-Hall, Englewood Cliffs, N.J., 1972.

CHAPTER
4

ACID-BASE CHEMISTRY

4.1. INTRODUCTION

A thorough examination of acid-base chemistry is important in the study of aqueous chemistry. It is essential for understanding the carbonate system that has an important influence on the pH of natural waters and may govern the solubility of certain metal ions. Many precipitation-dissolution, oxidation-reduction, and complexation reactions involve acid-base reactions. The concentrations of metal ions in water are governed to a great extent by acid-base phenomena. Thus the concentration of hydroxide ion may determine the concentration of many metal ions. Some metal ions behave as acids, for example, ferric ion has an acid strength comparable to phosphoric acid and much of its chemistry is governed by its acidic properties. In water treatment, the dosages of chemicals required to reduce hardness (calcium and magnesium) are partially governed by the acid-base properties of the solution being treated.

In this chapter we will consider in detail acid-base reactions using relevant examples from water treatment processes and natural water chemistry to illustrate the various principles. Special attention will be given to techniques for calculating the composition of solutions of acids and bases. The chapters that follow and deal with precipitation, coordination chemistry, and oxidation-reduction chemistry will make use of many of the principles developed in this chapter.

4.2. DEFINITION OF TERMS

The most commonly used definitions of an acid and a base are those of Brönsted and Lowry. These definitions state that an acid is a substance which can donate a hydrogen ion, H^+, or a proton as it is commonly called, and that a base is a substance which can accept a proton. In the general reaction,

$$B^- + HA \rightleftharpoons HB + A^-$$
$$\text{base}_1 \quad \text{acid}_1 \quad \text{acid}_2 \quad \text{base}_2$$

(4-1)

a proton is transferred from HA to the anion B^-. The substance HA is an acid because it can donate a proton; the substance B^- is a base because it can accept the proton. Note that the reaction is reversible so that a new acid, HB, and a new base, A⁻, are formed as the reaction proceeds to the right. In this reaction, HA, the proton donor, and A⁻, the base that forms when the proton is given up, are called an *acid-conjugate base pair*. Similarly, HB and B⁻ are an acid-conjugate base pair.

Water can be both an acid and a base. For example, in the following reaction,

$$HCl + H_2O \rightleftharpoons H_3O^+ + Cl^-$$
$$\text{acid}_1 \qquad \text{base}_1 \qquad \text{acid}_2 \qquad \text{base}_2$$

(4-2)

water is a base because it accepts a proton from HCl. In the reaction between the carbonate ion and water,

$$CO_3^{2-} + H_2O \rightleftharpoons OH^- + HCO_3^-$$
$$\text{base}_1 \qquad \text{acid}_1 \qquad \text{base}_2 \qquad \text{acid}_2$$

(4-3)

the water molecule donates a proton to the carbonate ion and is therefore an acid. Substances that can function as both an acid and a base, such as HCO_3^- and H_2O, are called *ampholytes*; they are *amphoteric*.

In aqueous solution the proton is bonded quite strongly to a water molecule so that it is more correct to use the symbol H_3O^+, the *hydrated proton* or *hydronium ion*, rather than H^+ to represent it in equations. Furthermore, there is good evidence that at least three additional water molecules are somewhat more loosely attached to the hydrated proton so that another representation of the proton in water would be $H_9O_4^+$ or $H^+ \cdot 4H_2O$. The hydroxyl ion, OH^-, also is hydrated, having three water molecules bonded to it, that is, $H_7O_4^-$, or $OH^- \cdot 3H_2O$. It is common to use H^+ as a shorthand notation for H_3O^+ or $H_9O_4^+$, and OH^- to represent $H_7O_4^-$, although the formula H_3O^+ often is used to emphasize that acid-base reactions are proton exchange reactions. Because of its amphoteric properties water undergoes auto-ionization thus,

$$H_2O \rightleftharpoons H^+ + OH^-$$

(4-4)

The equilibrium constant for this reaction is

$$K = \frac{\{H^+\}\{OH^-\}}{\{H_2O\}}$$

But because $\{H_2O\} = 1$ in dilute solution (see Section 3-3), by neglecting ionic strength effects we can write

$$K = [H^+][OH^-] = K_w$$

The auto-ionization reaction can be written also as

$$H_2O + H_2O \rightleftharpoons H_3O^+ + OH^-$$

(4-5)

with the resulting equilibrium constant,

$$K = \frac{\{H_3O^+\}\{OH^-\}}{\{H_2O\}^2}$$

or when $\{H_2O\} = 1$ and neglecting ionic strength effects,

$$K = [H_3O^+][OH^-] = K_w$$

We can derive the value of the equilibrium constant, K_w, for these reactions from free energy calculations. In doing this, we should recall that the free energy of the proton is zero. The free energy of the hydrated proton is the free energy of the proton, zero, plus the free energy of the water molecules bonded to it. Similarly, the free energy of the hydrated hydroxyl ion is the free energy of OH^- plus the free energy of the water molecules hydrating it. In Example 3-1, a $\Delta G°$ of $+19.09$ kcal was calculated for Eq. 4-4, and making the same calculation for Eq. 4-5, using $\Delta \overline{G_f}°$ values from Table 3-1 yields the identical value for $\Delta G°$.

$$\Delta G° = \Delta \overline{G}°_{f,H_3O^+} + \Delta \overline{G}°_{f,OH^-} - 2\Delta \overline{G}°_{f,H_2O}$$

$$\Delta G° = -56.69 + (-37.60) - 2(-56.69) = +19.09 \, kcal$$

(4-6)

Thus using the shorthand notation of H^+ for H_3O^+ and omitting water molecules from both sides of Eqs. 4-5 does not affect the free energy *change* for the reaction. As shown in Example 3-1, $K = 1 \times 10^{-14}$ at 25°C when $\Delta G° = +19.09$ kcal.

In this text we will follow common practice and use H^+ and OH^- to represent the hydrated proton and hydroxyl ion, respectively. On occasion we will use H_3O^+ to emphasize that acid-base reactions are truly proton exchange reactions.

The equilibrium constant, K_a, for an acid refers to the reaction in which *an acid donates a proton to a water molecule* (e.g., Eq. 4-2). The equilibrium constant, K_b, refers to the reaction in which *a base accepts a proton from a water molecule* (e.g., Eq. 4-3). *Large values of K_a indicate that the acid has a strong tendency to donate a proton to water*, i.e., it is a strong acid. Large values of K_b indicate that a substance has a strong tendency to accept a proton from water, i.e., it is a strong base. Conversely, small values of K_a and K_b denote weak acids and bases, respectively.

Table 4-1 is a listing of some K_a and K_b values of relevance to water chemistry. The acids are listed in the order of decreasing strength while the bases are in the order of increasing strength. The dividing line between what the chemist refers to as a strong acid and a weak acid lies somewhere in the region of iodic acid ($pK_a = 0.8$). Thus, of the acids listed, perchloric, hydrochloric, nitric, and H_3O^+ would be classified as strong acids while the remainder of the list are weak acids. Strong acids tend toward complete dissociation while weak acids are incompletely ionized. For acids with pK_a values near the dividing line, dissociation is incomplete for high concentrations but tends toward completion for low concentrations. This effect is known as the *leveling effect* of the solvent, water, and is dependent upon the relative affinities of the conjugate base and H_2O for the proton. Similarly, the dividing line between strong and weak bases is in the neighborhood of dihydrogen silicate, ($pK_b = 1.4$). Thus, of the bases listed, sulfide, hydroxide, amide, and oxide are strong bases whereas the remaining entries are weak bases. As illustrated in Table 4-1, the conjugate base of a strong acid is a weak base (small K_b) and the conjugate base of a weak acid is a strong base (large K_b).

An important relationship exists between the K_a value of an acid and the K_b value for its conjugate base. Using hydrocyanic acid, HCN, and the cyanide ion, CN^-, as an example of an acid-conjugate base pair, we can write equations for the donation of a proton to water by HCN and for the acceptance of a proton from water by CN^-.

1.
$$\underset{\text{base}}{CN^-} + H_2O \rightleftharpoons \underset{\substack{\text{conjugate}\\\text{acid}}}{HCN} + OH^-; \qquad K_b = \frac{\{HCN\}\{OH^-\}}{\{CN^-\}\{H_2O\}}$$

2.
$$\underset{\text{acid}}{HCN} + H_2O \rightleftharpoons H_3O^+ + \underset{\substack{\text{conjugate}\\\text{base}}}{CN^-}; \qquad K_a = \frac{\{H_3O^+\}\{CN^-\}}{\{HCN\}\{H_2O\}}$$

By adding (1) and (2), we obtain the overall equation,

$$2H_2O \rightleftharpoons H_3O^+ + OH^-; \qquad K = K_aK_b \qquad\qquad (4\text{-}7)$$

$$K = K_aK_b = \frac{\{H_3O^+\}\{CN^-\}\{HCN\}\{OH^-\}}{\{HCN\}\{H_2O\}\{CN^-\}\{H_2O\}} = \frac{\{H_3O^+\}\{OH^-\}}{\{H_2O\}^2}$$

Because $\{H_2O\} = 1$,

$$K_aK_b = \{H_3O^+\}\{OH^-\} = K_w \qquad\qquad (4\text{-}8)$$

where K_w is the equilibrium constant for ionization of water. Equation 4-8 enables us to calculate K_b for the conjugate base of an acid if K_a is known, or conversely, K_a for an acid if K_b for the conjugate base is known.

The K_w equation, which describes the dissociation of water, forms the basis for the definition of the pH scale. At 25°C, $K_w = 1.0 \times 10^{-14} = \{H^+\}\{OH^-\}$. Using the notation pX to signify $-\log X$, we can write

$$pH + pOH = pK_w = 14$$

where

$$pH = -\log \{H^+\}$$
$$pOH = -\log \{OH^-\}$$
$$pK_w = -\log K_w$$

When ionic strength effects are negligible,

$$pH = -\log [H^+]$$
$$pOH = -\log [OH^-]$$

TABLE 4-1 Acidity and Basicity Constants for Substances in Aqueous Solution at 25°C

Acid		$-\log K_a$ $= pK_a$	Conjugate Base		$-\log K_b$ $= pK_b$
$HClO_4$	Perchloric acid	-7	ClO_4^-	Perchlorate ion	21
HCl	Hydrochloric acid	~ -3	Cl^-	Chloride ion	17
H_2SO_4	Sulfuric acid	~ -3	HSO_4^-	Bisulfate ion	17
HNO_3	Nitric acid	~ 0	NO_3^-	Nitrate ion	14
H_3O^+	Hydronium ion	0	H_2O	Water	14
HIO_3	Iodic acid	0.8	IO_3^-	Iodate ion	13.2
HSO_4^-	Bisulfate ion	2	SO_4^{2-}	Sulfate ion	12
H_3PO_4	Phosphoric acid	2.1	$H_2PO_4^-$	Dihydrogen phosphate ion	11.9
$Fe(H_2O)_6^{3+}$	Ferric ion	2.2	$Fe(H_2O)_5OH^{2+}$	Hydroxo iron(III) complex	11.8
HF	Hydrofluoric acid	3.2	F^-	Fluoride ion	10.8
HNO_2	Nitrous acid	4.5	NO_2^-	Nitrite ion	9.5
CH_3COOH	Acetic acid	4.7	CH_3COO^-	Acetate ion	9.3
$Al(H_2O)_6^{3+}$	Aluminum ion	4.9	$Al(H_2O)_5OH^{2+}$	Hydroxo aluminum(III) complex	9.1

$H_2CO_3^*$	Carbon dioxide and carbonic acid	6.3	HCO_3^-	Bicarbonate ion	7.7
H_2S	Hydrogen sulfide	7.1	HS^-	Bisulfide ion	6.9
$H_2PO_4^-$	Dihydrogen phosphate	7.2	HPO_4^{2-}	Monohydrogen phosphate ion	6.8
$HOCl$	Hypochlorous acid	7.5	OCl^-	Hypochlorite ion	6.4
HCN	Hydrocyanic acid	9.3	CN^-	Cyanide ion	4.7
H_3BO_3	Boric acid	9.3	$B(OH)_4^-$	Borate ion	4.7
NH_4^+	Ammonium ion	9.3	NH_3	Ammonia	4.7
H_4SiO_4	Orthosilicic acid	9.5	$H_3SiO_4^-$	Trihydrogen silicate ion	4.5
C_6H_5OH	Phenol	9.9	$C_6H_5O^-$	Phenolate ion	4.1
HCO_3^-	Bicarbonate ion	10.3	CO_3^{2-}	Carbonate ion	3.7
HPO_4^{2-}	Monohydrogen phosphate	12.3	PO_4^{3-}	Phosphate ion	1.7
$H_3SiO_4^-$	Trihydrogen silicate	12.6	$H_2SiO_4^{2-}$	Dihydrogen silicate ion	1.4
HS^-	Bisulfide ion	14	S^{2-}	Sulfide ion	0
H_2O	Water	14	OH^-	Hydroxide ion	0
NH_3	Ammonia	~23	NH_2^-	Amide ion	−9
OH^-	Hydroxide ion	~24	O^{2-}	Oxide ion	−10

When $\{H^+\} = \{OH^-\}$, or $[H^+] = [OH^-]$, pH = pOH and pH = $pK_w/2 = 7$. In an aqueous solution where $[H^+] = [OH^-]$, the solution is said to be *neutral*. At 25°C, neutrality is attained at pH = 7. Values of pH below 7 signify that $[H^+] > [OH^-]$, and the solution is said to be *acidic*. Conversely, pH values above 7 indicate that $[OH^-] > [H^+]$ and the solution is said to be *basic* or *alkaline*. It is important to realize, however, that pH 7 only represents neutrality at 25°C because the pH scale has its origin in the K_w equation and K_w is a function of temperature.

For example, the reaction $H_2O \rightleftharpoons H^+ + OH^-$ has an enthalpy of reaction (from Table 3-1),

$$\Delta H° = \Delta \overline{H}°_{f,H^+} + \Delta \overline{H}°_{f,OH^-} - \Delta \overline{H}°_{f,H_2O}$$

$$\Delta H° = 0 + (-54.96) - (-68.32)$$

$$\Delta H° = +13.36 \, \text{kcal}$$

Using the Van't Hoff relationship, Eq. 3-27, we can calculate K_w at 50°C as follows,

$$\ln K_{w,50°} = \ln K_{w,25°} + \frac{\Delta H°}{R}\left(\frac{1}{298} - \frac{1}{323}\right)$$

$$= \ln(1.0 \times 10^{-14}) + \frac{13.36}{1.987 \times 10^{-3}}\left(\frac{1}{298} - \frac{1}{323}\right)$$

$$K_{w,50°} = 5.73 \times 10^{-14}$$

Thus, at 50°C, neutrality is at pH = $-\log(K_{w,50°})^{1/2} = 6.62$. Table 4-2 is a tabulation of K_w as a function of temperature and of the corresponding pH values of a neutral solution.

As previously indicated, the equilibrium constant, K_a, of an acid gives an indication of its strength. It is, however, quite helpful to think of acid strength in terms of the standard free energy change of the reaction in which 1 mole of protons is transferred from the acid to water. For example, for nitric acid, using values given in Table 3-1,

$$HNO_3 + H_2O \rightleftharpoons H_3O^+ + NO_3^-$$

$$\Delta G° = \Delta \overline{G}°_{f,NO_3^-} + \Delta \overline{G}°_{f,H_3O^+} - \Delta \overline{G}°_{f,HNO_3} - \Delta \overline{G}°_{f,H_2O}$$

$$\Delta G° = -26.43 - 56.69 - (-26.41) - (-56.69)$$

$$\Delta G° = -0.02 \, \text{kcal}$$

Thus, under standard conditions, 0.02 kcal of free energy are *released* for each mole of protons transferred from nitric acid to water. Because this proton transfer enables the system to attain a lower free energy, the reaction tends to proceed readily in the direction written and accordingly nitric acid is a very strong acid. From the relationship, $\Delta G° = -2.3RT \log K$

TABLE 4-2   Ion Product of Water

°C	K_w	pK_w	pH of a "neutral" solution ($\{H^+\} = \{OH^-\}$)
0	0.12×10^{-14}	14.93	7.47
15	0.45×10^{-14}	14.35	7.18
20	0.68×10^{-14}	14.17	7.08
25	1.01×10^{-14}	14.00	7.00
30	1.47×10^{-14}	13.83	6.92
40	2.95×10^{-14}	13.53	6.76

Source. H. S. Harned and B. B. Owen, *The Physical Chemistry of Electrolyte Solutions*, 3rd ed., Reinhold, New York, 1958.

$$\log K = \frac{-0.02}{-2.3 \times 1.98 \times 10^{-3} \times 298} = 1.47 \times 10^{-2}; \qquad K = 1.03$$

By comparison the free energy change for dissociation of the bisulfide ion,

$$HS^- + H_2O \rightleftharpoons H_3O^+ + S^{2-}$$

is + 19.1 kcal, a large positive number; accordingly, there is little tendency for this reaction to proceed as written and bisulfide is a weak acid ($K_a = 10^{-14.0}$).

4.3.   RATE OF REACTION

Acid-base reactions in aqueous solutions generally proceed extremely rapidly.[1] The reaction,

$$H^+ + OH^- \underset{k_2}{\overset{k_1}{\rightleftharpoons}} H_2O \qquad (4\text{-}9)$$

with $k_1 = 1.4 \times 10^{11}$ liter/mole, sec. and $k_2 = 2.5 \times 10^{-5}$/sec. at 25°C,[2] has the fastest reaction rate known in aqueous solution. Reaction rates of this magnitude are consistent with a diffusion-controlled reaction in which the ions diffuse to the point of contact and then react instantaneously. The H^+ and OH^- ions have exceptionally high mobilities.

Since the reaction in Eq. (4-9) is elementary, we can write

$$\text{Forward rate} = k_1 [H^+][OH^-] \qquad (4\text{-}10)$$

$$\text{Reverse rate} = k_2 [H_2O] \qquad (4\text{-}11)$$

---

[1] For an exception see the discussion of the reaction $H_2CO_3^* \rightleftharpoons H^+ + HCO_3^-$ in Section 4-13.

[2] D. Benson, *Mechanisms of Inorganic Reactions in Solution: An Introduction*, McGraw-Hill, London, 1968.

At equilibrium, forward rate = reverse rate; therefore,

$$[H^+][OH^-] k_1 = k_2 [H_2O] \tag{4-12}$$

or

$$\frac{k_2[H_2O]}{k_1} = [H^+][OH^-] \; (\text{mole}^2/\text{liter}^2) \tag{4-13}$$

Substituting for k_1 and k_2 and using 55.5 moles/liter as the concentration of water,[3] we obtain the concentration product of water,

$$\frac{(2.5 \times 10^{-5}/\text{sec})(55.5 \text{ mole/liter})}{1.4 \times 10^{+11} \text{ liter/mole, sec.}} = 0.99 \times 10^{-14} \cong 10^{-14}$$

When ionic strength effects are negligible the concentration product equals K_w.

The rate of reaction of H^+ with bases other than OH^- also is extremely rapid, although not quite as fast as its reaction rate with OH^-, since other bases diffuse more slowly.

The rate at which acids dissociate or give up a proton is a function of the affinity of the acid for the proton. Weak acids, with a high affinity for protons, dissociate relatively slowly. Similarly, weak bases have slower reaction rates than strong bases. Even though the reaction rates of weak acids and bases are slower, they can be considered essentially instantaneous as is shown by the following example for the weak acid H_2S, $pK_a = 7.1$.

Example 4-1

The dissociation of H_2S is given by

$$H_2S \underset{k_2}{\overset{k_1}{\rightleftharpoons}} H^+ + HS^-$$

where $k_1 = 4.3 \times 10^3 \text{ sec}^{-1}$.[4] Assuming that the reaction is irreversible, how long will it take for 50 percent of the H_2S to dissociate?

Solution

The rate law

$$\frac{d[H_2S]}{dt} = -k_1[H_2S]$$

---

[3] It is necessary to use 55.5 moles/liter here because in the rate law, the concentration of water in moles/liter has been used whereas in the equilibrium constant, K_w, we use the mole fraction for the concentration of water.

[4] D. Benson, *Mechanisms of Inorganic Reactions in Aqueous Solution: An Introduction*, McGraw-Hill, London, 1968.

can be integrated between time $= 0$ and time $= t$ to give

$$\ln \frac{[H_2S]_t}{[H_2S]_0} = -k_1 t$$

where

$[H_2S]_0 =$ concentration at $t = 0$, and
$[H_2S]_t =$ concentration at $t = t$.

Thus

$$\frac{\ln \frac{1}{2}}{-4.3 \times 10^{+3}} = t = 1.6 \times 10^{-4} \text{ sec}$$

From this value we would expect the dissociation of H_2S into HS^- and H^+ to be at equilibrium in a well-mixed solution only a fraction of a second after it is added. If H_2S were bubbled into a solution, the rate at which HS^- would be produced would be controlled by the efficiency of the mixing device, even for the most efficient mixing system.

4.4. EQUILIBRIUM CALCULATIONS—A GENERAL APPROACH

Because acid-base reactions in solution generally are so rapid, we can concern ourselves primarily with the determination of species concentrations at equilibrium. Usually, we desire to know $[H^+]$, $[OH^-]$, and the concentration of the acid and its conjugate base that result when an acid or a base is added to water. As we shall see later in this text, acid-base equilibrium calculations are of central importance in the chemistry of natural waters and in water and wastewater treatment processes. The purpose of this section is to develop a general approach to the solution of acid-base equilibrium problems and to apply this approach to a variety of situations involving strong and weak acids and bases.

Let us consider first the equations that describe a solution which results when an acid, HA, or a salt of its conjugate base, MA (where M is a cation) is added to water.

4.4.1. Mass Balances

In acid-base reactions the reacting species are conserved. When HA is added to water, the acid ionizes partially or completely,

$$HA + H_2O \rightleftharpoons A^- + H_3O^+$$

Let us assume that the system is homogeneous and closed (i.e., no species containing A can enter from, or leave to, the atmosphere, and that precipitation or dissolution of such species cannot occur). A mass balance on all species containing A gives

$$C_{T,A} = [HA] + [A^-] \tag{4-14}$$

where $C_{T,A}$ is the number of moles of "species containing A" per liter and is equal to the *analytical concentration* of HA or the number of moles of HA added per liter. [HA] and [A$^-$] are the molar concentrations of the acid and conjugate base in solution at equilibrium.

When C moles of the salt MA are added per liter, it dissociates

$$MA \rightleftharpoons M^+ + A^-$$

and the base, A$^-$, reacts with water,

$$A^- + H_2O \rightleftharpoons HA + OH^-$$

A mass balance on M gives,[5]

$$C = C_{T,M} = [M^+] + [MA] \tag{4-15}$$

or, given that MA dissociates completely,

$$C_{T,M} = [M^+]$$

where $C_{T,M}$ is the number of moles of M per liter. A mass balance on A yields

$$C_{T,A} = [HA] + [A^-] \tag{4-16}$$

where $C_{T,A} = C$.

When MA and HA both are added to a solution,

$$C_{T,A} = [HA] + [A^-] \tag{4-17}$$

where $C_{T,A}$ = sum of the moles of HA and MA added per liter of solution.

4.4.2. Equilibrium Relationships

The second group of equations we need to consider describe equilibrium relationships. For the example of HA added to pure water the following equilibria are pertinent. In aqueous solution, neglecting ionic strength effects, we obtain for the dissociation of water,

$$K_w = [H^+][OH^-] = 10^{-14} \text{ at } 25°C \tag{4-18}$$

The dissociation of HA is described by

$$K_a = \frac{[H^+][A^-]}{[HA]} \tag{4-19}$$

The same equations are valid if the salt MA is added to solution. The basicity constant,

$$K_b = \frac{[HA][OH^-]}{[A^-]} \tag{4-20}$$

---

[5] This mass balance assumes that M$^+$ does not form complexes with A$^-$ or other solutes—an assumption that is legitimate for the alkali metals but may not be valid for alkali earth and transition metals (see Chapter 5).

could be used in place of Eq. 4-19. Since $K_w = K_a K_b$, Eqs. 4-19 and 4-20 cannot both be used because they are not independent.

4.4.3. The Proton Condition

The proton condition is a special type of mass balance equation on protons. It is an essential component of equilibrium problem solving if either H^+ or OH^- are involved in the equilibria. It is used in this section only to solve problems related to solutions made up in the laboratory. In later sections and chapters it is used to solve natural water problems.

The proton mass balance is established with reference to a "zero level" or a "reference level" for protons (the proton reference level, PRL). The species having protons in excess of the PRL are equated with the species having less protons than the PRL. The PRL is established as the species with which the solution was prepared.

Example 4-2

Determine the proton condition when HA is added to water.

Solution

The species present are H_3O^+, H_2O, OH^-, HA, and A^-. All species are involved in reactions with H^+ or OH^-.

PRL = HA, H_2O,
Species with protons > PRL = H_3O^+,
Species with protons < PRL = OH^-, A^-.

Proton Condition

$$[H_3O^+] = [OH^-] + [A^-]$$

Note that HA and H_2O, the species at the PRL do not appear in the proton condition.

Example 4-3

Determine the proton condition when the salt MA is added to water.

Solution

The species present are H_3O^+, H_2O, OH^-, M^+, A^-, and HA. The species that involve reactions with H^+ or OH^- are H_3O^+, H_2O, OH^-, A^-, and HA.

PRL = H_2O and MA (or, equivalently, A^-, since M^+ completely dissociates from A^- and does not affect the proton condition).
Species with protons > PRL = HA, H_3O^+,
Species with protons < PRL = OH^-.

Proton Condition

$$[HA] + [H_3O^+] = [OH^-]$$

Again the species at the PRL, A^-, and H_2O, do not appear in the proton condition.

Example 4-4

State the proton condition for a solution of phosphoric acid, H_3PO_4, in pure water.

Solution

The species present are H_3O^+, H_2O, OH^-, H_3PO_4, $H_2PO_4^-$, HPO_4^{2-}, and PO_4^{3-}.
PRL = H_3PO_4, H_2O,
Species with protons > PRL = H_3O^+,
Species with protons < PRL = OH^-, $H_2PO_4^-$, HPO_4^{2-}, and PO_4^{3-}.

Proton Condition

$$[H_3O^+] = [OH^-] + [H_2PO_4^-] + 2[HPO_4^{2-}] + 3[PO_4^{3-}]$$

Note that the concentration of $[HPO_4^{2-}]$ is counted twice because it has two less protons per molecule than the PRL. Similarly, $[PO_4^{3-}]$ is counted three times, since PO_4^{3-} has three less protons per molecule than the PRL.

Example 4-5

State the proton condition for sodium bicarbonate, $NaHCO_3$, added to water.

Solution

The species present are H^+, H_2O, OH^-, Na^+, $H_2CO_3^*$, HCO_3^-, and CO_3^{2-}.
PRL = HCO_3^-, H_2O,
Species with protons > PRL = H^+, $H_2CO_3^*$,
Species with protons < PRL = OH^-, CO_3^{2-}.

Proton Condition

$$[H^+] + [H_2CO_3^*] = [OH^-] + [CO_3^{2-}]$$

A *proton* or *hydroxyl mass balance* also may be used to arrive at the proton condition.[6] This approach can be applied to the more complicated situation that arises when both MA and HA are added to the same solution. However, these more-complicated situations are more readily solved by obtaining the proton condition from a combination of the charge balance and the mass balance, as is shown in Section 4-4-4.

4.4.4. The Charge Balance or Electroneutrality Equation

The basis of the charge balance is that all solutions must be electrically neutral. Ions of one charge cannot be added to, formed in, or removed from a solution without the addition, formation, or removal, of an equal number of ions of the opposite charge. In a solution the total number of positive charges must equal the total number of negative charges.

---

[6] The details of this approach are given in J. N. Butler, *Ionic Equilibrium, A Mathematical Approach*, Addison-Wesley, Reading, Mass., 1964.

Example 4-6

Determine the charge balance when HA is added to water.

Solution

The species present are HA, A^-, H_2O, OH^-, and H^+.
Total positive charges = $[H^+]$,
Total negative charges = $[OH^-] + [A^-]$.

Charge Balance

$$\text{Total positive charges} = \text{total negative charges}$$

or

$$[H^+] = [OH^-] + [A^-]$$

Example 4-7

Determine the charge balance when MA is added to water.

Solution

The species present are M^+, A^-, HA, H^+, OH^-, and H_2O.
Total positive charges = $[M^+] + [H^+]$,
Total negative charges = $[OH^-] + [A^-]$.

Charge Balance

$$[M^+] + [H^+] = [OH^-] + [A^-]$$

Example 4-8

Write a charge balance statement for Na_2HPO_4 dissolved in water.

Solution

The species present are, Na^+, H^+, OH^-, H_2O, H_3PO_4, $H_2PO_4^-$, HPO_4^{2-}, and PO_4^{3-}.
Total positive charges = $[Na^+] + [H^+]$,
Total negative charges = $[OH^-] + [H_2PO_4^-] + 2[HPO_4^{2-}] + 3[PO_4^{3-}]$.

Charge Balance

$$[Na^+] + [H^+] = [OH^-] + [H_2PO_4^-] + 2[HPO_4^{2-}] + 3[PO_4^{3-}]$$

Note that the molar concentration of HPO_4^{2-} is multiplied by two and the molar concentration of PO_4^{3-} is multiplied by three in the charge balance, since HPO_4^{2-} carries two negative charges per mole while PO_4^{3-} carries three negative charges per mole.

Besides being essential to the solution of equilibrium problems, the charge balance equation has significant practical application in the analysis of waters. Since all waters must be electrically neutral, we can deduce that a complete water analysis must produce a result in which the total number of positive charges = the total number of negative charges. An acceptable water analysis will have this equation agreeing to within ± 5 percent for wastewater or seawater analyses and ± 2 percent for water analyses. Larger deviations than this indicate either errors in analysis or an overlooked species.

Example 4-9

An analysis of water from Well No. 9 of the Manhattan Beach, California, Water Co. showed: pH = 7.8, silica = 19 mg SiO_2/liter, calcium = 65 mg Ca^{2+}/liter (162.5 mg as $CaCO_3$/liter), magnesium = 18.2 mg Mg^{2+}/liter (75 mg as $CaCO_3$/liter), sodium = 76 mg Na^+/liter, bicarbonate = 286.7 mg HCO_3^-/liter, sulfate = 28 mg SO_4^{2-}/liter, and chloride = 98 mg Cl^-/liter. Is the water analysis satisfactory from the point of view of analytical accuracy?

Solution

Arrange anions and cations in separate columns as in Table 4-3. Compute millimoles/liter (mmoles/liter) of individual anions and cations by dividing the concentrations by the appropriate molecular weight. Then multiply by the charge/ion to obtain mmoles of charge units/liter, also called the milliequivalents/liter (meq/liter) of charge units. Sum the meq/liter of positive and negative charges and compare sums for agreement.

The analyses check with excellent accuracy. Note that SiO_2 does not enter into the computation, since it is not charged at pH = 7.8 (the pH of the water sample), and that H^+ and OH^- are negligible.

The charge balance and mass balance equations can be combined to give the proton condition. For complex solutions this is generally the best way to determine proton conditions. For example, when C moles of MA are added per liter of solution, assuming complete dissociation of MA, we obtain the following results.

Mass Balances

(1) $C_{T,M} = [M^+] = C$

(2) $C_{T,A} = [HA] + [A^-] = C$

TABLE 4-3   Anion-Cation Balance

Cations	Conc. mg/ liter	mg/ mmole	Charge/ ion	Meq Charge/ liter	Anions	Conc. mg/ liter	mg/ mmole	Charge/ ion	Meq Charge/ liter
Na^+	76	23	1	3.30	Cl^-	98	35.5	1	2.76
Ca^{2+}	65	40	2	3.25	SO_4^{2-}	28	96	2	0.58
Mg^{2+}	18.2	24.3	2	1.49	HCO_3^-	286.7	61	1	4.70
Σ (+ charge)				8.04	Σ (− charge)				8.04

Charge Balance

(3) $[M^+] + [H^+] = [A^-] + [OH^-]$

We can combine (1), (2), and (3) to eliminate M^+, C, and $C_{T,A}$ to arrive at the proton condition,

$$[HA] + [H^+] = [OH^-]$$

Using the equilibrium equations in combination with mass and charge balances and the proton condition, it is now possible for us to proceed with the solution of acid-base problems and to arrive at the concentration of each species present in such systems.

Example 4-10

State the equations necessary to arrive at the concentration of each species in

1. 10^{-3} *M* solution of HA.
2. 10^{-3} *M* solution of NaA.

Solution

a. The species present are $[H^+]$, $[OH^-]$, $[HA]$, and $[A^-]$. Since we have four unknowns, four independent equations are required.

Mass Balance on A

(1) $C_{T,A} = [HA] + [A^-] = 10^{-3}$

Equilibrium Constant, H_2O

(2) $K_w = 10^{-14} = \{H^+\}\{OH^-\}$

Equilibrium Constant, HA

(3) $K_a = \{H^+\}\{A^-\}/\{HA\}$

Proton Condition (or Charge Balance)

(4) $[H^+] = [A^-] + [OH^-]$

If we make the assumption that activity coefficients of all species are unity, that is, ionic strength effects are negligible, then activities, { }, are equal to molar concentrations, [ ], and knowing K_a and K_w we can solve equations (1) to (4) simultaneously, or graphically, to yield the concentration of each species present. If we cannot assume that activities and molar concentrations are identical, then the activity coefficient must be calculated using the relationships given in Section 3-6. The equilibrium constants then can be converted from constants based on activities to constants based on concentrations, and the solution procedure is the same.

b. The species present are Na^+, A^-, HA, OH^-, and H^+. We have five unknowns; therefore, five independent equations are required.

Mass Balance on Na^+

(1) $C_{T,Na} = [Na^+] = 10^{-3}$

Mass Balance on A

(2) $C_{T,A} = [HA] + [A^-] = 10^{-3}$

Equilibrium Constant, H_2O

(3) $K_w = 10^{-14} = \{H^+\}\{OH^-\}$

Equilibrium Constant, HA

(4) $K_a = \dfrac{\{H^+\}\{A^-\}}{\{HA\}}$

Charge Balance

(5) $[Na^+] + [H^+] = [A^-] + [OH^-]$

The proton condition, $[HA] + [H^+] = [OH^-]$, may be used instead of (5)

Analytical and graphical procedures for solving these equations are given in the sections that follow.

4.5. STRONG ACID–STRONG BASE CALCULATIONS

We can assume that when a strong acid is added to water, it is completely ionized. For example, when HCl ($K_a = 10^3$) is added to pure water, a mass balance on Cl gives

$$C_{T,Cl} = [HCl] + [Cl^-]$$

Since $K_a = 10^3$, $= [H^+][Cl^-]/[HCl]$ when activity coefficients have a value of unity,

$$[Cl^-] \gg [HCl]$$

or [HCl] is approximately 0, and

$$C_{T,Cl} = [Cl^-]$$

Although this assumption is valid for most strong acid or strong base calculations of interest, HCl is present in solution at all pH values. When the pH is very low (< -2), the number of HCl molecules becomes a significant fraction of $C_{T,Cl}$ and must be taken into account. However, it is extremely rare in water chemistry that we are concerned with pH values less than about 1. The following three examples illustrate how the equations describing a strong acid or a strong base solution can be solved.

Example 4-11

10^{-2} moles of HCl are added to 1 liter of distilled water at 25°C. Find $[H^+]$, $[OH^-]$ and $[Cl^-]$ for this solution. What is the pH of the solution? Demonstrate that $[HCl]$ is negligible. Assume that activity coefficients are unity.

Solution

The unknowns are H^+, OH^-, HCl, and Cl^-.

Equilibria

1. $HCl \rightleftharpoons H^+ + Cl^-$,

   $$K_a = 10^3 = \frac{[H^+][Cl^-]}{[HCl]}$$

2. $H_2O \rightleftharpoons H^+ + OH^-; K_w = 10^{-14} = [H^+][OH^-]$

Mass balance on Cl

3. $C_{T,Cl} = [Cl^-] + [HCl] = 10^{-2}$

Proton Condition (Charge Balance)

4. $[H^+] = [Cl^-] + [OH^-]$

Assumptions

5. In mass balance:
   Because HCl is a strong acid, assume that

   $$[Cl^-] \gg [HCl]$$

   Thus

   $$C_{T,Cl} = [Cl^-] = 10^{-2}$$

6. In proton condition:
   Solution is acid, so assume

   $$[H^+] \gg [OH]$$

   Thus

   $$[H^+] = [Cl^-]$$

7. From (5) and (6)

   $$[H^+] = [Cl^-] = 10^{-2}$$

8. From (7) and (2)

   $$[OH^-] = \frac{K_w}{[H^+]} = \frac{10^{-14}}{10^{-2}} = 10^{-12}$$

9. Check the proton condition to find whether the concentrations that have been determined, that is, $[H^+] = 10^{-2}$, $[Cl^-] = 10^{-2}$, $[OH^-] = 10^{-12}$, satisfy it to within 5 percent.

   $$[H^+] = [Cl^-] + [OH^-]$$
   $$10^{-2} = 10^{-2} + 10^{-12}$$

The proton condition is satisfied since the right-hand side and the left-hand side are virtually identical.

10. The pH of the solution is

$$pH = 2$$

11. To demonstrate that [HCl] is negligible compared to [Cl$^-$] and so justify our simplifying assumption in the mass balance:

$$K_a = 10^3 = \frac{[H^+][Cl^-]}{[HCl]}$$

$$10^3 = \frac{[10^{-2}][10^{-2}]}{[HCl]}$$

$$[HCl] = 10^{-7}$$

$$[Cl^-] = 10^{-2} \gg 10^{-7}$$

If the proton condition, or any equations for which assumptions were made, did not check within 5 percent, it would be necessary either to make a different simplifying assumption or to solve the equations without making an assumption.

The criterion of agreement to within 5 percent is arbitrary. However, in view of the variation in reported equilibrium constants (reported values for a single reaction usually vary by more than 5 percent) and because of the purpose for which the calculations are generally made, accuracy better than 5 percent is rarely needed or justified.

Based on the result of Example 4-11, we will make the assumption, without verification, in subsequent calculations that strong acids and strong bases are dissociated completely.

Example 4-12

10^{-7} moles of the strong acid HNO_3 are added to 1 liter of distilled water at 25°C. What is the pH of the solution? Assume that activity coefficients are unity.

Solution

The unknowns are H^+, OH^-, and NO_3^-.

Equilibria

1. $H_2O \rightleftharpoons H^+ + OH^-$; $K_w = 10^{-14} = [H^+][OH^-]$

Mass Balance on NO_3 (assuming complete dissociation of HNO_3)

2. $C_{T,NO_3} = [NO_3^-] = 10^{-7}$

Proton Condition (Charge Balance)

3. $[H^+] = [NO_3^-] + [OH^-]$

Assumptions

4. In proton condition:
    The solution is acid, so assume that

    $$[NO_3^-] \gg [OH^-]$$

    which is equivalent to assuming that $[H^+] \gg [OH^-]$. Thus

    $$[H^+] = [NO_3^-]$$

5. From (3) and (4)

    $$[H^+] = [NO_3^-] = 10^{-7}$$

6. From (5) and (1)

    $$[OH^-] = \frac{K_w}{[H^+]} = \frac{10^{-14}}{10^{-7}} = 10^{-7}$$

7. Check the assumption in the proton condition:

    $$[H^+] = [NO_3^-] + [OH^-]$$
    $$10^{-7} = 10^{-7} + 10^{-7}$$

    The assumption does not check since the right-hand side is 100 percent greater than the left-hand side.

8. We now must try to make another assumption. The best approach is to assume the opposite situation from the assumption that did not work. Therefore, we will now assume that $[OH^-] \gg [NO_3^-]$, having failed with our assumption that $[NO_3^-] \gg [OH^-]$.
    Using our new assumption in the proton condition, we find that

    $$[H^+] = [OH^-]$$

9. From (1) and (8)

    $$K_w = [H^+][OH^-] = [H^+]^2 = 10^{-14}$$

    $$[H^+] = 10^{-7} \quad \text{and} \quad [OH^-] = 10^{-7}$$

10. Check this new assumption in the proton condition:

    $$[H^+] = [NO_3^-] + [OH^-]$$
    $$10^{-7} = 10^{-7} + 10^{-7}$$

    The new assumption does not check any better than the original one, so we must resign ourselves to solving the equations without making an assumption in the proton condition. The usual approach to this solution, and indeed to solutions where assumptions are not made, is to solve first for unknown concentrations in the proton condition (charge balance) in terms of $[H^+]$ and constants. We then can solve for $[H^+]$ and use its value to solve for other species concentrations.

11. Substituting (1) and (2) in (3) to eliminate $[OH^-]$ and $[NO_3^-]$, we obtain

    $$[H^+] = 10^{-7} + \frac{10^{-14}}{[H^+]}$$

or

$$[H^+]^2 - 10^{-7}[H^+] - 10^{-14} = 0$$

which can be solved using the quadratic formula

$$[H^+] = \frac{-b \pm \sqrt{b^2 - 4ac}}{2a}$$

For this problem, $a = 1$, $b = -10^{-7}$, and $c = -10^{-14}$. Solving, we obtain

$$[H^+] = 1.618 \times 10^{-7}; \text{ or pH} = 6.79$$

A similar set of assumptions and a similar calculation procedure can be used to calculate the species concentrations in the solution of a strong base as shown in the following example.

Example 4-13

Calculate pH and pOH in a 10^{-4} M NaOH solution at 25°C. Assume that activity coefficients are unity.

Solution

The unknowns are $[H^+]$, $[OH^-]$, and $[Na^+]$.

Equilibria

1. $H_2O \rightleftharpoons H^+ + OH^-$; $K_w = [H^+][OH^-] = 10^{-14}$

2. Mass balance on Na^+ (assuming complete dissociation of NaOH),

$$C_{T,Na} = [Na^+] = 10^{-4}$$

Charge Balance

3. $[H^+] + [Na^+] = [OH^-]$

Assumptions

4. In charge balance, the solution is basic, so assume

$$[Na^+] \gg [H^+]$$

(This is equivalent to assuming $[OH^-] \gg [H^+]$). Thus

$$[Na^+] = [OH^-]$$

5. From (2) and (4)

$$[Na^+] = [OH^-] = 10^{-4}$$
$$\text{pOH} = 4$$

6. $[H^+] = \dfrac{K_w}{[OH^-]} = \dfrac{10^{-14}}{10^{-4}} = 10^{-10}$

$$\text{pH} = 10$$

7. Check assumption in the charge balance

$$10^{-10} + 10^{-4} = 10^{-4}$$

The assumption checks, since the left-hand side of the equation is only a fraction of a percent larger than the right-hand side of the equation.

In each of the preceding examples, ionic strength effects were neglected. Let us now recalculate the pH for Examples 4-11 and 4-13 taking ionic strength into account. In these examples, the mass balance and proton condition (charge balance) equations are unaffected because these are based on concentration rather than activity. However, we must make changes in all of the equilibrium expressions we use. Since the K_a equation is eliminated by a simplifying assumption, that is, strong acids and strong bases are completely dissociated, we are concerned only in these examples with the K_w equation. This becomes

$$10^{-14} = \{H^+\}\{OH^-\} = \gamma_{H^+}[H^+]\gamma_{OH^-}[OH^-]$$

Using the Güntelberg approximation of the DeBye-Hückel law (Eq. 3-36) to express activity coefficients as a function of ionic strength,

$$-\log \gamma = \frac{0.5Z^2\mu^{1/2}}{1 + \mu^{1/2}} \tag{3-36}$$

We can calculate γ if μ is known. For Example 4-11, $C_{T,Cl} = 10^{-2} \, M$

$$\mu = 0.5 \sum_i C_i Z_i^2$$

$$= 0.5(1^2[H^+] + (-1)^2[Cl^-] + (-1)^2[OH^-]) \tag{3-32}$$

$[OH^-] \, (= 10^{-12})$ can be neglected because it is approximately 10 orders of magnitude less than $[H^+] \, (= 10^{-2})$ and $[Cl^-] \, (= 10^{-2})$.

$$\mu = \tfrac{1}{2}(1 \times 10^{-2} + 1 \times 10^{-2}) = 10^{-2} \, M$$

$$-\log \gamma_{H^+} = -\log \gamma_{OH^-} = \frac{0.5(1)^2(10^{-2})^{1/2}}{1 + (10^{-2})^{1/2}} = 0.045$$

$$\gamma_{H^+} = \gamma_{OH^-} = 0.90$$

The value, $[H^+] = 10^{-2}$, does not change because it was calculated using only the mass balance and proton conditions, which are based on concentrations rather than activities.

$$pH = -\log \{H^+\} = -\log (\gamma_{H^+}[H^+]) = -\log (0.9 \times 10^{-2}) = 2.05$$

and from K_w,

$$[OH^-] = \frac{10^{-14}}{\gamma_{H^+}\gamma_{OH^-}[H^+]} = \frac{10^{-14}}{(0.9)(0.9)(10^{-2})} = 1.2 \times 10^{-12} \, M$$

These values can be compared with pH = 2 and $[OH^-] = 1.0 \times 10^{-12}$ when the effect of ionic strength was neglected. An error on the order of 10 percent in $[H^+]$ and 20 percent in $[OH^-]$ resulted from neglecting ionic strength effects. We conclude that in solutions with an ionic strength of 10^{-2} moles/liter and equilibria involving only monovalent ions, the effect of ionic strength can be significant.

For Example 4-13, where $[Na^+] = 10^{-4}$

$$\mu = 0.5(1^2[H^+] + (-1)^2[OH^-] + 1^2[Na^+])$$

Using $[H^+] = 10^{-10}$, which was obtained by neglecting the effect of μ, we can see that $[H^+]$ is very small compared to $[OH^-]$ and $[Na^+]$, and can be neglected in the calculation of μ. Therefore,

$$\mu = 0.5(1 \times 10^{-4} + 1 \times 10^{-4}) = 10^{-4}\,M$$

$$-\log \gamma_{H^+} = -\log \gamma_{OH^-} = \frac{(0.5)(1^2)(10^{-4})^{1/2}}{1 + (10^{-4})^{1/2}} = 4.9 \times 10^{-3}$$

$$\gamma_{H^+} = \gamma_{OH^-} = 0.99$$

From Example 4-13 we found $[OH^-] = 10^{-4}$; this value does not change because only the mass balance and proton condition were used to calculate it. Therefore,

$$\{OH^-\} = \gamma_{OH^-}[OH^-] = 0.99 \times 10^{-4}$$

so that including the effect of ionic strength we obtain pOH = 4.004 compared to pOH = 4.0 when ionic strength effects were neglected. Also,

$$\{H^+\} = \frac{10^{-14}}{\{OH^-\}} = \frac{10^{-14}}{0.99 \times 10^{-4}} = 1.01 \times 10^{-10}$$

and pH = 9.996 compared to pH = 10 when ionic strength effects were neglected. We therefore might conclude that at ionic strengths of 10^{-4} mole/liter, we can safely neglect the effects of ionic strength on equilibria involving only monovalent ions.

4.6.   WEAK ACID–WEAK BASE CALCULATIONS

Weak acids and weak bases do not ionize (or protolyze) completely in aqueous solution. The approach used to solve for the concentrations of solution components for weak acid or base solutions is similar to that used for strong acids and strong bases, but we are not able to make the simplifying assumption in the K_a or K_b equilibrium equations that complete dissociation takes place. Typical calculations for weak acid and weak base systems are illustrated in the following example.

Example 4-14

Calculate the concentration of all species in a solution prepared by adding 10^{-2} mole of acetic acid, CH_3COOH (HAc), to 1 liter of water at 25°C. Neglect ionic strength effects.

Solution

The unknowns are [H⁺], [OH⁻], [HAc], and [Ac⁻]. Thus four equations are required.

Equilibria

1. $HAc \rightleftharpoons H^+ + Ac^-$; $K_a = 10^{-4.7} = [H^+][Ac^-]/[HAc]$
2. $H_2O \rightleftharpoons H^+ + OH^-$; $K_w = 10^{-14} = [H^+][OH^-]$

Mass Balance on Ac

3. $C_{T,Ac} = [HAc] + [Ac^-] = 10^{-2}$

Charge Balance (or Proton Condition)

4. $[H^+] = [Ac^-] + [OH^-]$
   To solve equations (1) to (4), we can make the following assumptions: HAc is an acid so it dissociates to produce [H⁺] and [Ac⁻].

5. In the mass balance assume [Ac⁻] ≫ [HAc]. Then

$$C_{T,Ac} = [Ac^-] = 10^{-2}$$

The solution is acid and therefore [OH⁻] is small.

6. In the charge balance assume that [Ac⁻] ≫ [OH⁻].

$$[H^+] = [Ac^-]$$

7. From (5) and (6),

$$[H^+] = [Ac^-] = 10^{-2}$$

8. From (2) and (7),

$$[OH^-] = \frac{K_w}{[H^+]} = \frac{10^{-14}}{10^{-2}} = 10^{-12}$$

9. From (1) and (8),

$$[HAc] = \frac{[H^+][Ac^-]}{K_a} = \frac{[10^{-2}]^2}{10^{-4.7}}$$

$$[HAc] = 10^{+0.7} = 5\ M$$

10. Checking our assumptions, we find, in the proton condition,

$$[H^+] = [Ac^-] + [OH^-]$$
$$10^{-2} = 10^{-2} + 10^{-12}$$

This assumption is acceptable because the two sides of the equation are virtually equal.
   In the mass balance we find

$$10^{-2} = [Ac^-] + [HAc]$$
$$10^{-2} = 10^{-2} + 5$$

This equation is hopelessly unbalanced so that the assumption of [Ac⁻] ≫ [HAc] is not acceptable. When we think about this assumption a little more, we can see that it was a poor one because we know that HAc is a weak acid so that only a small amount of Ac^- will be present compared to the quantity of HAc. So let us now make the opposite assumption in the mass balance equation that $[HAc] \gg [Ac^-]$.

The mass balance becomes

11. $C_{T,HAc} = 10^{-2} = [HAc]$

12. Substituting (6) and (11) into (1) to eliminate [HAc] and $[Ac^-]$ to solve for $[H^+]$, we find that

$$10^{-4.7} = \frac{[H^+][H^+]}{10^{-2}}$$

$$[H^+] = 10^{-3.35} = 4.47 \times 10^{-4}$$

13. From (6) and (12)

$$[H^+] = [Ac^-] = 4.47 \times 10^{-4}$$

14. From (2) and (12)

$$[OH^-] = \frac{10^{-14}}{4.47 \times 10^{-4}} = 2.24 \times 10^{-11}$$

15. Checking our new assumptions, in the mass balance we obtain

$$10^{-2} = [HAc] + [Ac^-]$$
$$10^{-2} = 10^{-2} + 4.47 \times 10^{-4}$$

The error of this balance is $[(10^{-2} + 4.47 \times 10^{-4} - 10^{-2})/10^{-2}] \times 100 = 4.5$ percent, which by our criterion of < 5 percent error is just acceptable.

In the proton condition, we obtain

$$[H^+] = [Ac^-] + [OH^-]$$
$$4.47 \times 10^{-4} = 4.47 \times 10^{-4} + 2.24 \times 10^{-11}$$

This assumption is acceptable because the right-hand side of the equation is virtually equal to the left-hand side.

From this example we can see that even though one of our first assumptions was not valid, the approach of "making assumptions, obtaining a solution, and checking the assumptions" allowed us to correct the assumptions to obtain a solution with the desired degree of accuracy. If the second assumption in the mass balance had not yielded the desired accuracy, our approach would have been to obtain a solution without making any assumptions in the mass balance. This would have led to a quadratic equation in $[H^+]$, which could have been solved as in Example 4-12.

Example 4-15

Sodium hypochlorite (NaOCl), household bleach, is a commonly used disinfectant.

a. Neglecting ionic strength effects, determine the concentrations of the "free chlorine" species, HOCl and OCl⁻, in a freshly prepared solution of 10^{-3} M NaOCl in distilled water.

b. What is the concentration of HOCl, the most effective disinfectant species, if ionic strength effects are taken into account?

Solution

The unknowns are H^+, OH^-, Na^+, OCl^-, and HOCl. Thus five equations are required.

a. Neglecting ionic strength effects, the necessary equations are

Equilibria

1. $NaOCl \rightarrow Na^+ + OCl^-$, dissociation is complete[7] so [NaOCl] = 0.

2. $OCl^- + H_2O \rightleftharpoons HOCl + OH^-$; $K_b = [HOCl][OH^-]/[OCl^-] = 10^{-6.5}$ (Table 4-1).

Alternatively,

3. $HOCl \rightleftharpoons H^+ + OCl^-$; $K_a = [H^+][OCl^-]/[HOCl] = 10^{-7.5}$ (Table 4-1).

4. $H_2O \rightleftharpoons H^+ + OH^-$; $K_w = [H^+][OH^-] = 10^{-14}$

5. Mass balance on Na^+:

$$C_{T,Na} = [Na^+] = 10^{-3}$$

(Assume that NaOCl is completely dissociated.)

6. Mass balance on OCl:

$$C_{T,OCl} = [HOCl] + [OCl^-] = 10^{-3}$$

7. Proton condition:

$$[HOCl] + [H^+] = [OH^-]$$

Equations (3) to (7) will be used to solve this problem.

Assumptions

8. Arbitrarily assuming that $[OCl^-] \gg [HOCl]$, the mass balance on OCl becomes

$$[OCl^-] = 10^{-3}$$

9. Assuming that $[HOCl] \gg [H^+]$, the proton condition becomes

$$[HOCl] = [OH^-]$$

10. Substituting (4), (8) and (9) in (3) and solving for $[H^+]$, we obtain

$$10^{-7.5} = \frac{[H^+](10^{-3})}{(10^{-14}/[H^+])}$$

$$[H^+] = 10^{-9.25} = 5.6 \times 10^{-10} \, M$$

$$pH = 9.25$$

[7] Complete dissociation can be assumed in most instances for salts of the alkali and alkaline earth metals. Exceptions are discussed in Chapter 5.

11.   From (4) and (10)

$$[OH^-] = \frac{K_w}{[H^+]} = 10^{-4.75} = 1.78 \times 10^{-5}\,M$$

$$pOH = 4.75$$

12.   From (9) and (11)

$$[HOCl] = 10^{-4.75} = 1.78 \times 10^{-5}\,M$$

13.   Check the assumption in the mass balance on OCl:

$$10^{-3} = [HOCl] + [OCl^-]$$

$$10^{-3} = 1.78 \times 10^{-5} + 10^{-3}$$

$$error = \left(\frac{1.78 \times 10^{-5}}{10^{-3}}\right) \times 100 = 1.78\,percent$$

The assumption is acceptable.

14.   Check the assumption in the proton condition:

$$1.78 \times 10^{-5} + 5.6 \times 10^{-10} = 1.78 \times 10^{-5}$$

$$error = \left(\frac{5.6 \times 10^{-10}}{1.78 \times 10^{-5}}\right) \times 100 = 0.003\,percent$$

The assumption is acceptable.

b.   To account for ionic strength effects, we need the following five equations: (5), the mass balance on Na, (6), the mass balance on OCl, (7), the proton condition, and the two equilibrium relationships,

$$K_a = 10^{-7.5} = \frac{\{H^+\}\{OCl^-\}}{\{HOCl\}}$$

$$= \frac{\gamma_{H^+}[H^+]\gamma_{OCl^-}[OCl^-]}{\gamma_{HOCl}[HOCl]}$$

15. $^cK_a = 10^{-7.5}\gamma_{HOCl}/(\gamma_{H^+}\gamma_{OCl^-}) = [H^+][OCl^-]/[HOCl]$. Then

$$K_w = 10^{-14} = \{H^+\}\{OH^-\}$$

$$= \gamma_{H^+}[H^+]\gamma_{OH^-}[OH^-]$$

16. $^cK_w = 10^{-14}/(\gamma_{H^+}\gamma_{OH^-}) = [H^+][OH^-]$.

These equations are identical to the five equations used in part (a), except that ^cK_a and ^cK_w have replaced K_a and K_w, respectively. The approach to the solution of the equations is identical to that used in part (a).

17.   The ionic strength of the solution can be calculated to a good approximation[8] using the ionic concentrations computed in part (a):

$$\mu = 0.5 \sum_i C_i Z_i^2 \tag{3-32}$$

---

[8] It is an approximation because ionic strength affects the degree of dissociation of HOCl. For a precise calculation, after the solution to part (b) is obtained, a new ionic strength should be calculated and a new solution should be obtained. In this problem the degree of dissociation is very close to that calculated in part (a), making the iterative calculation unnecessary.

$$\mu = \tfrac{1}{2}([H^+](1)^2 + [OH^-](-1)^2 + [OCl^-](-1)^2 + [Na^+](1)^2)$$

$$= \tfrac{1}{2}(5.6 \times 10^{-10} + 1.78 \times 10^{-5} + 10^{-3} + 10^{-3})$$

$$= 10^{-3}\,M$$

18. Activity coefficients for the ions H^+, OH^-, OCl^-, and Na^+ can be calculated using the Güntelberg approximation of the DeBye-Hückel law,

$$-\log \gamma = \frac{0.5 Z^2 \mu^{1/2}}{1 + \mu^{1/2}}$$

$$= \frac{0.5(10^{-3})^{1/2}}{1 + (10^{-3})^{1/2}} = 1.53 \times 10^{-2} \qquad (3\text{-}36)$$

$$\gamma = 0.96$$

19. It is assumed that the activity coefficient of HOCl is unity, since the ionic strength at which activity coefficients for neutral molecules deviate significantly from unity is of the order of 0.1 M.

20. Using the activity coefficients and equations (15) and (16) yields

$$^{c}K_a = 10^{-7.46}$$

$$^{c}K_w = 10^{-13.96}$$

21. Solving equations (5), (6), (7), (15), and (16) as in part (a) yields the values shown in Table 4-4. It can be seen that there is very little change in [HOCl].

The chlorine solution in Example 4-15 has an ionic strength of 10^{-3} moles/liter, and again we see that the ionic strength correction had little effect and could have been neglected. Thus we can narrow down even further the range at which ionic strength effects become significant, since we notice significant effects at 10^{-2} moles/liter but not at 10^{-3} moles/liter. It should be remembered that in each of these situations we have been dealing with equilibria involving only monovalent ions—for ions of higher charge the ionic strength value at which activity coefficient corrections

TABLE 4-4   Effect of Ionic Strength of Species Concentrations in 10^{-3} M NaOCl Solution.

	Not Accounting for Ionic Strength Effects	Accounting for Ionic Strength Effects ($\mu = 10^{-3}$ mole/liter)
Na^+, M	10^{-3}	10^{-3}
OCl^-, M	10^{-3}	10^{-3}
HOCl, M	1.78×10^{-5}	1.81×10^{-5}
H^+, M	5.6×10^{-10}	6.02×10^{-10}
OH^-, M	1.78×10^{-5}	1.81×10^{-5}
pH	9.25	9.23
pOH	4.75	4.77

become significant falls to about 10^{-3}. The activity coefficient of a trivalent ion in a solution with an ionic strength of 10^{-3} M is approximately the same for a monovalent ion in a solution with an ionic strength of 10^{-2} M (see Fig. 3-4).

Note that the free chlorine at pH 9.25 is almost entirely present in the form of OCl^-. The distribution of the species $HOCl$ and OCl^- and their variation with pH is a key factor in disinfection capability because the effectiveness of these two species as a bactericide is markedly different ($HOCl$ being about 80 to 100 times more effective than OCl^- for some organisms).

4.7.   GRAPHICAL PROCEDURES FOR EQUILIBRIUM CALCULATIONS— pC–pH DIAGRAMS

Graphical procedures are available that allow equilibrium calculations to be carried out much more readily than by the computations we have been discussing. However, it is important to master the computation method because the basis of the graphical procedure lies in the computation method. Graphical procedures for acid-base systems consist of setting up equilibrium and mass balance equations on a double logarithmic plot of $-\log$ concentration (pC) and pH. The diagram so produced is examined to determine the pH at which the proton condition (or charge balance) is satisfied. All of the species concentrations then can be read off from the graph at this pH value. To develop the graphical procedure, let us first consider the situation when a weak acid is added to water. The approach we use, and the diagram that results, are valid equally for addition of the conjugate base of the weak acid, or for a mixture of the acid and its conjugate base. For this example, HCN, 10^{-3} moles, is added to 1 liter of distilled water at 25°C in a closed system with no atmosphere. We make the last stipulation because $HCN_{(aq)}$ is volatile and will tend to equilibrate with $HCN_{(g)}$ in the atmosphere above a solution. At this stage we wish to eliminate such equilibria from consideration. We will also neglect ionic strength effects. The equations that describe the system are

Equilibria

$$HCN \rightleftharpoons H^+ + CN^-;$$

$$K_a = \frac{[H^+][CN^-]}{[HCN]} = 10^{-9.3} \qquad \text{(Table 4-1)} \qquad (4\text{-}21)$$

$$H_2O \rightleftharpoons H^+ + OH^-;$$

$$K_w = [H^+][OH^-] = 1 \times 10^{-14} \qquad (4\text{-}22)$$

Mass Balance

$$C_{T,CN} = [HCN] + [CN^-] = 10^{-3} \qquad (4\text{-}23)$$

Proton Condition (Charge Balance)

$$[H^+] = [CN^-] + [OH^-] \tag{4-24}$$

The first equation we plot on the diagram is the mass balance, which in logarithmic form is

$$-\log C_{T,CN} = pC_{T,CN} = 3 \tag{4-25}$$

Equation 4-25 is not a function of pH so it plots as a horizontal line at pC = 3 (see line 1, Fig. 4-1).

The second equation plotted on the diagram is the K_w expression, which in logarithmic form is

$$\log K_w = \log [H^+] + \log [OH^-]$$

or

$$pOH = 14 - pH \tag{4-26}$$

The slope of this line is

$$\frac{dpOH}{dpH} = -1$$

and when pH = 0, pOH = 14.

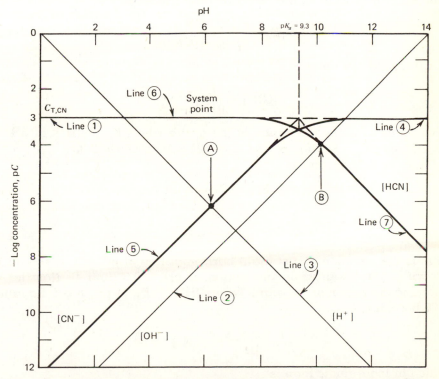

Fig. 4-1. The pC-pH diagram for cyanide-hydrocyanic acid; $C_{T,CN} = 10^{-3} M$; $pK_a = 9.3$ at 25°C.

From this slope and intercept we can plot this line on Fig. 4-1 as line 2.

Now, by definition,

$$-\log [H^+] = pH$$

so

$$\frac{d\,(-\log [H^+])}{d pH} = 1$$

and when pH = 0, $-\log [H^+] = 0$. From the slope and intercept we can plot this line on Fig. 4-1 as line 3.

The H^+ and OH^- concentration lines are common to the graphical solution of all acid-base problems (and many other types of equilibrium problems) so that it would be handy to prepare pC-pH diagrams with these lines already plotted on them. The first step would then be to plot pC_T on the diagram.

The next lines to be plotted are for $[CN^-]$ and $[HCN]$. To plot these quantities in the pC-pH diagram, first we must derive expressions relating each of them to pH and constants. Tackling $[CN^-]$ first, we can combine Eqs. 4-21 and 4-23 to eliminate $[HCN]$ and produce

$$K_a = \frac{[H^+][CN^-]}{(C_{T,CN} - [CN^-])} \tag{4-27}$$

Rearranging, we obtain

$$[CN^-] = \frac{K_a C_{T,CN}}{([H^+] + K_a)} \tag{4-28}$$

Now let us examine the region of the graph where $K_a \gg [H^+]$, or pH $\gg$ pK_a. Since $pK_a = 9.3$ this region is at pH values above 9.3.

Equation 4-28 becomes

$$[CN^-] = C_{T,CN} = 10^{-3}$$

and

$$p[CN^-] = 3 \tag{4-29}$$

This is the equation of a straight horizontal line at pC = 3. It should be drawn on the diagram in the region of pH > 9.3 (line 4). Now let us examine the region of the graph where $[H^+] \gg K_a$, or $pK_a > pH$. Equation 4-28 becomes

$$\frac{K_a C_{T,CN}}{[H^+]} = [CN^-]$$

which in logarithmic form is

$$\log C_{T,CN} + \log K_a - \log [H^+] = + \log [CN^-]$$

or

$$p[CN^-] = pC_{T,CN} + pK_a - pH \qquad (4\text{-}30)$$

The slope is

$$\frac{d(p[CN^-])}{dpH} = -1$$

When $pH = pK_a = 9.3$

$$p[CN^-] = pC_{T,CN} = 3$$

With this slope and coordinate we can plot the $[CN^-]$ line in the region $pH < 9.3$ (line 5).

Since approximations were made to derive both sections of the $p[CN^-]$ line, we may not draw either section of the line through pH 9.3, the so-called "system point." In the region around pH 9.3 we must draw a dashed line. We will return to the exact behavior of the lines in this region after development of the [HCN] line.

We can solve Eqs. 4-21 and 4-23 to eliminate $[CN^-]$ with the result that

$$K_a = \frac{[H^+](C_{T,CN} - [HCN])}{[HCN]}$$

and rearranging,

$$[HCN] = \frac{C_{T,CN}[H^+]}{[H^+] + K_a} \qquad (4\text{-}31)$$

In the pH region where $[H^+] \gg K_a$, $pK_a > pH$, this equation becomes

$$[HCN] = C_{T,CN}$$
$$p[HCN] = pC_{T,CN} = 3 \qquad (4\text{-}32)$$

We can draw a horizontal straight line at $pC = 3$ for HCN in the pH region below pK_a (line 6).

In the region where $[H^+] \ll K_a$, $pH > pK_a$, Eq. 4-31 becomes,

$$[HCN] = \frac{C_{T,CN}[H^+]}{K_a}$$
$$p[HCN] = pH + pC_{T,CN} - pK_a$$
$$\text{slope} = \frac{d(p[HCN])}{dpH} = +1 \qquad (4\text{-}33)$$

When $pH = pK_a = 9.3$

$$p[HCN] = pC_{T,CN} = 3$$

With this slope and coordinate we can plot the p[HCN] line in the region $pH > 9.3$ (line 7). Again we must forego drawing in the p[HCN] lines in the immediate vicinity of pK_a.

We now have lines for [HCN] and [CN⁻] plotted as a function of pH for the entire pH range, except in the immediate vicinity of pH = pK_a because of the assumptions that were made. To complete the graph in Fig. 4-1, let us examine the position of these two graphs at the point pH = pK_a. From Eq. 4-27, when pH = pK_a, or when [H⁺] = K_a, we obtain

$$[CN^-] = \frac{C_{T,CN}}{2}$$

or

$$p[CN^-] = pC_{T,CN} + 0.3 \qquad (4\text{-}34)$$

Similarly, when considering Eq. 4-31 at pH = pK_a we find

$$[HCN] = \frac{C_{T,HCN}}{2}$$

or in logarithmic form

$$p[HCN] = pC_{T,CN} + 0.3 \qquad (4\text{-}35)$$

From Eqs. 4-34 and 4-35 we can see that at the point (pC = ($pC_{T,CN}$ + 0.3), pH = pK_a) and [HCN] = [CN⁻]. Therefore we have a common point through which both the [HCN] and the [CN⁻] lines must pass, which is located 0.3 log units vertically below $pC_{T,CN}$ at pH = pK_a = 9.3. This point can be entered on the diagram. The [HCN] and the [CN⁻] lines both pass through this point. They can be drawn through this point as a smooth curve that starts at about one pH unit either above or below pK = pK_a.

If more accuracy is desired in this region, additional points can be calculated by substituting pH values in Eqs. 4-27 and 4-31. For example, at pH = pK_a − 1,

$$p[HCN] = pC_{T,CN} + 0.04$$

and at pH = pK_a + 1,

$$p[CN^-] = pC_{T,CN} + 0.04$$

The diagram in Fig. 4-1 now is complete. To use the diagram for determining equilibrium concentrations of various species, we must examine it (as illustrated in the following examples) to find the pH where the proton condition or charge balance is satisfied. It is only at points where proton conditions are satisfied that the diagram can be used to obtain equilibrium concentrations.

Example 4-16

What is the pH, pOH, [CN⁻], and [HCN] in a 10^{-3} M HCN solution at 25°C which is closed to the atmosphere so that no HCN can volatilize? Neglect ionic strength effects.

Solution

The proton condition for HCN in H_2O is

$$[H^+] = [CN^-] + [OH^-]$$

We must use the same technique of approximation that was employed in the computation procedure to find the point in the diagram where the proton condition is satisfied. We are not, however, flying blind in this instance because we have the diagram to guide us in our selection of assumptions. Examining point A in Fig. 4-1, we see that $[CN^-] \gg [OH^-]$ and $[H^+] = [CN^-]$. This point satisfies our simplified proton condition, which when neglecting $[OH^-]$ becomes $[H^+] = [CN^-]$. Point A occurs at pH = 6.2 and we can read off all the other species concentrations at pH 6.2: $[HCN] = 10^{-3}$, $[H^+] = 10^{-6.2}$, $[CN^-] = 10^{-6.2}$, and $[OH^-] = 10^{-7.8}$. Our assumption that $[CN^-] \gg [OH^-]$ is justified, since $10^{-6.2} \gg 10^{-7.8}$.

The same diagram (Fig. 4-1) can be used to determine the pH of 10^{-3} M solutions of sodium cyanide, NaCN, and of mixtures of HCN and NaCN.

Example 4-17

Find the pH, pOH, [HCN], [CN$^-$], and [Na$^+$] in a 10^{-3} M NaCN solution. The temperature is 25°C; neglect ionic strength effects.

Solution

The proton condition for NaCN in H_2O is

$$[HCN] + [H^+] = [OH^-]$$

Examining Fig. 4-1 at point B, we see that $[HCN] \gg [H^+]$ and the approximate proton condition $[HCN] = [OH^-]$, is satisfied. This point occurs at pH = 10.1. At this pH, $[H^+] = 10^{-10.1}$ M, $[OH^-] = 10^{-3.9}$, M $[HCN] = 10^{-3.9}$ M, $[CN^-] = 10^{-3.05}$ M. From a mass balance on sodium, $[Na^+] = C_{T,Na} = 10^{-3}$ M.

Checking our assumption in the proton condition, we find that

$$10^{-10.1} + 10^{-3.9} = 10^{-3.9}$$

The assumption is acceptable, since the error is a small fraction of 1 percent.

Example 4-18

Find the pH, [HOCl], and [OCl$^-$] in a solution to which 10^{-4} mole of sodium hypochlorite, NaOCl, and 5×10^{-5} mole of hypochlorous acid, HOCl, $pK_a = 7.5$, have been added per liter of solution. The temperature is 25°C; neglect ionic strength effects.

Solution

1. Construct a pC-pH diagram for HOCl following the procedure described previously. In the diagram, Fig. 4-2,

$$C_{T,OCl} = [HOCl] + [OCl^-] = 5 \times 10^{-5} + 10^{-4} = 1.5 \times 10^{-4}$$

$$pC_{T,OCl} = 3.8$$

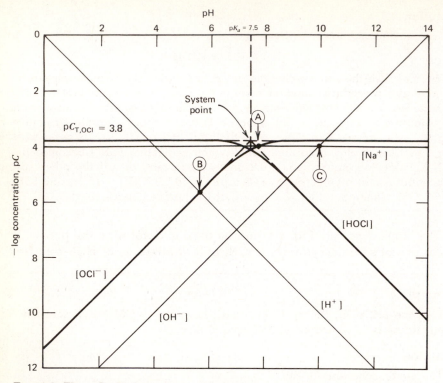

Fig. 4-2. The pC-pH diagram for hypochlorite-hypochlorous acid: $C_{T,OCl}$ = $1.5 \times 10^{-4}\ M$, 25°C.

2. Because we added both an acid and a conjugate base we must account for the cation (Na^+) accompanying the conjugate base in our diagram.

A mass balance on Na^+ yields

$$C_{T,Na} = [Na^+] = 10^{-4}\ M$$

$$pC_{T,Na} = 4$$

We can draw the $[Na^+]$ line in the diagram as a horizontal line at $pC_T = 4$.

To solve for the species present, we must examine the diagram for the point at which electroneutrality is met. For situations where an acid and its conjugate base are added, the charge balance is preferred over the proton condition for finding the point of neutrality because it is easier to write.

The charge balance for our mixture is

$$[Na^+] + [H^+] = [OH^-] + [OCl^-]$$

Examining Fig. 4-2, we can see that at point A, $[Na^+] \gg [H^+]$ and $[OCl^-] \gg [OH^-]$ so that the charge balance simplifies to $[Na^+] = [OCl^-]$, a condition that is met at pH 7.7. At this pH, $[Na^+] = 10^{-4}\ M$, $[OCl^-] = 10^{-4}\ M$, $[H^+] = 10^{-7.7} = 1.99 \times 10^{-8}\ M$, $[OH^-] = 10^{-6.3} = 5.0 \times 10^{-7}\ M$, and $[HOCl] = 10^{-4.2} = 6.3 \times 10^{-5}\ M$.

Checking the charge balance, we obtain,

$$1 \times 10^{-4} + 1.99 \times 10^{-8} = 1 \times 10^{-4} + 5.0 \times 10^{-7}$$

$$\text{error} = \frac{(1 \times 10^{-4} + 5.0 \times 10^{-7}) - (1 \times 10^{-4} + 1.99 \times 10^{-8})}{1 \times 10^{-4} + 1.99 \times 10^{-8}} \times 100$$

$$= 0.5 \, \text{percent}$$

so that the assumptions are acceptable.

We can prove that these are the only viable set of assumptions to make in the charge balance by using the pC-pH diagram to check the validity of other possible assumptions. For example, if we assume $[H^+] \gg [Na^+]$ and $[OCl^-] \gg [OH^-]$, then the charge balance becomes $[H^+] = [OCl^-]$, a condition that is met at point B in Fig. 4-2. However, at this pH of 5.8, $[H^+] = 10^{-5.8}$ and $Na^+ = 10^{-4}$ so that $[Na^+] \gg [H^+]$ and our assumption is not valid. If we make the assumptions that $[Na^+] \gg [H^+]$ and $[OH^-] \gg [OCl^-]$, then the charge balance reduces to $[Na^+] = [OH^-]$, a condition that is satisfied at point C in Fig. 4-2. At this pH of 10, $[OH^-] = 10^{-4}$ and $[OCl^-] = 10^{-3.8}$ so that the assumption of $[OH^-] \gg [OCl^-]$ is not acceptable. We have now made all possible combinations of assumptions in the charge balance and could by this process of elimination eventually find the correct set of assumptions.

There are instances when assumptions in the proton condition are not possible. One of the more common situations encountered is when the analytical concentration of an acid is close to its K_a value or the K_b value of its conjugate base, or the analytical concentration of a base is close to its K_b value or the K_a value of its conjugate acid. It also becomes difficult to make simplifying assumptions in the proton condition for both acids and bases when their analytical concentration approaches $10^{-7} \, M$. The approach to this type of problem is illustrated in the following example.

Example 4-19

Determine all species concentrations present in a $10^{-4.5} \, M$ NH_4Cl solution $(K_{a,NH_4^+} = 10^{-9.3})$.

Solution

Note that $pC_{T,NH_3} = 4.5$, which is close to $pK_b = pK_w - pK_a = 14 - 9.3 = 4.7$. We can expect trouble here. Construction of the pC-pH diagram follows previously described procedures (see Fig. 4-3). The proton condition is

$$[H^+] = [NH_{3(aq)}] + [OH^-]$$

Examination of Fig. 4-3 reveals that in the region where the proton condition is satisfied (an approximate pH of 6.7 to 7), it is not possible to neglect $[OH^-]$ or $[NH_{3(aq)}]$ with respect to each other. We have several alternatives: (1) use a computational procedure, (2) guess the point in the diagram (e.g., just to the left-hand side of $[H^+] = [NH_{3(aq)}]$) where $[H^+] = [NH_{3(aq)}] + [OH^-]$, (3) read off values of

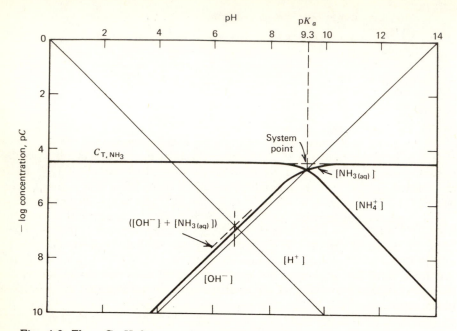

Fig. 4-3. The pC-pH diagram for a $10^{-4.5}$ M NH$_4$Cl solution at 25°C.

[OH$^-$] and [NH$_{3(aq)}$] from the diagram at several pH values in the critical region and then construct a pH versus p([NH$_{3(aq)}$] + [OH$^-$]) line so that the proton condition can be solved exactly in the diagram. With experience, alternative 2 is acceptable and speedy. Alternative 3 is illustrated in Table 4-5 and by the dashed line in Fig. 4-3. Using this alternative, we find that pH = p([NH$_{3(aq)}$] + [OH$^-$]) at pH = 6.8 and [NH$_{3(aq)}$] = $10^{-6.75}$, [NH$_4^+$] = $10^{-4.5}$, [H$^+$] = $10^{-6.8}$, and [OH$^-$] = $10^{-7.2}$.

Example 4-26 later in this chapter deals with another situation in which approximations in the charge balance are not possible.

4.8. DETERMINATION OF TEMPERATURE AND IONIC STRENGTH EFFECTS ON EQUILIBRIA USING pC-pH DIAGRAMS

The pC-pH diagrams can be used for systems having temperatures other than 25°C and ionic strengths other than zero. The modification of a 25°C-based pC-pH diagram for temperatures other than 25°C merely entails the use of equilibrium constants for the appropriate temperature in the equilibrium expressions used to construct the diagram. The Van't Hoff relationship (Eq. 3-27) can be used to convert 25°C equilibrium constants to the values at the desired temperature.

Construction of pC-pH diagrams for ionic strengths other than zero requires the use of activities rather than concentrations in all equilibrium

TABLE 4-5 Proton Condition Solution for a $10^{-4.5} M$ NH$_4$Cl Solution (See Fig. 4-3)

pH	pOH	[OH$^-$]	p[NH$_{3(aq)}$]	[NH$_{3(aq)}$]	([OH$^-$] + [NH$_{3(aq)}$])	p([OH$^-$] + [NH$_{3(aq)}$])
5.5	8.5	3.1×10^{-9}	8.3	5×10^{-9}	8.1×10^{-9}	8.1
6.0	8.0	10^{-8}	7.8	1.6×10^{-8}	2.6×10^{-8}	7.6
6.5	7.5	3.2×10^{-8}	7.3	5×10^{-8}	8.2×10^{-8}	7.1
7.0	7.0	10^{-7}	6.8	1.6×10^{-7}	2.6×10^{-7}	6.6
7.5	6.5	3.2×10^{-7}	6.3	5×10^{-7}	8.2×10^{-7}	6.1
8.0	6.0	10^{-6}	5.8	1.6×10^{-6}	2.6×10^{-6}	5.6

expressions. Mass balance and charge balance (or proton condition) equations are based on concentration and therefore are not corrected for activity.

As an example of these two types of corrections, let us construct the pC-pH diagram for a $10^{-3} M$ hypochlorous acid solution at 15°C with an ionic strength of 0.1 M. In this example, equilibrium constants and pH values based on concentration measurements will be denoted as ^cK and p^{c}H while the constants and pH values based on activity will be referred to as K and pH. The constant K_a for HOCl at 15°C is $10^{-7.6}$; K_w at 15°C is 4.5×10^{-15} (see Table 4-2). The master variable in the pC-pH diagram is p^{c}H, defined as

$$p^cH = -\log [H^+] \tag{4-36}$$

This is plotted as line 1 in Fig. 4-4.

The line for [OH$^-$] is derived as follows:

$$K_w = \{H^+\}\{OH^-\} = \gamma_{H^+} [H^+] \gamma_{OH^-}[OH^-]$$

$$^cK_w = [H^+][OH^-] = \frac{K_w}{\gamma_{H^+} \gamma_{OH^-}}$$

$$-\log [OH^-] = p^cK_w - p^cH = -\log K_w + \log \gamma_{H^+} + \log \gamma_{OH^-} - p^cH$$

At $\mu = 0.1$ we find from Fig. 3-4, that $\gamma_{H^+} = 0.83$ and $\gamma_{OH^-} = 0.75$; we then can compute that $^cK_w = 7.2 \times 10^{-15}$ and

$$-\log [OH^-] = 14.14 - p^cH \tag{4-37}$$

This equation plots as line 2 in Fig. 4-4.

The mass balance equations are defined in terms of concentrations and do not need correction for μ or temperature. Thus $C_{T,OCl} = 10^{-3} = $ [HOCl] + [OCl$^-$] and

$$pC_{T,OCl} = 3 \tag{4-38}$$

Equation 4-38 plots as line 3 in Fig. 4-4.

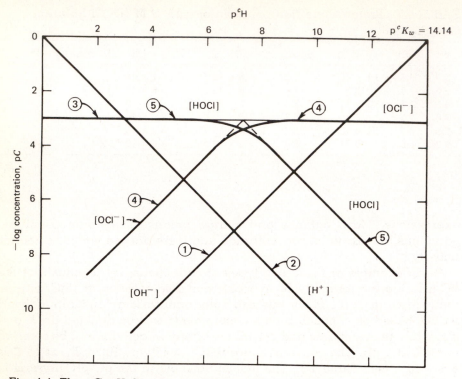

Fig. 4-4. The pC-p^{c}H diagram for HOCl—OCl$^-$ at 15°C, $\mu = 0.1$.

The lines for [OCl$^-$] and [HOCl] are constructed using the ^cK_a equation and therefore will reflect the modification of this equation to account for temperature and ionic strength effects on K_a

$$K_a = \frac{\{H^+\}\{OCl^-\}}{\{HOCl\}} = \frac{\gamma_{H^+}[H^+]\gamma_{OCl^-}[OCl^-]}{\gamma_{HOCl}[HOCl]}$$

$$^cK_a = \frac{[H^+][OCl^-]}{[HOCl]} = \frac{K_a\,\gamma_{HOCl}}{\gamma_{H^+}\gamma_{OCl^-}}$$

(4-39)

When $\mu = 0.1$, from Fig. 3-4, $\gamma_{H^+} = 0.83$, $\gamma_{OCl^-} = 0.76$, and $\gamma_{HOCl} = 1$. Since K_a at 15°C $= 10^{-7.6}$, we can substitute these values in Eq. 4-39 to obtain ^cK_a $= 10^{-7.4}$. The [OCl$^-$] and [HOCl] lines (lines 4 and 5, respectively, in Fig. 4-4) can then be derived from the equations

$$[OCl^-] = \frac{C_{T,OCl}\,^cK_a}{[H^+] + \,^cK_a}$$

(4-40)

and

$$[HOCl] = \frac{C_{T,OCl}\,[H^+]}{[H^+] + \,^cK_a}$$

(4-41)

The system point is (p^cH = pK_a, pC = $-\log C_{T,OCl}$), instead of (pH = pK_a, pC = $-\log C_{T,OCl}$) in the absence of corrections for ionic strength.

The remaining equation necessary to describe the system is either the proton condition or the charge balance. Since these equations are based on concentration rather than activity, no correction for ionic strength is necessary. The pC-pH diagram in Fig. 4-4 can now be used to obtain concentrations as previously demonstrated.

4.9. MULTIPROTIC ACID-CONJUGATE BASE EQUILIBRIUM CALCULATIONS

A multiprotic acid can donate more than one proton. Consider, for example, sulfuric acid, H_2SO_4, which can donate two protons,

$$H_2SO_4 \rightleftharpoons HSO_4^- + H^+; \qquad K_{a,1} \cong 10^3$$
$$\text{bisulfate}$$
$$\text{ion}$$

$$HSO_4^- \rightleftharpoons SO_4^{2-} + H^+; \qquad K_{a,2} = 10^{-2}$$
$$\text{sulfate}$$
$$\text{ion}$$

H_2SO_4 is classified as a strong acid while HSO_4^- is a weak acid.

For the general case of a multiprotic acid added to water, we can construct a pC-pH diagram in a similar fashion to that used for a monoprotic acid. The procedure is somewhat more complex because we have additional species to consider as well as additional equilibrium statements.

To illustrate the procedure, we will determine the pH and concentrations of all species present ($[H_2A]$, $[HA^-]$, $[A^{2-}]$, $[H^+]$, $[OH^-]$) when 10^{-2} moles of the weak acid H_2A ($pK_{a,1} = 4$ and $pK_{a,2} = 8$) are added to 1 liter of water at 25°C. We will neglect ionic strength effects. The equilibria are

$$H_2O \rightleftharpoons H^+ + OH^-; \quad K_w = [H^+][OH^-] = 10^{-14} \qquad (4\text{-}42)$$

$$H_2A \rightleftharpoons H^+ + HA^-; \quad K_{a,1} = \frac{[H^+][HA^-]}{[H_2A]} = 10^{-4} \qquad (4\text{-}43)$$

$$HA^- \rightleftharpoons H^+ + A^{2-}; \quad K_{a,2} = \frac{[H^+][A^{2-}]}{[HA^-]} = 10^{-8} \qquad (4\text{-}44)$$

The mass balance on species containing A is

$$C_{T,A} = [H_2A] + [HA^-] + [A^-] = 10^{-2} \qquad (4\text{-}45)$$

and is plotted in Fig. 4-5 as a horizontal straight line at $pC_{T,A} = 2$.

We will next develop lines for each of the A-containing species: $[H_2A]$, $[HA^-]$, and $[A^{2-}]$. First for H_2A, by combining the mass balance, Eq. 4-45, and the two equilibrium relationships, Eqs. 4-43 and 4-44, we obtain $[H_2A]$ as a function of $[H^+]$, $C_{T,A}$, and equilibrium constants,

$$C_{T,A} = [H_2A]\left(1 + \frac{K_{a,1}}{[H^+]} + \frac{K_{a,1}K_{a,2}}{[H^+]^2}\right)$$

or

$$[H_2A] = C_{T,A}\left(\frac{1}{1 + (K_{a,1}/[H^+]) + (K_{a,1}K_{a,2}/[H^+]^2)}\right) \qquad (4\text{-}46)$$

We must now examine the behavior of Eq. 4-46 at various pH values so that we can plot an $[H_2A]$ line in the pC-pH diagram. In the region where $[H^+] \gg K_{a,1} \gg K_{a,2}$, or $\text{pH} < pK_{a,1} < pK_{a,2}$,

$$1 > \frac{K_{a,1}}{[H^+]} > \frac{K_{a,1}K_{a,2}}{[H^+]^2}$$

so that Eq. 4-46 becomes

$$C_{T,A} = [H_2A] = 10^{-2}$$

or

$$pC_{T,A} = p[H_2A] = 2 \qquad (4\text{-}47)$$

This is plotted in Fig. 4-5 as line 1.

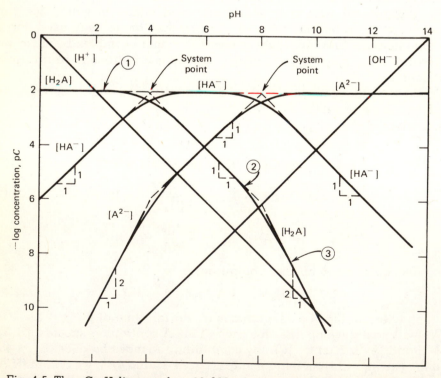

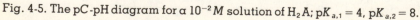

Fig. 4-5. The pC-pH diagram for a $10^{-2}\,M$ solution of H_2A; $pK_{a,1} = 4$, $pK_{a,2} = 8$.

In the region $K_{a,1} > [H^+] > K_{a,2}$, or $pK_{a,1} < pH < pK_{a,2}$, Eq. 4-46 becomes

$$[H_2A] = C_{T,A} \frac{[H^+]}{K_{a,1}}$$

or in logarithmic form

$$\log [H_2A] = \log C_{T,A} + \log [H^+] - \log K_{a,1}$$

or

$$p[H_2A] = pC_{T,A} + pH - pK_{a,1} \tag{4-48}$$

Since $d(p[H_2A])/d\ pH = +1$, this is the equation of a straight line with a slope of $+1$ that passes through the point ($pH = pK_{a,1}$, $p[H_2A] = pC_{T,A}$), the first system point. This line is plotted as line 2 in Fig. 4-5 between pH 4 ($= pK_{a,1}$) and pH 8 ($= pK_{a,2}$). In the vicinity of $pH = pK_{a,1}$ and $pH = pK_{a,2}$ dashed lines are plotted; we must make adjustments here, since our assumptions break down in these regions.

In the region $K_{a,1} \gg K_{a,2} \gg [H^+]$ or $pK_{a,1} < pK_{a,2} < pH$, Eq. 4-46 becomes

$$[H_2A] = C_{T,A} \left(\frac{1/K_{a,1} K_{a,2}}{[H^+]^2} \right)$$

or in logarithmic form

$$\log [H_2A] = \log C_{T,A} + 2 \log [H^+] - \log K_{a,1} - \log K_{a,2}$$

or

$$p[H_2A] = pC_{T,A} + 2pH - pK_{a,1} - pK_{a,2} \tag{4-49}$$

Since $d(p[H_2A])/d\ pH = +2$, this is the equation of a straight line with a slope of $+2$ that passes through the point ($pH = pK_{a,2}$, $p[H_2A] = pC_{T,A} + pK_{a,2} - pK_{a,1}$). This line is plotted in Fig. 4-5 as line 3 in the region $pH > pK_{a,2}$. Dashed lines are again plotted in the region where pH approaches $pK_{a,2}$.

To complete the curve for $p[H_2A]$ versus pH, we must determine its behavior in the pH regions near $pK_{a,1}$ and $pK_{a,2}$. From Eq. 4-43 we can show that when $pH = pK_{a,1}$, $[H^+] = K_{a,1}$ and $[HA^-] = [H_2A]$. Since, at pH values in the region of $pK_{a,1}$, $[A^{2-}]$ is negligible, we can write

$$[H_2A] = \frac{C_{T,A}}{2}$$

or in logarithmic form,

$$p[H_2A] = pC_{T,A} + 0.3 \tag{4-50}$$

Thus at $pH = pK_{a,1}$, $p[H_2A]$ plots at a point 0.3 units below the intersection of lines 1 and 2.

When $[H^+] = K_{a,2} = 10^{-8}$, we can determine the $p[H_2A]$ value by solving Eq. 4-46 exactly

$$[H_2A] = C_{T,A}\left(\frac{1}{1 + (10^{-4}/10^{-8}) + (10^{-12}/10^{-16})}\right)$$

$$= C_{T,A}\left(\frac{1}{2 \times 10^{+4}}\right) = C_{T,A}(5 \times 10^{-5})$$

$$\log[H_2A] = \log C_{T,A} + \log(5 \times 10^{-5})$$

$$p[H_2A] = 6.3$$

Thus at pH 8 = $pK_{a,2}$, $p[H_2A]$ plots at a point 0.3 units below the intersection of lines 2 and 3 in Fig. 4-5.

The position of the $[HA^-]$ and $[A^{2-}]$ lines can be determined in a similar fashion by making approximations in the following equations developed by substituting the equilibrium expressions in the mass balance and solving for $[HA^-]$ and $[A^{2-}]$, respectively, in terms of $C_{T,A}$, $[H^+]$ and constants,

$$[HA^-] = C_{T,A}\left(\frac{1}{([H^+]/K_{a,1}) + 1 + (K_{a,2}/[H^+])}\right) \tag{4-51}$$

and

$$[A^{2-}] = C_{T,A}\left(\frac{1}{([H^+]^2/K_{a,1}K_{a,2}) + ([H^+]/K_{a,2}) + 1}\right) \tag{4-52}$$

It would be a useful exercise for the student to construct lines for pH versus $p[HA^-]$ and $p[A^{2-}]$ using these equations and appropriate assumptions. These lines plot as shown in Fig. 4-5. This pC-pH diagram differs in two major ways from that for a monoprotic acid:

1. The diprotic acid diagram has two system points, $(pK_{a,1}, pC_{T,A})$ and $(pK_{a,2}, pC_{T,A})$.

2. The lines for $[H_2A]$ and $[A^-]$ have regions where the slopes $(d\ pC/d\ pH)$ are $+2$ and -2, respectively.

The use of the diagram is illustrated in the following example.

Example 4-20

Find the pH of the 25°C solution resulting from the addition of (1) 10^{-2} moles of H_2A to 1 liter of distilled water and (2) 10^{-2} moles of Na_2A to 1 liter of distilled water. Here $pK_{a,1} = 4$ and $pK_{a,2} = 8$.

Solution

1. To find the desired pH, we can use Fig. 4-5 in conjunction with the proton condition for H_2A,

$$[H^+] = [HA^-] + 2[A^{2-}] + [OH^-] \tag{4-53}$$

Examining the diagram, we see that at pH 3.05, the approximate proton condition $[H^+] = [HA^-]$ is satisfied. At this pH, $2[A^{2-}]$ and $[OH^-]$ are negligible compared to $[HA^-]$. Therefore, the proton condition is satisfied only at pH 3.05. At this pH, $[H_2A] = 10^{-2.05}$, $[HA^-] = 10^{-3.05}$, $[A^{2-}] = 10^{-8}$, and $[OH^-] = 10^{-10.95}$. Checking the assumptions in the proton condition,

$$10^{-3.05} = 10^{-3.05} + 10^{-8} + 10^{-10.95}$$

The equation is valid to within a fraction of 1 percent.

2. We can use Fig. 4-5 together with the proton condition for Na_2A to find the pH. It is written as,

$$2[H_2A] + [HA^-] + [H^+] = [OH^-] \tag{4-54}$$

Examining the diagram, we see that at pH 10, $[HA^-] = [OH^-]$, and that $2[H_2A]$ and $[H^+]$ are negligible in comparison to $[HA^-]$. Therefore, the proton condition is satisfied only at pH 10. At this pH, $[OH] = 10^{-4}$, $[H^+] = 10^{-10}$, $[H_2A] = 10^{-10}$, $[HA^-] = 10^{-4}$, and $[A^{2-}] = 10^{-2}$. Checking the proton condition, we see that there is much less than 5 percent error and therefore our assumptions are acceptable.

Example 4-21

The dibasic acid succinic acid is an important metabolic intermediate formed in plant and animal cells. The structure of succinic acid ($pK_{a,1} = 4.2$ and $pK_{a,2} = 5.5$ at 25°C) is

$$\begin{aligned} &CH_2 - COOH \\ &| \\ &CH_2 - COOH \end{aligned}$$

which we will represent as H_2Su. Find the concentrations of all species in a 10^{-2} molar solution of succinic acid (H_2Su).

Solution

Figure 4-6 is a pC-pH diagram for H_2Su showing variations of $[H^+]$, $[OH^-]$, $[H_2Su]$, $[HSu^-]$, and $[Su^{2-}]$ with pH. Note that because of the close proximity of the pK_a values the lines for $[H_2Su]$ and $[Su^{2-}]$ have no straight line portions between $pK_{a,1}$ and $pK_{a,2}$. The approach to obtaining the position of these lines in the diagram is similar to that shown previously for the acid H_2A, except that the same simplifying assumptions cannot be made in the general equation obtained by combining the mass balance and equilibrium equations for $pK_{a,1} < pH < pK_{a,2}$ (e.g., see Eq. 4-46 for the general acid H_2A). The approach to obtaining a solution for concentrations of H_2Su, HSu^-, and Su^{2-} in this region is to substitute values of H^+ into the general equation for each of these species and to solve for the concentration at each pH.

For H_2Su the proton condition is

$$[H^+] = [HSu^-] + 2[Su^{2-}] + [OH^-]$$

Examining the diagram, we can see that at pH 3.2, $[H^+] = [HSu^-]$, and that $2[Su^{2+}]$ and $[OH^-]$ are negligible compared to $[HSu^-]$. Thus the proton condition is satisfied only at pH 3.2. At this pH, $[H^+] = 10^{-3.2}$, $[OH^-] = 10^{-10.8}$, $[H_2Su] = 10^{-2}$,

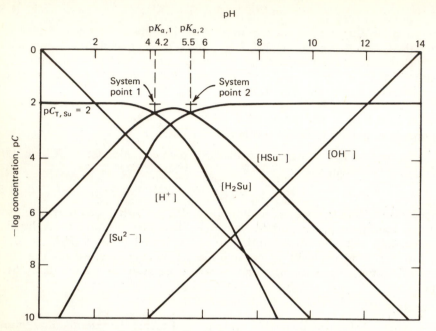

Fig. 4-6. The pC-pH diagram for $10^{-2}\,M$ succinic acid (H_2Su); $pK_{a,1} = 4.2$, $pK_{a,2} = 5.5$ at 25°C.

$[HSu^-] = 10^{-3.2}$, and $[Su^{2-}] = 10^{-5.3}$. Checking our assumptions in the proton condition, we obtain

$$10^{-3.2} = 10^{-3.2} + 2 \times 10^{-5.3} + 10^{-10.8}$$

The assumptions are acceptable.

The pC-pH diagram for a triprotic acid can be developed by extending the procedure used to plot the diagrams for monoprotic and diprotic acids to multiprotic acids. The advantage of the graphical procedure over computational method becomes greater and greater as the number of equilibria increases. As an example, the pC-pH diagram for phosphoric acid, H_3PO_4, is shown in Fig. 4-7 for $C_{T,PO_4} = 10^{-3}$, ($pK_{a,1} = 2.1$, $pK_{a,2} = 7.2$ and $pK_{a,3} = 12.3$ at 25°C). Applicable equations are, neglecting ionic strength effects,

Equilibria

$$K_w = 10^{-14} = [H^+][OH^-] \qquad (4\text{-}55)$$

$$H_3PO_4 \rightleftharpoons H^+ + H_2PO_4^-; \qquad K_{a,1} = \frac{[H^+][H_2PO_4^-]}{[H_3PO_4]} = 10^{-2.1} \qquad (4\text{-}56)$$

$$H_2PO_4^- \rightleftharpoons H^+ + HPO_4^{2-}; \qquad K_{a,2} = \frac{[H^+][HPO_4^{2-}]}{[H_2PO_4^-]} = 10^{-7.2} \qquad (4\text{-}57)$$

$$HPO_4^{2-} \rightleftharpoons H^+ + PO_4^{3-}; \qquad K_{a,3} = \frac{[H^+][PO_4^{3-}]}{[HPO_4^{2-}]} = 10^{-12.3} \qquad (4\text{-}58)$$

Mass Balance on PO₄ Species

$$C_{T,PO_4} = 10^{-3} = [H_3PO_4] + [H_2PO_4^-] + [HPO_4^{2-}] + [PO_4^{3-}] \qquad (4\text{-}59)$$

The equilibria and mass balance can be combined and solved for $[H_3PO_4]$, $[H_2PO_4^-]$, $[HPO_4^{2-}]$, and $[PO_4^{3-}]$ in terms of C_{T,PO_4}, $[H^+]$, and constants to yield

$$[H_3PO_4] = C_{T,PO_4}\left(\frac{1}{1 + (K_{a,1}/[H^+]) + (K_{a,1}K_{a,2}/[H^+]^2) + (K_{a,1}K_{a,2}K_{a,3}/[H^+]^3)} \right) \qquad (4\text{-}60)$$

$$[H_2PO_4^-] = C_{T,PO_4}\left(\frac{1}{([H^+]/K_{a,1}) + 1 + (K_{a,2}/[H^+]) + (K_{a,2}K_{a,3}/[H^+]^2)} \right) \qquad (4\text{-}61)$$

$$[HPO_4^{2-}] = C_{T,PO_4}\left(\frac{1}{([H^+]^2/K_{a,1}K_{a,2}) + ([H^+]/K_{a,2}) + 1 + (K_{a,3}/[H^+])} \right) \qquad (4\text{-}62)$$

$$[PO_4^{3-}] = C_{T,PO_4}\left(\frac{1}{([H^+]^3/K_{a,1}K_{a,2}K_{a,3}) + ([H^+]^2/K_{a,1}K_{a,2}) + ([H^+]/K_{a,3}) + 1} \right) \qquad (4\text{-}63)$$

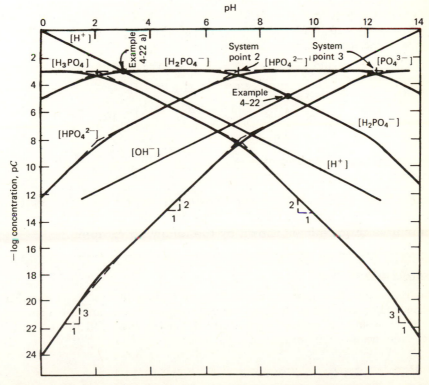

Fig. 4-7. The pC-pH diagram for a $10^{-3} M$ phosphate solution; $pK_{a,1} = 2.1$, $pK_{a,2} = 7.2$, $pK_{a,3} = 12.3$ at 25°C.

These equations can be examined for their behavior with respect to pH in various regions. For example, the slope of the $[H_3PO_4]$ versus pH line is 0 when pH $<$ $pK_{a,1}$, $+1$ when $pK_{a,1}$ $<$ pH $<$ $pK_{a,2}$, $+2$ when $pK_{a,2}$ $<$ pH $<$ $pK_{a,3}$, and $+3$ when pH $>$ $pK_{a,3}$. The system points are $(pK_{a,1}, pC_{T,PO_4})$, $(pK_{a,2}, pC_{T,PO_4})$ and $(pK_{a,3}, pC_{T,PO_4})$. The use of Fig. 4-7 is illustrated in the following example.

Example 4-22

Determine the concentration of each phosphate species, OH^- and H^+ in a 10^{-3} M Na_2HPO_4 solution.

Solution

Using Fig. 4-7, the mass balance on Na^+,

$$C_{T,Na} = [Na^+] = 2C_{T,PO_4} = 2 \times 10^{-3} \tag{4-64}$$

and the proton condition,

$$[H^+] + [H_2PO_4^-] + 2[H_3PO_4] = [OH^-] + [PO_4^{3-}] \tag{4-65}$$

we see that at pH 9.1, $[H_2PO_4^-] = [OH^-]$, and that $[H^+]$ and $2[H_3PO_4]$ are much less than $[H_2PO_4^-]$, and $[PO_4^{3-}]$ is much less than $[OH^-]$. Also, $[H_2PO_4^-]$ and $[PO_4^{3-}]$ are much less than $[OH^-]$. At this pH, $[H^+] = 10^{-9.1}$ M, $[OH^-] = 10^{-4.9}$ M, $[Na^+] = 2 \times 10^{-3}$ M, $[H_3PO_4] = 10^{-11.7}$ M, $[H_2PO_4^-] = 10^{-4.9}$ M, $[HPO_4^{2-}] = 10^{-3}$ M, and $[PO_4^{3-}] = 10^{-6.2}$ M. Checking the proton condition,

$$10^{-9.1} + 10^{-4.9} + 2 \times 10^{-1.7} = 10^{-4.9} + 10^{-6.2}$$

$$1.26 \times 10^{-5} = 1.32 \times 10^{-5}$$

$$\text{error} = \left(\frac{(1.32 \times 10^{-5} - 1.26 \times 10^{-5})}{1.26 \times 10^{-5}} \right) \times 100$$

$$= 4.9 \, \text{percent}$$

The assumption is just acceptable.

The pC-pH diagrams have so far been used only for solutions containing one acid and its conjugate base(s). These diagrams are equally applicable to mixtures of acids as is shown by the following example.

Example 4-23

Determine the pH of a solution prepared by adding 10^{-3} moles of ammonium acetate, NH_4Ac, per liter of distilled water. (Note that this solution is identical to that which would be obtained by adding 10^{-3} moles of NH_4OH ($NH_{3(aq)}$) and 10^{-3} moles HAc, acetic acid, per liter of water.)

Solution

NH_4Ac dissociates to NH_4^+ and Ac^-, and these species equilibrate as follows:

$$NH_4^+ \rightleftharpoons H^+ + NH_{3(aq)}; \qquad K_a = 10^{-9.3} = \frac{[H^+][NH_{3(aq)}]}{[NH_4^+]}$$

$$HAc \rightleftharpoons H^+ + Ac^-; \qquad K_a = 10^{-4.7} = \frac{[H^+][Ac^-]}{[HAc]}$$

Mass Balances

$$C_{T,NH_3} = 10^{-3} = [NH_4^+] + [NH_{3(aq)}]$$

$$C_{T,Ac} = 10^{-3} = [HAc] + [Ac^-]$$

We can plot the pC-pH diagrams for these two acid-conjugate base pairs on the same diagram (Fig. 4-8). Because $C_{T,NH_3} = C_{T,Ac}$ is the same for each pair, some of the lines are superimposed. The proton condition is

$$[HAc] + [H^+] = [NH_{3(aq)}] + [OH^-]$$

Examining the diagram, we see that at pH 7, $[HAc] = [NH_{3(aq)}]$ and that $[HAc] \gg$ $[H^+]$ and $[NH_3] \gg [OH^-]$ so that the proton condition is satisfied.

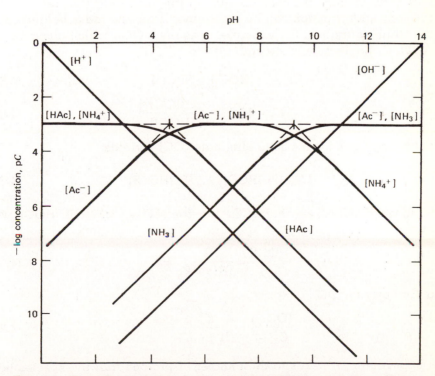

Fig. 4-8. The pC-pH diagram for a $10^{-3}M$ NH_4Ac solution at 25°C.; $pK_{a,HAc} = 4.7$, $pK_{a,NH_4^+} = 9.3$.

4.10. IONIZATION FRACTIONS AND DISTRIBUTION DIAGRAMS

On several occasions in the previous section we developed equations that express the amount of an acid or its conjugate base as a fraction of the total analytical concentration (e.q., Eqs. 4-28 and 4-31 for the HCN–CN⁻ system; Eqs. 4-46, 4-51 and 4-52 for the diprotic acid H_2A and Eqs. 4-60 to 4-63, for the phosphoric acid system). We used these equations, with simplifying assumptions, to develop pC-pH diagrams and noted that there were pH regions (close to the pK values) where our assumptions broke down. The reader should also have noted, probably with some trepidation, that the equations were increasing in complexity as the number of protons in the acid increased. To simplify the situation, we can introduce the concept of an ionization fraction (or α value). Tables of α values as a function of pH may be used to aid computations. Graphical presentation of α values as a function of pH—the so-called distribution diagrams—are helpful in graphical solutions of equilibrium problems. A table of α versus pH is given in Appendix 1. The major advantage of the ionization fraction, in addition to its simplicity of expression, is that it is independent of the analytical concentration. Therefore, multiplication of the analytical concentration by the appropriate ionization fraction at a particular pH value will directly produce the concentration of the species at that pH.

The ionization fraction can be determined from the mass balance and equilibrium equations. For example, consider HOCl, which has a pK_a of 7.5 at 25°C. Neglect ionic strength effects.

$$C_{T,OCl} = [HOCl] + [OCl^-] \tag{4-66}$$

$$K_a = 10^{-7.5} = \frac{[H^+][OCl^-]}{[HOCl]} \tag{4-67}$$

Substituting Eq. 4-67 in 4-66 to eliminate $[OCl^-]$ yields

$$C_{T,OCl} = [HOCl] + \frac{K_a}{[H^+]} [HOCl] \tag{4-68}$$

Solving for $[HOCl]/C_{T,OCl}$, that is, the fraction of $C_{T,OCl}$ which is [HOCl], we obtain

$$\frac{[HOCl]}{C_{T,OCl}} = \frac{[H^+]}{[H^+] + K_a} = \alpha_0 \tag{4-69}$$

and similarly for OCl^-,

$$\frac{[OCl^-]}{C_{T,OCl}} = \frac{K_a}{[H^+] + K_a} = \alpha_1 \tag{4-70}$$

The subscript on α refers to the number of protons that the species has released. If ionic strength effects are important, $^cK_a = [H^+][OCl^-]/[HOCl]$ $(= K_a/(\gamma_{H^+}\gamma_{OCl^-}))$ should be used in place of K_a in Eqs. 4-69 and 4-70.

Using Eqs. 4-69 and 4-70, or alternatively the table of α versus pH in Appendix 1, we can determine α_0 and α_1 at various pH values and plot them in a distribution diagram as α versus pH (Fig. 4-9). Distribution diagrams for the multiprotic acids, H_2S and H_3PO_4, are shown in Fig. 4-10; such curves can be obtained using either the table or the general equations for α in Appendix 1.

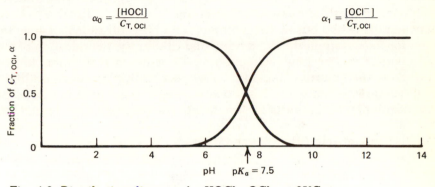

Fig. 4-9. Distribution diagram for HOCl—OCl⁻ at 25°C.

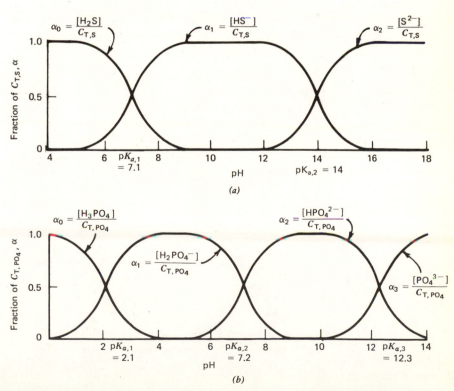

Fig. 4-10. Distribution diagrams for (a) H_2S, and (b) H_3PO_4 at 25°C.

4.11. MIXTURES OF ACIDS AND BASES—pH CALCULATION

Until now we have concerned ourselves with situations in which acids, and salts of conjugate bases, have been added singly to distilled water. We must now address situations in which we have to predict the nature and extent of composition *changes* that occur when acids, conjugate bases, and salts are added to each other. Many important analytical methods, natural processes, and treatment schemes involve reactions between acids and bases, for example, acid-base titrations such as those used for the determination of acidity and alkalinity; the addition of metal ion coagulants, or lime for softening or water stabilization in water treatment; the neutralization of acid wastes by in-plant base addition or by discharge and dilution in surface waters; the production of acid rain; the dissolution of minerals by groundwaters bearing CO_2, and so on.

4.11.1. Strong Acid-Strong Base Mixtures

The calculation of solution pH after the addition of a strong acid to a solution containing a strong base, or vice versa, is very similar to the calculations used in Section 4-5 for determining pH in solutions of strong acid or strong base. The general approach is to write equations that describe the solution after the acid and base are mixed, make simplifying assumptions, find the solution to the equations, and then check the assumptions. The following example will serve as an illustration.

Example 4-24

A 0.1 M NaOH solution is added to 1 liter of 10^{-3} M HCl solution. What is the pH after addition of (1) 5 ml, (2) 10 ml and (3) 20 ml 0.1 M NaOH? The solution temperature is 25°C. Neglect ionic strength effects.

Solution

Because HCl is a strong acid and NaOH is a strong base, their dissociation in aqueous solution is assumed to be complete. The equations needed to solve the problem are

Equilibrium

1. $K_w = 10^{-14} = [H^+][OH^-]$

Mass Balances on Cl^- and Na^+

2. $C_{T,Cl} = [Cl^-] = \dfrac{\text{number of moles}}{\text{total volume}} = \dfrac{1 \text{ liter} \times 10^{-3} \text{ moles/liter}}{1 \text{ liter} + \text{volume added}}$

3. $C_{T,Na} = [Na^+] = \dfrac{\text{number of moles}}{\text{total volume}} = \dfrac{0.1 \text{ moles/liter} \times \text{volume added}}{1 \text{ liter} + \text{volume added}}$

Charge Balance

4. $[Na^+] + [H^+] = [Cl^-] + [OH^-]$

a. The volume added is 0.005 liter (5 ml). Substituting (1), (2), and (3) in (4), we obtain

$$\frac{(0.1)(0.005)}{1.005} + [H^+] = \frac{10^{-3}}{1.005} + \frac{10^{-14}}{[H^+]}$$

or

5. $$\frac{[H^+] - 10^{-14}}{[H^+]} = \frac{10^{-3} - 5 \times 10^{-4}}{1.005} = 5 \times \frac{10^{-4}}{1.005}$$

This equation, which is a quadratic in $[H^+]$, could be solved exactly. However, we can make a simplifying assumption. Since the amount of strong acid present (1 liter $\times$ 10^{-3} mole/liter = 10^{-3} moles) exceeds the amount of base added (0.005 liter $\times$ 10^{-1} mole/liter = 5×10^{-4} moles) we would expect the solution to be acid so that $[H^+] \gg [OH^-]$. Neglecting $[OH^-]$ (= $10^{-14}/[H^+]$) in (5), we obtain

$$[H^+] = \frac{5 \times 10^{-4}}{1.005} = 5 \times 10^{-4} M \qquad pH = 3.3$$

At this pH, $[OH^-] = 10^{-10.7}$ and the charge balance is satisfied to well within 5 percent error.

b. Substituting (2) and (3) into (4) and using a volume of NaOH added of 0.01 liter (10 ml),

$$\frac{(0.1)(0.01)}{1.01} + [H^+] = [OH^-] + \frac{10^{-3}}{1.01}$$

or

$$[H^+] = [OH^-]$$

and from equation (1) pH = 7.0.

At this point we have added a quantity of base equal to the amount of acid present (the stoichiometric amount to satisfy the reaction, NaOH + HCl $\rightarrow$ NaCl + H_2O); we therefore have a salt solution with a pH of 7.

c. Substituting (2) and (3) in (4) and using a volume of NaOH added of 0.02 liter (20 ml), we find that

$$\frac{(0.1)(0.02)}{1.02} + [H^+] = [OH^-] + \frac{10^{-3}}{1.02}$$

Since the amount of strong base added exceeds the amount of acid present, we would expect the solution to be basic with $[OH^-] \gg [H^+]$. Thus

$$[OH^-] = \frac{2 \times 10^{-3} - 10^{-3}}{1.02} = 9.80 \times 10^{-4}$$

where

$$pOH = 3.01$$
$$pH = 10.99$$

Example 4-25

Spent regenerant solution from an ion exchange process contains 10^{-1} moles of NaOH per liter. It must be partially neutralized by the addition of H_2SO_4 prior to discharge. The required final pH is 10. What volume of a 1 M H_2SO_4 solution must be added per liter of waste? The temperature is 25°C and the ionic strength is approximately 0.1.

Solution

Equations

1. $C_{T,Na} = [Na^+] = (10^{-1}\,M)(1\text{ liter})/(1\text{ liter} + V)$.

2. $C_{T,SO_4} = [SO_4^{2-}] = (1\,M)(V)/(1\text{ liter} + V)$.

3. $10^{-14} = \{H^+\}\{OH^-\}$; $\{H^+\} = 10^{-10}$; $\{OH^-\} = 10^{-4}$; $[H^+] = 10^{-10}/\gamma_{H^+} = 10^{-10}/0.83$ $= 1.20 \times 10^{-10}$; $[OH^-] = 10^{-4}/\gamma_{OH^-} = 10^{-4}/0.75 = 1.33 \times 10^{-4}$ (activity coefficients from Fig. 3-4).

4. $[Na^+] + [H^+] = 2[SO_4^{2-}] + [OH^-]$.

(Note HSO_4^- need not be included in the charge balance because at pH 10 its concentration is negligible.) Substituting from (1), (2), and (3) in (4), we obtain

$$\frac{0.1}{(1+V)} + 1.20 \times 10^{-10} = \frac{2V}{(1+V)} + 1.33 \times 10^{-4}$$

Assume that

$$1.2 \times 10^{-10} \ll \frac{0.1}{(1+V)}$$

then

$$0.1 - 2V - 1.33 \times 10^{-4}(1+V) = 0$$

$$2.0001V = 0.0999$$

$$V = 0.0499 \text{ liter} = 49.9 \text{ ml}$$

We should note that the stoichiometric statement

$$2NaOH + H_2SO_4 \rightarrow Na_2SO_4 + 2H_2O$$

is satisfied and the pH = 7, when

$$V_{acid} \times N_{acid} = V_{base} \times N_{base}$$

$$V_{acid} \times 2 = 1000 \text{ ml} \times 0.1$$

$$V_{acid} = 50 \text{ ml}$$

We can see then that it took 49.9 ml acid to lower the pH from the initial value of approximately 13 to pH 10, but a further increment of only 0.1 ml would lower the pH by 3 units to pH 7. If ionic strength effects had not been taken into account, the same answer would have been obtained for this problem.

We can perform the calculation shown in Example 4-25 many times over and from the data construct a curve of pH versus acid added, expressed either as ml added or as the equivalent fraction, f, the number of equivalents of H^+ added per mole of base present. This curve is known as a titration curve. In Fig. 4-11 the titration curve for addition of H_2SO_4 to the spent regenerant solution, as in Example 4-25, is plotted.

It is apparent that in the region of the equivalence point, or the point of stoichiometric addition, the pH of a strong base solution to which a strong acid is being added (or vice versa) is very sensitive to the amount of acid added. The solution, which is a mixture of a strong base and its salt, is poor at resisting changes in pH for incremental additions of acid. The solution is thus poorly *buffered* for pH. Although we will take up the subject of buffers in more detail in Section 4-12, it is worthwhile to note that a strong acid alone would be a poor choice of neutralizing agent for a waste which is primarily a strong base. Because of the wide fluctuation of pH for small additions of acid near the equivalence point, process control to any desired pH in this region would be very difficult.

4.11.2. Weak Acid-Strong Base and Strong Acid-Weak Base Mixtures

The situation with these mixtures becomes slightly more complicated than for strong acid-strong base mixtures because we cannot always assume complete dissociation of the weak acid or weak base. However, the importance of these types of systems far outweighs that of the strong acid-strong base system. Most natural waters indeed behave like mixtures of weak acids and the salts of strong bases. A consideration of such systems will lead us to an understanding of pH buffers and eventually to one of the important pH buffering systems in natural water, that is, the carbonate system. At first we will examine some simple situations

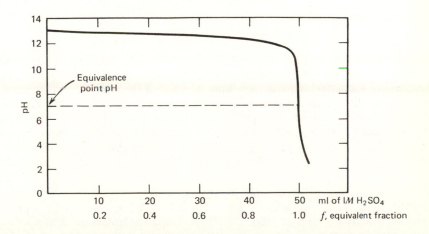

Fig. 4-11. Titration of 1 liter of $0.1\,M$ NaOH with $1\,M$ H_2SO_4 at 25°C.

illustrated by examples. We will use two approaches: the first using pC-pH diagrams and the second a trial-and-error method.

Example 4-26

A solution contains 10^{-3} moles of acetic acid, HAc. What is the solution pH after 0.0025, 0.005, and 0.01 liters of 0.1 M NaOH and 0.03 liter of 1 M NaOH have been added per liter? From these data sketch a titration curve (equivalent fraction, f, added versus pH, and ml added versus pH) and show the relationship between the titration curve and the pC-pH diagram. The temperature is 25°C; neglect ionic strength effects.

Solution

Since the volume of base added per liter is very small, it is reasonable to assume that its addition causes no volume change.[9] The general equations are as follows,

1. $C_{T,Ac} = [HAc] + [Ac^-] = 10^{-3} M$.

2. $C_{T,Na}$ = number of moles added/sample volume = $[Na^+] = V(1)(0.1)/1$ liter = $0.1V$ M.

3. $K_a = 10^{-4.7} = [H^+][Ac^-]/[HAc]$.

4. $K_w = 10^{-14} = [H^+][OH^-]$.

5. $[Na^+] + [H^+] = [OH^-] + [Ac^-]$.

We use equations (1), (3), and (4) to plot a pC-pH diagram for the acetic acid-acetate system ($C_{T,Ac} = 10^{-3} M$) (Fig. 4-12). To determine the solution pH after each NaOH addition, we must first determine Na^+ from equation (2) and then substitute this value in the charge balance, equation (5). The solution for the charge balance is obtained by inspection of the pC-pH diagram. For $V = 0.0025$ liter: From (2)

$$[Na^+] = 0.1V = 2.5 \times 10^{-4} M$$

The charge balance becomes

$$2.5 \times 10^{-4} + [H^+] = [OH^-] + [Ac^-]$$

Now assuming $[H^+] \ll 2.5 \times 10^{-4}$ and $[OH^-] \ll [Ac^-]$, the charge balance becomes $[Ac^-] = 2.5 \times 10^{-4} M$. The charge balance solution in the pC-pH diagram can be found at pH 4.3. Checking our assumption, we find that $[Na^+] = 10^{-3.6}$, which is not much greater than $[H^+] = 10^{-4.3}$. The approximate solution of the charge balance does not work because $[Na^+]$ and $[H^+]$ are the same order of magnitude. We must resort to finding values of $([Na^+] + [H^+])$ from the diagram in the region around pH 4.3 and then plotting $-\log ([Na^+] + [H^+])$ versus pH. We then can determine the pH where $-\log ([Na^+] + [H^+]) = -\log [Ac^-]$, which is the solution

---

[9] If the volume change had not been neglected, $C_{T,Ac} = [HAc] + [Ac^-] = (10^{-3})/(1 \text{ liter} + V)$, and $C_{T,Na} = [Na^+] = (0.1 \ V)/(1 \text{ liter} + V)$. $C_{T,Ac}$ and $C_{T,Na}$ values would both change when NaOH is added. The graphical procedure could still be used, but a different pC-pH diagram would have to be constructed after each addition.

to the charge balance. From the diagram we see that this occurs at pH = 4.4, and at this pH our second assumption that $[OH^-] \ll [Ac^-]$ is acceptable. For $V = 0.005$ liter: From (2)

$$[Na^+] = (0.005)(0.1) = 5 \times 10^{-4}\ M$$

The charge balance becomes

$$5 \times 10^{-4} + [H^+] = [OH^-] + [Ac^-]$$

If we assume that $[H^+] \ll 5 \times 10^{-4}$ and $[OH^-] \ll [Ac^-]$, the charge balance becomes $[Ac^-] = 5 \times 10^{-4}$, a solution that is met at pH = 4.7. Our assumptions are valid at this pH.

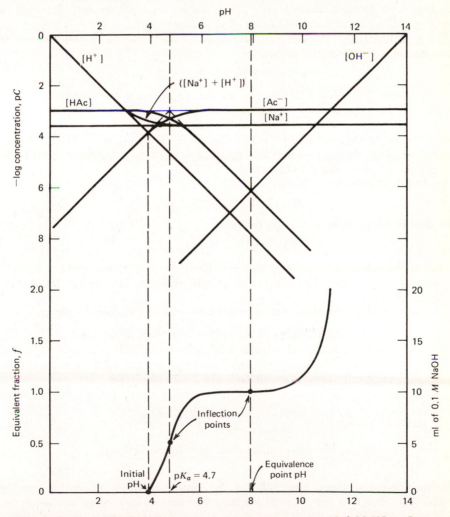

Fig. 4-12. The pC-pH diagram and titration curve for $10^{-3}\ M$ HAc—Ac$^-$ at 25°C.

The addition of $5 \times 10^{-4}\,M$ NaOH has produced $5 \times 10^{-4}\,M$ NaAc so that the solution is an equimolar mixture of undissociated HAc and Ac^-. Note that this occurs at a pH value equal to the pK_a of the acid. Finally, note that because our assumptions were acceptable, the influence of $[H^+]$ and $[OH^-]$ on pH changes caused by the addition of base are negligible. In other words all of the NaOH added to such a mixture can be assumed to react with HAc; and all of any strong acid added to such a mixture can be assumed to react with Ac^-, that is,

$$HAc + OH^- \rightarrow Ac^- + H_2O$$

or

$$Ac^- + H^+ \rightarrow HAc$$

so that for strong base addition the $[Ac^-]$ concentration is equal to the analytical concentration of strong base added. As we shall see later, all of these observations are of great importance to the subject of pH buffers.

For $V = 0.010$, the number of equivalents of OH^- added = the number of equivalents of HAc initially present.

$$[Na^+] = (0.01)(0.1) = 10^{-3} = C_{T,Ac}$$

Substituting $[HAc] + [Ac^-]$ for $[Na^+]$ in the charge balance, we obtain

$$[HAc] + [H^+] = [OH^-]$$

which is the proton condition for a solution containing $C_{T,Ac}$ moles/liter of NaAc. Assuming $[HAc] \gg [H^+]$, we find that the charge balance is satisfied at pH 7.8 in the pC-pH diagram where $[HAc] = [OH^-]$.

For $V = 0.03$ liter of $1\,M$ NaOH, from (2)

$$[Na^+] = (0.03)(1) = 3 \times 10^{-2}$$

The charge balance, equation (5), becomes

$$3 \times 10^{-2} + [H^+] = [OH^-] + [Ac^-]$$

If we assume that $[H^+] \ll 3 \times 10^{-2}$ and $[OH^-] \gg [Ac^-]$, the charge balance becomes $[OH^-] = 3 \times 10^{-2}$. From the pC-pH diagram the charge balance is satisfied at pH 12.5.

We now have sufficient data to sketch the titration curve, summarizing:

For $V = 0.0025$ liter, $f = (0.1)(0.0025)/1 \times 10^{-3} = 0.25$; pH $= 4.4$
For $V = 0.005$ liter, $f = (0.1)(0.005)/1 \times 10^{-3} = 0.50$; pH $= 4.7$
For $V = 0.010$ liter, $f = (0.1)(0.01)/1 \times 10^{-3} = 1.0$; pH $= 7.8$
For $V = 0.03$ liter, $f = (1)(0.03)/1 \times 10^{-3} = 30$; pH $= 12.5$

where

$$f = \frac{\text{number of equivalents NaOH added}}{\text{moles of (HAc} + Ac^-)}$$

$$= \frac{\text{number of equivalents NaOH added/liter}}{C_{T,Ac}}$$

The titration curve was plotted in Fig. 4-12, together with the pC-pH diagram. By comparing these two plots, the following important observations can be made.

1. When $f = 0.5$, pH $= pK_a$. Any addition of base corresponding to $f < 0.5$ will produce a solution pH between that of the original HAc solution and the pK_a.

2. At $f = 0.5$ the titration curve has an inflection point. At this pH ($= pK_a$) the change of pH for a given addition of strong base is small—the solution is very well buffered.

3. When $f = 1.0$, the pH is that of a 10^{-3} M NaAc solution. This pH is the equivalence point. At this pH value the curve has an inflection point in the opposite direction to that at $f = 0.5$. Here the resistance of the solution to pH change on the addition of strong base (or strong acid) is small—the solution is very poorly buffered.

4. The pH increases very slowly as it approaches the pH of the solution being added.

5. It is useful to compare the strong acid-strong base titration curve in Fig. 4-11 with the curve for a weak acid-strong base in Fig. 4-12. Note that the strong acid-strong base system has no good pH region of buffering corresponding to that of the weak acid-strong base system.

We will now examine a strong acid-weak base system using a trial-and-error method of solution.

Example 4-27

What is the pH of a 10^{-3} M NH$_{3(aq)}$ solution following the addition of (1) 0.4 liter of 10^{-3} M HCl and (2) 0.8 liter of 10^{-3} M HCl? Estimate the pH before attempting to calculate it exactly. Neglect ionic strength effects. The pK_a of NH$_4^+$ at 25°C is 9.3.

Solution

pH Estimate

Before HCl addition, this will be the pH of a 10^{-3} M NH$_{3(aq)}$ solution. This can either be read from a pC-pH diagram or calculated as in Example 4-14. Roughly sketching the pC-pH diagram (Fig. 4-13), we can rapidly estimate the pH; neglecting dilution, we obtain

PRL

$$NH_{3(aq)}, H_2O.$$

Proton Condition

$$[NH_4^+] + [H^+] = [OH^-]$$

Assuming $[NH_4^+] \gg [H^+]$, we find $[NH_4^+] = [OH^-]$ at pH $= 10.5$ before HCl is added.

After 1 equivalent fraction of strong acid has been added, we will have a 10^{-3} M NH$_4$Cl solution.[10] Again using the pC-pH diagram (Fig. 4-13) to solve for the approximate pH, we find

[10] The reader should compare the proton condition for a 10^{-3} M NH$_4$Cl solution with that for a 10^{-3} M NH$_{3(aq)}$ solution to which 10^{-3} mole HCl/liter have been added to verify this statement.

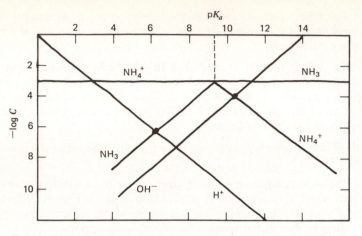

Fig. 4-13. A sketch of the pC-pH diagram for $10^{-3}\,M$ NH$_3$.

PRL

$$NH_4^+, H_2O$$

Proton condition

$$[NH_{3(aq)}] + [OH^-] = [H^+]$$

Assuming that $[NH_{3(aq)}] \gg [OH^-]$
Find $[NH_{3(aq)}] = [H^+]$ at pH $\cong 6.1$.
  When 0.4 liter of $10^{-3}\,M$ HCl have been added,

$$f = \frac{0.4 \times 10^{-3}}{10^{-3}} = 0.4$$

so that the pH will be between the initial pH and pH $= pK_a$, or between 10.5 and 9.3, and fairly close to 9.3.
  When 0.8 liter of $10^{-3}\,M$ HCl has been added,

$$f = \frac{(0.8 \times 10^{-3})}{(10^{-3})} = 0.8$$

and the pH will be between 9.3 and 6.1, and fairly close to 6.1.
  The pH will now be calculated exactly.

Mass Balances

1. $C_{T,NH_3} = [NH_{3(aq)}] + [NH_4^+] =$ number of moles/total volume $=$
   10^{-3} moles/(1 liter $+ V$ liter).

2. $C_{T,Cl} =$ number of moles added/total volume $=$
   $(10^{-3}$ moles/liter$)(V)$/(1 liter $+ V$ liter).

where $V =$ the volume of HCl solution added.

Equilibria

3. $K_a = 10^{-9.3} = [NH_{3(aq)}][H^+]/[NH_4^+]$

4. $K_w = 10^{-14} = [H^+][OH^-]$

Charge Balance

5. $[NH_4^+] + [H^+] = [OH^-] + [Cl^-]$

Knowing $\alpha_0 = [NH_4^+]/C_{T,NH_3}$ and $\alpha_1 = [NH_3]/C_{T,NH_3}$, we can substitute in the charge balance to obtain

$$\alpha_0 C_{T,NH_3} + [H^+] = \frac{10^{-14}}{[H^+]} + \frac{10^{-3}V}{(1+V)}$$

Since we know V, and α_0 is a function only of $[H^+]$ and constants, we have an equation with one unknown which we can solve by trial and error using the table of α values versus pH in Appendix 1.

For $V = 0.4$ liter, $C_{T,NH_3} = 0.71 \times 10^{-3}$ from (1).

$$\alpha_0 (0.71 \times 10^{-4}) + [H^+] = \frac{10^{-14}}{[H^+]} + 0.285 \times 10^{-4}$$

Try pH = 9.5, $[H^+] = 10^{-9.5}$. From Appendix 1, $\alpha_0 = 0.387$

$$(0.387)(0.71 \times 10^{-3}) + 10^{-9.5} = \frac{10^{-14}}{10^{-9.5}} + 0.285 \times 10^{-3}$$

$$0.275 \times 10^{-3} = 0.31 \times 10^{-4}$$

The equation is not satisfied.

Try pH = 9.4 and $[H^+] = 10^{-9.4}$. From Appendix 1, $\alpha_0 = 0.443$.

$$(0.443)(0.71 \times 10^{-3}) + 10^{-9.4} = \frac{10^{-14}}{10^{-9.4}} + 0.285 \times 10^{-3}$$

$$0.31 \times 10^{-4} = 0.31 \times 10^{-4}$$

The equation is satisfied.

For $V = 0.8$ liter, $C_{T,NH_3} = 0.556 \times 10^{-3}$ M (from (1)) and the charge balance is

$$\alpha_0 (0.556 \times 10^{-3}) + [H^+] = \frac{10^{-14}}{[H^+]} + 0.444 \times 10^{-3}$$

By trial-and-error, we find this equation to be satisfied with a reasonable degree of accuracy when $\alpha_0 = 0.799$ and pH = 8.7.

The charge balance in Example 4-27 could also have been solved using the quadratic formula. Although this formula may have worked well for this example, the procedure used here applies equally well to higher-order polynomials for which the quadratic formula is not applicable.

In the next section we shall see that there is an even easier approximate solution to such problems, based on the assumption of complete reaction, that can be used with acceptable accuracy in most instances.

4.12.   pH BUFFERS AND BUFFER INTENSITY

As we have previously stated, a buffer solution is a solution that in some way has the ability to maintain a stable composition when various components are added or removed. Although pH buffers (solutions that resist change in pH on the addition of strong acid, H^+, or strong base, OH^-) are the most commonly talked about buffers, they are by no means the only type. Other examples are metal ion buffers, and oxidation-reduction potential buffers. It is now thought that the mineral composition of many natural waters is regulated by a buffer system involving silicates (clay minerals such as kaolinite). The pH of about 95 percent of the naturally occurring waters is within the range of 6 to 9; constituents such as SiO_2 and K^+ are also regulated within very narrow limits. Constant and well-regulated pH is essential for the proper functioning of the many biological and chemical processes involved in water and wastewater treatment. For example, the efficiency of the activated sludge process decreases markedly when the pH falls below about 6. In most situations this does not present a problem because, as we shall see later, most waters and wastewaters are usually well buffered in the pH range of 6.0 to 6.6 by the carbonate system. There are instances when this is not so. For example, in extended aeration activated sludge plants located in the low alkalinity, poorly buffered water areas of the northeastern United States, significant operation problems can be caused by low pH values. In treatment plants in this region the oxidation of ammonia to nitrate (nitrification) by the reaction

$$NH_4^+ + 2O_2 \rightleftharpoons NO_3^- + H_2O + 2H^+$$

produces 2 moles of strong acid (H^+) for every mole of NH_4^+ oxidized. With wastewater containing 25 mg NH_4^+–N/liter, this amounts to the production of

$$\frac{25 \text{ mg N/liter}}{14 \text{ mg N/m}M} \times 2\frac{\text{m}M \text{ H}^+}{\text{m}M \text{ N}} = 3.57 \text{ m}M/\text{liter}$$

of strong acid, which will destroy 3.57 mM/liter of HCO_3^- alkalinity (or 179 mg/liter of alkalinity as $CaCO_3$).[11]

Similar problems are encountered in chemical processes where acids (e.g., alum, ferric chloride) are used as coagulants and precipitants in water and wastewater treatment. If any of these acid-consuming processes consume most of the alkalinity in a water or wastewater, severe depressions of pH (to values of 4 to 5) can occur.

Just as the buffering of water results in resistance to pH depression upon the addition of strong acid, we can show that pH elevation is resisted upon the addition of strong base. In treatment processes such as

[11] See Section 4-13 for a discussion of alkalinity.

water softening and phosphate removal by calcium phosphate precipitation, it is often necessary to raise the pH of a solution to about 11. Field scale operating data have shown that the amount of the base, hydrated lime ($Ca(OH)_2$), required to affect this pH elevation is a direct function of the alkalinity and buffer capacity of the solution.

In this section we will look at the procedure for determining the pH of a buffer solution and then consider methods to predict and determine the buffer intensity, or the degree of resistance of a buffered solution to pH change.

4.12.1. Buffer pH

A solution buffered at a particular pH value will contain an acid that reacts with any strong base added to the solution and a base that reacts with any strong acid added to the solution. The most common method of preparing a buffered solution is to add a mixture of a weak acid and its conjugate base; the pK_a of the acid should be very close to the desired pH of the buffer solution for maximum buffer intensity.

Let us examine the general case when C_{HA} moles/liter of the weak acid HA and C_{NaA} moles/liter of its conjugate base A^- are added to the distilled water. The equations describing the system, neglecting ionic strength, are

Mass Balances

$$C_{T,A} = C_{HA} + C_{NaA} = [HA] + [A^-] \tag{4-71}$$

$$C_{T,Na} = [Na^+] = C_{NaA} \tag{4-72}$$

Equilibria

$$K_a = \frac{[A^-][H^+]}{[HA]} \tag{4-73}$$

$$K_w = [H^+][OH^-] \tag{4-74}$$

Charge Balance

$$[Na^+] + [H^+] = [A^-] + [OH^-] \tag{4-75}$$

Combining Eqs. 4-72 and 4-75 to eliminate $[Na^+]$, we obtain

$$[A^-] = [H^+] - [OH^-] + C_{NaA} \tag{4-76}$$

and combining Eqs. 4-71 and 4-76, we arrive at a similar expression for [HA],

$$[HA] = C_{HA} + [OH^-] - [H^+] \tag{4-77}$$

Equations 4-76 and 4-77 can be substituted into Eq. 4-73 to obtain a general expression for pH of the buffer,

$$[H^+] = K_a \frac{(C_{HA} + [OH^-] - [H^+])}{(C_{NaA} - [OH^-] + [H^+])} \tag{4-78}$$

For most buffer solutions, and almost invariably for prepared buffers, $C_{HA} \gg ([OH^-] - [H^+])$ and $C_{NaA} \gg (-[OH^-] + [H^+])$. (The pC-pH diagram can be used to determine when this condition is met.) Equation 4-78 then becomes

$$[H^+] = K_a \frac{C_{HA}}{C_{NaA}} \tag{4-79}$$

where $C_{HA} = [HA]$ and $C_{NaA} = [A^-]$, or in logarithm form,

$$pH = pK_a + \log\left(\frac{C_{NaA}}{C_{HA}}\right)$$

or

$$pH = pK_a + \log \frac{[salt]}{[acid]} \tag{4-80}$$

This equation is commonly used for the preparation of buffer solutions at a given pH. When ionic strength effects are important, ^cK_a should be used instead of K_a in Eqs. 4-78 to 4-80.

Example 4-28

1. From an examination of the equilibrium constants in Table 4-1, determine which acid-conjugate base pair should be used to buffer a solution at (i) pH 4.7 and (ii) pH 7. What ratio of acid to conjugate base should be used in each of these buffers?

2. What is the pH of each of these buffers after addition of 10^{-4} moles of NaOH per liter of solution if $C_T = 10^{-3}\,M$ in each case?

Solution

1. (a) Examination of Table 4-1 shows that acetic acid, HAc, with a pK_a of 4.7 should be a good buffer at pH 4.7. Assuming that $[H^+]$ and $[OH^-]$ are small compared to C_{HAc} and C_{NaAc}, we can use Eq. 4-79 or 4-80 to determine the ratio of C_{HAc}/C_{NaAc}:

$$\frac{[H^+]}{K_a} = \frac{10^{-4.7}}{10^{-4.7}} = \frac{C_{HAc}}{C_{NaAc}} = 1$$

Note that establishing the buffer pH determines the *ratio* but not the *concentration* of acetic acid and sodium acetate to use. The concentrations of acid and conjugate base are determined by the required ability of the buffer to resist pH changes upon addition of strong acid or base, or the buffer intensity.

(b) Examination of Table 4-1 shows that both $H_2S_{(aq)}$ and $H_2PO_4^-$ have pK_a values near 7.0. $H_2S_{(aq)}$ is volatile and would be difficult to keep in solution, besides which it is toxic and evil-smelling. Therefore, $H_2PO_4^-$

is the better choice. Phosphoric acid is a triprotic acid and we are interested in the $pK_{a,2}$ value describing the equilibrium,

$$H_2PO_4^- \rightleftharpoons HPO_4^{2-} + H^+$$

We can consider this equilibrium in isolation from the two other equilibria if the pK_a values differ by more than about 2 pH units. This condition is met for phosphoric acid where $pK_{a,1} = 2.1$, $pK_{a,2} = 7.2$, and $pK_{a,3} = 12.3$. For $pK_{a,2} = 7.2$, Eq. 4-79 becomes

$$\frac{C_{Na_2HPO_4}}{C_{NaH_2PO_4}} = \frac{10^{-7.2}}{10^{-7.0}} = 0.63$$

2. (a) The initial solution with $C_{T,Ac} = 10^{-3}$ was prepared using $C_{HAc} = C_{NaAc} = 5 \times 10^{-4}$. Because $[H^+]$ and $[OH^-]$ are very small with respect to $[HAc]$ and $[Ac^-]$ in this region, we can assume that the small amount of OH^- added reacts completely with HAc. Thus $[HAc]$ decreases, and $[Ac^-]$ increases by the amount of OH^- added per liter. (Our check to determine whether this approach is valid is to determine if the assumptions used to develop Eq. 4-80, that is, $[HAc] \gg ([OH] - [H^+])$ and $[Ac^-] \gg ([OH^-] + [H^+])$, are valid at the final pH of the solution.) On this basis $[HAc] = 5 \times 10^{-4} - 10^{-4} = 4 \times 10^{-4}$, and $[Ac^-] = 5 \times 10^{-4} + 10^{-4} = 6 \times 10^{-4}$.

Applying Eq. (4-80), we obtain

$$pH = pK_a + \log \frac{[Ac^-]}{[HAc]}$$

$$= 4.7 + \log \frac{6 \times 10^{-4}}{4 \times 10^{-4}}$$

$$= 4.7 + 0.2 = 4.9$$

Because $[H^+]$ and $[OH^-]$ are much smaller than $[HAc]$ and $[Ac^-]$ at pH $= 4.9$, the assumption of complete reaction is valid.

(b) Knowing that $C_{T,PO_4} = 10^{-3} M$ and $C_{Na_2HPO_4}/C_{NaH_2PO_4} = 0.63$, we find that $C_{NaH_2PO_4} = [H_2PO_4^-] = 6.1 \times 10^{-4} M$ and $C_{Na_2HPO_4} = [HPO_4^{2-}] = 3.9 \times 10^{-4} M$. After the addition of 10^{-4} mole NaOH/liter, $[HPO_4^{2-}] = 3.9 \times 10^{-4} + 10^{-4} = 4.9 \times 10^{-4} M$, and $[H_2PO_4^-] = 6.1 \times 10^{-4} - 10^{-4} = 5.1 \times 10^{-4} M$.

$$pH = 7.2 + \log \frac{4.9 \times 10^{-4}}{5.1 \times 10^{-4}}$$

$$= 7.2 - 0.02 = 7.18$$

Again $[H_2PO_4^-]$ and $[HPO_4^{2-}]$ are much greater than $[H^+]$ and $[OH^-]$ at this pH so our assumption of complete reaction is valid. The reader may wish to prove this using the approach given in Example 4-27.

4.12.2. Buffer Intensity

Buffer intensity, β, or buffer capacity, as it is often called, is defined as the moles/liter of strong base, C_B, (or OH^-) which when added to a solution causes a unit change in pH. Thus

$$\beta = \frac{d\,C_B}{d\,\mathrm{pH}} = -\frac{d\,C_A}{d\,\mathrm{pH}} \qquad (4\text{-}81)$$

where C_A is the moles/liter of strong acid (or H$^+$) added. The buffer intensity can be determined experimentally and by computation at individual pH values. Experimental measurement involves determining a titration curve that shows the course of pH change with moles of strong acid or base added. The slope of a titration curve is $d\,\mathrm{pH}/dV$ where $V =$ ml of X normal acid or base added. The slope of the titration curve can be converted into $1/\beta$ by multiplying by the factor C', where

$C' = ((\text{sample volume, liters})/(\text{normality of titrant}))(10^3 \text{ ml/liter})$

Thus

$$C'\,(d\,\mathrm{pH}/dV) = d\,\mathrm{pH}/dC_B = 1/\beta$$

Therefore, the reciprocal of the slope of the titration curve is proportional to the buffer intensity. For example, the titration curve for $10^{-3}\,M$ HAc (Fig. 4-12) is reproduced in Fig. 4-14 together with a plot of β derived from slope measurements on this curve.

We can generally conclude that the buffer intensity is a maximum at pH $=$ pK_a, that is, when [Ac$^-$] $=$ [HAc] and that β is at a minimum when $f = 1$, when [HAc]$_{\text{initial}} = C_B$, at the equivalence point of the titration of HAc with a strong base. Although we have learned all of these facts previously, a knowledge of the absolute value of β helps us to design buffers at a given pH with the desired capacity to resist pH change. It is in this aspect of buffer design that the computational method for determining β is most useful.

To develop a general equation for β, let us consider a solution containing

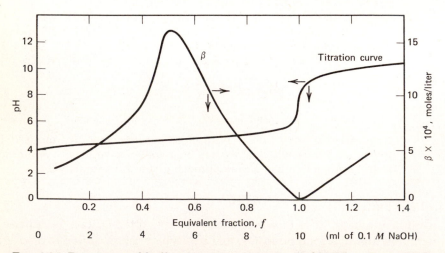

Fig. 4-14. Titration and buffer intensity curves for $10^{-3}\,M$ HAc—Ac$^-$ at 25°C.

a monoprotic acid, HA, and the salt of its conjugate base NaA to which C_A moles/liter of strong acid (HCl) are added.

If we neglect ionic strength effects, the equilibria are

$$K_a = \frac{[H^+][A^-]}{[HA]} \tag{4-82}$$

$$K_w = [H^+][OH^-] \tag{4-83}$$

Mass Balances

$$C_A = [Cl^-] \quad \text{and} \quad C_{T,A} = [HA] + [A^-] \tag{4-84}$$

Charge Balance

$$[Na^+] + [H^+] = [A^-] + [OH^-] + [Cl^-]$$

or using $C_A = [Cl^-]$

$$C_A = [Na^+] + [H^+] - [A^-] - [OH^-] \tag{4-85}$$

Since

$$\beta = \frac{-d\,C_A}{d\,pH} = \frac{-d\,C_A}{d\,[H^+]}\frac{d\,[H^+]}{d\,pH} \tag{4-86}$$

Substituting in Eq. 4-86 for C_A from Eq. 4-85, we obtain

$$\beta = -\left(\frac{d\,[Na^+]}{d\,[H^+]} + \frac{d\,[H^+]}{d\,[H^+]} - \frac{d\,[A^-]}{d\,[H^+]} - \frac{d\,[OH^-]}{d\,[H^+]}\right)\left(\frac{d\,[H^+]}{d\,pH}\right) \tag{4-87}$$

Evaluating each of the differentials, we obtain

$$\frac{d\,[Na^+]}{d\,[H^+]} = 0$$

$$\frac{d\,[H^+]}{d\,[H^+]} = 1$$

From Eqs. 4-82 and 4-84,

$$[A^-] = \frac{K_a C_{T,A}}{[H^+] + K_a} \tag{4-88}$$

and

$$[HA] = \frac{[H^+] C_{T,A}}{[H^+] + K_a} \tag{4-89}$$

Then

$$-\frac{d\,[A^-]}{d\,[H^+]} = \frac{d\,((-K_a C_{T,A})/([H^+] + K_a))}{d\,[H^+]} = +\frac{C_{T,A}K_a}{([H^+] + K_a)^2} \tag{4-90}$$

From Eq. 4-83

$$[OH^-] = \frac{K_w}{[H^+]}$$

Then

$$-\frac{d\,[OH^-]}{d\,[H^+]} = \frac{d\,(-K_w/[H^+])}{d\,[H^+]} = \frac{K_w}{[H^+]^2} \qquad (4\text{-}91)$$

$$\frac{d\,[H^+]}{d\,pH} = \frac{d\,[H^+]}{d\,(-\log[H^+])} = \frac{d\,[H^+]}{d\,(\ln[H^+])/-2.3} = \frac{-2.3\,d\,[H^+]}{d\,[H^+]/[H^+]} = -2.3[H^+] \quad (4\text{-}92)$$

Substituting the values of the differentials in Eq. 4-87, we find that

$$\beta = -\left(0 + 1 + \frac{C_{T,A}K_a}{([H^+] + K_a)^2} + \frac{K_w}{[H^+]^2}\right)(-2.3[H^+])$$

$$= 2.3[H^+] + 2.3[H^+]\left(\frac{C_{T,A}K_a}{([H^+] + K_a)^2}\right) + 2.3[H^+]\frac{K_w}{[H^+]^2} \qquad (4\text{-}93)$$

$$= 2.3\left([H^+] + [OH^-] + \frac{[H^+]\,C_{T,A}K_a}{([H^+] + K_a)^2}\right)$$

In the event that ionic strength effects are important, ^cK_a should be used in place of K_a and ^cK_w in place of K_w.

Substituting from Eqs. 4-84, 4-88, and 4-89 gives an alternative expression for β:

$$\beta = 2.3\left([H^+] + [OH^-] + \frac{[HA][A^-]}{[HA] + [A^-]}\right) \qquad (4\text{-}94)$$

Knowing that $\alpha_0 = [HA]/C_{T,A}$ and $\alpha_1 = [A^-]/C_{T,A}$, substituting into Eq. 4-94, we can derive a third way of expressing β:

$$\beta = 2.3\,([H^+] + [OH^-] + \alpha_0\alpha_1 C_{T,A}) \qquad (4\text{-}95)$$

This equation, together with the table of α values in Appendix 1 allows the rapid computation of β as a function of pH.

If no weak acid-conjugate base pair were present, derivation by the above procedure would have yielded the buffer intensity attributable to water:

$$\beta_{H_2O} = 2.3\,([H] + [OH^-]) \qquad (4\text{-}96)$$

so that Eq. 4-95 can be written

$$\beta = \beta_{H_2O} + \beta_{HA} \qquad (4\text{-}97)$$

We can also use a similar derivation of β for a multiprotic acid, H_nA. If the pK_a values are separated by more than 2 pH units, the derived equation produces results within 5 percent error of the correct value of β.[12]

Thus

$$\beta = 2.3\left([H^+] + [OH^-] + \frac{[H_nA][H_{n-1}A^-]}{[H_nA] + [H_{n-1}A^-]} \right. $$
$$\left. + \frac{[H_{n-1}A^-][H_{n-2}A^{2-}]}{[H_{n-1}A^-] + [H_{n-2}A^{2-}]} + \cdots \right) \qquad (4\text{-}98)$$

or in terms of α values,

$$\beta = 2.3([H^+] + [OH^-] + \alpha_0\alpha_1 C_{T,A} + \alpha_1\alpha_2 C_{T,A} + \cdots) \qquad (4\text{-}99)$$

where

$$\alpha_0 = \frac{[H_nA]}{C_{T,A}}, \; \alpha_1 = \frac{[H_{n-1}A^-]}{C_{T,A}}, \; \alpha_2 = \frac{[H_{n-2}A^{2-}]}{C_{T,A}}, \cdots$$

and

$$C_{T,A} = [H_nA] + [H_{n-1}A^-] + [H_{n-2}A^{2-}] + \cdots$$

More than one acid-conjugate base pair may be used to buffer a solution. A derivation similar to the previous equations yields the equations applicable to such a solution,

$$\beta = 2.3\left([H^+] + [OH^-] + \frac{[HA][A^-]}{[HA] + [A^-]} + \frac{[HB][B^-]}{[HB] + [B^-]} + \cdots \right) \qquad (4\text{-}100)$$

or

$$\beta = \beta_{H_2O} + \beta_{HA} + \beta_{HB} + \cdots \qquad (4\text{-}101)$$

or in terms of α values,

$$\beta = 2.3([H^+] + [OH^-] + \alpha_{0,A}\alpha_{1,A}C_{T,A} + \alpha_{0,B}\alpha_{1,B}C_{T,B} + \cdots) \qquad (4\text{-}102)$$

where HA and HB are of total concentration $C_{T,A}$ and $C_{T,B}$ and $\alpha_{0,A}$ and $\alpha_{1,A}$ are the α values for HA and A^-, and so forth.

Example 4-29

Compute the buffer intensity of a 10^{-3} M acetic acid solution at several pH values to determine the variation of β with pH. Determine the individual contributions of the acetic acid (β_{HAc}) and of water (β_{H_2O}) to the total buffer intensity (β) and plot their variation with pH. How do the computed values of β compare with the values measured from the titration curve in Fig. 4-14? ($K_{a,HAc} = 4.7$; neglect ionic strength effects.)

[12] J. E. Ricci, *Hydrogen Ion Concentration, New Concepts in a Systematic Treatment*, Princeton University Press, Princeton, N.J., 1952.

Solution

For selected pH values, determine α_0 and α_1 from Appendix 1. Using these α values determine

$$\beta = \beta_{H_2O} + \beta_{HAc}$$

$$= 2.3([H^+] + [OH^-]) + 2.3\,\alpha_0\alpha_1 C_{T,HAc}$$

At pH 4.7

$$\beta_{H_2O} = 2.3\,(10^{-4.7} + 10^{-9.3}) = 4.58 \times 10^{-5}\text{mole/liter}$$

$$\beta_{HAc} = 2.3\,(0.5 \times 0.5 \times 10^{-3}) = 5.75 \times 10^{-4}\text{mole/liter}$$

$$\beta = 0.046 \times 10^{-4} + 5.75 \times 10^{-4} = 6.21 \times 10^{-4}\text{mole/liter}$$

See Table 4-6. The values of β_{H_2O} and β_{HAc} are plotted in Fig. 4-15.

Examination of Fig. 4-15 shows that β_{HAc} has a maximum at pH $=$ pK_a and is symmetrical about this pH. This same phenomenon is also observed for other acid-conjugate base systems, including multiprotic systems. There will be a maximum at each pK_a for the multiprotic species. Also, β_{H_2O} has a minimum at pH $=$ p$K_w/2$ and is symmetrical about this pH. When β_{H_2O} and β_{HAc} are summed to give β, Fig. 4-15 shows that β_{H_2O} is predominant at very high and very low pH but that it contributes very little to β in the pH range 5 to 9 for a $10^{-3}\,M$ buffer solution. It should be noted that β_{H_2O} is the buffer intensity of a strong acid-strong base system.

Example 4-30

A strongly acidic pharmaceutical waste, $[H^+] = 10^{-1.8}\,M$, is to be neutralized to pH 6 prior to discharge. Continuous neutralization with 50 percent caustic soda, NaOH, resulted in an effluent pH that varied widely in the range 4 to 11 because of the low buffer intensity of the water. A laboratory study showed that the addition of sodium bicarbonate, NaHCO$_3$, or soda ash, Na$_2$CO$_3$, could be used to increase the buffer intensity of the solution so that the desired final pH could be

TABLE 4-6   Computation of β, Example 4-29

pH	α_0	α_1	β_{H_2O}	β_{HAc}	β
4.7	0.50	0.50	4.58×10^{-5}	5.75×10^{-4}	6.2×10^{-4}
4.4	0.66	0.33	0.1×10^{-5}	5.0×10^{-4}	5.9×10^{-4}
4.1	0.80	0.20	1.8×10^{-4}	3.7×10^{-4}	5.5×10^{-4}
3.7	0.91	0.09	4.6×10^{-4}	1.9×10^{-4}	6.5×10^{-4}
2.7	0.99	0.01	4.6×10^{-3}	2.3×10^{-5}	4.6×10^{-3}
2.0	1.00	0.00	2.3×10^{-2}	0	2.3×10^{-2}
5.0	0.33	0.66	2.3×10^{-5}	5.1×10^{-4}	5.3×10^{-4}
5.3	0.20	0.80	1.1×10^{-5}	3.7×10^{-4}	3.8×10^{-4}
5.7	0.09	0.91	4.6×10^{-6}	1.9×10^{-4}	1.9×10^{-4}
7.7	0.01	0.99	4.61×10^{-6}	2.3×10^{-5}	2.8×10^{-5}
9.0	0.00	1.00	2.3×10^{-5}	0	2.3×10^{-5}
11.0	0.00	1.00	2.3×10^{-3}	0	2.3×10^{-3}

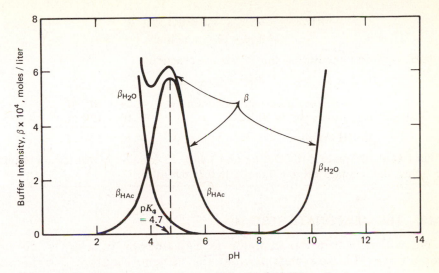

Fig. 4-15. Buffer intensity versus pH for 10^{-3} M HAc—Ac⁻.

achieved in the neutralization process. Given that a β of 0.75 mM/liter is necessary if the neutralization system is to continuously produce an effluent with a pH sufficiently close to 6, calculate the NaOH and NaHCO₃ required to achieve the neutralization. Neglect ionic strength effects.[13]

Solution

Calculate the NaHCO₃ required using Eq. 4-99:

$$\beta = 2.3([H^+] + [OH^-] + \alpha_0\alpha_1 C_{T,CO_3} + \alpha_1\alpha_2 C_{T,CO_3})$$

From Appendix 1 at pH = 6, α_0 = 0.666 = $[H_2CO_3^*]/C_{T,CO_3}$, α_1 = 0.334 = $[HCO_3^-]/C_{T,CO_3}$, and α_2 = 0 = $[CO_3^{2-}]/C_{T,CO_3}$.

$$0.75 \times 10^{-3} = 2.3(10^{-6} + 10^{-8} + (0.666)(0.334)C_{T,CO_3} + (0.334)(0)C_{T,CO_3})$$

$$C_{T,CO_3} = 1.42 \times 10^{-3} M$$

Thus 1.42×10^{-3} mole NaHCO₃ must be added per liter of waste to achieve the desired buffer intensity.

Next we calculate the required NaOH. In the waste there are $10^{-1.8}$ moles H⁺/liter (1.58×10^{-2} mole/liter). Adjustment of the solution to pH 6 requires that $(1.58 \times 10^{-2} - 1 \times 10^{-6}) = 1.58 \times 10^{-2}$ mole H⁺/liter be neutralized. The H⁺ concentration neutralized by NaHCO₃ via the reaction

$$HCO_3^- + H^+ \rightarrow H_2CO_3^*$$

is equal to $[H_2CO_3^*]$ at pH = 6

[13] This example is based upon information given by R. W. Okey, K. Y. Chen, and A. Z. Sycip, "Techniques for Continuous Neutralization of Strong Acid Wastes," presented at the 32nd Purdue Industrial Waste Conference, Lafayette, Ind., 1978.

$$[H_2CO_3^*] = \alpha_0 C_{T,CO_3} = (0.666)(1.42 \times 10^{-3})$$
$$= 9.46 \times 10^{-4}\, \text{mole/liter}$$

NaOH required to neutralize the H^+ via the reaction,

$$OH^- + H^+ \rightarrow H_2O$$

Total H^+ requiring neutralization	$1.58 \times 10^{-2}\ M$
Neutralization by $NaHCO_3$	$-9.46 \times 10^{-4}\ M$
NaOH required	$\simeq 1.48 \times 10^{-2}\ M$

Thus 1.42×10^{-3} mole $NaHCO_3$/liter and 1.48×10^{-2} mole NaOH/liter are required to adjust the solution pH to 6.0 and β to 0.75 mM/liter.

4.13.   THE CARBONATE SYSTEM

We will now turn our attention to a discussion of what is the most important acid-base system in water: the carbonate system. The chemical species that make up the carbonate system—gaseous CO_2, $CO_{2(g)}$; aqueous or dissolved CO_2, $CO_{2(aq)}$; carbonic acid, H_2CO_3; bicarbonate, HCO_3^-; carbonate, CO_3^{2-}; and carbonate-containing solids—comprise one of the major acid-conjugate base systems in natural waters. Because of this the analytical techniques for measuring the capacity of a water to neutralize strong acid and strong base (alkalinity and acidity, respectively) are based largely on the specific properties of the carbonate system. The carbonate species also buffer natural waters. In addition, individual species are of interest to us because they participate in important reactions other than strictly acid-base interactions. Carbon dioxide is a participant in the biological processes of respiration (CO_2 produced) and biosynthesis by autotrophs or photosynthetic organisms (CO_2 consumed). The dissolution of CO_2 from the atmosphere into water and the release of CO_2 from supersaturated waters involves a heterogeneous reaction between gas and liquid phases. Carbonate ion participates in hetero-geneous equilibria with solids containing carbonate, notably calcium carbonate. The formation of $CaCO_3$ precipitates is one of the bases of the precipitation process for water softening. The dissolution of carbonate-bearing minerals is a major source of inorganic carbon-containing species in natural waters. Carbonates and bicarbonates form complexes and ion pairs with metals such as Ca^{2+} and Mg^{2+}. Because of their almost universal occurrence and involvement in reactions, the carbonate species almost always has a bearing on the type and quantity of chemicals required to treat water and wastewater.

4.13.1.   The Carbonate Species and Their Acid-Base Equilibria

The various components of the carbonate system are interrelated by the following equilibria. (For the moment we will neglect the equilibria between carbonate species and metal ions and the involvement of carbonate-containing solids (see Chapters 5 and 6)). The temperature dependence of the constants we will use most frequently is presented in Table 4-7.

$$CO_{2(g)} \rightleftharpoons CO_{2(aq)}; \quad K = K_H = 10^{-1.5} \tag{4-103}$$

where K_H is Henry's constant.

$$CO_{2(aq)} + H_2O \rightleftharpoons H_2CO_3; \quad K_m = 10^{-2.8} \tag{4-104}$$

$$H_2CO_3 \rightleftharpoons H^+ + HCO_3^-; \quad K_1' = 10^{-3.5} \tag{4-105}$$

$$H_2CO_3^* \rightleftharpoons H^+ + HCO_3^-; \quad K_{a,1} = 10^{-6.3} \tag{4-106}$$

$$HCO_3^- \rightleftharpoons H^+ + CO_3^{2-}; \quad K_{a,2} = 10^{-10.3} \tag{4-107}$$

When we are presented with a problem involving the carbonate system, we must first make a decision concerning the nature of the system. This step is important because carbonate species can be involved in homogeneous solution equilibria as well as heterogeneous gas/liquid and liquid/solid equilibria. We must examine our system to see whether it can be treated as

1. An open system with no solid present.

2. An open system with solid present.

3. A closed system with no solid present.

4. A closed system with solid present.

The use of simple sketches helps us to decide the nature of our system. For example,

1. A water sample containing carbonate species is titrated with a strong acid. During titration the sample is warmed and vigorously bubbled with air to equilibrate the sample with the atmosphere.

TABLE 4-7   Temperature Dependence of Some Important Carbonate Equilibrium Constants

	Temperature, °C						
Reaction	5	10	15	20	25	40	60
1. $CO_{2(g)} + H_2O \rightleftharpoons CO_{2(aq)}$; pK_H	1.20	1.27	1.34	1.41	1.47	1.64	1.8
2. $H_2CO_3^* \rightleftharpoons HCO_3^- + H^+$; $pK_{a,1}$	6.52	6.46	6.42	6.38	6.35	6.30	6.30
3. $HCO_3^- \rightleftharpoons CO_3^{2-} + H^+$; $pK_{a,2}$	10.56	10.49	10.43	10.38	10.33	10.22	10.14
4. $CaCO_{3(s)} \rightleftharpoons Ca^{2+} + CO_3^{2-}$; pK_{so}	8.09	8.15	8.22	8.28	8.34	8.51	8.74
5. $CaCO_{3(s)} + H^+ \rightleftharpoons Ca^{2+} + HCO_3^-$; $p(K_{so}/K_{a,2})$	−2.47	−2.34	−2.21	−2.10	−1.99	−1.71	−1.40

Source: From T. E. Larson and A. M. Buswell, "Calcium Carbonate Saturation Index and Alkalinity Interpretations," J. Am. Water Works Assoc., 34: 1664 (1942).

The system can be sketched as in the following figure. The system is *open* to the atmosphere and not in equilibrium with a carbonate-containing solid.

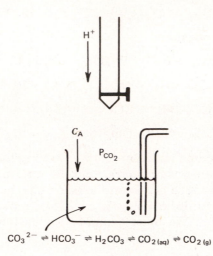

$$CO_3^{2-} \rightleftharpoons HCO_3^- \rightleftharpoons H_2CO_3 \rightleftharpoons CO_{2\,(aq)} \rightleftharpoons CO_{2\,(g)}$$

2. A water sample containing carbonate species is titrated with strong acid. The titration is conducted rapidly in a nearly full container with little shaking. The system can be sketched as in the following figure. The system is considered to be *closed* to the atmosphere (although some CO_2 may escape) and is not in equilibrium with any carbonate-containing solid.

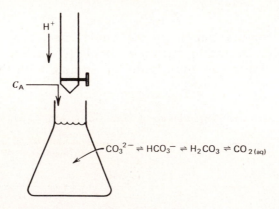

$$CO_3^{2-} \rightleftharpoons HCO_3^- \rightleftharpoons H_2CO_3 \rightleftharpoons CO_{2\,(aq)}$$

3. The hypolimnion of a stratified lake is in contact with sediment containing calcite, $CaCO_{3(s)}$. The system can be sketched as in the following figure. The system is *closed* to the atmosphere and tends to be in equilibrium with a carbonate-containing solid.

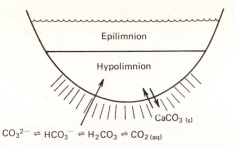

$$CO_3^{2-} \rightleftharpoons HCO_3^- \rightleftharpoons H_2CO_3 \rightleftharpoons CO_{2\,(aq)}$$

4. Chemicals (e.g., lime, soda ash, strong acid, strong base, and CO_2) are added to the water in a water treatment plant to soften the water and adjust its pH. The system can be sketched as in the following figure. Because the surface area to volume ratio of the treatment basin is small, the residence time is short, and solid $CaCO_3$ is in abundance, the system is considered to be *closed* to the atmosphere and it tends toward equilibrium with carbonate-containing solids. (A closed system assumption is not entirely accurate because some atmospheric CO_2 exchange may occur.)

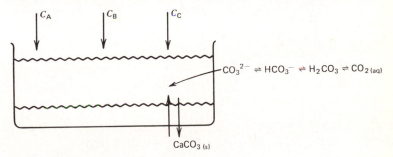

5. Nitrification (oxidation of ammonia to nitrate) and oxidation of organic matter to CO_2 and water take place in the vigorously aerated aeration basin of an activated sludge plant. The system can be sketched as in the following figure. The system is *open* and in equilibrium with the atmosphere with a P_{CO_2} of the aerating gas. It is not in equilibrium with a carbonate-containing solid.

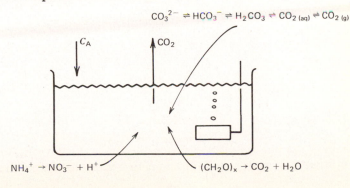

From Eq. 4-104 we can conclude that the concentration of hydrated carbon dioxide, $CO_{2(aq)}$, predominates over the concentration of carbonic acid, H_2CO_3.

$$K_m = 10^{-2.8} = \frac{[H_2CO_3]}{[CO_{2(aq)}]} = 1.6 \times 10^{-3}$$

Thus $[H_2CO_3]$ is only 0.16 percent of $[CO_{2(aq)}]$. Because it is difficult to distinguish between $CO_{2(aq)}$ and H_2CO_3 by analytical procedures such as acid-base titration, a hypothetical species $(H_2CO_3^*)$ is used to represent H_2CO_3 plus $CO_{2(aq)}$. We will follow this convention. Using Eqs. 4-103 and 4-104, we can determine the ionization constant, $K_{a,1}$ for $H_2CO_3^*$ as follows: Since

$$[H_2CO_3] + [CO_{2(aq)}] = [H_2CO_3^*] \tag{4-108}$$

where $H_2CO_3^*$ is a hypothetical species only, and neglecting ionic strength effects, the equilibrium constants for ionization of $H_2CO_3^*$ and H_2CO_3 can be written as

$$\frac{[H^+][HCO_3^-]}{[H_2CO_3^*]} = K_{a,1} \quad \text{and} \quad \frac{[H^+][HCO_3^-]}{[H_2CO_3]} = K_1'$$

Substituting Eq. 4-108 in the $K_{a,1}$ equation, we obtain

$$K_{a,1} = \frac{[H^+][HCO_3^-]}{([H_2CO_3] + [CO_{2(aq)}])} \tag{4-109}$$

Now multiplying the numerator and denominator of Eq. 4-109 by $[H_2CO_3]$ and knowing that $[CO_{2(aq)}]/[H_2CO_3] = K_m$, we find that

$$\frac{([H^+][HCO_3^-])/[H_2CO_3]}{([CO_{2(aq)}] + [H_2CO_3])/[H_2CO_3]} = \frac{K_1'}{(1/K_m) + 1} = K_{a,1}$$

Since

$$\frac{1}{K_m} = \frac{1}{10^{-2.8}} = 631$$

$$\frac{1}{K_m} \gg 1$$

therefore,

$$K_{a,1} = \frac{K_1'}{(1/K_m)} = K_1' K_m$$

$$K_{a,1} = 10^{-3.5}\, 10^{-2.8} = 10^{-6.3}$$

The implications of this development are important because they point out that H_2CO_3 is a fairly strong acid, $(K_1 = 10^{-3.5})$, but there is very little of it in solution. This point is emphasized because the composite first dissociation constant, $K_{a,1}$, for $H_2CO_3^*$ is $10^{-6.3}$, a value that would give the impression that carbonic acid is a very weak acid. As we shall see later when we examine the dissolution of minerals by CO_2, carbonic acid has some of the properties of a strong acid. Because of the preceding equilibrium calculations and because $[CO_{2(aq)}] \gg [H_2CO_3]$, we can write

$$[H_2CO_3^*] \cong [CO_{2(aq)}] \tag{4-110}$$

Although equilibrium calculations justify the use of the hypothetical species, $H_2CO_3^*$, there are some kinetic considerations that must be addressed. A simple experiment will illustrate the problem.[14]

Take 200 ml of chilled distilled water and add a piece of dry ice (solid CO_2) to it to increase its dissolved CO_2 content. Add 3 to 4 drops of bromothymol blue indicator, which is yellow below pH 6, green at pH 7, and blue at pH 8 and above. The indicator will turn yellow in the chilled CO_2 solution. Now add 1 to 2 ml of 0.1 M NaOH (strong base) and swirl the flask to mix. The solution will be momentarily (seconds) blue, then return to its original yellow color through an intermediate shade of green. These color changes indicate that after the addition of strong base the pH of the solution was at pH above 8 for a significant time and then gradually returned to below pH 6. For some reason then, the strong base that was added to the solution was not neutralized, or remained unreacted, for a measurable time period. If we recall our earlier discussions on kinetics, we realize that slow reaction rates are not characteristic of acid-base reactions. The explanation for this behavior lies in the slow (relative to other deprotonation reactions) rate of hydration of aqueous CO_2. The reaction, $CO_{2(aq)} + H_2O \rightleftharpoons H_2CO_3$ has a rate constant of 0.0025 to 0.03 at 20 to 25°C. By chilling the solution as we did in our experiment, the rate is slowed even further. From our equilibrium discussions we know that $CO_{2(aq)}$ is the major dissolved CO_2 species. However, it is the minor species H_2CO_3 that participates in acid-base reactions with OH^-,

$$H_2CO_3 + OH^- \rightarrow HCO_3^- + H_2O$$

This reaction, like most acid-base reactions, is extremely rapid. Therefore, on the addition of excess OH^-, the available H_2CO_3 is consumed at a faster rate than it can be produced. The result is that the pH increases momentarily until the slower CO_2 hydration reaction can supply more H_2CO_3 to react with the OH^-.

The reason for the slow hydration rate lies in the change of molecular configuration that must occur in CO_2 hydration. The $CO_{2(aq)}$ molecule is

---

[14] This experiment is taken from D. M. Kern, "The Hydration of Carbon Dioxide," *J. Chem. Educ.*, *37*: 14 (1960).

linear, $O\!=\!C\!=\!O$, with waters of hydration attached, while H_2CO_3 has the form,

with associated waters of hydration.

Throughout this text we will use the hypothetical species $H_2CO_3^*$ to represent $CO_{2(aq)}$ plus H_2CO_3. We will treat it as a diprotic acid, but we should keep in mind its true nature.

4.13.2. Calculation of Carbonate Species Concentrations in Open and Closed Systems

In working problems concerned with carbonate equilibria in waters, we can take one of two general approaches. The first method is to treat the problem rigorously. This procedure often leads to complex equations. The second method is to make simplifying assumptions to provide an easy but approximate solution. We intend to present both ways of attacking the problem because it is only by having a fundamental knowledge of the system that one can develop the ability to make rational simplifications.

First, let us examine homogeneous (no precipitates), open systems. In these systems the carbonate species in solution are in equilibrium with the CO_2 gas in the atmosphere above the solution. The normal atmosphere contains $10^{-3.5}$ atm of CO_2, and when this is in equilibrium with $CO_{2(aq)}$ from Eqs. 4-103 and 4-110, we obtain

$$[H_2CO_3^*] \cong [CO_{2(aq)}] = K_H P_{CO_2} = 10^{-1.5} \times 10^{-3.5} = 10^{-5} M \qquad (4\text{-}111)$$

Let us now construct a pC-pH diagram for the carbonate system open to an atmosphere containing $P_{CO_2} = 10^{-3.5}$ atm (Fig. 4-16). Our first step is to determine the concentration of each carbonate species as a function of pH. From Eq. 4-111, $[H_2CO_3^*] = 10^{-5}$,

$$-\log [H_2CO_3^*] = 5$$

Therefore, $[H_2CO_3^*]$ versus pH plots as a horizontal straight line at $-\log C = 5$ as shown in Fig. 4-16.
For $[HCO_3^-]$, we know that

$$K_{a,1} = \frac{[H^+][HCO_3^-]}{[H_2CO_3^*]} \qquad (4\text{-}112)$$

Because $[H_2CO_3^*] = 10^{-5}$,

$$[HCO_3^-] = \frac{10^{-5}K_{a,1}}{[H^+]}$$

and

$$-\log [HCO_3^-] = 5 + pH + 6.3 = pH + 11.3 \qquad (4\text{-}113)$$

This equation yields the line for $[HCO_3^-]$ shown in Fig. 4-16. Similarly, for CO_3^{2-} we know that

$$K_{a,2} = \frac{[H^+][CO_3^{2-}]}{[HCO_3^-]} \qquad (4\text{-}114)$$

Combining Eq. 4-111 and the equations for $K_{a,1}$ and $K_{a,2}$ yield

$$[CO_3^{2-}] = \frac{10^{-5}K_{a,1}K_{a,2}}{[H^+]^2}$$

$$-\log [CO_3^{2-}] = 5 + pK_{a_1} + pK_{a_2} + pH = 21.6 + 2pH \qquad (4\text{-}115)$$

This equation yields the line for $[CO_3^{2-}]$ shown in Fig. 4-16. Knowing that

$$C_{T,CO_3} = [H_2CO_3^*] + [HCO_3^-] + [CO_3^{2-}] \qquad (4\text{-}116)$$

we can obtain the concentration of the individual species at various pH

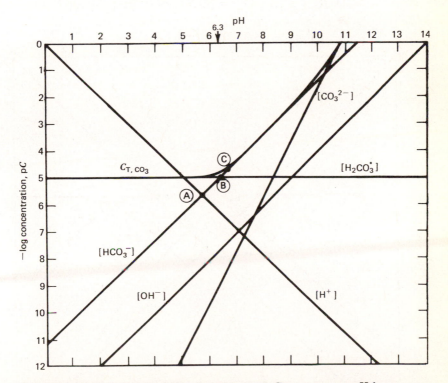

Fig. 4-16. Total concentration of carbonates, C_{T,CO_3}, versus pH for an open system at 25°C.

from Fig. 4-16, determine C_{T,CO_3} as a function of pH using Eq. 4-116, and plot $-\log C_{T,CO_3}$ versus pH. The resulting curve is shown in Fig. 4-16. Note that $[H_2CO_3^*]$ predominates for pH $< pK_{a,1}$, $[HCO_3^-]$ predominates for $pK_{a,1} < pH < pK_{a,2}$, and $[CO_3^{2-}]$ predominates for pH $> pK_{a,2}$. Examination of Fig. 4-16 also reveals that C_{T,CO_3} increases rapidly as pH increases above $pK_{a,1}$.

Now let us tackle the closed system, with a $C_{T,CO_3} = 10^{-5}M$. Since the constraint that the system be in equilibrium with a certain partial pressure of $CO_{2(g)}$ has been removed, we can treat $H_2CO_3^*$ as a nonvolatile diprotic acid. Thus construction of the pC-pH diagram follows the same steps taken in Section 4-8 for the diagram of the hypothetical diprotic acid H_2A (Fig. 4-5). The diagram is shown in Fig. 4-17.

Let us first determine the composition of $10^{-5} M$ solutions of $H_2CO_3^*$, sodium bicarbonate, $NaHCO_3$, and sodium carbonate, Na_2CO_3 in both systems. The proton conditions for the closed system are

$H_2CO_3^*$: $[H^+] = [OH^-] + [HCO_3^-] + 2[CO_3^{2-}]$

or approximately

$$[H^+] = [HCO_3^-]$$

$NaHCO_3$: $[H^+] + [H_2CO_3^*] = [CO_3^{2-}] + [OH^-]$

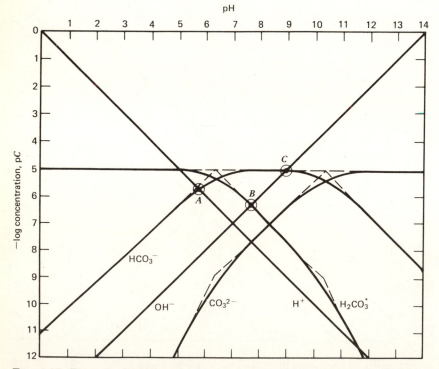

Fig. 4-17. The pC-pH diagram for a $10^{-5} M$ carbonate solution at 25° C.

or approximately

$$[H_2CO_3^*] = [OH^-]$$

Na$_2$CO$_3$: $$[H^+] + 2[H_2CO_3^*] + [HCO_3^-] = [OH^-]$$

or approximately

$$[HCO_3^-] = [OH^-]$$

These proton conditions are met at points A, B, and C, respectively, in Fig. 4-17. The concentrations of the various species in these solutions as taken from Fig. 4-17 are tabulated in Table 4-8.

The calculations are slightly more complex for the open system. Because CO$_2$ can enter the solution or be evolved, C_{T,CO_3} is an unknown. To make comparisons with the closed system solution, let us consider (1) distilled water, (2) distilled water + 10^{-5} mole NaHCO$_3$/liter, and (3) distilled water + 10^{-5} mole Na$_2$CO$_3$/liter, each in equilibrium with the atmosphere. When atmospheric CO$_2$ is in equilibrium with distilled water, Henry's law gives [H$_2$CO$_3^*$].

$$[H_2CO_3^*] = K_H P_{CO_2}$$

$$10^{-1.5} \times 10^{-3.5} = 10^{-5}\ M$$

The proton condition is

$$[H^+] = [HCO_3^-] + 2[CO_3^{2-}] + [OH^-]$$

or approximately [H$^+$] = [HCO$_3^-$], and is satisfied at point A on Fig. 4-16. Examination of the figure shows that $C_{T,CO_3} = 10^{-4.9}$. The solution composition is given in Table 4-8.

TABLE 4-8   Comparison of Composition and pH of Open and Closed Carbonate Systems

Solution	Closed System (Fig. 4-22)						
	[H$^+$]	[OH$^-$]	[H$_2$CO$_3^*$]	[HCO$_3^-$]	[CO$_3^{2-}$]	pH	C_{T,CO_3}
10^{-5} M H$_2$CO$_3^*$	$10^{-5.7}$	$10^{-8.3}$	$10^{-5.1}$	$10^{-5.7}$	$10^{-10.4}$	5.7	$10^{-5.0}$
10^{-5} M NaHCO$_3$	$10^{-7.6}$	$10^{-6.4}$	$10^{-6.4}$	10^{-5}	$10^{-7.6}$	7.6	$10^{-5.0}$
10^{-5} M Na$_2$CO$_3$	10^{-9}	10^{-5}	$10^{-7.6}$	10^{-5}	$10^{-6.3}$	9.0	$10^{-5.0}$

Solution	Open System						
	[H$^+$]	[OH$^-$]	[H$_2$CO$_3^*$]	[HCO$_3^-$]	[CO$_3^{2-}$]	pH	C_{T,CO_3}
10^{-5} M H$_2$CO$_3^*$	$10^{-5.7}$	$10^{-8.3}$	10^{-5}	$10^{-5.7}$	$10^{-10.4}$	5.7	$10^{-4.9}$
10^{-5} M NaHCO$_3$	$10^{-6.3}$	$10^{-7.7}$	10^{-5}	10^{-5}	$10^{-9.5}$	6.3	$10^{-4.7}$
10^{-5} M Na$_2$CO$_3$	$10^{-6.7}$	$10^{-7.3}$	10^{-5}	$10^{-4.7}$	$10^{-8.3}$	6.6	$10^{-4.5}$

The addition of 10^{-5} mole $NaHCO_3$/liter of distilled water followed by equilibration with the atmosphere at 25°C results in the following charge balance

$$[H^+] + [Na^+] = [HCO_3^-] + 2[CO_3^{2-}] + [OH^-] \qquad (4\text{-}117)$$

where $[Na^+] = 1 \times 10^{-5}$. Figure 4-16, which is applicable to open systems, shows that this equation is satisfied at pH 6.3 (point B) where $[Na^+] = 1 \times 10^{-5} = [HCO_3^-]$, $[H^+] \ll 10^{-5}$, and $[HCO_3^-] \gg 2[CO_3^{2-}] + [OH^-]$. The concentrations of the various species at this pH are presented in Table 4-8.

When $10^{-5} M$ Na_2CO_3 is added to distilled water, the same charge balance, Eq. 4-117 results but $[Na^+]$ is now $2 \times 10^{-5} M$. Using Fig. 4-16, we find the charge balance is satisfied at pH 6.6, where $[Na^+] = 2 \times 10^{-5} = [HCO_3^-]$, $[Na^+] \gg [H^+]$ and $[HCO_3^-] \gg 2[CO_3^{2-}] + [OH^-]$. The species concentrations at this pH, taken from Fig. 4-16, are also given in Table 4-8.

Comparison of the closed and open systems in Table 4-8 reveals the following important distinctions

1. In the open system $[H_2CO_3^*]$ remains constant.

2. The pH of a $10^{-5} M$ $CO_{2(aq)}$ solution and of a distilled water solution that is equilibrated with the atmosphere is equal. It would be a useful exercise for the reader to experimentally determine the pH value of distilled water immediately after preparation and then again after standing the distilled water overnight in a beaker open to the atmosphere. For freshly distilled water the pH should be only slightly below neutrality while after contact with the atmosphere containing $10^{-3.5}$ atm (P_{CO_2}), the pH will drift down to the neighborhood of 5.7.

3. The C_{T,CO_3} concentration is constant in the closed system but varies with pH in the open system.

4. Even though the $NaHCO_3$ and Na_2CO_3 solutions in the open system contain higher C_{T,CO_3} than the $NaHCO_3$ and Na_2CO_3 solutions in the closed system, their pH values are significantly lower. We can conclude that equilibration of closed solutions of $NaHCO_3$ and Na_2CO_3 with the atmosphere will cause a depression in pH. If we were to bubble a $10^{-5} M$ Na_2CO_3 solution with air containing $P_{CO_2} = 0.0003$ atm ($10^{-3.5}$ atm), then over a prolonged period its pH would drift down from 9 to about 6.6.

Let us now turn our attention to the calculation of solution composition following the addition of strong acid or strong base to the closed carbonate system. The results and methods that we derive from this exercise will prove useful in (1) calculating solution composition during water treatment operations where strong acids or bases are added to carbonate-bearing waters in processes such as coagulation and stabilization, (2) assessing pH and solution composition when acidic wastes are discharged to receiving waters, and (3) calculating the pH and solution composition at various stages of the alkalinity and acidity titrations. We will demonstrate methods of calculating the solution pH and composition following strong-acid or strong-base addition when the analytical concentration of carbonates, C_{T,CO_3}, is known.

Example 4-31

Calculate the solution pH, solution composition, and the equivalent fractions, f, of titrant for a solution containing $10^{-3} M$ Na_2CO_3 to which are added

1. 10^{-3} mole HNO_3 per liter of solution.

2. 2×10^{-3} mole HNO_3 per liter of solution. Assume that a negligible volume change occurs as a result of the addition of HNO_3 and that the system is closed.

Solution

1. The equivalent fraction is calculated as

$$f = \frac{\text{number of equivalents of } HNO_3 \text{ added/liter}}{C_{T,CO_3}} = \frac{1 \times 10^{-3}}{1 \times 10^{-3}} = 1$$

To determine the solution composition, we need the mass balance, the charge balance, and the appropriate pC-pH diagram.

Mass Balances

$$C_{T,Na} = [Na^+] = 2 \times 10^{-3}$$

$$C_{T,NO_3} = [NO_3^-] = 10^{-3}$$

$$C_{T,CO_3} = [H_2CO_3^*] + [HCO_3^-] + [CO_3^{2-}] = 10^{-3}$$

Charge Balance

$$[Na^+] - [NO_3^-] = [HCO_3^-] + 2[CO_3^{2-}] + [OH^-] - [H^+]$$

After substituting from mass balances on $[Na^+]$ and $[NO_3^-]$, we can write

$$10^{-3} = [HCO_3^-] + 2[CO_3^{2-}] + [OH^-] - [H^+]$$

which is the charge balance for a $10^{-3} M$ $NaHCO_3$ solution. We can also substitute $[H_2CO_3^*] + [HCO_3^-] + [CO_3^{2-}]$ for 10^{-3} in the charge balance to obtain

$$[H_2CO_3^*] + [H^+] = [CO_3^{2-}] + [OH^-]$$

which is the proton condition for a solution prepared by adding only $NaHCO_3$ to distilled water. From the pC-pH diagram for a closed system with $C_{T,CO_3} = 10^{-3}$

M, Fig. 4-18, this proton condition is met at pH 8.3 where $[H_2CO_3^*] = [CO_3^{2-}]$, $[H_2CO_3^*] \gg [H^+]$, and $[CO_3^{2-}] \gg [OH^-]$. The solution composition is

$$[H^+] = 10^{-8.3}, [OH^-] = 10^{-5.7}, [H_2CO_3^*] = 10^{-5},$$
$$[HCO_3^-] = 10^{-3}, [CO_3^{2-}] = 10^{-5}$$

2. The equivalent fraction is calculated as

$$f = \frac{\text{number of equivalents of HNO}_3 \text{ added/liter}}{C_{T,CO_3}} = \frac{2 \times 10^{-3}}{1 \times 10^{-3}} = 2$$

We determine the solution composition as in part (1).

Mass Balances

$$C_{T,Na} = [Na^+] = 2 \times 10^{-3}$$
$$C_{T,NO_3} = 2 \times 10^{-3}$$
$$C_{T,CO_3} = [H_2CO_3^*] + [HCO_3^-] + [CO_3^{2-}]$$

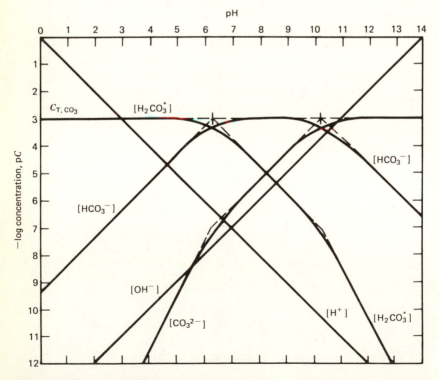

Fig. 4-18. The pC-pH diagram for a carbonate solution, $C_{T,CO_3} = {}^{-3} M$ at 25°C.

Charge Balance

$$[Na^+] - [NO_3^-] = [HCO_3^-] + 2[CO_3^{2-}] + [OH^-] - [H^+]$$

Substituting mass balances on Na^+ and NO_3^- in the charge balance, we can write

$$0 = [HCO_3^-] + 2[CO_3^{2-}] + [OH^-] - [H^+]$$

or

$$[H^+] = [HCO_3^-] + 2[CO_3^{2-}] + [OH^-]$$

This is the proton condition of a 10^{-3} M $H_2CO_3^*$ solution. It is satisfied at pH 4.7 in the pC-pH diagram (Fig. 4-23), where $[H^+] = [HCO_3^-]$ and $[HCO_3^-] \gg 2[CO_3^{2-}] + [OH^-]$. The solution composition is $[H^+] = 10^{-4.7}$, $[OH^-] = 10^{-9.3}$, $[H_2CO_3^*] = 10^{-3}$, $[HCO_3^-] = 10^{-4.7}$, and $[CO_3^{2-}] = 10^{-10.3}$.

Similarly, it can be shown that if one equivalent fraction of strong base is added to a solution containing C_{T,CO_3} moles/liter of $H_2CO_3^*$, the pH of a solution with the same C_{T,CO_3} but made up with only $NaHCO_3$ results, $pH_{HCO_3^-}$; if two equivalent fractions of strong base are added, the pH of a C_{T,CO_3}-molar solution of Na_2CO_3, $pH_{CO_3^{2-}}$, results.

The application of pC-pH diagrams for closed system problems is most useful when the initial solution pH corresponds to a pure carbonate solution (i.e., pH_{CO_2} for an $H_2CO_3^*$ solution, $pH_{HCO_3^-}$ for an $NaHCO_3$ solution and $pH_{CO_3^{2-}}$ for an Na_2CO_3 solution) and the amount of strong acid or strong base added is equal to one or two equivalent fractions. Such cases are the exception; however, our coverage of these simple examples has hopefully instilled into the reader's mind the nature of the assumptions that can be made in various pH regions of the carbonate system. Generally, when strong acid or strong base is added to solutions containing carbonates, the initial pH does not correspond to pH_{CO_2}, $pH_{HCO_3^-}$, or $pH_{CO_3^{2-}}$ and the amount of acid or base added does not correspond to $f = 1$ or 2.

Example 4-32

A solution contains 10^{-3} M C_{T,CO_3} and has a pH of 7.0. No weak acids or weak bases are present other than the carbonate species, but other anions and cations are present. Neglecting ionic strength effects, calculate the pH after addition of:

1. 10^{-4} mole HCl/liter.

2. 3×10^{-4} mole NaOH/liter.

Solution

1. We will solve this problem using one exact solution and one approximate method of solution.

Equilibria

$$[H^+][OH^-] = K_w = 10^{-14}$$

$$\frac{[HCO_3^-][H^+]}{[H_2CO_3^*]} = K_{a,1} = 10^{-6.3}$$

$$\frac{[CO_3^{2-}][H^+]}{[HCO_3^-]} = K_{a,2} = 10^{-10.3}$$

Mass Balances

$$C_{T,CO_3} = 10^{-3} = [H_2CO_3^*] + [HCO_3^-] + [CO_3^{2-}]$$

C_A = moles/liter of negative charge units other than those of the carbonates.[15]
C_B = moles/liter of positive charge units other than $[H^+]$.[16]

Charge Balance

$$C_B + [H^+] = [HCO_3^-] + 2[CO_3^{2-}] + [OH^-] + C_A$$

or

$$C_B - C_A = [HCO_3^-] + 2[CO_3^{2-}] + [OH^-] - [H^+]$$

From Appendix 1, $\alpha_0 = 0.1667$ and $\alpha_1 = 0.8337$. Thus $[H_2CO_3^*] = \alpha_0 C_{T,CO_3} = 1.7 \times 10^{-4} M$ and $[HCO_3^-] = \alpha_1 C_{T,CO_3} = 8.3 \times 10^{-4} M$.

To calculate $[CO_3^{2-}]$, we enter the values for $[HCO_3^-]$ and $[H^+]$ in the $K_{a,2}$ equation to find

$$[CO_3^{2-}] = \frac{K_{a,2}[HCO_3^-]}{[H^+]} = \frac{10^{-10.3} \times 8.3 \times 10^{-4}}{10^{-7}} = 4.1 \times 10^{-7} M$$

The initial charge balance is then

$$\begin{aligned} C_B - C_A &= [HCO_3^-] + 2[CO_3^{2-}] + [OH^-] - [H^+] \\ &= 8.3 \times 10^{-4} + 2(4.1 \times 10^{-7}) + 10^{-7} - 10^{-7} \\ &= 8.3 \times 10^{-4} \end{aligned}$$

After the addition of 10^{-4} mole HCl/liter:

Exact Solution

The final charge balance after adding $10^{-4} M$ HCl,

$$C_B - C_A = 8.3 \times 10^{-4} - 10^{-4} = 7.3 \times 10^{-4}$$

$$7.3 \times 10^{-4} = [HCO_3^-] + 2[CO_3^{2-}] + [OH^-] - [H^+]$$

Assuming that $7.3 \times 10^{-4} \gg [H^+]$, and $[HCO_3^-] \gg 2[CO_3^{2-}] + [OH^-]$, we obtain

$$[HCO_3^-] = 7.3 \times 10^{-4}$$

In the mass balance,

$$C_{T,CO_3} = 10^{-3} = [H_2CO_3^*] + [HCO_3^-] + [CO_3^{2-}]$$

[15] For example, if SO_4^{2-}, Cl^-, Br^-, . . . are in solution, $C_A = 2[SO_4^{2-}] + [Cl^-] + [Br^-]$ + · · ·

[16] For example, if Na^+, K^+, Ca^{2+} . . . are in solution, $C_B = [Na^+] + [K^+] + 2[Ca^{2+}]$ + · · ·

assume that $[CO_3^{2-}]$ is negligible

$$10^{-3} = [H_2CO_3^*] + 7.3 \times 10^{-4}$$

$$[H_2CO_3^*] = 2.7 \times 10^{-4}$$

Using these values in the $K_{a,1}$ equation because it includes both $[H_2CO_3^*]$ and $[HCO_3^-]$, we obtain

$$[H^+] = \frac{10^{-6.3} \times 2.7 \times 10^{-4}}{7.3 \times 10^{-4}} = 10^{-6.73} \, M$$

From K_w, $[OH^-] = 10^{-7.27}$. From $K_{a,2}$, $[CO_3^{2-}] = 10^{-10.3} \times 7.3 \times 10^{-4}/10^{-6.73} = 10^{-6.71}$. Checking our assumptions,

$$[HCO_3^-] = 7.3 \times 10^{-4} \gg 2[CO_3^{2-}] + [OH^-]$$
$$= 2 \times 10^{-6.71} + 10^{-7.27}$$

and

$$[H_2CO_3^*] + [HCO_3^-] = 10^{-3} \gg [CO_3^{2-}]$$

since

$$[CO_3^{2-}] = 10^{-6.71}$$

We can see that the assumptions are valid.

Approximate Solution

The strong acid added (10^{-4} mole HCl/liter) is assumed to react completely with only the strongest base, HCO_3^-, which is present at significant concentration. (The bases OH^- and CO_3^{2-} are stronger, but their concentrations are at least 2 orders of magnitude lower.)

$$HCO_3^- + H^+ \rightarrow H_2CO_3^*$$

Final $[HCO_3^-]$ = initial $[HCO_3^-] - [H^+]$ added

$$= 8.3 \times 10^{-4} - 10^{-4} = 7.3 \times 10^{-4} \, M$$

Final $[H_2CO_3^*]$ = initial $[H_2CO_3^*] + [H^+]$ added

$$= 1.7 \times 10^{-4} + 10^{-4} = 2.7 \times 10^{-4} \, M$$

From $K_{a,1}$, $[H^+] = 10^{-6.3} \times 2.7 \times 10^{-4}/7.3 \times 10^{-4} = 10^{-6.73}$.

$$pH = 6.73$$

This approximate solution is in excellent agreement with the exact solution. The assumption of complete reaction can be made when the concentration of the predominant acid and its conjugate base pair is much greater than the H^+ or OH^- concentrations. Whether this estimate will be valid can generally be judged very easily using the appropriate pC-pH diagram and the estimated pH; alternatively, the solution obtained in this way can be checked using the set of equations that describe the system.

2. Addition of 3×10^{-4} moles NaOH/liter.
   This problem can be solved in the same fashion as given in (1). The approximate solution is slightly different, however, and it is as follows:

Approximate Solution

The strong base is added in an amount (3×10^{-4} *M*) greater than the initial $[H_2CO_3^*]$ (= 1.7×10^{-4} *M*). We assume that the strong base first reacts stoichiometrically with all of the $H_2CO_3^*$ (the $[H^+]$ is negligible), then any remaining OH^- reacts with HCO_3^-, the weaker acid; thus the reactions,

$$H_2CO_3^* + OH^- \rightarrow HCO_3^- + H_2O \quad \text{and} \quad HCO_3^- + OH^- \rightarrow CO_3^{2-} + H_2O$$

are both important. Also we can deduce that our final solution will not contain a significant amount of $H_2CO_3^*$, since we assume that it has all reacted with the added OH^-. Therefore, we should examine the $K_{a,2}$ equation for relationships between concentrations of the major species, that is, HCO_3^- and CO_3^{2-}.

Final $[H_2CO_3^*]$ = initial $[H_2CO_3^*]$ − equivalent amount of NaOH
$$= 1.7 \times 10^{-4} - 1.7 \times 10^{-4} = 0$$

$[OH^-]$ remaining after reaction with $H_2CO_3^*$ = $[OH^-]$ added
$$- [OH^-] \text{ reacted with } H_2CO_3^* = 3 \times 10^{-4} - 1.7 \times 10^{-4} = 1.3 \times 10^{-4}.$$

Final $[HCO_3^-]$ = initial $[HCO_3^-]$ + $[HCO_3^-]$ formed from $H_2CO_3^*$
$$- [HCO_3^-] \text{ reacted with } OH^- \text{ to form } CO_3^{2-}$$

Final $[HCO_3^-]$ = $8.3 \times 10^{-4} + 1.7 \times 10^{-4} - 1.3 \times 10^{-4}$
$$= 8.7 \times 10^{-4} M$$

Final $[CO_3^{2-}]$ = initial $[CO_3^{2-}]$ + $[CO_3^{2-}]$ formed = 4.1×10^{-7}
$$+ 1.3 \times 10^{-4} = 1.3 \times 10^{-4}$$

From $K_{a,2}$, $[H^+] = 10^{-10.3} \times 8.7 \times 10^{-4}/1.3 \times 10^{-4} = 10^{-9.47}$.

$$pH = 9.47$$

From $K_{a,1}$ and K_w, respectively, $[H_2CO_3^*] = 10^{-6.23}$ and $[OH^-] = 10^{-4.53}$. Substitution into the charge balance and mass balance shows that the values obtained are satisfactory.

The approximate solution approach is especially useful in certain practical problems, such as the one illustrated in the following example.

Example 4-33

A stream has a pH of 8.3 and a C_{T,CO_3} of 3×10^{-3} moles/liter. A wastewater containing 1×10^{-2} *M* H_2SO_4 is to be discharged to the stream. What is the most waste that can be discharged per liter of stream water, and the corresponding dilution ratio, if the pH may not drop below 6.7?

Solution

At pH 8.3, $pH_{HCO_3^-}$, essentially all of the carbonates are in the form of HCO_3^-. At pH 6.7, the ratio of $[HCO_3^-]$ to $[H_2CO_3^*]$ can be determined from the $K_{a,1}$ equation,

$$\frac{[HCO_3^-]}{[H_2CO_3^*]} = \frac{K_{a,1}}{[H^+]} = \frac{10^{-6.3}}{10^{-6.7}} = 2.5$$

Since $[CO_3{}^{2-}]$ is very small, the mass balance becomes

$$C_{T,CO_3} = 3 \times 10^{-3} = [H_2CO_3^*] + [HCO_3^-]$$

Combining the ratio $[HCO_3^-]/[H_2CO_3^*]$ with the mass balance, at pH 6.7,

$$[H_2CO_3^*] = \left(\frac{1}{3.5}\right)(3 \times 10^{-3}) = 8.6 \times 10^{-4}M$$

$$[HCO_3^-] = \left(\frac{2.5}{3.5}\right)(3 \times 10^{-3}) = 2.14 \times 10^{-3}M$$

Assuming a complete reaction between H^+ and HCO_3^-, we obtain

$$H^+ + HCO_3^- \rightarrow H_2CO_3^*$$

8.6×10^{-4} mole H^+ can be added per liter of stream water, since this will convert 8.6×10^{-4} mole of HCO_3^- to $H_2CO_3^*$. The waste is 10^{-2} M H_2SO_4, or 2×10^{-2} M H^+. Thus

$$\frac{8.6 \times 10^{-4}}{2 \times 10^{-2}} = 0.043$$

or 0.043 liter of waste may be added/liter of stream water.

4.13.3. Alkalinity and Acidity

Alkalinity is a measure of the capacity of a water to neutralize strong acid. In natural waters this capacity is attributable to bases such as HCO_3^-, $CO_3{}^{2-}$ and OH^- as well as to species often present in small concentrations such as silicates, borates, ammonia, phosphates, and organic bases. *Acidity* is a measure of the capacity of a water to neutralize strong base. In natural waters this capacity is usually attributable to acids such as $H_2CO_3^*$ and HCO_3^- and sometimes to strong acids, namely, H^+. We can define alkalinity and acidity in two ways: operationally and analytically (or mathematically). Let us first examine the operational definition of alkalinity and then proceed to see how this is related to the mathematical definition when the components of the solution contributing to the alkalinity are only carbonate species and OH^-.

In the determination of total alkalinity a known volume of sample is titrated with a standard solution of a strong acid to a pH value in the approximate range of 4 to 5 and usually in the range 4.5 to 4.8. This endpoint is commonly indicated by the color change of the indicator methyl orange; therefore, the total alkalinity is often referred to as the methyl orange alkalinity. The H^+ added is the stoichiometric amount required for the following reactions:

$$H^+ + OH^- \rightleftharpoons H_2O$$

$$H^+ + HCO_3^- \rightleftharpoons H_2CO_3^*$$

$$2H^+ + CO_3{}^{2-} \rightleftharpoons H_2CO_3^*$$

The pH at the true endpoint of the total alkalinity titration should be that of a solution of $H_2CO_3^*$ and H_2O. Since we will assume that the alkalinity titration is a closed system (no atmosphere), the solution at the endpoint of the titration should be an $H_2CO_3^*$ solution with a C_{T,CO_3} that equals the C_{T,CO_3} of the solution being titrated. We refer to the pH of such a solution as pH_{CO_2}, and we call the solution a C_{T,CO_3}-molar solution of CO_2.

For the carbonate system we can theoretically identify two more significant pH values that occur during the course of an alkalinity titration. These are $pH_{HCO_3^-}$ and $pH_{CO_3^{2-}}$. They represent, respectively, (1) the pH of solution to which has been added the stoichiometric amount of H^+ required to complete the following two reactions:

$$H^+ + OH^- \rightleftharpoons H_2O$$

$$H^+ + CO_3^{2-} \rightleftharpoons HCO_3^-$$

(i.e., the solution will have the pH of a C_{T,CO_3}-molar HCO_3^- solution), and (2) the pH of a solution to which has been added the stoichiometric amount of H^+ required to complete only the following reaction:

$$H^+ + OH^- \rightleftharpoons H_2O$$

In the latter case, the carbonate species will be present all as CO_3^{2-}, or as the form that results when C_{T,CO_3} moles of Na_2CO_3 is added to H_2O (i.e., the pH of a C_{T,CO_3}-molar CO_3^{2-} solution). The value of $pH_{HCO_3^-}$ is about pH 8.3 while $pH_{CO_3^{2-}}$ is generally between pH 10 and 11 and varies with C_{T,CO_3}.

It is possible to determine experimentally the amount of acid required per liter of solution to lower the pH of a solution to $pH_{HCO_3^-}$. This is the carbonate alkalinity. This endpoint can be determined with a pH meter or by the color change of the indicator phenolphthalein so that it is often referred to as the phenolphthalein alkalinity. The amount of acid required to reach $pH_{CO_3^{2-}}$ (the caustic alkalinity) cannot be determined readily in the laboratory because of the poorly defined endpoint, caused by the masking effect of the buffering of water (i.e., the reaction $H^+ + OH^- \rightarrow H_2O$). Caustic alkalinity can be determined by calculation if the carbonate and total alkalinity are known.

Example 4-34

A sample of water from the overflow of the recarbonation basin that follows a precipitation/softening process has a pH of 9.0; 200 ml of the water require 1.1 ml of 0.02 N H_2SO_4 to titrate it to the phenolphthalein endpoint and 22.9 ml of 0.02 N H_2SO_4 to titrate it further to the methyl orange endpoint. Assuming the sample contains no calcite particles, what are the total and carbonate alkalinities of the sample in meq/liter and the total alkalinity in mg/liter as $CaCO_3$?

Carbonate alkalinity = the meq of acid/liter of sample required to reach the
phenolphthalein endpoint
= 1.1 ml × 0.02 eq/liter × 1000 meq/eq ×
1/(200 ml sample volume)
= 0.11 meq/liter

Total alkalinity = the meq of acid required/liter of sample to reach the methyl
orange endpoint
= 24 ml × 0.02 eq/liter × 1000 meq/eq ×
1/(200 ml sample volume)
= 2.4 meq/liter

2.4 meq/liter × 50 mg $CaCO_3$/meq = 120 mg/liter as $CaCO_3$

If we make the previous approximation of complete reaction and
combine it with the assumption that the alkalinity of a water is due solely
to carbonate species and OH^-, we can, to a good approximation, make
some deductions about the initial composition of a solution from the
alkalinity titrations. Thus we assume that by pH 8.3 the reactions

$$OH^- + H^+ \rightarrow H_2O$$

and

$$CO_3^{2-} + H^+ \rightarrow HCO_3^-$$

are complete, and by about pH 4.3 to 4.7 the reaction

$$HCO_3^- + H^+ \rightarrow H_2CO_3^*$$

is complete. We can also deduce that each mole of CO_3^{2-} present will
consume one H^+ when the solution is titrated to pH 8.3 and another H^+ as
it is titrated from pH 8.3 to approximately pH 4.3. If the volume of acid to
reach pH 8.3 (V_p) was equal to the volume of acid required to proceed from
pH 8.3 to pH 4.3 (V_{mo}), we would know that the original solution contained
only CO_3^{2-} as the major alkalinity species.

If, when we put phenolphthalein indicator into the solution, it immediately became colorless (or if the pH was below about 8.3) and we then
required an acid volume of V_{mo} to reach about pH 4.3, the original solution
would have contained only HCO_3^- as the major species contributing to
alkalinity.

If the initial solution required V_p ml of acid to reach pH 8.3 but no
further acid addition was required to reach pH 4.3, the alkalinity was due
to OH^- alone. By similar reasoning we can show that if $V_p > V_{mo}$, then the
major alkalinity species are OH^- and CO_3^{2-}; if $V_{mo} > V_p$, then the major
alkalinity species are CO_3^{2-} and HCO_3^-. The third possible combination
of major alkalinity-causing species, OH^- and HCO_3^-, does not exist
because there is no pH range over which these two species are concurrently
the major alkalinity species (see Fig. 4-18). It should be emphasized that

the concentrations obtained by this procedure are approximate and significant errors are introduced if the solution pH is greater than about 9.5. These relationships are tabulated in Table 4-9.

Example 4-34 (continued)

We can now determine the major alkalinity species from the titration values.

$$V_p = 1.1 \text{ ml}$$

$$V_{mo} = 22.9 \text{ ml}$$

Since $V_{mo} > V_p$, the major species present are HCO_3^- and CO_3^{2-}. V_p = ml acid required to titrate the CO_3^{2-} according to the reaction: $H^+ + CO_3^{2-} \rightarrow HCO_3^-$. When V_p is added, 1 H^+ is added for each CO_3^{2-}; thus the acid added in eq/liter = the concentration of CO_3^{2-} in moles/liter. From Table 4-9,

$$[CO_3^{2-}] = V_p \text{ ml} \times \text{normality of titrant in eq/liter} \times \frac{1}{(\text{sample volume, ml})}$$

$$\times \frac{1 \text{ mole } CO_3^{2-}}{\text{eq } H^+}$$

$$= 1.1 \times 0.02 \times 1/200$$

$$[CO_3^{2-}] = 0.11 \times 10^{-3} \text{mole/liter}$$

Since $H^+ + HCO_3^- \rightarrow H_2CO_3^*$, the HCO_3^- originally present plus the HCO_3^-

TABLE 4-9   Approximate Concentrations of Carbonate Species and Hydroxide from Alkalinity Titration Data

Condition	Predominant Form of Alkalinity	Approximate Concentration, M
$V_p = V_{mo}$	CO_3^{2-}	$[CO_3^{2-}] = V_p \times N \times V^{-1}$
$V_p = 0$	HCO_3^-	$[HCO_3^-] = V_{mo} \times N \times V^{-1}$
$V_{mo} = 0$	OH^-	$[OH^-] = V_p \times N \times V^{-1}$
$V_{mo} > V_p$	CO_3^{2-} and HCO_3^-	$[CO_3^{2-}] = V_p \times N \times V^{-1}$
		$[HCO_3^-] = (V_{mo} - V_p) \times N \times V^{-1}$
$V_p > V_{mo}$	OH^- and CO_3^{2-}	$[CO_3^{2-}] = V_{mo} \times N \times V^{-1}$
		$[OH^-] = (V_p - V_{mo}) \times N \times V^{-1}$

Where:

V_p = volume of titrant to the phenolphthalein endpoint, ml
V_{mo} = volume of titrant from the phenolphthalein endpoint to the methyl orange endpoint, ml
V = sample volume, ml
N = normality of titrant, eq/liter

Note: The accuracy of the OH^- and CO_3^{2-} concentrations obtained from this table depend upon the value of C_{T,CO_3}. Accuracy is acceptable for $C_{T,CO_3} > 2 \times 10^{-3} M$; however, results will be significantly in error if $C_{T,CO_3} < 1 \times 10^{-3} M$.

formed by titration of the sample to the phenolphthalein endpoint will be converted to $H_2CO_3^*$ when the sample is titrated to the methyl orange endpoint. Thus, from Table 4-9,

$$[HCO_3^-] = (V_{mo} - V_p)\text{ml} \times \text{normality of titrant in eq/liter}$$
$$\times 1 \text{ mole } HCO_3^-/eq \times 1/\text{sample volume in ml}$$
$$= (22.9 - 1.1) \times 0.02 \times \frac{1}{200}$$
$$= 2.18 \times 10^{-3} \text{mole/liter}$$

We can make similar definitions and calculations concerning acidity measurements. For a water in which acidity is contributed solely by H^+ and carbonate species, the endpoint of the total acidity titration is at a pH where a stoichiometric amount of OH^- has been added to complete the following three reactions:

$$OH^- + H^+ \rightleftharpoons H_2O$$
$$OH^- + HCO_3^- \rightleftharpoons CO_3^{2-} + H_2O$$
$$2OH^- + H_2CO_3^* \rightleftharpoons CO_3^{2-} + 2H_2O$$

The species present at the endpoint pH are those which result when Na_2CO_3 equal to the C_{T,CO_3} of the sample, is added to pure water. This pH is $pH_{CO_3^{2-}}$ and is generally between 10 and 11. The carbon dioxide acidity endpoint is at $pH_{HCO_3^-}$, about 8.3. The mineral acidity endpoint is at pH_{CO_2}, about pH 4 to 5. In the analysis of waters the two endpoints used in acidity determinations are pH 8.3 ($pH_{HCO_3^-}$) and pH 4–5 (pH_{CO_2}). As in the caustic alkalinity determination the $pH_{CO_3^{2-}}$ endpoint is ill defined. Since the endpoints at pH 8.3 and pH 4 to 5 correspond, respectively, to the color changes of the indicators phenolphthalein and methyl orange, carbon dioxide acidity is referred to as phenolphthalein acidity and mineral acidity is often called methyl orange acidity.[17]

Using the assumption of complete reactions, we can deduce that adding base to achieve a pH of 4 to 5 will allow the completion of the reaction

$$H^+ + OH^- \rightarrow H_2O$$

and the carbonate species will be in the form that exists when CO_2 is added to pure water. Addition of strong base to pH 8.3 will complete the reactions

$$H^+ + OH^- \rightarrow H_2O$$

and

$$H_2CO_3^* + OH^- \rightarrow HCO_3^- + H_2O$$

[17] For wastewaters in which weak acids and bases other than the carbonate system are present, pH 3.7 is used instead of pH 4 to 5. See *Standard Methods*, 14th ed., 1975.

Figure 4-19 shows the relationships between the acidity and alkalinity titration curves and the pC-pH diagram. It shows that 1 equivalent fraction of strong acid, ($f = 1$), must be added to lower the pH from $pH_{CO_3{}^{2-}}$ to $pH_{HCO_3{}^-}$, the carbonate alkalinity endpoint. An additional equivalent fraction must be added to lower the pH from $pH_{HCO_3{}^-}$ to pH_{CO_2}, the total

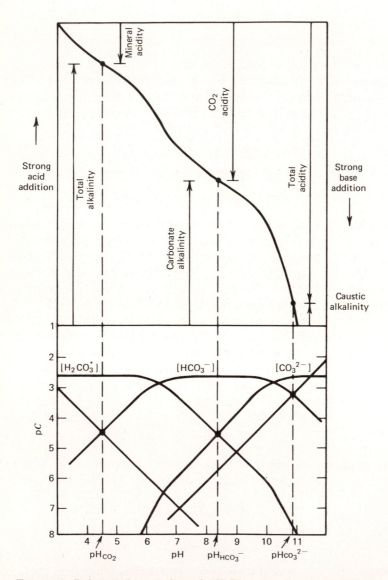

Fig. 4-19. Relationship of the pC-pH Diagram to the titration curve for the carbonate system at 25° C.

alkalinity endpoint. These relationships show that for the carbonate system

$$\text{total alkalinity - carbonate alkalinity} = C_{T,CO_3}$$

Example 4-35

Two hundred and fifty ml of a freshly sampled well water (approximately pH 6.7) requires the addition of 5.8 ml 0.1 N NaOH to raise its pH to 8.3; the same volume of the water requires the addition of 12.2 ml (V_{mo}) of 0.1 N HCl to lower its pH to 4.3. What are (1) the total alkalinity and total acidity, (2) the carbon dioxide acidity and the carbonate alkalinity, and (3) the approximate concentrations of the major carbonate species?

Solution

1. Total alkalinity is based on the volume of titrant required to lower the pH of the sample to the methyl orange endpoint, which is approximated by pH 4.3 in this problem.

    Total alkalinity = 12.2 ml $\times$ 100 meq/liter $\times$ 1/250 ml sample volume
    = 4.88 meq/liter

    Total acidity is equivalent to the amount of strong base which must be added per liter to raise the pH to $pH_{CO_3^{2-}}$. In doing so the following reactions will essentially be completed,

    $$H^+ + OH^- \rightarrow H_2O$$

    $$H_2CO_3^* + OH^- \rightarrow HCO_3^- + H_2O$$

    and

    $$HCO_3^- + OH^- \rightarrow CO_3^{2-} + H_2O$$

    Although we are not given the volume of titrant to $pH_{CO_3^{2-}}$, we can deduce this from the information given. The pH of the sample is approximately 6.7; thus mineral acidity is absent and we can disregard the first reaction. It takes 5.8 ml of 0.1 N NaOH/250 ml sample to complete the second reaction and raise the pH to 8.3. To complete the third reaction, a further 5.8 ml/250 ml is required to convert the HCO_3^- formed via the second reaction to CO_3^{2-}. In addition, the HCO_3^- present in the sample (which titrated as total alkalinity) will need to be converted to CO_3^{2-} during the total acidity titration. This will require an additional 12.2 ml of 0.1 N NaOH/250 ml sample, an amount equivalent to the strong acid required in the total alkalinity titration. The total amount of base added then is 5.8 + 5.8 + 12.2 = 23.8 ml/250 ml sample.

    Total acidity = 23.8 ml $\times$ 100 meq/liter $\times$ $\dfrac{1}{250}$ ml sample volume
    = 9.52 meq/liter [18]

[18] Note that total acidity can also be calculated with the aid of Fig. 4-19. That is, total acidity = CO_2 acidity + 1 eq fraction = CO_2 acidity + C_{T,CO_3}, where C_{T,CO_3} = CO_2 acidity + total alkalinity. Since 5.8 ml of 0.1 N titrant was required to titrate to the CO_2 acidity endpoint and 12.2 ml of 0.1 N titrant was required to titrate to the total alkalinity endpoint, total acidity = (5.8 + 5.8 + 12.2) $\times$ 100 eq/liter $\times$ (1/250 sample volume) = 9.52 meq/liter.

2. Carbon dioxide acidity corresponds to the amount of base required to raise pH to 8.3.

$$CO_2 \text{ acidity} = 5.8 \text{ ml} \times 100 \text{ meq/liter} \times \frac{1}{250} \text{ ml sample volume}$$

$$= 2.32 \text{ meq/liter}$$

3. Since $V_p = 0$ (initial pH < 8.3) and $V_{mo} = 12.2$ ml, the major total alkalinity species is HCO_3^-. Thus

$$[HCO_3^-] = V_{mo} \text{ ml} \times \text{normality of titrant in meq/liter}$$

$$\times 1/\text{ml sample} \times 1 \text{ mole } HCO_3^-/\text{eq of } H^+$$

$$= 12.2 \times 100 \times \frac{1}{250}$$

$$= 4.88 \text{ m}M/\text{liter}$$

Since the mineral acidity $= 0$ (initial pH > 4.3) and base (8 ml) is required to reach pH 8.3, the major acidity species is $H_2CO_3^*$. Also note that 1 OH^- is required to convert each $H_2CO_3^*$ to HCO_3^-.

$$[H_2CO_3^*] = 5.8 \text{ ml} \times \text{normality of titrant in meq/liter} \times 1/\text{ml sample volume}$$

$$\times 1 \text{ mole } H_2CO_3^*/\text{eq of } OH^-$$

$$= 5.8 \times 100 \times \frac{1}{250}$$

$$= 2.32 \text{ m}M$$

4.13.4.   Alkalinity and Acidity Endpoints

We have been careful in the preceding discussion of alkalinity and acidity titrations and calculations to refer to the pH of the various endpoints[19] as "approximate values." The actual values that correspond to pH_{CO_2}, $pH_{HCO_3^-}$, and $pH_{CO_3^{2-}}$ are not truly fixed values, rather they vary with the C_{T,CO_3} in solution. If we treat the titrations as closed systems, the C_{T,CO_3} at the endpoint will be the same as the C_{T,CO_3} in the initial solution. This appears to be a reasonable approach if the solution is not shaken vigorously and if the titration is conducted rapidly. The variation in equivalence point pH values is shown in Fig. 4-20.

Reference to Fig. 4-20 and the statement in *Standard Methods* (14th ed.) that "the following pH values are suggested as the endpoints for the corresponding alkalinity concentrations as calcium carbonate: pH 5.1 for total alkalinities of about 50 mg/liter (1.0 meq/liter), pH 4.8 for 150 mq/liter (3 meq/liter) and pH 4.5 for 500 mg/liter (10 meq/liter)" shows that this variation is significant indeed for pH_{CO_2}. Figure 4-20 shows that $pH_{CO_3^{2-}}$ also is strongly dependent on C_{T,CO_3} while $pH_{HCO_3^-}$ is much less dependent.

---

[19] The reader should keep in mind the difference between endpoint pH and equivalence point pH. The equivalence point pH is the theoretical pH at which a titration is complete while the endpoint pH is an approximation of the equivalence point pH.

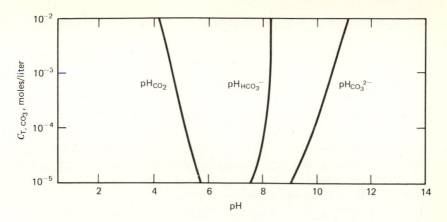

Fig. 4-20. The pH values for alkalinity and acidity titration endpoints in a closed system as a function of the initial total carbonate concentration.

The equivalence point pH values can be calculated exactly using previously presented methods, especially the one employing the pC-pH diagram (e.g., see Section 4-12-2). Here we will examine approximate methods for their calculation. For pH_{CO_2}, we will check the rationale behind the *Standard Methods* statement quoted previously.

At the total alkalinity (and mineral acidity) equivalence point, we assume that there is present an $H_2CO_3^*$ solution of concentration C_{T,CO_3} which is equal to C_{T,CO_3} of the sample prior to titration. Therefore, at the endpoint,

$$[H_2CO_3^*] = C_{T,CO_3}$$

Considering only the $K_{a,1}$ equation

$$K_{a,1} = \frac{[H^+][HCO_3^-]}{[H_2CO_3^*]}$$

and assuming $[H^+] = [HCO_3^-]$, which is the approximate proton condition for an $H_2CO_3^*$ solution,

$$[H^+] = (K_{a,1}C_{T,CO_3})^{1/2}$$

For a total alkalinity of 50 mg/liter as $CaCO_3$ (1 meq/liter) and an original sample pH between 7.5 and 9,

$$[HCO_3^-] = 1 \times 10^{-3} \ M$$

Titration to pH_{CO_2} will produce a $10^{-3} \ M \ H_2CO_3^*$ solution with an $[H^+]$ of

$$[H^+] = (10^{-6.3} \times 10^{-3})^{1/2} = 10^{-4.65}$$

$$pH = 4.65$$

Similarly, for a total alkalinity of 150 mg/liter as $CaCO_3$ (3 meq/liter), we obtain

$$pH_{CO_2} = 4.5$$

and for total alkalinity of 500 mg/liter as $CaCO_3$ (10 meq/liter),

$$pH_{CO_2} = 4.2$$

The *Standard Methods* endpoint values are higher than the closed system pH_{CO_2} values, perhaps indicating that some allowance has been made for the escape of CO_2 during the titration that would result in a lower $H_2CO_3^*$ concentration at the endpoint.

The variation of pH_{CO_2} with C_{T,CO_3} has led to alkalinity titration methods (employed in Europe) in which the majority of the acid is added, the solution is boiled to expel CO_2, and then, after cooling, the small amount of remaining acid is added. The pH of the endpoint is then at about pH 5.1 and is independent of alkalinity, since it is governed solely by the small amount of CO_2 generated from the last small acid addition plus the CO_2 in the atmosphere. If a sample is warmed and bubbled with nitrogen gas (free of CO_2) during titration, the equivalence point would be the pH, where $[H_2CO_3^*] = 0$ and $[H^+] = [OH^-]$ which at 25°C is pH 7.

The $pH_{HCO_3^-}$ equivalence point (for carbonate alkalinity and carbon dioxide acidity) is at about pH 8.3 and varies relatively less than pH_{CO_2} with C_{T,CO_3}.

Since at $pH_{HCO_3^-}$ we have an HCO_3^- solution, the proton condition is

$$[H^+] + [H_2CO_3^*] = [OH^-] + [CO_3^{2-}]$$

or approximately, since at pH > 7, $[H^+] < [OH^-]$,

$$[H_2CO_3^*] = [OH^-] + [CO_3^{2-}]$$

Substituting for these terms from $K_{a,1}$, K_w, and $K_{a,2}$ respectively, we find

$$\frac{[H^+][HCO_3^-]}{K_{a,1}} = \frac{K_w}{[OH^-]} + \frac{[HCO_3^-]K_{a,2}}{[H^+]}$$

Solving for $[H^+]$ and recognizing $[HCO_3^-] = C_{T,CO_3}$, we find that

$$[H^+] = \left(\frac{K_w K_{a,1}}{C_{T,CO_3}}\right)^{1/2} + (K_{a,1} K_{a,2})^{1/2}$$

The first term becomes insignificant compared to the second term when $C_{T,CO_3} > 10^{-3}M$. Thus for $C_{T,CO_3} = 10^{-2} M$,

$$[H^+] = (10^{-6.3} \times 10^{-10.3})^{1/2} = 10^{-8.3}$$

$$pH = 8.3$$

For the range of alkalinities in which most natural waters fall (10^{-3} to

10^{-2} M or 50 to 500 mg/liter as $CaCO_3$), $pH_{HCO_3^-}$ will be very close to 8.3. At extremely low alkalinities, as Fig. 4-20 shows, the pH will be less than 8.3.

The $pH_{CO_3^{2-}}$ endpoint (for caustic alkalinity and total acidity) is between pH 9–11 and varies significantly with C_{T,CO_3} as shown in Fig. 4-20. Since at $pH_{CO_3^{2-}}$ we have a CO_3^{2-} solution, the proton condition is

$$2[H_2CO_3^*] + [HCO_3^-] + [H^+] = [OH^-]$$

At pH $\gg$ 7, $[H_2CO_3^*]$ and $[H^+]$ are negligible so that $[HCO_3^-] = [OH^-]$. Substituting from $K_{a,2}$ and K_w and recognizing $[CO_3^{2-}] = C_{T,CO_3}$ when $C_{T,CO_3} > 10^{-2}$, we find

$$[H^+] = \left(\frac{K_w K_{a,2}}{C_{T,CO_3}}\right)^{1/2}$$

For values of $C_{T,CO_3} < 10^{-2}$, $pH_{CO_3^{2-}}$ can most easily be determined by using the pC-pH diagram.

4.13.5. Analytical Definition of Alkalinity and Acidity

Alkalinity and acidity are defined on the basis of the constituent bases and acids, respectively, which contribute to the alkalinity and acidity. The various alkalinity and acidity measurements are defined in terms of the proton conditions of the equivalence points to which the alkalinity and acidity determinations are measured. Alkalinity requires the addition of strong acid (H^+). Thus the presence of alkalinity with respect to a particular endpoint indicates an excess of species containing less protons than the reference (or endpoint) species over species containing more protons than the reference species. Acidity requires the addition of strong base (OH^-), which is tantamount to removing protons. Thus the presence of acidity with respect to a particular endpoint indicates an excess of species containing more protons than the reference species over species containing less protons than the reference species. Table 4-10 summarizes the various equivalence points, their proton conditions, and the associated mathematical definitions of acidity and alkalinity for a water containing only the carbonate species, OH^- and H^+ as buffering components. Each of the mathematical definitions of alkalinity and acidity can be expressed in terms of α values; see Table 4-11.

From these expressions we can see that for systems which are buffered by carbonate species, H^+, and OH^- alone, there is a definite relationship between total alkalinity, pH, and C_{T,CO_3} and between total acidity, pH, and C_{T,CO_3}. For such systems, if two of these quantities are measured, the third can be calculated. Since many natural waters can be treated as though they were solely buffered by carbonate species, H^+, and OH^-, the determination of initial pH and either total alkalinity or total acidity will allow the computation of C_{T,CO_3} and all species.

TABLE 4-10 Alkalinity and Acidity Equations and Equivalence Points

Equivalence Point	Reference Species	Proton Condition	Definition (Alkalinity and Acidity in eq/liter)
pH_{CO_2}	$H_2CO_3^*$ H_2O	$[H^+] = [HCO_3^-] + 2[CO_3^{2-}] + [OH^-]$	Total alkalinity $= [HCO_3^-] + 2[CO_3^{2-}] + [OH^-] - [H^+]$ Mineral acidity $= [H^+] - [HCO_3^-] - 2[CO_3^{2-}] - [OH^-]$
$pH_{HCO_3^-}$	HCO_3^- H_2O	$[H^+] + [H_2CO_3^*] = [CO_3^{2-}] + [OH^-]$	Carbonate alkalinity $= [CO_3^{2-}] + [OH^-] - [H_2CO_3^*] - [H^+]$ Carbon dioxide acidity $= [H^+] + [H_2CO_3^*] - [CO_3^{2-}] - [OH^-]$
$pH_{CO_3^{2-}}$	CO_3^{2-} H_2O	$[H^+] + [HCO_3^-] + 2[H_2CO_3^*] = [OH^-]$	Caustic alkalinity $= [OH^-] - [HCO_3^-] - 2[H_2CO_3^*] - [H^+]$ Total acidity $= [H^+] + [HCO_3^-] + 2[H_2CO_3^*] - [OH^-]$

TABLE 4-11 Alkalinity and Acidity in Terms of C_{T,CO_3} and α Values

Parameter	
Total alkalinity	$C_{T,CO_3}(\alpha_1 + 2\alpha_2) + \dfrac{K_w}{[H^+]} - [H^+]$
Carbonate alkalinity	$C_{T,CO_3}(\alpha_2 - \alpha_0) + \dfrac{K_w}{[H^+]} - [H^+]$
Caustic alkalinity	$\dfrac{K_w}{[H^+]} - [H^+] - C_{T,CO_3}(\alpha_1 + 2\alpha_0)$
Total acidity	$C_{T,CO_3}(\alpha_1 + 2\alpha_0) + [H^+] - \dfrac{K_w}{[H^+]}$
Carbon dioxide acidity	$C_{T,CO_3}(\alpha_0 - \alpha_2) + [H^+] - \dfrac{K_w}{[H^+]}$
Mineral acidity	$[H^+] - \dfrac{K_w}{[H^+]} - C_{T,CO_3,2}$

Note: If ionic strength effects are important, ^cK_w should be used in place of K_w, and the α values should be calculated using ^cK_a rather than K_a.

Example 4-36

What is the total inorganic carbon concentration (in mg carbon/liter) in a water buffered by the carbonate system, H^+, and OH^-? The pH on sampling is 7.8 and the water requires 13.7 ml 0.02 N HCl to lower the pH of 100 ml to 4.5. What is the major component of the alkalinity? (Neglect ionic strength effects.)

$$\text{Total alkalinity (eq/liter)} = 13.7 \text{ ml} \times 0.02 \text{ eq/liter} \times 1/100 \text{ ml sample volume}$$
$$= 2.74 \times 10^{-3} \text{eq/liter}$$

From Table 4-10,

$$2.74 \times 10^{-3} = C_{T,CO_3}(\alpha_1 + 2\alpha_2) + \frac{K_w}{[H^+]} - [H^+]$$

$$\cong C_{T,CO_3}(\alpha_1 + 2\alpha_2)$$

From Appendix 1, $\alpha_0 = 0.031$, $\alpha_1 = 0.969$, and $\alpha_2 \cong 0$. Thus

$$C_{T,CO_3} = \frac{2.74 \times 10^{-3}}{\alpha_1}$$

$$= \frac{2.74 \times 10^{-3}}{0.969}$$

$$= 2.83 \times 10^{-3}$$

Therefore,

$$\text{Total inorganic carbon concentration} = 2.83 \times 10^{-3} \text{ mole/liter}$$
$$\times 12,000 \text{ mg carbon/mole}$$
$$= 33.9 \text{ mg C/liter}$$

To determine the major component of alkalinity, we must determine the concentrations of each of the alkalinity species in the original sample. Knowing pH = 7.8, we obtain

$$[H^+] = 10^{-7.8}$$

$$[OH^-] = 10^{-6.2}$$

$$[H_2CO_3^*] = C_{T,CO_3}\alpha_0 = 2.83 \times 10^{-3} \times 0.031 = 8.8 \times 10^{-5} M$$

$$[HCO_3^-] = C_{T,CO_3}\alpha_1 = 2.83 \times 10^{-3} \times 0.969 = 2.74 \times 10^{-3} M$$

From $K_{a,2}$,

$$[CO_3^{2-}] = \frac{2.74 \times 10^{-3} \times 10^{-10.3}}{10^{-7.8}}$$

$$= 8.7 \times 10^{-6} M$$

The major species contributing to alkalinity is thus HCO_3^-.

Example 4-37

We can use the mathematical definition of alkalinity to provide an exact solution to the Example 4-34 worked previously by the approximate method of complete reactions. We found that the total alkalinity = 2.4×10^{-3} eq/liter and the initial pH was 9.0.

From Table 4-11,

$$\text{Total alkalinity} = C_{T,CO_3}(\alpha_1 + 2\alpha_2) + \frac{K_w}{[H^+]} - [H^+]$$

At pH 9.0, from Appendix 1, $\alpha_1 = 0.952$ and $\alpha_2 = 0.048$. Thus

$$\text{Total alkalinity} = 2.4 \times 10^{-3} = C_{T,CO_3}(0.952 + 2(0.048)) + \frac{10^{-14}}{10^{-9}} - 10^{-9}$$

$$C_{T,CO_3} = 2.28 \times 10^{-3}$$

Now

$$[CO_3^{2-}] = \alpha_2 C_{T,CO_3} = (0.048)(2.28 \times 10^{-3}) = 1.1 \times 10^{-4}$$

$$[HCO_3^-] = \alpha_1 C_{T,CO_3} = (0.952)(2.28 \times 10^{-3}) = 2.17 \times 10^{-3}$$

and from $K_{a,1}$,

$$[H_2CO_3^*] = 10^{-9} \times 2.17 \times \frac{10^{-3}}{10^{-6.3}} = 4.3 \times 10^{-6}$$

These values check well with those obtained previously using the approximate equations in Table 4-9, where

$$[H_2CO_3^*] = 0$$

$$[HCO_3^-] = 2.18 \times 10^{-3} M$$

$$[CO_3^{2-}] = 1.1 \times 10^{-4} M$$

The relationship between alkalinity, C_{T,CO_3} and pH has been used by Deffeyes to construct a diagram relating alkalinity to C_{T,CO_3} at a series of constant pH values. This diagram, shown in Fig. 4-21, can be used for a wide variety of calculations involving the addition and removal of various species from the carbonate system.

If two of the three characteristics, pH, C_{T,CO_3}, and total alkalinity, of a water are given, the third characteristic can be determined using the diagram.

Example 4-38

A water has a pH of 8.5 and a total alkalinity of 1.8 meq/liter. What is its C_{T,CO_3}? The temperature is 25°C.

Solution

From Fig. 4-26, at the intersection of the pH 8.5 line and total alkalinity = 1.8 meq/liter, $C_{T,CO_3} = 1.8 \times 10^{-3} M$ (see point A).

In the Deffeyes diagram coordinate system, the addition of strong acid (C_A, moles/liter) and strong base (C_B, moles/liter) can be represented by

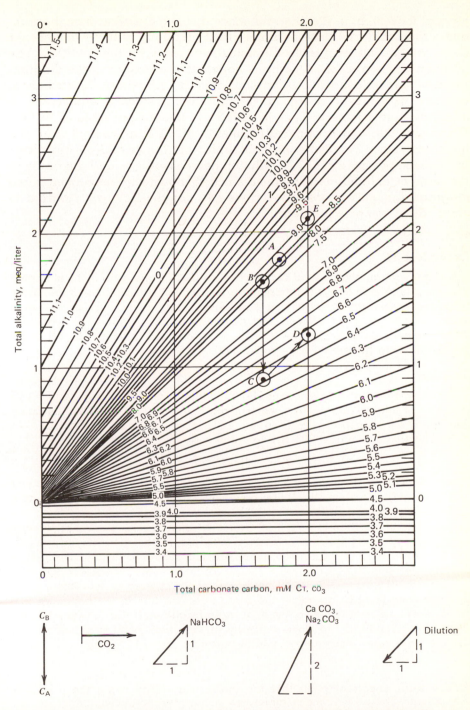

Fig. 4-21. The Deffeyes diagram for 25°C. From K. S. Deffeyes, Lim. & Ocean., 10: 412(1965). Reproduced by permission of *Limnology and Oceanography*.

vertical movement in the diagram, since this is a closed system diagram and strong acid and base addition affect alkalinity without influencing C_{T,CO_3}, given that negligible dilution takes place by the addition. The addition or removal of CO_2 causes horizontal movement in the diagram because this influences C_{T,CO_3} without affecting alkalinity.[20] Bicarbonate addition or removal causes a 45° movement, since 1 equivalent of alkalinity and 1 mole of C_{T,CO_3} is added or removed per mole of HCO_3^-. Carbonate addition or removal causes a 60° (2/1) movement from the horizontal, since 2 equivalents of alkalinity and 1 mole of C_{T,CO_3} are added or removed per mole of CO_3^{2-}. Dilution decreases C_{T,CO_3} and alkalinity to the same extent and therefore results in a 45° downward movement in the diagram. Armed with these tools, we can use the Deffeyes diagram to solve rather complex problems.

Example 4-39

A water with an alkalinity of 82.5 mg/liter as $CaCO_3$ and a pH of 8.0 is treated with $Cl_{2(g)}$ to oxidize $NH_{3(aq)}$ to nitrogen gas (breakpoint chlorination). If the initial ammonia concentration is 3.5 mg NH_3-N/liter, and the product of ammonia oxidation is N_2 gas, what is the pH, total alkalinity, and C_{T,CO_3} following treatment? What combination of NaOH and $NaHCO_3$ must be used to adjust the pH of the treated water to 9.0 and the C_{T,CO_3} to 2.0 mM?

Solution

The original water sample contains (82.5 mg/liter as $CaCO_3$/(50 mg/meq)) = 1.65 meq/liter of total alkalinity. The water characteristics, 1.65 meq/liter alkalinity and pH 8.0 are located at point B on Fig. 4-21.

The reaction of chlorine with NH_3 to produce $N_{2(g)}$ is

$$2NH_3 + 3Cl_2 \rightarrow N_{2(g)} + 6H^+ + 6Cl^-$$

Thus 1 mole of NH_3–N produces 3 moles of H^+ (strong acid) or 14 g NH_3–N produce 3 moles H^+. Thus 3.5 mg NH_3–N/liter produces $3 \times (3.5 \times 10^{-3}$ g/liter)/(14 g/mole) or 0.75×10^{-3} mole of H^+/liter.

The effect of addition of 0.75×10^{-3} mole H^+/liter of solution, C_A, is to move vertically downward from point B, 0.75 meq/liter in the diagram to point C. The characteristics of the treated water are thus pH = 6.4, total alkalinity = 0.9 meq/liter, and $C_{T,CO_3} = 1.65$ mM.

The point pH 9.0 and $C_{T,CO_3} = 2.0$ mM lies above an upward 45° movement in the diagram from point C. To adjust the sample composition to the point, we first satisfy the C_{T,CO_3} requirement of the final water by adding $NaHCO_3$ to point D and then adding NaOH (C_B) to move vertically upward from point D to point E, the desired final water. The required additions are

---

[20] This is an important point that the reader can verify using the alkalinity equations, Table 4-8. It is also an important consideration in sampling because CO_2 release or uptake after a sample is taken but before total alkalinity is measured will not affect the total alkalinity value.

$$NaHCO_3 (C \text{ to } D) = 0.35 \, mM$$

$$NaOH (D \text{ to } E) = 0.75 \, mM$$

Since $NaOH + NaHCO_3 \rightarrow Na_2CO_3$, 0.35 mM Na_2CO_3 and $(0.75 - 0.35) = 0.4 \, mM$ NaOH could be added instead to produce the same effect.

Further examination of the expressions in Table 4-10 allows the following useful relationships to be established:

$$\text{Total alkalinity} - \text{carbonate alkalinity} = C_{T,CO_3} \qquad (4\text{-}118)$$

Proof

From Table 4-10,

$$[HCO_3^-] + 2[CO_3^{2-}] + [OH^-] - [H^+] + [H_2CO_3^*] + [H^+] - [OH^-] - [CO_3^{2-}]$$
$$= [H_2CO_3^*] + [HCO_3^-] + [CO_3^{2-}]$$
$$= C_{T,CO_3}$$

Similarly,

$$\text{Total alkalinity} + \text{carbonate dioxide acidity} = C_{T,CO_3} \qquad (4\text{-}119)$$

and

$$\text{Carbon dioxide acidity} - \text{mineral acidity} = C_{T,CO_3} \qquad (4\text{-}120)$$

In each of these equations both of the capacity parameters can be readily determined by titration with a strong acid or base. The relationship in Eq. 4-118 is applicable for waters with pH > 8.3; Eq. 4-119 is useful for waters in the pH range 8.3 to 4.3 and Eq. 4-120 applies to waters with pH below 4.3. The diagram in Fig. 4-24 also shows these relationships.

By employing these relationships, we can compute C_{T,CO_3}, the initial pH, and individual alkalinity or acidity species based solely on the two appropriate titration values.

Example 4-40

A 25°C water sample requires 10 ml 0.02 N H_2SO_4 to lower the pH of 100 ml to 4.3. The same volume of water requires 4 ml 0.02 N NaOH to raise its pH to 8.3. Compute the total inorganic carbon content, C_{T,CO_3}, the initial pH, and the concentrations of $[H^+]$, $[OH^-]$, $[H_2CO_3^*]$, $[HCO_3^-]$, and $[CO_3^{2-}]$ in the initial water sample assuming that alkalinity and acidity are solely due to the carbonate system, H^+ and OH^-.

Solution

Titrations are for total alkalinity and carbon dioxide acidity.

$$\text{Total alkalinity} = 10 \text{ ml} \times 0.02 \text{ eq/liter} \times \frac{1}{100} \text{ ml sample volume}$$
$$= 2 \times 10^{-3} \text{ eq/liter}$$
$$\text{Carbon dioxide acidity} = 4 \text{ ml} \times 0.02 \text{ eq/liter} \times \frac{1}{100} \text{ ml sample volume}$$
$$= 8 \times 10^{-4} \text{ eq/liter}$$

From Eq. 4-119,

$$\text{Total alkalinity} + \text{carbon dioxide acidity} = C_{T,CO_3}$$

$$C_{T,CO_3} = 2.8 \times 10^{-3} M$$

From Table 4-11,

$$\text{Total alkalinity} = C_{T,CO_3}(\alpha_1 + 2\alpha_2) + \frac{K_w}{[H^+]} - [H^+]$$

$$2 \times 10^3 = 2.8 \times 10^{-3}(\alpha_1 + 2\alpha_2) + \frac{10^{-14}}{[H^+]} - [H^+]$$

Assume $2\alpha_2 \ll \alpha_1$, since the pH is less than 8.3, and $C_{T,CO_3} \gg [OH^-] - [H^+]$.

$$\alpha_1 = \frac{2 \times 10^{-3}}{2.8 \times 10^{-3}} = 0.71$$

From Appendix 1, the pH = 6.7. At pH 6.7, from Appendix 1, $\alpha_0 = 2.85$, $\alpha_1 = 0.715$. Thus

$$[H_3CO_3^*] = \alpha_0 C_{T,CO_3} = 0.8 \times 10^{-3} M$$

$$[HCO_3^-] = \alpha_1 C_{T,CO_3} = 2 \times 10^{-3} M$$

Within the limits of accuracy of Appendix 1, $\alpha_2 = 0$. However, we can use $[HCO_3^-]$ and $[H^+]$ as calculated together with $K_{a,2}$ to find $[CO_3^{2-}] = 5 \times 10^{-7}$.

Alternatively, knowing the total alkalinity ($= 2 \times 10^{-3}$) and having calculated C_{T,CO_3} ($= 2.8 \times 10^{-3}$), we can use the Deffeyes diagram, Fig. 4-21, to obtain the pH of 6.75.

If bases other than the carbonates and OH^- are present that react with protons in the pH range of the alkalinity and acidity titrations, we cannot use these calculations and diagrams without modifications. In addition the analytical definition of acidity and alkalinity must be changed to incorporate these bases. The following reactions are important in this regard in waters and wastewaters.

	$-\log K_a$ (25°C)
Silica (silicic acid):	
$H_4SiO_4 \rightleftharpoons H_3SiO_4^- + H^+$	
Boric acid:	9.3
$H_3BO_3 + H_2O \rightleftharpoons B(OH)_4^- + H^+$	
Dihydrogen phosphate:	7.2
$H_2PO_4^- \rightleftharpoons HPO_4^{2-} + H^+$	
Hydrogen sulfide:	7.1
$H_2S_{(aq)} \rightleftharpoons HS^- + H^+$	
Ammonium ion:	9.3
$NH_4^+ \rightleftharpoons NH_3 + H^+$	
Acetic acid:	4.7
$HAc \rightleftharpoons Ac^- + H^+$	

Many other organic acids important in water chemistry have pK_a values similar to that of acetic acid, for example, propionic acid and butyric acid. Many natural waters contain silicates and organic bases that contribute to the total alkalinity. Wastewaters contain substantial quantities of organic bases, ammonia, and phosphates. Anaerobic digester supernatant often contains high concentrations of bases similar to acetate, as well as carbonates, ammonia, and phosphates. For very complex systems for which a detailed chemical analysis is not available, no attempt is made to work with the mathematical definition of alkalinity. However, if the system is chemically defined we can show that these substances will contribute to alkalinity if, during the titration, some of the base is converted to the conjugate acid. We can modify the total alkalinity definition to include these species so that

$$\text{Total alkalinity, eq/liter} = [OH^-] + [HCO_3^-] + 2[CO_3^{2-}] + [H_3SiO_4^-] \\ + [B(OH)_4^-] + [HPO_4^{2-}] + [NH_3] + f'[Ac^-] - [H^+] \quad (4\text{-}121)$$

where f' is the fraction of $[Ac^-]$ titrated at the total alkalinity endpoint. $H_3SiO_4^-$, $B(OH)_4^-$, HPO_4^{2-}, and NH_3 will each be completely converted to their conjugate acids. In terms of α values this expression becomes

$$\text{Total alkalinity, eq/liter} = [OH^-] + \alpha_{1,CO_3}C_{T,CO_3} + 2\alpha_{2,CO_3}C_{T,CO_3} \\ + \alpha_{1,Si}C_{T,Si} + \alpha_{1,B}C_{T,B} + \alpha_{2,PO_4}C_{T,PO_4} \quad (4\text{-}122) \\ + \alpha_{1,NH_3}C_{T,NH_3} + (\alpha_{1,Ac} - \alpha'_{1,Ac})C_{T,Ac} - [H^+]$$

where $\alpha'_{1,Ac}$, is the fraction of $C_{T,Ac}$ present as Ac^- at the endpoint of the titration.

Example 4-41

The East Bay Municipal Utility District, California, obtains water from the Mokelumne River. In January 1968 the total alkalinity of the treated and distributed water was 20 mg/liter as $CaCO_3$, the pH was 9.65, and the silica content was 8 mg/liter as SiO_2. What fraction of the total alkalinity is contributed by the silicate, by the carbonate system, and by the hydroxide? Here $pK_{a,Si} = 9.5$.

Solution

$$\text{Total alkalinity (eq/liter)} = (20 \text{ mg/liter as } \frac{CaCO_3}{50} \text{ mg } CaCO_3/\text{meq}) \times 10^{-3} \text{ eq/meq}$$
$$= 4 \times 10^{-4} \text{ eq/liter}$$

From Eq. 4-122,
$$\text{Total alkalinity} = \alpha_{1,CO_3}C_{T,CO_3} + 2\alpha_{2,CO_3}C_{T,CO_3} + [OH^-] + \alpha_{1,Si}C_{T,Si} - [H^+]$$
$$C_{T,Si} = 8 \text{ mg/liter as } SiO_2/60 \times 10^{-3} \text{ mg } SiO_2/\text{mole}$$
$$= 1.33 \times 10^{-4} M$$

At pH 9.65, from Appendix 1, $\alpha_{1,Si} = 0.58$,

$$[OH^-] = \frac{K_w}{[H^+]} = \frac{10^{-14}}{10^{-9.65}} = 10^{-4.35} = 4.44 \times 10^{-5} M$$

Thus

$$\text{Silica in the form of } H_3SiO_4{}^- = \alpha_1 C_{T,Si} = 0.58 \times 1.33 \times 10^{-4}$$
$$= 0.77 \times 10^{-4} \text{ eq/liter}$$

Alkalinity due to silicate $= 0.77 \times 10^{-4}$ eq/liter

$$= \left(\frac{0.77 \times 10^{-4}}{4 \times 10^{-4}} \right) \times 100 = 19.2 \text{ percent}$$

Alkalinity due to hydroxide $= 0.44 \times 10^{-4}$ eq/liter

$$= \left(\frac{0.44 \times 10^{-4}}{4 \times 10^{-4}} \right) \times 100 = 11.0 \text{ percent}$$

Alkalinity due to carbonate and bicarbonate $= 100 - (19.2 + 11.0)$
$$= 69.8 \text{ percent}$$

Therefore, in this water a significant fraction of the total alkalinity (almost 20 percent) is due to the silicate system.

Many interesting aspects of the carbonate system occur in connection with heterogeneous systems containing carbonate solids and soluble carbonate complexes. These topics are dealt with in the next two chapters.

4.14. PROBLEMS

1. Calculate the pH of the following perchloric acid, $HClO_4$, solutions:
   (a) 10^{-3} M.
   (b) 10^{-8} M.

2. Calculate the pH of the following potassium hydroxide, KOH, solutions:
   (a) 10^{-4} M.
   (b) 10^{-6} M.

3. Arsenious acid is a weak acid

$$HAsO_2 \rightleftharpoons H^+ + AsO_2{}^-; \qquad K_a = 10^{-9.22}$$

   If 10^{-3} moles of the herbicide $NaAsO_2$, are added to 1 liter of distilled water at 25°C, what is the resulting pH? Determine by solving the appropriate set of equations analytically.

4. (a) A solution of 10^{-3} mole of sodium propionate ($CH_3CH_2COO\ Na$) is added to 1 liter of distilled water. If the $pK_a = 4.87$ for propionic acid, what is the solution pH? Solve the appropriate set of equations analytically and graphically and compare the results.
   (b) Propionic acid is a common constituent of anaerobic digesters. If the pH of the digester contents is 7.6, what percentage of the total amount of propionic acid is ionized at 25°C?

5. Aqueous free chlorine is hypochlorous acid, HOCl, and hypochlorite ion, OCl^-, with the relative amounts depending on the pH of the solution.

(a) Given that $pK_a = 7.5$ for HOCl at 25°C, what is the pH of a solution prepared by adding 10^{-3} mole of NaOCl to 1 liter of water?

(b) What is the pH at which 50 percent of the total free chlorine is present as HOCl?

(c) What is the pK_b for OCl^- and to what reaction does this constant apply? Neglect ionic strength effects; temperature = 25°C.

6. A solution of 10^{-2} mole of Na_2HPO_4 is added to 1 liter of solution.

   (a) Write the mass balances for Na and PO_4, $C_{T,Na}$, and C_{T,PO_4}.

   (b) Write the proton condition.

   (c) Write the charge balance.

7. What pH results when 10^{-4} mole of NaH_2PO_4 are added to 1 liter of distilled water? The temperature = 25°C; assume that $\mu = 10^{-2}$ after the addition. Compare the value obtained when ionic strength effects are taken into account with that calculated when it is neglected. (See Fig. 3-4 for activity coefficients.)

8. What is the pH of a solution that is prepared by adding 10^{-5} mole of Na_2CO_3 to 1 liter of water? The temperature = 15°C; $\mu = 0$. (See Tables 4-2 and 4-7 for equilibrium constants.) Compare this value with that obtained if the temperature is 25°C.

9. What is the pH of a 10^{-3} M $KHSO_4$ solution at 25°C; $\mu \cong 0$?

10. Using either the table or the equations in Appendix 1, calculate

    (a) α_0 and α_1 for hydrofluoric acid, HF, at pH 4.2.

    (b) α_0, α_1, and α_2 for $H_2CO_3^*$ at pH 6.6. [Include both approximate (± 0.05) and exact values.]

    (c) α_0, α_1, α_2, and α_3 for H_3PO_4 at pH 8.0. Assume that activity effects are negligible and the temperature is 25°C.

11. Calculate the α values as required in Problem 4-10 given that the ionic strength is 0.01.

12. 2.5×10^{-7} mole of $NH_{3(aq)}$ and 2.5×10^{-7} mole NaOH are added to 1 liter of distilled water at 25°C. What is the pH if $\mu = 0$?

13. 10^{-2} moles NH_4Ac and 10^{-2} moles NaOH are added to 1 liter of water at 25°C. What is the pH if $\mu \cong 0$?

14. What is the pH of a 25°C solution containing 10^{-3} moles $NaHCO_3$ and 2×10^{-3} mole NH_3 per liter if $\mu = 10^{-2}$?

15. A nitric acid solution, pH = 2.7, results from NO_x removal from a stack gas. Neglect ionic strength effects and the temperature is 25°C.

    (a) How much Na_2CO_3 must be added to neutralize this solution prior to discharge? (The final pH is 8.3. Assume that no weak acids are present in the scrubber water.)

    Hint: What is the predominant carbonate species at pH 8.3?

    (b) What is the buffer intensity of the final solution?

16. How much soda ash, Na_2CO_3, in moles/liter is required to neutralize a pickle liquor solution containing 10^{-1} mole H_2SO_4/liter? Assume that the H_2SO_4 will react only with the Na_2CO_3 and that the pH after neutralization is 8.3.

17. An experiment is to be conducted on breakpoint chlorination (the oxidation of ammonia by chlorine) in which it is desired to maintain the pH constant within 0.5 units of the initial pH of 8. Make the assumption that all of the ammonia is in the NH_4^+ form (a reasonable assumption, since the pK_a of NH_4^+ is 9.3 and the pH of interest is pH 8). The breakpoint reaction between Cl_2 and NH_3 proceeds as follows:

$$3Cl_2 + 2NH_4^+ \rightleftharpoons N_{2(g)} + 8H^+ + 6Cl^-$$

The maximum amount of ammonia that will be used in the experiments is 12.5 mg as N/liter (0.89×10^{-3} moles/liter). Select an appropriate acid-conjugate base pair and determine the concentration of it that will control the pH to within 0.5 units of 8.0 during the reaction. Neglect ionic strength effects; the temperature $= 25°C$. Determine the buffer intensity of this solution. (*Note:* There is a detailed discussion of the breakpoint chlorination reaction in Chapter 7.)

18. An industrial wastewater is to be discharged to a receiving stream with a pH of 8.3 and a total alkalinity $= 2 \times 10^{-3}$ eq/liter. The wastewater contains 5×10^{-3} M H_2SO_4, and the pH of the stream should not be permitted to drop below 6.3.
    (a) What is the maximum dilution ratio (volume waste/volume stream water) that can be used for discharge of the wastewater?
    (b) What is the buffer intensity of the solution at pH 6.3?

19. A sample of natural water contains 1×10^{-3} M CO_3^{2-} and 3×10^{-3} M HCO_3^-.
    (a) As the pH is lowered during the alkalinity titration, at what pH is the CO_2 in solution in equilibrium with atmospheric CO_2?
    (b) What is the pH of the total alkalinity equivalence point, pH_{CO_2}? (Neglect dilution effects.) Give your answer to nearest 0.1 pH unit.
    (c) If 50 percent of the CO_2 formed during the titration "escapes," what is the "new" pH of the total alkalinity equivalence point?

20. A sample of natural water that has been equilibrated with $CaCO_{3(s)}$ is isolated from its surroundings. Indicate whether the addition of small quantities of the following will increase, decrease, or have no effect on the total alkalinity or total acidity and state very briefly why. Neglect ionic strength effects.
    (a) HCl.   (b) $FeCl_3$.
    (c) Na_2SO_4.   (d) CO_2.
    (e) Na_2CO_3.

21. A natural water has the following *partial* analysis:

    pH $= 8.3$ $[Ca^{2+}] = 5 \times 10^{-4}$ M
    $[HCO_3^-] = 3 \times 10^{-3}$ M $[Mg^{2+}] = 1 \times 10^{-4}$ M
    $[CO_{2(aq)}] = 3 \times 10^{-5}$ M $[SO_4^{2-}] = 1 \times 10^{-4}$ M

(a) What volume of $0.02 NH_2SO_4$ is required to titrate a 100-ml sample to the total alkalinity endpoint? What is the total alkalinity in eq/liter and in mg/liter as $CaCO_3$?

(b) A waste containing 10^{-2} moles NaOH/liter is to be discharged to this water. The pH cannot be raised above 9.5. What is the maximum number of liters of waste that can be added to each liter of the natural water?

22.   A *partial* water analysis is given as follows:

$CO_2 = 44$ mg/liter $[Cl^-] = 1 \times 10^{-3} M$
$[HCO_3^-] = 2 \times 10^{-3} M$ $[SO_4^{2-}] = 1 \times 10^{-4} M$

(a) What is the solution pH and CO_3^{2-} concentration?

(b) What is the caustic, carbonate, and total alkalinity (in eq/liter and in mg/liter as $CaCO_3$)?

(c) What is the mineral, CO_2, and total acidity (in eq/liter)?

(d) What is the pH if 5×10^{-4} mole OH^- (as NaOH) are added per liter of the above sample?

23.   Fifty ml of a natural water sample is titrated with $0.02 N$ H_2SO_4. The titrant volume required to titrate to pH 8.3 is 6 ml and an additional 8 ml is required to titrate to pH 4.3.

(a) What is the caustic, carbonate, and total alkalinity in meq/liter?

(b) What is the mineral, CO_2, and total acidity in meq/liter?

(c) What is C_{T,CO_3}?

(d) What is pH_{CO_2} and $pH_{CO_3^{2-}}$ (to the nearest 0.1 pH unit)?

(e) What is $[H^+]$, $[OH^-]$, $[CO_3^{2-}]$, $[HCO_3^-]$ and $[H_2CO_3^*]$ in the original sample? Calculate using both the approximate method described in Section 4.13.3 and the exact procedure and compare the results.

24.   A solution has a carbonate alkalinity of 1 meq/liter and a total alkalinity of 6 meq/liter. The ionic strength of the solution is $10^{-2} M$ and the temperature is 40°C. Calculate $[H^+]$, $[HCO_3^-]$, and $[CO_3^{2-}]$ neglecting ionic strength effects and compare with the values obtained when corrections are made for ionic strength effects. Use the constants given in Tables 4-2 and 4-4 for your calculations.

25.   In some regions carbonate-bearing minerals are lacking in the earth and the lakes in these regions have a low alkalinity. Acid rains, resulting from conversion of industrial SO_2 emissions to H_2SO_4, can cause significant pH depressions in such lakes and may result in fish kills. What volume of acid rain, pH = 4.0, is required to lower the pH of a lake to 6.7? The lake has a volume of 20×10^6 ft^3, a pH = 7.0, and an alkalinity of 30 mg/liter as $CaCO_3$?

4.15.   ADDITIONAL READING

ACID-BASE EQUILIBRIA

Butler, J. N., *Ionic Equilibrium—A Mathematical Approach*, Addison-Wesley, Reading, Mass., 1964.

Laitinen, H. A., *Chemical Analysis: An Advanced Text and Reference*, McGraw-Hill, New York, 1960.

Ricci, J. E., *Hydrogen Ion Concentration, New Concepts in a Systematic Treatment*, Princeton University Press, Princeton, N.J., 1952.

Sillen, L. G., "Graphic Presentation of Equilibrium Data," Chapter 8 in *Treatise on Analytical Chemistry*, Part I, Vol. 2, I. M. Kolthoff and P. J. Elving, eds., Wiley-Interscience, New York, 1959.

CARBONATE EQUILIBRIA
Garrels, R. M., and C. L. Christ, *Solutions, Minerals and Equilibria*, Harper, New York, 1965.

Stumm, W., and J. J. Morgan, *Aquatic Chemistry*, Wiley-Interscience, New York, 1970

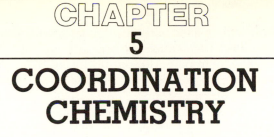

CHAPTER 5

COORDINATION CHEMISTRY

5.1. INTRODUCTION

Complex formation is important in the chemistry of natural waters and wastewaters from several standpoints. Complexes modify metal species in solution, generally reducing the free metal ion concentration so that effects and properties which depend on free metal ion concentration are altered. These effects include such things as the modification of solubility (see Section 6-5), the toxicity and possibly the biostimulatory properties of metals, the modification of surface properties of solids, and the adsorption of metals from solution.

Complex formation is used extensively in water analysis. The determination of hardness, Ca^{2+} and Mg^{2+} concentration, employs the complexation of these metals with the chelating agent ethylenediaminetetra-acetic acid (EDTA); the titrimetric finish of the chemical oxygen demand (COD) test uses the complexing agent 1,10-phenanthroline to detect the presence of Fe^{2+} and so indicate the endpoint; chloride analysis by the mercurimetric method depends on the formation of the complex $HgCl_{2(aq)}^{\circ}$ between Hg^{2+} and the chloride ion.

The objective of this chapter is to present sufficient background information to enable the student to understand the principles of complex formation and stability. Methods for calculating the concentrations of complexes in fairly simple systems are presented; the approaches to calculating species distribution in more complicated systems are discussed together with the results of such computations. A discussion of complex formation in several dilute aqueous systems is presented to illustrate important points. The effect of complexation on solubility is the major topic in Chapter 6.

5.2. NOMENCLATURE AND DEFINITION OF TERMS

Coordination compounds or *complexes* consist of one or more *central atoms* or *central ions*, usually metals, with a number of ions or molecules, called *ligands*, surrounding them and attached to them. The complex can be nonionic, cationic, or anionic, depending upon the charges of the central ions and the ligands. Usually, the central ions and ligands can exist individually as well as combined in complexes. The total possible number of attachments to a central atom or central ion or the total possible number of coordinated species is referred to as the *coordination number.* Ligands are attached to the central species by coordinate covalent bonds in which both of the electrons participating in the bond are derived from the ligand. Thus we can regard the central species as an electron acceptor and the ligand as an electron donor. The central species is a Lewis acid and the ligands are Lewis bases.[1] Since metal ions have an affinity for accepting electrons, they all form coordination compounds, with a tendency that increases as the electron-accepting affinity of the metal ion increases. Molecules and ions with free electron pairs tend to form complexes whose strength is a function of their ability to donate or share that pair of electrons.[2]

There are a few basic rules to follow in naming complexes.[3] Ligands are named as molecules if they are neutral. If they have a positive charge, the suffix "-ium" is used; if the group is negative, the suffix "-o" is used. Exceptions to this are the terms "aquo" for H_2O and "ammine" for NH_3. The prefixes "di-, tri-, tetra-," and so forth are used to indicate the number of ligands. The characteristic suffix for anionic complexes is "-ate"; there are no characteristic suffixes for cationic and neutral complexes. The oxidation state of the central metal ion is generally given in parentheses following the name of the complex, (I), (II), (-I), and so forth. These rules are illustrated by the following examples:

$Fe(CN)_6^{4-}$, hexacyanoferrate(II) ion.

$Al(H_2O)_6^{3+}$, hexaaquoaluminum(III) ion.

$Cu(NH_3)_4^{2+}$, tetraamminecopper(II) ion.

$MgCO_3^{\circ}$, carbonatomagnesium(II).

$CaSO_4^{\circ}$, sulfatocalcium(II).

The superscript "o" on the magnesium and calcium complexes indicates the complex is uncharged.

The coordination number of a metal ion may vary from one ion to another. For example, Ni(II) has a coordination number of 4 in its complex with carbonyl (CO), $Ni(CO)_4^{2+}$; with 1,10-phenanthroline in the complex $Ni(phenanthroline)_3^{2+}$ the coordination number is 6. Iron(III) has coordination numbers of 6 for water, $Fe(H_2O)_6^{3+}$, and cyanide, $Fe(CN)_6^{3-}$, but 4 for chloride, $FeCl_4^{-}$.

---

[1] Substances that can accept a pair of electrons are called Lewis acids and substances that can donate a pair of electrons are called Lewis bases.

[2] F. Basolo and R. G. Pearson, *Mechanisms of Inorganic Reactions,* 2nd ed., John Wiley, New York, 1967.

[3] See F. Basolo and R. G. Pearson for a more complete discussion of nomenclature.

Ligands that attach to a central metal ion at only one point, such as H_2O, OH^-, Cl^-, and CN^-, are called *monodentate* ligands. Ligands that attach at two or more sites are *multidentate* ligands or *chelating agents*. The complex formed by a chelating agent and a central metal ion is known as a *chelate*. Chloride, for example, forms a monodentate complex with mercuric ion, while carbonate and sulfate can occasionally be bidentate—attaching to a central metal ion at two sites:

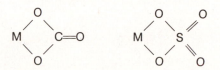

The chelating agent $EDTA^{4-}$ can attach at six sites, since each of the acetate groups and the two nitrogen atoms have free electron pairs necessary for coordinate bond formation. For example, the EDTA complex with Ca^{2+} that forms during the titrimetric determination of water hardness is

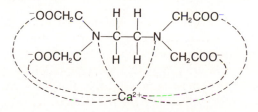

Each mole of the chelating agent 1,10-phenanthroline, which is used as an indicator for Fe^{2+} in the titrimetric finish of the COD test, can satisfy two coordination sites with the electron pairs on the two nitrogen atoms in each molecule. With Fe^{2+} a red-colored complex is formed with three molecules of 1,10-phenanthroline satisfying the coordination number of 6 for the Fe^{2+} ion:

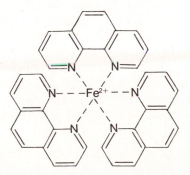

Complexes containing one central ion are called *mononuclear* complexes; when there is more than one central ion or molecule the complexes

are *polynuclear* complexes. When aluminum ions are added to water buffered at about a neutral pH, there is evidence that polynuclear hydroxoaluminum(III) complexes form from the condensation (dehydration) of mononuclear hydroxoaluminum(III) complexes:

$$Al(H_2O)_6^{3+} + H_2O \rightleftharpoons Al(H_2O)_5OH^{2+} + H_2O$$

hexaquoaluminum(III) ion   monohydroxopentaaquoaluminum(III) ion

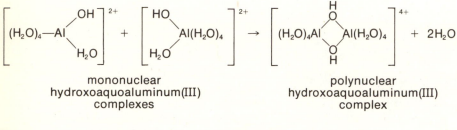

mononuclear
hydroxoaquoaluminum(III)
complexes

polynuclear
hydroxoaquoaluminum(III)
complex

or

$$2Al(H_2O)_5OH^{2+} \rightleftharpoons Al_2(H_2O)_8(OH)_2^{4+} + 2H_2O$$

The bond strength of metal ion complexes varies considerably. Weakly associated complexes that have one or more layers of water between the ligand and the central ion are known as *ion pairs*, for example, $CaCO_3°$, $CaHCO_3^+$, $CaSO_4°$, $CaOH^+$, $MgCO_3°$, and $MgSO_4°$. In these compounds, the degree of interaction is only slightly greater than the electrostatic interactions accounted for by the DeBye-Hückel theory.

5.3.   REACTION RATE

The rates of coordination reactions are characterized by the terms *labile* (very fast reactions) and *inert* (very slow reactions). It is important to understand that these terms do not, in any way, indicate the stability of a complex. They describe the kinetics of the interaction of ligands and a central ion; complex stability is described by the magnitude of the equilibrium constant of the reaction by which a complex dissociates into its component central ion and ligands. An inert complex is not necessarily stable, that is, it does not necessarily have little tendency to dissociate. For example, the tetracyanomercurate(II) complex, $Hg(CN)_4^{2-}$, is labile but very stable; while the reverse is true for the tetracyanonickelate(II) complex, $Ni(CN)_4^{2-}$, which is inert and unstable.[4]

The substitution reaction in which a ligand replaces a coordinated water molecule to form a complex in aqueous solution,

$$M(H_2O)_n + L \rightleftharpoons ML(H_2O)_{n-1} + H_2O$$

---

[4] R. G. Wilkins, *The Study of Kinetics and Mechanism or Reactions of Transition Metal Complexes*, Allyn and Bacon, Boston, 1974.

is virtually complete within seconds to minutes at typical natural water concentrations (10^{-3} M) for the following combinations of central metal ion (M) and ligand (L):

METAL ION	LIGAND
Ni^{2+}	CH_3COO^-
Mn^{2+}, Fe^{2+}, Co^{2+}, Cu^{2+}, Zn^{2+}, Cd^{2+}	H_2O, F^-
Mn^{2+}, Co^{2+}, Cu^{2+}, Zn^{2+}, Cd^{2+}	$EDTAH^{3-}$
Co^{2+}, Ni^{2+}, Cu^{2+}, Zn^{2+}	NH_3

Substances that are not complexes, for example, methylmercury, $HgCH_3^+$, will not undergo ligand replacement reactions.

For the same reaction, if Fe^{3+} is the metal and Cl^- is the ligand several hours are required, but if Fe^{3+} has an OH^- attached to it, that is, $FeOH^{2+}$, only minutes are required. The reaction between Fe^{3+} and SO_4^{2-} or SCN^-, on the other hand, requires only minutes.

Notable exceptions to the statement that coordination reactions occur rapidly are the reactions involving replacement of H_2O from $Cr(H_2O)_6^{3+}$ by ligands such as SO_4^{2-}, which take days to years. Changes in the structure of polymeric metal hydroxide species may take weeks.

5.4.   COMPLEX STABILITY AND EQUILIBRIUM CALCULATIONS

Equilibrium constants for complexes are usually stated for reactions written in the direction of complex formation, that is,

$$\text{Ligand} + \text{central metal ion} \rightleftharpoons \text{complex}$$

For example, for the ammine copper complex,

$$NH_{3(aq)} + Cu^{2+} \rightleftharpoons Cu(NH_3)^{2+}$$

and

$$\frac{\{Cu(NH_3)^{2+}\}}{\{Cu^{2+}\}\{NH_{3(aq)}\}} = 10^{+4} = K = \text{stability constant}$$

When stated for a complex formation reaction, the equilibrium constant is called a *formation* or a *stability constant*. Conversely, if the equilibrium constant is stated for the dissociation of the complex it is called a *dissociation or an instability constant*. Large values of stability constants indicate stable complexes.

It is difficult to formulate general rules, or scales of strength, for complex stability because so many exceptions to the rules exist. However, some general statements can be made that aid in predicting what will take place in aqueous solution. A-metal cations, including the alkali metals, Na^+ and K^+; alkaline earth metals, Mg^{2+} and Ca^{2+}; and other ions such as Al^{3+} and Si^{4+} preferentially coordinate with ligands containing oxygen as the electron donor, such as carbonate, hydroxide, borate, and so forth. Other metal ions, of the so-called B-metals, which include Ag^+,

Zn^{2+}, Hg^{2+}, Pb^{2+}, and Sn^{2+}, show a stronger tendency to form complexes with ligands containing S, P, or N as donor atoms, such as ammonia, sulfide, and phosphite, rather than with oxygen-containing electron donors. Within these groups of metal ions, complex stability is proportional to the charge of the metal ion. Alkali metals form the least stable complexes and indeed rarely complex with any ligand except H_2O, while the transition metal ions, for example, Cr^{2+}, Cr^{3+}, Fe^{2+}, Fe^{3+}, and Ni^{2+} have a strong tendency to form complexes with a wide variety of ligands.

Ligands also differ in their ability to form stable complexes. The ligands phosphate, hydroxide, carbonate, and so forth, are potent complex formers while the perchlorate, ClO_4^-, and nitrate, NO_3^-, ions show very little tendency to form complexes. It is for this reason that nitrate or perchlorate salts are used as swamping electrolytes in experiments where it is desirable to have a constant ionic strength.[5]

We can also make some general statements about the stability of chelates. Coordination compounds containing five- and six-membered rings are the most stable. For example, $EDTA^{4-}$ and 1,10-phenanthroline complexes with five-membered rings are more stable than some carbonate and sulfate complexes that have four-membered rings. The stability of a chelate generally increases with an increase in the number of points of attachment between the chelating agent and the central metal ion. For example, the complexes formed between Cu^{2+} and the compounds ethylenediamine,

$$H_2N—CH_2—CH_2—NH_2$$
$$Cu^{2+}$$

diethylenetriamine,

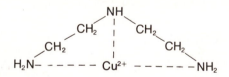

and triethylenetetramine,

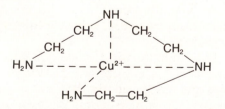

[5] For a more extensive discussion of the complexing ability of metal ions and ligands, see the following references: S. Arhland, S. J. Clatt, and W. R. Davies, *Quart. Rev.* (London) *12:* 165 (1958); and W. Stumm and J. J. Morgan, *Aquatic Chemistry*, Wiley-Interscience, New York, 1970.

have stability constants, respectively, of $\sim 10^{11}$, $\sim 10^{16}$, and $\sim 10^{20}$. The phenomenon of increasing complex stability with the increasing number of attachments is called the *chelate effect*.

If we introduce a ligand that forms complexes with two metal ions into a solution containing equal concentrations of the two metal ions, the ligand will form complexes preferentially with the metal ion that produces the complex with the larger stability constant. The titration of hardness with EDTA illustrates this point. The chelating agent, $EDTA^{4-}$, is gradually added (by titration) to a solution containing Ca^{2+} and Mg^{2+}. The following complex-forming reactions can occur:

$$Ca^{2+} + EDTA^{4-} \rightleftharpoons (EDTA\text{-}Ca)^{2-} \quad K = 10^{+10.7}$$

$$Mg^{2+} + EDTA^{4-} \rightleftharpoons (EDTA\text{-}Mg)^{2-} \quad K = 10^{+8.7}$$

Because the stability constant of the $(EDTA\text{-}Ca)^{2-}$ complex is two orders of magnitude greater than the stability constant of the $(EDTA\text{-}Mg)^{2-}$ complex, the reaction between $EDTA^{4-}$ and Ca^{2+} is virtually complete (i.e., all Ca^{2+} is complexed) prior to any reaction between $EDTA^{4-}$ and Mg^{2+}.

Another aspect of complex equilibria that can be understood using a knowledge of stability constant values is the converse of the previous situation, that is, the competition of a number of ligands for a single metal ion rather than the competition of a number of metal ions for a single ligand. The complex with the largest stability constant again forms preferentially. When a ligand that forms a very stable complex with a metal ion is added to a solution of the metal ion complexed with another ligand in a less stable complex, the stronger complexing agent (ligand) will tend to decompose the existing complex and withdraw the metal ion into a complex with itself. The weaker complex must be sufficiently labile for this to occur. This fact is used in the titration of hardness with EDTA. The solution to be titrated (containing Ca^{2+} and Mg^{2+}) is dosed with the complexing agent Eriochrome Black T (EBT), or comparable substance, which forms a complex with Mg^{2+} as follows:

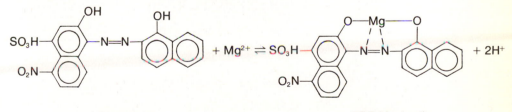

EBT (blue)   EBT-Mg (red)

$$\frac{[EBT\text{-}Mg][H^+]^2}{[EBT][Mg^{2+}]} = 10^{+7}$$

We can note two things from the above equations:

1. The EBT-Mg complex $(K = 10^{+7})$ is weaker than the EDTA-Mg complex $(K = 10^{+8.7})$.

2. The color of the EBT-Mg complex is red and the color of uncomplexed EBT is blue. Thus when the EBT-Mg complex is destroyed by, for example, adding EDTA, the color of the solution will change from red to blue. It should be possible to detect the endpoint when all of the EBT-Mg complex has been destroyed.

In the total hardness analysis, EDTA is added from a buret into the solution containing Ca^{2+} and the EBT-Mg complex. We have seen previously that EDTA forms a stronger complex with Ca^{2+} than with Mg^{2+} so that the EDTA will first react to form an EDTA-Ca(II) complex. Only when virtually all the Ca^{2+} has been complexed will the EDTA start to decompose the EBT-Mg(II) complex. The completion of the reaction with EBT-Mg is heralded by the color change from the red of the EBT-Mg complex to the blue of free EBT. It signifies the endpoint of the titration at which all of both the Ca^{2+} and Mg^{2+} (total hardness) have reacted with the EDTA titrant.

For complexes that contain more than one ligand or central metal ion, there are two ways of writing stability constants. Stepwise formation constants are equilibrium constants for the reactions in which the central metal ion consecutively adds one ligand. An overall formation constant is the equilibrium constant for the reaction in which the central metal ion combines with all of the ligands necessary to form a specific complex. We can illustrate these two types of constants using the formation constants of the chloromercury(II) complexes as an example. For stepwise formation we have

$$Hg^{2+} + Cl^- \rightleftharpoons HgCl^+; \qquad \log K_1 = 7.15 \qquad (5\text{-}1)$$

$$HgCl^+ + Cl^- \rightleftharpoons HgCl_2^{\circ}; \qquad \log K_2 = 6.9 \qquad (5\text{-}2)$$

$$HgCl_2^{\circ} + Cl^- \rightleftharpoons HgCl_3^-; \qquad \log K_3 = 2.0 \qquad (5\text{-}3)$$

$$HgCl_3^- + Cl^- \rightleftharpoons HgCl_4^{2-}; \qquad \log K_4 = 0.7 \qquad (5\text{-}4)$$

and for the overall formation reactions we can write

$$Hg^{2+} + Cl^- \rightleftharpoons HgCl^+; \qquad \log \beta_1 = 7.15 \qquad (5\text{-}5)$$

$$Hg^{2+} + 2Cl^- \rightleftharpoons HgCl_2^{\circ}; \qquad \log \beta_2 = 14.05 \qquad (5\text{-}6)$$

$$Hg^{2+} + 3Cl^- \rightleftharpoons HgCl_3^-; \qquad \log \beta_3 = 15.15 \qquad (5\text{-}7)$$

$$Hg^{2+} + 4Cl^- \rightleftharpoons HgCl_4^{2-}; \qquad \log \beta_4 = 15.75 \qquad (5\text{-}8)$$

where K_i is the equilibrium constant for the reaction in which the complex containing "i" ligands is formed from the complex containing "$i-1$" ligands and β_i is the equilibrium constant for the reaction in which the complex containing i ligands is formed from the metal ion and i ligands. Although either type of constant (but not both) may be used to determine the concentrations of the various complexes in solution, it is somewhat safer to use the stepwise formation constants. The inexperienced user may not remember all of the complex species that can be formed when using the overall formation constants. Such an omission is not possible when the stepwise formation constants are used.

When one or more of the intermediate species is unknown (or their stability constant value is not defined), we are forced to use the overall formation constant. For example, for the series of complexes formed between the hydrated ferric ion, $Fe(H_2O)_6^{3+}$, and the hydroxide ion we have

$$Fe(H_2O)_6^{3+} + OH^- \rightleftharpoons Fe(H_2O)_5OH^{2+} + H_2O; \qquad K_1$$

$$Fe(H_2O)_5(OH)^{2+} + OH^- \rightleftharpoons Fe(H_2O)_4(OH)_2^+ + H_2O; \qquad K_2$$

$$Fe(H_2O)_4(OH)_2^+ + OH^- \rightleftharpoons Fe(H_2O)_3(OH)_3^\circ{}_{(aq)}; \qquad K_3$$

$$Fe(H_2O)_3(OH)_3^\circ{}_{(aq)} + OH^- \rightleftharpoons Fe(H_2O)_2(OH)_4^- + H_2O; \qquad K_4$$

The concentration of the species $Fe(H_2O)_3(OH)_3^\circ{}_{(aq)}$ which exists in solution is unknown so that while K_1 and K_2 are defined, K_3 and K_4 are not.

$$K_1 = \frac{[Fe(H_2O)_5(OH)^{2+}]}{[Fe(H_2O)_6^{3+}][OH^-]} = 10^{+10.95}$$

$$K_2 = \frac{[Fe(H_2O)_4(OH)_2^+]}{[Fe(H_2O)_5(OH)^{2+}][OH^-]} = 10^{+9.42}$$

$$K_3 = \frac{[Fe(H_2O)_3(OH)_3^\circ{}_{(aq)}]}{[Fe(H_2O)_4(OH)_2^+][OH^-]} = ?$$

$$K_4 = \frac{[Fe(H_2O)_2(OH)_4^-]}{[Fe(H_2O)_3(OH)_3^\circ{}_{(aq)}][OH^-]} = ?$$

However, for the overall formation constants, we have a defined relationship involving the species $Fe(H_2O)_2(OH)_4^-$.

$$\beta_1 = \frac{[Fe(H_2O)_5(OH)^{2+}]}{[Fe(H_2O)_6^{3+}][OH^-]} = 10^{11.84}$$

$$\beta_2 = \frac{[Fe(H_2O)_4(OH)_2^+]}{[Fe(H_2O)_6^{3+}][OH^-]^2} = 10^{21.26}$$

$$\beta_3 = \frac{[Fe(H_2O)_3(OH)_3^\circ{}_{(aq)}]}{[Fe(H_2O)_6^{3+}][OH^-]_3} = ?$$

$$\beta_4 = \frac{[Fe(H_2O)_2(OH)_4^-]}{[Fe(H_2O)_6^{3+}][OH^-]^4} = 10^{33.0}$$

Note that it is common to omit the H_2O ligands when writing equations such as those given here. For example, $Fe(H_2O)_5(OH)^{2+}$ is usually written as $Fe(OH)^{2+}$. We generally follow this practice throughout the remainder of this text.

The procedures for calculating species concentration in a solution containing metal ions and ligands are similar to that used for acid-base problems in Chapter 4; these procedures are illustrated in the following example.

Example 5-1

The COD analysis employs mercuric sulfate, $HgSO_4$, as a source of Hg^{2+} to complex Cl^- ion and so prevent its oxidation to $Cl_{2(aq)}$ by $K_2Cr_2O_7$ during the analysis. Typically, 0.4 g $HgSO_4$ is added to 20 ml of sample along with 40 ml of other reagents. If a sample contains 1000 mg Cl^-/liter, what will be the concentration of each chloromercury(II) complex and of the free Cl^- ion in the solution? Recall that in the COD analysis the sulfuric acid concentration is 50 percent so that the pH is very low (< 1) and hydroxomercury(II) complexes are not important. Neglect ionic strength effects and assume a temperature of 25°C.

Solution

The unknowns are $[Hg^{2+}]$, $[Cl^-]$, $[HgCl^+]$, $[HgCl_2^0{}_{(aq)}]$, $[HgCl_3^-]$, and $[HgCl_4^{2-}]$: $[H^+]$ and $[OH^-]$ are not of importance because they do not enter into reactions with Hg^{2+} or Cl^- under the conditions of the COD test. Six equations are needed. These are the mass balances on Hg and Cl and the four equilibrium equations.

(1) $C_{T,Hg} = \dfrac{0.4\,g\,HgSO_4}{60\,ml\,total\,volume/1000\,ml/liter} \times \dfrac{1}{297\,g\,HgSO_4/mole}$

$= 2.24 \times 10^{-2}\,M\,(= 10^{-1.65}\,M)$

$= [Hg^{2+}] + [HgCl^+] + [HgCl_2^0] + [HgCl_3^-] + [HgCl_4^{2-}]$

(2) $C_{T,Cl} = \dfrac{1000\,mg/liter \times (20\text{-}ml\,sample/1000\,ml/liter)}{(60\,ml\,total\,volume/1000\,ml/liter)} \times \dfrac{1}{35,500\,mg\,Cl^-/mole}$

$= 9.39 \times 10^{-3}\,M\,(= 10^{-2.02}\,M)$

$= [Cl^-] + [HgCl^+] + 2[HgCl_2^0] + 3[HgCl_3^-] + 4[HgCl_4^{2-}]$

(3) $\beta_1 = 10^{7.15} = \dfrac{[HgCl^+]}{[Hg^{2+}][Cl^-]}$

(4) $\beta_2 = 10^{14.05} = \dfrac{[HgCl_2^0{}_{(aq)}]}{[Hg^{2+}][Cl^-]^2}$

(5) $\beta_3 = 10^{15.05} = \dfrac{[HgCl_3^-]}{[Hg^{2+}][Cl^-]^3}$

(6) $\beta_4 = 10^{15.75} = \dfrac{[HgCl_4^{2-}]}{[Hg^{2+}][Cl^-]^4}$

We will use a trial and error approach to solve this problem. We will make certain assumptions, find a solution, and then check the assumptions against the results. If the assumptions do not check out, we will make new assumptions and try again.

Assume that the concentrations of all complexes are insignificant in the mass balance equations so that

$$[Cl^-] = C_{T,Cl} = 10^{-2.02} \quad \text{and} \quad [Hg^{2+}] = C_{T,Hg} = 10^{-1.65}$$

Then, from Eqs. (3) to (6),

$$[HgCl^+] = (10^{+7.15})(10^{-2.02})(10^{-1.65}) = 10^{+3.48}$$

$$[HgCl_2^{\circ}{}_{(aq)}] = (10^{+14.05})(10^{-1.65})(10^{-2.02})^2 = 10^{+8.36}$$

$$[HgCl_3^-] = (10^{+15.05})(10^{-1.65})(10^{-2.02})^3 = 10^{+7.34}$$

$$[HgCl_4{}^{2-}] = (10^{+15.75})(10^{-1.65})(10^{-2.02})^4 = 10^{+6.02}$$

We can see immediately that the assumption is not valid, since the concentration of each complex far exceeds the values of both $C_{T,Hg}$ and $C_{T,Cl}$. Even though the calculated concentrations of the chloromercury complexes are far too high, we can glean some useful information from this otherwise disappointing result. We see that the concentration of the $HgCl^{\circ}{}_{2(aq)}$ complex is larger than the concentrations of the other complex species and that Cl^- is much less than the concentrations of all the complexes. Using this information, let us assume that the $HgCl^{\circ}{}_{2(aq)}$ complex is the sole important Cl-containing species. The Cl mass balance now becomes

$$C_{T,Cl} = 10^{-2.02} = 2[HgCl_2^{\circ}{}_{(aq)}]$$

or

$$[HgCl_2^{\circ}{}_{(aq)}] = 10^{-2.32}$$

Since $C_{T,Hg} > \frac{1}{2}C_{T,Cl}$, we must assume that some free $[Hg^{2+}]$ is present—there is just not enough Cl^- to complex it all. We therefore assume that the important Hg-containing species are Hg^{2+} and $HgCl^{\circ}{}_{2(aq)}$.

The Hg mass balance becomes

$$C_{T,Hg} = 10^{-1.65} = [Hg^{2+}] + [HgCl_2^{\circ}{}_{(aq)}]$$

Then

$$[Hg^{2+}] = 10^{-1.65} - 10^{-2.32} = 10^{-1.75}$$

From Eq. (4)

$$[Cl^-]^2 = \frac{[HgCl_2^{\circ}]}{10^{14.05}[Hg^{2+}]}$$

$$= \frac{10^{-2.32}}{10^{14.05}\,10^{-1.75}}$$

$$= 10^{-14.62}$$

$$[Cl^-] = 10^{-7.31}$$

From Eqs. (3), (5), and (6),

$$[HgCl^+] = 10^{+7.15}[Hg^{2+}][Cl^-] = 10^{+7.15}10^{-1.75}10^{-7.31}$$

$$[HgCl^+] = 10^{-1.91}$$

$$[HgCl_3^-] = 10^{+15.05}[Hg^{2+}][Cl^-]^3$$

$$[HgCl_3^-] = 10^{-8.63}$$

$$[HgCl_4{}^{2-}] = 10^{+15.75}[Hg^{2+}][Cl^-]^4 = 10^{+15.75}10^{-1.75}(10^{-7.31})^4$$

$$[HgCl_4{}^{2-}] = 10^{-15.24}$$

Substituting this set of values into the mass balance equations for $C_{T,Hg}$ and $C_{T,Cl}$ shows again that the calculated concentrations are still too high. Examination of the new values shows $[HgCl^+]$ to have a value close to that of $[HgCl_2^{\circ}{}_{(aq)}]$. Therefore, it should not be neglected. Taking this into account, the mass balances become

(1') $C_{T,Hg} = [Hg^{2+}] + [HgCl^+] + [HgCl_2^{\circ}] = 10^{-1.65}$

(2') $C_{T,Cl} = [HgCl^+] + 2[HgCl_2^{\circ}] = 10^{-2.02}$

From Eqs. (1'), (2'), (3), and (4),

(3') $10^{-1.65} = [Hg^{2+}](1 + 10^{7.15}[Cl^-] + 10^{14.05}[Cl^-]^2)$

(4') $[Hg^{2+}] = \dfrac{10^{-1.65}}{1 + 10^{7.15}[Cl^-] + 10^{14.05}[Cl^-]^2}$

From Eqs. (2'), (3), and (4')

(5') $10^{-2.02} = [Hg^{2+}](10^{7.15}[Cl^-] + 2 \times 10^{14.05}[Cl^-]^2)$

and

(6') $[Hg^{2+}] = \dfrac{10^{-2.02}}{10^{7.15}[Cl^-] + 2 \times 10^{14.05}[Cl^-]^2}$

Combining Eqs. (4') and (6') to eliminate $[Hg^{2+}]$ and solve for $[Cl^-]$, we find $[Cl^-]$ $= 3 \times 10^{-8} M$. Substituting this value in Eq. (5') and solving for $[Hg^{2+}]$ yields $[Hg^{2+}]$ $= 1.42 \times 10^{-2} M$.

Using these values for $[Cl^-]$ and $[Hg^{2+}]$ in Eqs. (3), (4), (5), and (6), we obtain

$$[HgCl^+] = 6 \times 10^{-3}$$

$$[HgCl_2^{\circ}{}_{(aq)}] = 1.59 \times 10^{-3}$$

$$[HgCl_3^-] = 4.8 \times 10^{-10}$$

$$[HgCl_4^{2-}] = 8 \times 10^{-17}$$

Checking our assumptions in Eqs. (1) and (2), we see that the values obtained satisfy these equations to within 3 percent accuracy.

This example shows that the majority of the chloride is tied up as $HgCl^+$ and $HgCl_2^{\circ}{}_{(aq)}$, with minor amounts present as other chloromercury(II) complexes and a very small amount ($3 \times 10^{-8} M$) as the free Cl^- ion.

The implication of this calculation for the COD analysis is that the oxidation of chloride can be almost completely excluded, since $K_2Cr_2O_7$ will not oxidize any of the chloromercury(II) complexes. If we were to conduct a COD analysis on the 1000 mg Cl^-/liter solution in the absence of $HgSO_4$ the following redox reaction would proceed:[6]

$$6e^- + 14H^+ + Cr_2O_7^{2-} \rightarrow 2Cr^{3+} + 7H_2O$$

$$(2Cl^- \rightarrow Cl_{2(aq)} + 2e^-) \times 3$$

$$Cr_2O_7^{2-} + 14H^+ + 6Cl^- \rightarrow 2Cr^{3+} + 3Cl_{2(aq)} + 7H_2O$$

[6] See Chapter 7 for an extensive discussion of oxidation-reduction reactions.

Since we express the COD result in terms of the equivalent amount of oxygen to complete the oxidation, and express the result as "mg O$_2$/liter," it is productive to write the redox reaction in terms of oxygen:

$$4Cl^- \rightarrow 2Cl_{2(aq)} + 4e^-$$

$$\frac{4e^- + 4H^+ + O_2 \rightarrow 2H_2O}{4H^+ + 4Cl^- + O_2 \rightarrow 2Cl_2 + 2H_2O}$$

Thus 32 mg of O$_2$ will oxidize $4 \times 35.5 (= 142)$ mg of Cl$^-$. If we assume that the entire 1000 mg Cl$^-$/liter is oxidized, a COD of $32 \times 1000/142 = 225$ mg/liter will result.

In the presence of sufficient HgSO$_4$ the COD of 1000 mg Cl$^-$/liter is very small, but not zero, because there is some oxidation of the Cl$^-$ present in equilibrium with chloromercury(II) complexes. This oxidation removes Cl$^-$ from solution and forces further dissociation of the complexes. For example,

$$HgCl_2{}^o{}_{(aq)} \rightleftharpoons Hg^{2+} + 2Cl^-$$

Cripps and Jenkins[7] found that the extent of this oxidation was a function of the duration of the COD digestion. The longer that the oxidation was allowed to proceed, the greater the extent of chloromercury(II) complex dissociation and the greater was the COD due to Cl$^-$ oxidation. Formulated as an empirical "chloride correction factor," they found that the COD due to Cl$^-$ oxidation was $0.00041 \times$ mg Cl$^-$/liter $\times$ time of digestion in hours.

5.5. HYDROLYSIS OF METAL IONS—H$_2$O AND OH$^-$ AS LIGANDS

In aqueous solution, *free* metal ions are complexed with water. The metal ions are said to be hydrated. The interaction of these hydrated metal ions with acids and bases is a ligand exchange reaction that is commonly called *hydrolysis*, or *protolysis*. These terms describe the general reaction in which a proton is transferred from an acid to water, or from water to a base. This type of reaction involving hydrated metal cations as the proton-donors or acids occurs readily and is of extreme importance in natural waters. For example, the stepwise hydrolysis of the aquoaluminum(III) ion can be represented by the following series of equations:

$$Al(H_2O)_6{}^{3+} + H_2O \rightleftharpoons Al(H_2O)_5OH^{2+} + H_3O^+ \tag{5-9}$$

$$Al(H_2O)_5OH^{2+} + H_2O \rightleftharpoons Al(H_2O)_4(OH)_2{}^+ + H_3O^+ \tag{5-10}$$

$$Al(H_2O)_4(OH)_2{}^+ + H_2O \rightleftharpoons Al(H_2O)_3(OH)_{3(s)} + H_3O^+ \tag{5-11}$$

$$Al(H_2O)_3(OH)_{3(s)} + H_2O \rightleftharpoons Al(H_2O)_2(OH)_4{}^- + H_3O^+ \tag{5-12}$$

[7] J. M. Cripps and D. Jenkins, "A COD Method Suitable for the Analysis of Highly Saline Waters," *J. Water Pollut. Control Fed.*, 36: 1240–1246 (1964).

From these equations it is easy to visualize that the hydrolysis of metal ions is a stepwise replacement of coordinated molecules of "water of hydration" by hydroxyl ions. In Eqs. 5-9 to 5-12 this replacement occurs by the transfer of protons from waters of hydration to free water molecules to form an hydronium ion. The hydrolysis reaction can also be written as the replacement of a water of hydration by a hydroxyl ion, as was done for the hydrolysis reactions of Fe^{3+} in Section 5-4. The two ways of writing the reaction are essentially the same because they can be interconverted by using the reaction for ionization of water,

$$Al(H_2O)_6{}^{3+} + OH^- \rightleftharpoons Al(H_2O)_5(OH)^{2+} + H_2O; \qquad K_1 = 10^9$$

$$2H_2O \rightleftharpoons H_3O^+ + OH^-; \qquad K_w = 10^{-14}$$

$$Al(H_2O)_6{}^{3+} + H_2O \rightleftharpoons Al(H_2O)_5(OH)^{2+} + H_3O^+; \qquad K_1 = 10^{-5}$$

Note that the equilibrium constants are different but that each defines the relationship between $[Al(H_2O)_6{}^{3+}]$ and $[Al(H_2O)_5OH^{2+}]$ as a function of pH. Equilibrium constants of the hydroxocomplexes of some metal ions of importance in water chemistry are presented in Table 5.1. This table shows that Fe(III) and many other multivalent ions behave similarly to the aquoaluminum(III) ion.

The hydrolysis reactions depicted in Eqs. 5-9 to 5-12 are acid-base (proton transfer) reactions. Because of this, the pH of the solution will influence the distribution of the various species. In general, the percentage of the hydrolyzed species increases as the pH increases, just as the concentration of a conjugate base would increase if the pH of a solution containing its conjugate acid were raised. All trivalent and most divalent metal ions are complexed to some extent with OH^- at the pH of natural waters. The alkaline earth metals hydrolyze significantly only at high pH

TABLE 5-1 Equilibrium Constants for Mononuclear Hydroxo Complexes

	log β_1[a]	log β_2	log β_3	log β_4	log K_{so}[b]
Fe^{3+}	11.84	21.26	—	33.0	−38
Al^{3+}	9	—	—	34.3	−33
Cu^{2+}	8.0	—	15.2	16.1	−19.3
Fe^{2+}	5.7	(9.1)[c]	10	9.6	−14.5
Mn^{2+}	3.4	(6.8)	7.8	—	−12.8
Zn^{2+}	4.15	(10.2)	(14.2)	(15.5)	−17.2
Cd^{2+}	4.16	8.4	(9.1)	(8.8)	−13.6

[a] β_i is the equilibrium constant for the reaction, $M^{n+} + iOH^- \rightarrow M(OH)^{(n-i)+}$.
[b] K_{so} is the equilibrium constant for the reaction $M(OH)_{n(s)} \rightleftharpoons M^{n+} + nOH^-$.
[c] ( ) indicates an estimated value.

(pH > 9 to 10). The pH of a solution to which metal ion is added can be calculated just as it was for acids in Chapter 4. The procedure is demonstrated by the following example.

Example 5-2

Calculate the pH of a 10^{-5} M $Hg(ClO_4)_2$ solution at 25°. Neglect ionic strength effects.

Solution

Because ClO_4^- is a very poor complex former, perchloratomercury(II) complexes can be neglected. The equilibrium of concern is

(1) $Hg^{2+} + 2H_2O \rightleftharpoons 2H^+ \; Hg(OH)_2{}^\circ$; $\log K = -6.3$

The species $HgOH^+$ can also form in solution by the following reaction

$$Hg^{2+}{}_{(aq)} + H_2O \rightleftharpoons H^+ + HgOH^+; \quad \log K = -3.7$$

but the $Hg(OH)_2{}^\circ$ complex generally predominates; to simplify this example, we neglected $HgOH^+$. If this species had not been neglected, our approach would have been similar to that used to determine the pH of a diprotic acid solution.

Mass Balances

(2) $C_{T,Hg} = [Hg^{2+}] + [Hg(OH)_2{}^\circ] = 10^{-5}$

(3) $C_{T,ClO_4} = 2 \times 10^{-5} = [ClO_4^-] = 2C_{T,Hg}$

Charge Balance

(4) $2[Hg^{2+}] + [H^+] = [ClO_4^-] + [OH^-]$

Dissociation of Water

(5) $K_w = 10^{-14} = [H^+][OH^-]$

There are five unknowns, $[H^+]$, $[OH^-]$, $[Hg^{2+}]$, $[Hg(OH)_2{}^\circ]$ and $[ClO_4^-]$, and we have stated five equations. We will combine these equations to arrive at one equation in $[H^+]$, which we will solve by trial and error.

First let us set

$$\alpha_2 = \frac{[Hg(OH)_2{}^\circ]}{C_{T,Hg}}$$

Combining Eqs. (1) and (2) yields

(6) $\alpha_2 = \dfrac{10^{-6.3}}{([H^+]^2 + 10^{-6.3})}$

Then, combining Eqs. (2), (3), and (4) to eliminate $[ClO_4^-]$, we obtain

(7) $[H^+] = 2[Hg(OH)_2{}^\circ] + [OH^-]$

and substituting for $[Hg(OH)_2{}^\circ]$ from Eq. (6), and for $[OH^-]$ from Eq. (5),

$$[H^+] - \frac{K_w}{[H^+]} - 2\alpha_2 C_{T,Hg} = 0$$

or

(8) $[H^+] - \dfrac{10^{-14}}{[H^+]} - 2 \times 10^{-5}\alpha_2 = 0$

Solving Eqs. (6) and (8) by trial and error yields a solution pH of 4.7.

The pH found in Example 5-2 is quite low, showing that the mercuric ion has the property of a weak acid of approximately the same strength as acetic acid. The acid strength of some metal ions is quite considerable. For example, the pH of equimolar H_3PO_4 and Fe^{3+} solutions are similar. The strong acid properties of hydrated metal ions become evident in processes such as coagulation/flocculation, where hydrated aluminum sulfate (alum) or ferric chloride is added to a natural water that is buffered with bicarbonate. The metal ion "titrates" (or reacts with) a stoichiometric amount of alkalinity, just as the addition of an equal amount of strong mineral acid would.

Example 5-3

A raw water with a pH of 8.0 and a total alkalinity of 80 mg/liter as $CaCO_3$ requires a dose of 25 mg/liter aluminum sulfate $(Al_2(SO_4)_3 \cdot 14H_2O)$ for effective coagulation/flocculation. What is the dose of sodium hydroxide (NaOH) in mg/liter required to maintain the alkalinity of the water at 80 mg/liter as $CaCO_3$ during coagulation/flocculation? Make the simplifying assumption that the only hydroxoaluminum(III) species to form is $Al(OH)_{3(s)}$; this assumption is valid at many surface water treatment plants.

Solution

We can treat this problem as a stoichiometric acid/base reaction. Since the alkalinity for the reaction will come entirely from the NaOH to be added, we use it as the base in the reaction.

$$Al_2(SO_4)_3 \cdot 14H_2O + 6NaOH \rightleftharpoons 2Al(OH)_{3(s)} + 6Na^+ + 3SO_4^{2-}$$

The molecular weight of alum is 594 and that of NaOH is 40. Thus

$$\frac{25 \text{ mg/liter}}{594 \text{ mg alum/mM}} \times \frac{6 \text{ mM NaOH}}{\text{mM alum}} = 0.252 \text{ mM NaOH/liter}$$

is required, or

$$\frac{0.252 \text{ mM}}{\text{liter}} \times \frac{40 \text{ mg NaOH}}{\text{mM}} = 10.1 \text{ mg NaOH/liter}$$

We assumed in Example 5-3 that when alum was added to water, there was a stoichiometric reaction with the hydroxide to form aluminum hydroxide. Although this assumption is good enough for estimating the amount of natural alkalinity that reacts, or the amount of hydroxide that must be added to restore that which reacted with the alum, it does not accurately represent the detailed equilibrium picture. As we have seen already, a variety of mononuclear hydroxoaluminum(III) complexes form. Aqueous solutions of aluminum ions at neutral pH values have also been reported to contain polynuclear hydroxoaluminum(III) complexes,[8] such as $Al_2(OH)_2^{4+}$, $Al_7(OH)_{17}^{4+}$ and $Al_{13}(OH)_{34}^{5+}$. It is thought that such species are formed by the condensation of mononuclear hydrolysis products (see Section 5-2).

Although the above species have been proposed as stable species, other polynuclear hydroxoaluminum(III) complexes are thought to exist as short-lived intermediates between the hydrated aluminum ion and aluminum hydroxide precipitate $(Al(OH)_{3(s)})$ in supersaturated solution.[9] We shall indeed see (Chapter 6) that when we consider the presence of the stable polynuclear aluminum hydrolysis species in the representation of heterogeneous equilibria, the stability region of the aluminum hydroxide solid is more confined than when their presence is disregarded. The polynuclear complexes generally have higher OH^- and less H_2O content per mole of aluminum than the mononuclear hydroxoaluminum complexes. Some workers suggest that the higher OH^- content of these species makes "hydrolyzed aluminum ion" a more effective coagulant in water treatment than the mononuclear hydrolysis species.[10] They argue that the polynuclear species act as short-chain polymers that adsorb to specific sites on particles, thereby destabilizing them so aggregation can take place. Although adsorption can take place even if the polymer and particle are of the same charge, destabilization appears more efficient if they have opposite charges. The classical "residual turbidity" versus "alum dose" plot, Fig. 5-1, illustrates how alum functions as a coagulant for a water of moderate alkalinity. At low alum doses there is no reduction in turbidity. At these low doses there is insufficient hydroxoaluminum(III) species to provide effective destabilization. The further addition of alum to the point at which complete destabilization occurs causes a reduction in turbidity to a minimum value. The further increase in alum dose will result in restabilization of the particles because of a near complete coverage of the particle with aluminum hydrolysis product. The further addition of alum to very high doses results in the formation of a precipitate of $Al(OH)_{3(s)}$ because the solubility product of $Al(OH)_{3(s)}$ is exceeded (see Chapter 6). This bulky precipitate enmeshes particles in it and settles rapidly to form the so-called "sweep floc" region of coagulation-flocculation. At most water treatment plants, coagulation/flocculation takes place in the "sweep floc" range because it is very difficult to vary the

[8] J. J. Morgan, chapter in "Equilibrium Concepts in Natural Water Systems," *Adv. in Chem. Series No. 67*, American Chemical Soc., Washington, D.C., 1967.

[9] W. Stumm and J. J. Morgan, *Aquatic Chemistry*, Wiley-Interscience, New York, 1970, p. 246.

[10] W. Stumm and J. J. Morgan, *Aquatic Chemistry*, Wiley-Interscience, New York, 1970, pp. 256–258.

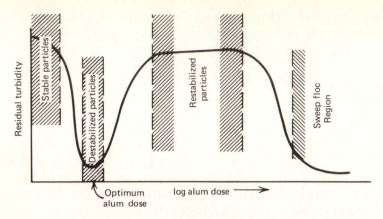

Fig. 5-1. Alum dose versus residual turbidity for water coagulation/flocculation. The residual turbidity is that which remains in a test solution to which alum was added. After mixing to simulate that which occurs in a water treatment plant, the sample is allowed to settle for 30 min before turbidity is measured.

coagulant dose to correspond to changing influent conditions as required to operate in the range of complete particle destabilization.[11]

Trivalent iron, Fe(III), which like aluminum(III) is used in water and wastewater treatment as a coagulating agent, as well as other multivalent metal cations (e.g., lead), form polynuclear hydroxocomplexes. These complexes tend to predominate at high metal concentrations and intermediate to high pH values. For example, Fig. 5-2 is a distribution diagram of the hydroxocomplexes of Pb^{2+} in aqueous solution. Below pH 6, Pb^{2+}

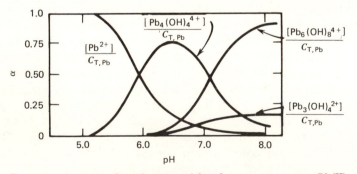

Fig. 5-2. Species distribution of lead in an aqueous Pb(II) solution $C_{T,Pb} = 0.04$ M. After Olin A., *Svensk Kim Tidskr.*, 73: 482 (1961). Reprinted by permission of *Svensk Kim. Tidskr.*

[11] See C. R. O'Melia, chapter in "Physiochemical Processes," by W. J. Weber, Jr., Wiley-Interscience, New York, 1971, for a complete review and discussion of coagulation/flocculation.

is the major lead-containing species; at higher pH values the polymeric hydroxocomplexes $Pb_4(OH)_4^{4+}$, $Pb_6(OH)_8^{4+}$, and $Pb_3(OH)_4^{2+}$ predominate. The equilibria on which the diagram is based are

$$4Pb^{2+} + 4H_2O \rightleftharpoons Pb_4(OH)_4^{4+} + 4H^+; \qquad \log K = -19.9$$

$$3Pb^{2+} + 4H_2O \rightleftharpoons Pb_3(OH)_4^{2+} + 4H^+; \qquad \log K = -23.2$$

$$6Pb^{2+} + 8H_2O \rightleftharpoons Pb_6(OH)_8^{4+} + 8H^+; \qquad \log K = -42.7$$

Similar distribution diagrams for concentrations of total iron(III) of $C_{T,Fe(III)} = 10^{-2}\ M$ and $C_{T,Fe(III)} = 10^{-4}\ M$ are presented in Fig. 5-3. The diagrams are based on equilibrium constants of mononuclear hydroxo complexes (see Table 5-1). The equilibrium constant for the dimeric $Fe_2(OH)_2^{4+}$ is $2Fe^{3+} + 2H_2O \rightleftharpoons Fe_2(OH)_2^{4+} + 2H^+$; $\log K = -2.91$. Like lead, the fraction of the total iron(III) present as hydroxo complexes, including the dimeric $Fe_2(OH)_2^{4+}$, increases as the pH increases. Also illustrated in Fig. 5-3 is the effect of total iron(III) concentration on the predominance of polymeric hydroxocomplexes of Fe(III). At pH 2, for example, virtually no $Fe_2(OH)_2^{4+}$ is present in the $10^{-4}\ M\ C_{T,Fe(III)}$ solution whereas a significant amount exists in the $10^{-2}\ M\ C_{T,Fe(III)}$ solution.

The previously discussed interaction of hydroxoaluminum(III) polymers with particle surfaces is but one example of the ways in which the types of species of metals, especially the complexes, influence the distribution of metals in the natural water environment. The bulk of the total metal content of natural waters and wastewaters is usually associated with particulates either as precipitates (solids) or adsorbed on particle surfaces, such as clays or organic detritus. In general, hydroxo, carbonato, and

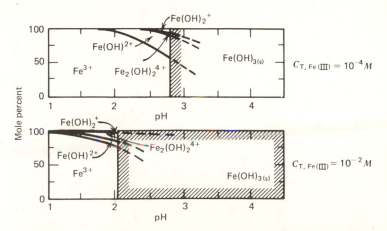

Fig. 5-3. Species distribution in an aqueous Fe(III) solution. After W. Stumm and J. J. Morgan, *Aquatic Chemistry*, Wiley-Interscience, New York, 1970. Reprinted by permission of John Wiley & Sons.

sulfato complexes of metals tend to sorb more strongly at clay surfaces than do the free metal ions. For example, James and Healy[12] have shown that the adsorption of cobalt on silica surfaces is a function of the ionic state of the cobalt (see Table 5-2 and Fig. 5-4). At pH values below 6, where the cobalt is present entirely as Co^{2+}, there is very little adsorption.

TABLE 5-2   Cobalt Equilibria

Equilibria	log K
$Co^{2+} + 2OH^- \rightleftharpoons Co(OH)_{2(s)}$	14.9
$Co(OH)^+ + H^+ \rightleftharpoons Co^{2+} + H_2O$	9.6
$Co(OH)_2^\circ + H^+ \rightleftharpoons Co(OH)^+ + H_2O$	9.2
$Co(OH)_2^\circ \rightleftharpoons Co(OH)_{2(s)}$	5.7
$Co(OH)_3^- + H^+ \rightleftharpoons Co(OH)_2^\circ + H_2O$	12.7

From L. G. Sillen and A. E. Martell. *Chemical Society of London, Special Bull., No. 17* (1964).

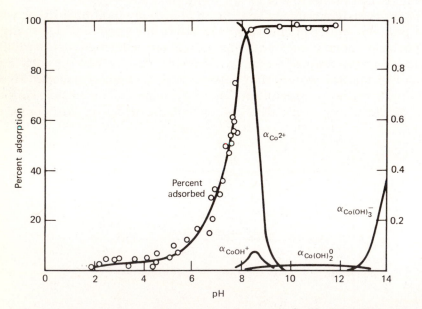

Fig. 5-4. Experimental adsorption isotherm for Co(II). Adsorption at $1.2 \times 10^{-4} M$ on silica at 25°C; 15 g/liter SiO_2. Computed hydrolysis data for this concentration are also shown as the percentage of each aquo complex as a function of pH. Reprinted from R. O. James and T. W. Healy, *J. Colloid and Interface Sci.*, 40: 42–52 (1972) by permission of Academic Press.

[12] R. O. James and T. W. Healy, "Adsorption of Hydrolyzable Metal Ions at the Oxide-Water Interface, I. Co(II) Adsorption on SiO_2 and TiO_2 as Model Systems," *J. Colloid Interfac. Sci.*, 40: 42 (1972).

As the pH increases to between 6 to 8.5, the removal of cobalt from solution by adsorption on the silica particles increases dramatically. In the pH range from 6 to 8.5 the free Co^{2+}, ion decreases from 100 percent of the total cobalt to zero, and the predominant cobalt species become the adsorbable hydroxocobalt(II) complexes, $CoOH^+$ and $Co(OH)_2^0$.

5.6.   COMPLEXES WITH OTHER INORGANIC LIGANDS

Natural waters contain significant concentrations of inorganic and organic ligands in addition to H_2O and OH^-. In this section we will look at some examples of the effects of inorganic ligands on the speciation and properties of metals in natural waters. The next section of this chapter will deal with organic ligands.

Reference to Table 1-5 shows that the predominant metal cations in typical fresh and ocean waters are Na^+, Ca^{2+}, and Mg^{2+}; the major ligands are HCO_3^-, Cl^-, and SO_4^{2-}. At the typical natural water pH values of between 6.5 to 8.5, these metals are not strong complex formers in comparison with metals such as Al^{3+} and Fe^{3+}.

For complicated systems such as natural waters it is necessary to use a computer to determine the degree of complexation of the various cations with the ligands available. To illustrate this technique, we determine the degree of complexation of the various cations present in the ocean and in average river water.[13]

Assume the following compositions for ocean and river water at 25°C:

	Ocean	Average River Water
$C_{T,Na}(M)$	0.47	2.7×10^{-4}
$C_{T,K}(M)$	10^{-2}	5.9×10^{-5}
$C_{T,Ca}(M)$	10^{-2}	3.8×10^{-4}
$C_{T,Mg}(M)$	5.4×10^{-2}	3.4×10^{-4}
$C_{T,Cl}(M)$	0.55	2.2×10^{-4}
$C_{T,SO_4}(M)$	3.8×10^{-2}	1.2×10^{-4}
Total alkalinity (eq/liter)	2.3×10^{-3}	9.6×10^{-4}
pH (units)	7.9	7.9

We neglect interactions with the atmosphere and consider the following species:

K^+, Na^+, Ca^{2+}, Mg^{2+}, H^+, Cl^-, SO_4^{2-}, HSO_4^-, CO_3^{2-}, HCO_3^-, $H_2CO_3^*$, OH^-, $CaOH^+$, $CaHCO_3^+$, $MgOH^+$, $MgHCO_3^+$, $NaCO_3^-$, $NaSO_4^-$, $CaCO_3^0$, $CaSO_4^0$, $MgCO_3^0$, $MgSO_4^0$, $NaOH^0$, $NaHCO_3^0$, and KSO_4^-

---

[13] These calculations were made by Dr. L. L. Russell, James M. Montgomery Engineers, Walnut Creek, Calif.

The equations that describe the system are the mass balances:

$$C_{T,K} = [K^+] + [KSO_4^-]$$

$$C_{T,Ca} = [Ca^{2+}] + [CaOH^+] + [CaHCO_3^+] + [CaCO_3^0] + [CaSO_4^0]$$

$$C_{T,Mg} = [Mg^{2+}] + [MgOH^+] + [MgHCO_3^+] + [MgCO_3^0] + [MgSO_4^0]$$

$$C_{T,Na} = [Na^+] + [NaOH^0] + [NaHCO_3^0] + [NaCO_3^-] + [NaSO_4^-]$$

$$C_{T,Cl} = [Cl^-]$$

$$C_{T,CO_3} = [H_2CO_3^*] + [HCO_3^-] + [CO_3^{2-}] + [CaHCO_3^+] + [CaCO_3^0]$$

$$C_{T,SO_4} = [SO_4^{2-}] + [HSO_4^-] + [CaSO_4^0] + [MgSO_4^0] + [NaSO_4^-] + [KSO_4^-]$$

Charge Balance

$$[K^+] + 2[Ca^{2+}] + [CaOH^+] + [CaHCO_3^+] + 2[Mg^{2+}] + [MgOH^+]$$
$$+ [MgHCO_3^+] + [H^+] + [Na^+] = [KSO_4^-] + [NaCO_3^-] + [NaSO_4^-]$$
$$+ [HCO_3^-] + 2[CO_3^{2-}] + [Cl^-] + [OH^-]$$

Alkalinity Equation

$$\text{Total alkalinity (eq/liter)} = [HCO_3^-] + 2[CO_3^{2-}] + [OH^-] - [H^+]$$

and the equilibrium relationships:

$$H_2O \rightleftharpoons H^+ + OH^-; \quad K_w = 10^{-14}$$

$$HSO_4^- \rightleftharpoons H^+ + SO_4^{2-}; \quad K_{a,2} = 10^{-2}$$

$$H_2CO_3^* \rightleftharpoons H^+ + HCO_3^-; \quad K_{a,1} = 10^{-6.3}$$

$$HCO_3^- \rightleftharpoons H^+ + CO_3^{2-}; \quad K_{a,2} = 10^{-10.3}$$

$$CaOH^- \rightleftharpoons Ca^{2+} + OH^-; \quad K = 10^{-1.3}$$

$$CaHCO_3^- \rightleftharpoons Ca^{2+} + HCO_3^-; \quad K = 10^{-1.26}$$

$$CaSO_4^0 \rightleftharpoons Ca^{2+} + SO_4^{2-}; \quad K = 10^{-2.31}$$

$$CaCO_3^0 \rightleftharpoons Ca^{2+} + CO_3^{2-}; \quad K = 10^{-3.2}$$

$$MgOH^+ \rightleftharpoons Mg^{2+} + OH^-; \quad K = 10^{-2.58}$$

$$MgHCO_3^+ \rightleftharpoons Mg^{2+} + HCO_3^-; \quad K = 10^{-1.16}$$

$$MgCO_3^0 \rightleftharpoons Mg^{2+} + CO_3^{2-}; \quad K = 10^{-3.4}$$

$$MgSO_4^0 \rightleftharpoons Mg^{2+} + SO_4^{2-}; \quad K = 10^{-2.36}$$

$$NaOH \rightleftharpoons Na^+ + OH^-; \quad K = 10^{+0.7}$$

$$NaHCO_3^0 \rightleftharpoons Na^+ + HCO_3^-; \quad K = 10^{+0.25}$$

$$NaCO_3^- \rightleftharpoons Na^+ + CO_3^{2-}; \quad K = 10^{-1.27}$$

$$NaSO_4^- \rightleftharpoons Na^+ + SO_4^{2-}; \quad K = 10^{-0.72}$$

$$KSO_4^- \rightleftharpoons K^+ + SO_4^{2-}; \quad K = 10^{-0.96}$$

Tables 5-3 to 5-6 are the results of the solution of these equations.

TABLE 5-3   Concentrations of Various Species in Ocean Water (moles/liter).

Cation	Uncomplexed	OH$^-$	HCO$_3^-$	CO$_3^{2-}$	SO$_4^{2-}$	Total
			Complexed with			
Calcium	7.8×10^{-3}	5.2×10^{-8}	6.8×10^{-5}	1.8×10^{-5}	2.1×10^{-3}	10^{-2}
Magnesium	4.1×10^{-2}	5.3×10^{-6}	2.8×10^{-4}	1.5×10^{-4}	1.2×10^{-2}	5.4×10^{-2}
Sodium	4.5×10^{-1}	5.4×10^{-8}	2.2×10^{-4}	3.8×10^{-5}	9.9×10^{-3}	4.7×10^{-1}
Potassium	9.6×10^{-3}	—	—	—	3.6×10^{-4}	10^{-2}
Hydrogen					7.0×10^{-8}	
Uncomplexed			1.5×10^{-3}	1.4×10^{-5}	1.3×10^{-2}	

TABLE 5-4   Percentage Distribution of Complex Species in Ocean Water.

Cation	Uncomplexed	OH$^-$	HCO$_3^-$	CO$_3^{2-}$	SO$_4^{2-}$
			Complexed with		
Calcium	78	—	0.7	0.2	21
Magnesium	76	—	0.5	0.3	22
Sodium	96				2
Potassium	96				4

TABLE 5-5   Concentration of Various Species in Average River Water (moles/liter).

Cation	Uncomplexed	OH$^-$	HCO$_3^-$	CO$_3^{2-}$	SO$_4^{2-}$	Total
			Complexed with			
Calcium	3.7×10^{-4}	5.1×10^{-9}	5.2×10^{-6}	1.7×10^{-6}	5.5×10^{-6}	3.8×10^{-4}
Magnesium	3.3×10^{-4}	8.6×10^{-8}	3.7×10^{-6}	2.4×10^{-6}	5.5×10^{-6}	3.4×10^{-3}
Sodium	2.7×10^{-4}	4.1×10^{-10}	1.3×10^{-7}	1.8×10^{-8}	1.3×10^{-7}	2.7×10^{-4}
Potassium	5.9×10^{-4}	—	—	—	4.8×10^{-7}	5.9×10^{-4}
Hydrogen	—	—	—	—	1.2×10^{-10}	—
Uncomplexed	—	—	9.4×10^{-4}	4.3×10^{-6}	1.1×10^{-4}	—

TABLE 5-6   Percentage Distribution of Complex Species in Average River Water.

Cation	Uncomplexed	OH$^-$	HCO$_3^-$	CO$_3^{2-}$	SO$_4^{2-}$
			Complexed with		
Calcium	97		1	1	1
Magnesium	97		1	1	1
Sodium	100				
Potassium	100				

The results show that only in ocean water are any of the metals in this model complexed to any significant extent by the inorganic ligands examined. In addition only sulfate is a significant complexing agent and only calcium and magnesium complexes are important. This type of calculation gives us confidence in ignoring the complexes of Ca^{2+}, Mg^{2+}, Na^+, and K^+ with OH^-, HCO_3^-, Cl^-, SO_4^{2-}, and CO_3^{2-}, when dealing with fresh waters of neutral pH values. Increasing the pH value or ligand concentration, especially sulfate, may require that complexes of these metals be taken into consideration.

Several analytical methods will differentiate the "free" (hydrated) metal ions from dissolved complexed metal ions. These methods include specific ion electrodes, polarographic, and other amperometric and voltammetric methods and various types of spectroscopy (see Section 7-10). Specific ion electrodes only respond to the free metal ion for which they are "specific." To determine the relative amounts of complexed and uncomplexed metal ion in a solution, we can use a "wet chemical" method to measure the total concentration of "free + complexed" ions, and then an ion-specific electrode to determine the free metal ion concentration (activity). Care must be taken to eliminate interferences that may affect these measurements. We deduce the concentration of the "complexed ions" by the difference between these two measurements. For example, in the EDTA titration method for hardness,[14] free and complexed calcium and magnesium ions are measured.

Methods such as anodic stripping voltammetry, polarography, and spectroscopy not only can differentiate between "free" and "complexed" metal ion species but in some instances can be used to identify the nature and determine the concentrations of individual complex species.

Inorganic ligands that complex metal ions can alter the effects that metal ions exert in natural waters. For example, the literature on fish toxicity contains references to the effects of water hardness on the toxicity of heavy metals. The general observation made is that heavy metals are much less toxic to fish in hard water than they are in soft water. We can provide an explanation of this result in terms of complex formation between the heavy metal copper, and the alkalinity ions (HCO_3^- and CO_3^{2-}), since the alkalinity usually increases together with the water hardness.

The distribution of copper species in a natural water buffered by the carbonate system is presented in Fig. 5-5. The diagram is based on the presence of the copper precipitate tenorite, $CuO_{(s)}$. If the solution were not in equilibrium with this solid, the species distribution would not be quite the same; however, radical changes would not occur. Therefore, the diagram can be used to determine regions where certain species are important. From Fig. 5-5 it can be seen that only at below pH 6.5 is free

[14] *Standard Methods for Analysis of Water and Wastewater*, 14th ed., Am. Publ. Health Assoc., New York, 1975.

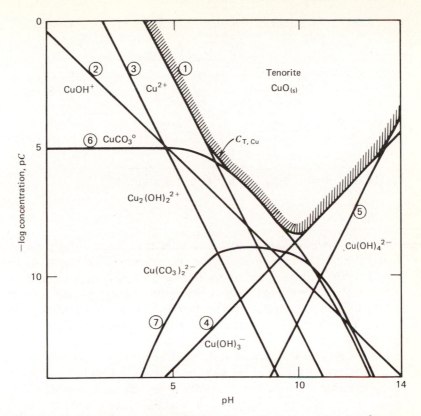

Fig. 5-5. Copper species distribution in a water containing total inorganic carbon, C_{T,CO_3}, $= 10^{-3}\ M$.

copper ion, Cu^{2+}, the predominant copper-containing species. At higher pH values the carbonatocopper(II) complex, $CuCO_3^\circ$, and the hydroxocopper(II) complexes, $Cu(OH)_3^-$ and $Cu(OH)_4^{2-}$, are the major copper-containing species. The diagram predicts that in the pH range 6.5 to 9.5 (a range that covers the pH span of most natural waters) the major copper-containing species is $CuCO_3^\circ$. Making the assumption that the complex formation reaction

$$Cu^{2+} + CO_3^{2-} \rightleftharpoons CuCO_3^\circ_{(aq)}; \qquad \log K = -6.8$$

controls the concentration of Cu^{2+} in this pH range, the Cu^{2+} concentration at various alkalinities can be calculated for a variety of pH values (Table 5-7).

It is apparent that alkalinity has a profound effect on the free copper ion concentration. For example, at pH 7 increasing the alkalinity from 50 to 250 mg/liter as $CaCO_3$ decreases the Cu^{2+} ion level from 25 to 9 percent of the total copper present. Returning to our original observation that

TABLE 5-7　Effect of Alkalinity and pH on Cu^{2+} Levels

pH	$\dfrac{[Cu^{2+}]}{C_{T,Cu}} \times 100(\%)$ at Alkalinity of		
	50 mg/liter as $CaCO_3$	100 mg/liter as $CaCO_3$	250 mg/liter as $CaCO_3$
7	25	14	9
7.5	11	6	3
8	—	2	0.9

For $C_{T,Cu}$ values from 8×10^{-7} M to 1.6×10^{-5} M.

copper is more toxic to fish in soft water than in hard water, we should note that high hardness is usually accompanied by high alkalinity. Therefore, in hard water a greater fraction of the total copper is complexed than in soft water. If we assume that the only copper species toxic to fish is the Cu^{2+} ion, then increasing the fraction of the carbonatocopper(II) complex by increasing the carbonate concentration will decrease the toxicity.

Complex formation influences the concentration of ligands (and their effects) as well as metal ions. An important example of this relates to the toxicity of "cyanide" to fish. Most toxicity researchers agree that the toxic component of a "cyanide" solution is undissociated hydrocyanic acid, HCN. When a complex-forming metal ion is added to a cyanide solution, the equilibrium between HCN and CN^- will be displaced in the direction of CN^- because the metal ion will remove free CN^- from solution, incorporating it into a cyano metal complex. It is not uncommon to encounter waste discharges containing both heavy metals and cyanide, since metal cyanide solutions are used in a variety of metal finishing processes. Most heavy metals form cyano complexes. For example, the following equilibria are applicable to the addition of nickel to a cyanide solution,

$$HCN \rightleftharpoons H^+ + CN^-; \qquad K_a = 6.17 \times 10^{-10}$$

$$Ni^{2+} + 4CN^- \rightleftharpoons Ni(CN)_4{}^{2-}; \qquad \beta_4 = 10^{30}$$

There is no evidence for the existence of the mono-, di-, and tricyanonickel(II) complexes; therefore, we need only consider the $Ni(CN)_4{}^{2-}$ complex. Figure 5-6 shows the profound effect that this complex has on the HCN concentration.

At pH 7.5, for example, the solution containing nickel has an HCN level of about 3×10^{-6} M (~0.08 mg/liter) while in the cyanide solution containing no nickel the HCN level is 4×10^{-5} (~1.1 mg/liter); that is, all of the cyanide present is HCN. Knowing that the component toxic to fish

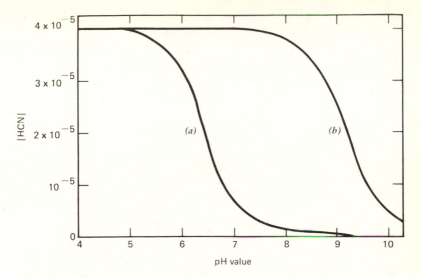

Fig. 5-6. Free HCN in equilibrium at 25°C with (a) $10^{-5} M$ nickel, and (b) $4 \times 10^{-5} M$ total cyanide, $C_{T,CN}$, with no correction for ionic strength. K_a for HCN $= 6.17 \times 10^{-10}$; β_4 for $Ni(CN)_4^{2-} = 10^{+30}$. Reprinted from H. A. C. Montgomery. "Some Analytical Aspects of the Behavior of Certain Pollutants in Aqueous Systems," *Progr. Water Technol.*, 3: 73 (1973) by permission of Pergamon Press.

is HCN, we can guess that the addition of Ni^{2+} to a cyanide solution would have a dramatic effect on its toxicity. Doudoroff et al.[15] have provided a striking illustration of this effect using the toxicity to fish of solutions of nickel, copper, zinc, cadmium, and silver cyanide complexes. When the median resistance (or immobilization) time of the fish is plotted against the total cyanide concentration of these solutions (log/log plot), no sensible relationship could be obtained. However, when the same plot was made against the concentration of molecular HCN in the complex cyanide solutions there was a linear relationship between HCN concentration and median resistance time (see Fig. 5-7).

5.7. COMPLEXES WITH ORGANIC LIGANDS

Complex formation involving organic ligands is important in two general areas of water chemistry. First, in the analysis of various constituents of water, organic compounds are used as complexing agents. For example, EDTA is used to determine water hardness by forming

[15] P. Doudoroff, G. Leduc, and C. R. Schneider, "Acute Toxicity to Fish of Solutions Containing Complex Metal Cyanides, in Relation to Concentrations of Molecular Hydrocyanic Acid," *Trans. Amer. Fish Soc.*, 95: 6–22 (1966).

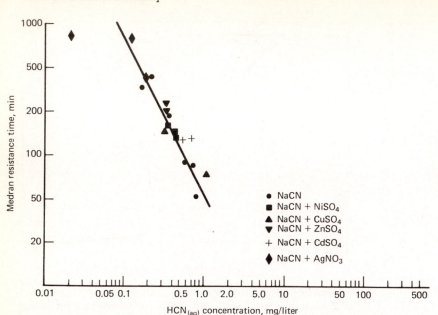

Fig. 5-7. Bluegill median resistance (immobilization) times plotted against free, HCN concentration in various simple and complex cyanide solutions. Adapted from P. Doudoroff, et al., *Trans. Amer. Fish. Soc.*, 95: 6–22 (1966). Reprinted by permission of the American Fisheries Society.

complexes with Ca^{2+} and Mg^{2+}; 1,10-phenanthroline is used as an indicator in the COD test, forming a complex with Fe^{2+}, and dimethylglyoxime,

$$CH_3—CNOH$$
$$|$$
$$CH_3—CNOH$$

can be used for the colorimetric determination of nickel by forming the red complex

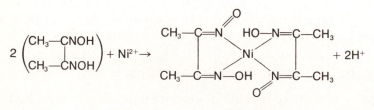

red complex

Second, a wide variety of organic compounds in natural waters and wastewater can act as complexing agents for metal ions. The nature and extent of metal ion complexation by natural water organics is not well-defined, probably because of the poorly defined nature of these organic compounds and also because of the staggering complexity of these multimetal, multiligand systems. Several examples will be given here to illustrate the importance of complexation by various organic compounds in determining free metal ion concentration in natural waters and wastewaters and to illustrate the complexity of the systems.

5.7.1. The Nature of Copper in Water and Wastewater

Copper forms complexes with numerous ligands, both inorganic and organic that are present in water and wastewater. Table 5-8 gives some examples of the types of soluble complexes detected by Stiff[16] in natural and treated waters and sewage. Beside complexes with inorganic ligands (i.e., carbonate and cyanide), the river waters contained significant amounts of copper complexed by organic ligands, that is, in the categories of "amino acid complex" and "inert humic complex." Amino acid complexes of copper are formed according to the following general equation:

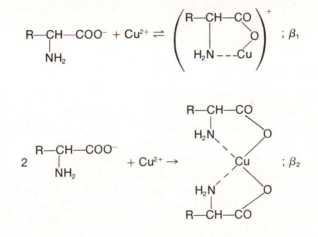

TABLE 5-8   Distribution of Soluble Copper in Natural and Treated Water and Wastewater

| | | | | Hardness, | | | Copper Concentration, μg/liter, in the form of | | | |
Sample	Copper added μg/liter	pH	HCO_3^- mmole/liter	mg/liter as $CaCO_3$	Cu^{2+}	$CuCO_3^\circ$	Amino Acid Complex	Inert Humic Complex	Hexanol Extractable	Cyanide complex, etc.
Tap water Stevenage	200	7.51	5.6	320	5.8	202	—	—	—	—
Settled sewage Stevenage	200	7.87	9	250	0.03	3.5	32	40	24	100
River Arrow	800	7.62	5.1	400	11	435	126	48	—	200
River Lee at Edmonton	800	8.19	5.7	340	0.9	148	480	—	170	—

Source: M. J. Stiff, *Water Research*, 5, 585–599 (1971). Reprinted by permission of Pergamon Press, Elmsford, N.Y.

[16] M. J. Stiff, "The Chemical States of Copper in Polluted Fresh Water and a Scheme of Analysis to Differentiate Them," *Water Research*, 5: 585–599 (1971).

For the amino acids leucine, valine, alanine, serine, glutamic acid, aspartic acid, and tyrosine, the values of log β_1 range from 7 to ~9; values of log β_2 are in the approximate range of 14 to 16. The amino acid cysteine, which contains the –SH group forms an extremely strong complex with cuprous ion (Cu^+) with a log β_1 value of 19.5.

As we will see later, "humic acids" are an extremely complex group of compounds. Because of this no one compound can act as a true model for this group. However, phenolic and carboxylic acid groups are common functional groups in humic acids. The substance 3,4-dihydroxybenzoic acid serves to illustrate one type of complexing group that may bind metals such as copper to humic acids:

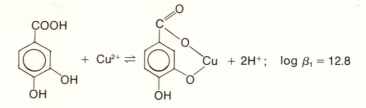

3,4 dihydroxybenzoic acid

From the values of the stability constants of these organic complexes of copper, it is evident that they must play a significant role in the complexation of copper in natural waters. Indeed, it is significant to note that in none of the waters listed in Table 5-8, is Cu^{2+}—the uncomplexed cupric ion—a major copper-containing species. The picture becomes even more complex when the distribution of copper between solution and particulate phases is considered. Beside the copper-containing precipitates (e.g., tenorite, $CuO_{(s)}$) we must also realize that the types of organic and inorganic ligand groups which complex with copper in solution also can exist on the surfaces of particulate solids such as organic detrital material; there they serve to bind the copper to a solid particle and transfer it from the "soluble copper" category to the "particulate copper" category. Stiff[17] analyzed some polluted river waters in England and found that between 43 to 88 percent (average 69 percent) of the total copper present was in the particulate phase.

5.7.2.   Complexation by NTA

Nitrilotriacetic acid (NTA) is a triprotic acid with the structure:

$$COOHCH_2{-}N\begin{array}{l}CH_2COOH \\ \\ CH_2COOH\end{array}$$

nitrilotriacetic acid, NTA

[17] M. J. Stiff, *Water Research*, 5: 585–599 (1971).

Its sodium salt, $N(CH_2COONa)_3$, can be used as a component of commercial laundry detergents. In these products the NTA is a so-called "builder." Its function is to complex the components of water hardness ($Ca^{2+} + Mg^{2+}$) and prevent them from reacting with the detergent molecule itself, thereby reducing the effectiveness of the detergent. This type of deliberate complexation is often called "sequestration," and compounds such as NTA are known as "sequestering agents." Commercial laundry detergents containing NTA were developed because some regions of the United States and Canada prohibited the use of detergents containing the usual sequestering agents for Ca^{2+} and Mg^{2+}—the polyphosphates—because of their suspected contribution to causing accelerated eutrophication of surface waters.

The introduction of NTA caused concern in some segments of the scientific community on several counts. It was thought that NTA introduced into detergents would find its way into domestic sewage. Being a strong complexing agent, it will keep toxic heavy metals in solution, thereby preventing their precipitation as hydroxides, carbonates, phosphates, sulfides, and so forth, and their subsequent removal by incorporation into wastewater-derived sludges. If the NTA itself and the NTA heavy metal complexes are not readily biodegradable, they will pass through the wastewater treatment processes, carrying into the receiving waters both their complexing ability and any heavy metals complexed by them. Once in the receiving water, free NTA will have the ability to transport heavy metals from sediments into solution by complexing them. In the course of time the NTA heavy metal complexes will be biodegraded, releasing the complexed heavy metal that will cause toxicity in the receiving water.

Some aspects of these concerns, such as the question of biodegradability of NTA and its heavy metal complexes, are beyond the scope of this text. However, NTA appears to be biodegradable in secondary, biological, waste treatment processes. The concerns dealing with complexation will be examined here.

Because NTA forms many metal complexes with a wide range of stabilities, it is important to examine a model or conduct experiments on systems where the concentrations of the metals and the amount of NTA available for complexing closely resemble that found in the natural water system of interest. In addition, the presence of other ligands, both inorganic and organic, should be considered. Because pH influences the speciation of NTA and the concentrations of other ligands such as CO_3^{2-}, OH^-, and so forth, the pH of the system under study should be carefully controlled in experimental and model systems.

Equilibrium models that consider many metals and many ligands are constructed using the same set of equations used for simple acid-base and complexation calculations. First, total concentrations of all components are stated; then all possible species are identified; and mass balances, a charge balance, and equilibrium equations are written. It is, however, tedious and virtually impossible to solve these equations manually so that a computer solution is needed. The reader is referred to published articles for descriptions of computer programs suitable for the solution of multimetal, multiligand systems.[18] Here we will examine the results of such a solution. Consider the system shown in Table 5-9 which represents the composition of a lake water at pH 8. The model does not consider the presence of organic ligands other than NTA. However, it is estimated that the stability constants of NTA metal complexes are significantly greater than other organic-matter metal complexes. Because of this the result should approximate what might be found in a water containing a low organic-matter concentration. The acid base equilibria of NTA, carbonic acid, and phosphoric acid and the complex formation of all metals with OH^-, HCO_3^-, CO_3^{2-}, PO_4^{3-}, SO_4^{2-}, and Cl^- as well as with NTA^{3-} were considered. The results of the calculations will be discussed here; for details of the equations used see Childs.[19] The results of the computations are presented in Table 5-10, which shows that at pH 8, NTA, just like any other complexing agent, complexes metals in order of the values of the stability constants of the NTA metal complex. At an NTA concentration of $10^{-7} M$, there is only sufficient NTA to result in minor complexation of only the metals with the largest NTA metal complex stability constants. Thus 4 percent of the total Cu(II) (log $K = 13$, where K is for the reaction $Cu^{2+} + NTA^{3-} \rightarrow CuNTA^-$), 2 percent of the total Pb(II)

TABLE 5-9

Major Components	Moles/liter	Minor Components	Moles/liter
$C_{T,Ca}$	10^{-3}	$C_{T,Cu(II)}$	2×10^{-6}
$C_{T,Mg}$	2.5×10^{-4}	$C_{T,Fe(III)}$	2×10^{-6}
$C_{T,Na}$	5×10^{-4}	$C_{T,Sr(II)}$	2×10^{-6}
C_{T,CO_3}	10^{-3}	$C_{T,Mn(II)}$	2×10^{-6}
C_{T,SO_4}	3×10^{-4}	$C_{T,Zn(II)}$	1.5×10^{-6}
$C_{T,Cl}$	7.5×10^{-4}	$C_{T,Pb(II)}$	3×10^{-7}
		$C_{T,Ba(II)}$	1.5×10^{-7}
		$C_{T,Ni(II)}$	10^{-7}
		C_{T,PO_4}	10^{-6}
		$C_{T,NTA}$	10^{-8} to 2×10^{-4}

[18] J. N. DeBoice and J. F. Thomas, "Chemical Treatment for Phosphate Control," J. Water Pollut. Control Fed., 47: 2246 (1975); F. Morel and J. J. Morgan, "A Numerical Method for Computing Equilibria in Aqueous Chemical System," Env. Sci. Technol., 6: 58 (1972); and D. D. Perrin and I. G. Sayce, "Computer Calculation of Equilibrium Concentrations in Mixtures of Metal Ions and Complexing Species," Talanta, 14: 833–842 (1967).

[19] C. W. Childs, "Chemical Equilibrium Models for Lake Water which Contains Nitrilotriacetate and for 'Normal' Lake Water," Proc. 14th Conf. Great Lakes Res., 198–210 (1971), Intl. Assoc. Great Lakes Res.

TABLE 5-10   Variation of Metal Complexation by NTA with NTA Concentration at pH 8

Total Concentration $C_{T,x}$ M	Complex Species	Log Formation Constant of	Percentage of Total Metal Present as Indicated Complex at Stated NTA Concentration		
			NTA = $10^{-7}\,M$	NTA = $3 \times 10^{-6}\,M$	NTA = $2 \times 10^{-4}\,M$
Cu(II) = 2×10^{-6}	CuNTA$^-$	13	4	82	100
Pb(II) = 3×10^{-7}	PbNTA$^-$	11.8	2	80	100
Ni(II) = 10^{-7}	NiNTA$^-$	11.3	1	60	100
Fe(III) = 2×10^{-6}	Fe(OH)NTA$^-$ Fe(OH)$_2$NTA^{2-}	10.9⎱ 3.1⎰	0.4	34	100
Zn(II) = 1.5×10^{-6}	ZnNTA$^-$	10.4	0.2	20	100
H$^+$ = 10^{-8}	HNTA^{2-}	10.3	0	0	9
Mn(II) = 2×10^{-6}	MnNTA$^-$	7.4	0	0	100
Ca(II) = 10^{-3}	CaNTA$^-$	6.4	0	<0.1	17
Mg(II) = 2.5×10^{-4}	MgNTA$^-$	5.4	0	0	2
Sr(II) = 2×10^{-6}	SrNTA$^-$	5.0	0	0	0
Ba(II) = 1.5×10^{-7}	BaNTA$^-$	4.8	0	0	0
Na(I) = 5×10^{-4}	NaNTA^{2-}	2.2	0	0	0

Source: C. W. Childs, *Proc. 14th Conf. Great Lakes Res.*, 198–210 (1971). *Intl. Assoc. Great Lakes Res.* (Reprinted by permission of the International Association for Great Lakes Research.)

(log K = 11.8), and 1 percent of the total Ni(II) (log K = 11.3) as well as fractional percentages of Fe(III) (log K = 10.9) and Zn(II) (log K = 10.4) are complexed with NTA.

As the NTA supply increases to $3 \times 10^{-6}\,M$, significant complexing of the above metals takes place; for example Cu(II), 82 percent; Pb(II), 80 percent; Ni(II), 60 percent; Fe(III), 34 percent; and Zn(II), 20 percent. Again, since the NTA is not present in a quantity in excess of the combined concentration of all the heavy metals, there is incomplete complexing of each of them. Moreover, the completeness of complexation is in strict order of values of the stability constants of the heavy metal NTA complexes. At a total NTA level of 2×10^{-4} all of the heavy metals are completely complexed by NTA. In addition, some of the major cations, Ca(II) and Mg(II), which form weaker NTA complexes than the heavy metals, are bound to NTA at levels of 17 and 2 percent, respectively.

It is interesting to note that although the log $(1/K_{a,3})$ value for NTA is 10.3, (an equilibrium constant value that is comparable to the log K of the Zn-NTA complex (10.4)), only 9 percent of the H$^+$ is bound to NTA at total NTA levels of $2 \times 10^{-4}\,M$, while the Zn(II) is completely complexed by NTA. The difference in degree of complexation lies in the relative concentrations of Zn(II) and H$^+$. $C_{T,Zn(II)}$ is two orders of magnitude greater than [H$^+$] at pH 8, that is, $1.5 \times 10^{-6}\,M$ versus $10^{-8}\,M$. Comparison of the equilibria

$$\frac{[\text{Zn}-\text{NTA}^-]}{[\text{Zn}^{2+}][\text{NTA}^{3-}]} = 10^{10.4}$$

and

$$\frac{[\text{HNTA}^{2-}]}{[\text{H}^+][\text{NTA}^{3-}]} = 10^{10.3}$$

shows that for the same $[\text{NTA}^{3-}]$ concentration, for $[\text{H}^+] = 10^{-8}$ and $[\text{Zn}^{2+}] = 10^{-6}$,

$$\frac{\dfrac{[\text{ZnNTA}^-]}{(10^{-6})[\text{NTA}^{3-}]}}{\dfrac{[\text{HNTA}^{2-}]}{(10^{-8})[\text{NTA}^{3-}]}} = \frac{10^{10.4}}{10^{10.3}} = 0.1$$

$$(10^{-8}) \quad [\text{ZnNTA}^-] = [\text{HNTA}^{2-}](10^{-6})(0.1)$$

Therefore,

$$[\text{HNTA}^{2-}] = \frac{[\text{ZnNTA}^-]}{10}$$

That is, higher concentrations of metal ions as well as large complex formation constant values tend to favor the formation of NTA-metal complexes.

If we examine the other side of the coin, we can see the way in which the concentration of NTA and the relative values of the stability constants of metal ion-NTA complexes determine the distribution of added NTA between the various metal ion NTA complexes. Figure 5-8 shows that at very low levels of added NTA the NTA is primarily present as the CuNTA^- complex with minor amounts present as PbNTA^- and CaNTA^- complexes in descending order of importance. As the quantity of added NTA increases to levels such as those typically encountered (i.e., $\sim 10^{-6}\ M$) in waters receiving secondary sewage effluent from communities using NTA based detergents, the $\text{Fe(OH)}_2\text{NTA}^{2-}$, Fe(OH)NTA^-, and ZnNTA^- complexes start to be more important but, although decreasing in percentage, the CuNTA complex is still by far the most significant repository of added NTA. At added NTA levels of $10^{-4}\ M$, CaNTA accounts for close to 85 percent of the NTA in solution, HNTA^{2-} for close to 8 percent, and MgNTA^- for about 2 percent. The heavy metal complexes of NTA account cumulatively for less than 5 percent of the added NTA. It should be kept in mind, however, that while these heavy metal NTA complexes only account for a minor fraction of the added NTA, the heavy metals are completely complexed by NTA.

Table 5-10 and Fig. 5-8 emphasize the importance of conducting model studies and experiments on complex multi-metal, multi-ligand systems at concentrations of metals and ligands that closely resemble the environment being studied.

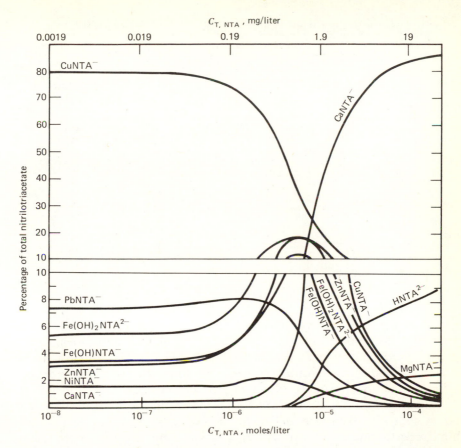

Fig. 5-8. Distribution of nitrilotriacetate in model lake water at pH 8. After C. W. Childs, *Proc. 14th Conf. Great Lakes Res.*, 198–210, 1971. Reprinted by permission of the International Association for Great Lakes Research.

The influence of NTA as a complexing agent on the distribution of metal species can be assessed by comparing the concentrations of various metal species both in the presence and absence of NTA.

Table 5-11 shows the calculated distribution of some selected metals in a water in the absence of NTA, but otherwise of the same composition as that used to compute the distribution of NTA complexes given in Table 5-10. NTA at $3 \times 10^{-6} M$ effectively draws the metals that are complexed by the inorganic ligands present from these complexes into NTA complexes. The metals that are poorly complexed by inorganic ligands are not significantly affected by the addition of this amount of NTA.

5.7.3. Metal Ion Association with Humic Substances

5.7.3.1. CHARACTERISTICS OF HUMIC SUBSTANCES Humic substances are an extremely complex and diverse group of organic materials

TABLE 5-11   Complexation of Metal Ions by Inorganic Ligands

| Metal Ion | Percent Complexed with Indicated Ligand | | | | | | | Percent Complexed by all Ligands | Percent Complexed by NTA³⁻ after addition of 3 × 10⁻⁶ M NTA |
	$(OH^-)_2$	(HCO_3^-)	(CO_3^{2-})	$(CO_3^{2-})_2$	$(OH^-)(CO_3^{2-})$	(SO_4^{2-})	(Cl^-)		
$Cu(II) = 2 \times 10^{-6}$		0.3	89	0.7				97	82
$Pb(II) = 3 \times 10^{-7}$		1	21			3	1	72	80
$Ni(II) = 10^{-7}$		2	40			3		47	60
$Fe(III) = 2 \times 10^{-6}$	98				2			100	34
$Zn(II) = 1.5 \times 10^{-6}$		2	29			3		39	20
$Mn(II) = 2 \times 10^{-6}$						5		5	0
$Ca(II) = 10^{-3}$		2				5		5	<0.1
$Mg(II) = 2.5 \times 10^{-4}$		1	1			5		5	0
$Sr(II) = 2 \times 10^{-6}$						5		5	0
$Ba(II) = 1.5 \times 10^{-7}$						5		5	0
$Na(I) = 5 \times 10^{-4}$								0	0

Source: C. W. Childs, Proc. 14th Conf. Great Lakes Res., 198–210 (1971). Intl. Assoc. Great Lakes Res. (Reprinted by permission of the International Association for Great Lakes Research.)

whose structure is not well-defined. They are a mixture of poorly biodegradable decomposition products and by-products of natural organic matter produced by both plants and animals. Humic substances have poorly defined physical and chemical characteristics (e.g., melting point, refractive index, and molecular weight) and have been characterized by Schnitzer and Khan[20] as "amorphous, brown or black, hydrophilic, acidic, polydisperse substances of molecular weights ranging from several hundreds to tens of thousands." Humic substances account for the bulk of the organic material in natural waters and soils. In dilute aqueous solution their "brown or black" color becomes the "yellow-brown" typical of natural water color and of the color of biologically treated sewage effluents.

Humic substances have been arbitrarily divided into three diverse groups of compounds on the basis of their solubility in dilute acid and dilute base. Fulvic acids are soluble in both dilute acid (pH 1) and dilute base. Humic acids are soluble in dilute base but are precipitated by dilute acid (pH 1). Humin is insoluble in both dilute acid and dilute base. Some authorities divide humic substances into only two groups: the humic acids and the fulvic acids, stating that the humins have the same characteristics as humic acids but that their solubility in base is hindered because they are associated with clay minerals in natural waters.[21] The gross chemical and physical properties of humic acids and fulvic acids are presented in Table 5-12. It is the fulvic acid fraction that appears to

[20] M. Schnitzer and S. U. Khan, Humic Substances in the Environment, Dekker, New York, 1972.

[21] R. F. Packham, "Studies on Organic Color in Natural Water," Proc. Soc. Water Treat. Exam. 13: 316–329 (1964).

be the predominant group of humic substances in natural waters. Fulvic acids appear to have lower molecular weights than humic acids; they contain a higher percentage of oxygen and the oxygen appears to be located in a greater percentage of carboxyl groups.

Although the chemical structure of both the humic acid and fulvic acid fractions of the humic substances is not precisely known, the nature of the major functional groups present is fairly well defined. The following types of functional groups have been reported:

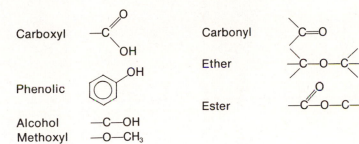

TABLE 5-12   Physical and Chemical Properties of Humic and Fulvic Acids

Property	Humic Acids	Fulvic Acids
Elemental Composition (% by weight)		
C	50–60	40–50
H	4–6	4–6
O	30–35	44–50
N	2–4	<1–3
S	1–2	0–2
Solubility in strong acid (pH 1)	Not soluble	Soluble
Molecular weight range	Few 100 → several million	180–10,000
Functional group distribution	Percent of oxygen in indicated functional group	
carboxyl —COOH	14–45	58–65
phenol —⟨O⟩—OH	10–38	9–19
alcohol —C—OH	13–15	11–16
carbonyl —C=O	4–23	4–11
methoxyl —O—CH$_3$	1–5	1–2

Source: M. Schnitzer and S. U. Khan, *Humic Substances in the Environment,* Dekker, New York, 1972.

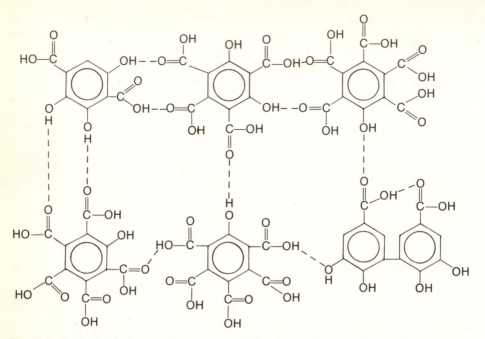

Fig. 5-9. Suggested chemical structure of fulvic acid. Reprinted from M. Schnitzer and S. U. Khan, Humic Substances in the Environment, p. 196, Marcel Dekker, New York, 1972, by courtesy of Marcel Dekker, Inc.

Schnitzer[22] has proposed that the structure in Fig. 5-9 represents a significant portion of the fulvic acid group. The various aromatic ring structures are bonded together by hydrogen bonds between the functional groups on adjacent rings.

This structure does not account for the nitrogen and sulfur content of humic substances. It has been suggested that these elements are derived from parts of other types of molecule, for example, proteins, which are associated with the humic substances. Indeed it has been proposed[23] that humic substances consist of an aromatic core to which peptides, carbohydrates, metals, and phenolic acids are chemically or physically attached. It can be seen that the structure in Fig. 5-9 is an "open" network. In fact, there have been suggestions that organic and inorganic materials associated with humic substances are trapped inside these "holes" in the humic substance structure.[24]

5.7.3.2. INTERACTION OF HUMIC SUBSTANCES WITH METALS   A discussion of this topic must always start with a statement of the difficulties

[22] M. Schnitzer, "Agron. Abstracts," Am. Soc. Agron, 1971, p. 71.
[23] M. Schnitzer and S. U. Khan, Humic Substances in the Environment, Dekker, New York, 1972.
[24] M. Schnitzer and S. U. Khan, Humic Substances in the Environment, Dekker, New York, 1972.

involved with determining the nature of the association between humic substances and metals. It is difficult to define the nature of the product of a chemical reaction when the nature of the reactants is not known. We are faced with this situation here, since the exact chemical structure of humic substances is not well defined and this structure can change from one water to another. For example, Christman[25] has found that several waters each with the same color intensity (measured as 100 cobalt-platinum units)[26] contained dissolved organic carbon concentrations of anywhere from 10 to 30 mg C/liter.

Humic substances undoubtedly have the ability to combine with (or bind) considerable quantities of metal ions. For example, many metals can be enriched into peat (which is a type of soil organic matter of a humic nature) from soil waters so that the concentration of the metal in the peat is up to 10,000 times that of its concentration in the water. It has also been observed that in many natural waters, Fe(III) is associated with the color causing organics. Schnitzer[27] has estimated that a solution containing 100 mg of fulvic acid/liter can maintain in solution 8.4 mg Fe(III) and 4.0 mg Al(III)/liter. The concentration of Fe(III) is more than two orders of magnitude greater than that expected in solution in a water at ~pH 7 in equilibrium with ferric hydroxide.

For this reason there is no question about the ability of humic materials to combine with metals; it is the nature of the association that is open to question. Two modes of binding appear to be significant. These are (1) the formation of complexes or chelates between the functional groups of the humic substance and the metal, and (2) an association between the humic substances and a colloidal particle of metal hydroxide (possibly through sorption on the surface of the particle) whereby the colloidal particles are peptized, that is, stabilized, in suspension and prevented from coagulating or settling. The formation of one or another of these associations between humic material and metals may depend on the initial states of humic matter and metal and their absolute and relative concentrations. High concentrations of metal (exceeding the solubility limits of the metal hydroxide) and high metal to humic substance ratios would seem to favor the formation of peptized particles over soluble complexes. Indeed Schnitzer[28] proposes that the transport and deposition of metals such as Fe(III) in soils is regulated by humic substance-iron compounds of various solubilities. Schnitzer suggests that humic sub-

[25] R. F. Christman, Symposium on Organic Matter in Natural Waters, D. W. Hood, ed. University of Alaska, College, Alaska, 1968.

[26] Standard Methods for the Examination of Water and Wastewater 14th ed. Am. Publ. Health Assoc. New York, 1975, p. 64.

[27] M. Schnitzer, "Metal-Organic Matter Interactions in Soils and Waters," in Organic Compounds in Aquatic Environments, S. D. Faust, and J. V. Hunter, eds. Dekker, New York, 1971, p. 297.

[28] M. Schnitzer, "Metal-Organic Matter Interactions in Soils and Waters," in Organic Compounds in Aquatic Environments, S. D. Faust and J. V. Hunter, eds., Dekker, New York, 1971, p. 297.

stances dissolve Fe(III) from iron-containing solids in the soil to form Fe(III)-humic substance complexes. At the surface of the soil these complexes contain a mole ratio[29] of Fe(III):humic substance of 1:1. As the soil water percolates down through the soil, the Fe(III)-humic substance gradually becomes more and more enriched with Fe(III) (mole ratios of up to 6:1 have been found) and concomitantly more and more insoluble. At very high Fe(III)-humic substance mole ratios, the complexes precipitate out to form iron-rich layers in the soil horizon. The high concentrations of metals such as iron, copper, and uranium in highly organic soils and deposits (such as peat and coal) are thought to arise also from the association of metals with natural organics in these materials.

The nature of the complexes formed between humic substances and metals are not known unequivocally. By blocking various functional groups on the humic substances and then exposing the modified humic material to metals in solution, it is possible to obtain an idea of which functional groups are important in the complexation of metals. Schnitzer and Khan[30] report that at pH 3.5, 75 percent of the complexed Fe(III), 66 percent of the Al(III), and 50 percent of the Cu(II) reacted simultaneously with strongly acidic carboxyl groups and with phenolic groups. A further 8 percent of the Fe(III), 20 percent of the Al(III), and 32 percent of the Cu(II) reacted with the more weakly acidic carboxyl groups alone. From these results we can account for in excess of 80 percent of the complexed Fe(III), Al(III), and Cu(II) by complexes such as

or

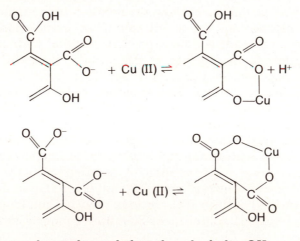

The experiments also indicated that the alcoholic OH groups were not involved in forming metal complexes.

[29] To determine mole ratios, it is necessary to know the average molecular weight of fulvic acid. For the fulvic acid studied by Schnitzer the average molecular weight was 670.

[30] M. Schnitzer and S. U. Khan, *Humic Substances in the Environment*, Dekker, New York, 1972.

Although it is impossible to write an exact complexation reaction for metal ions and humic substances, it is possible to derive stability constants on an operational basis, that is, for individual preparations of humic acids and fulvic acids from specific sources. The formation constant values in Table 5-13 were obtained by Schnitzer and Hansen[31] for a particular fulvic acid.

A comparison of the formation constants for metal-fulvic acid complexes shown in Table 5-13 with those values for the complexes of the same metals with NTA (Table 5-10) would seem to indicate that the metal-fulvic acid complexes are much weaker than the metal-NTA complexes. Indeed some of the values of the constants for metal-fulvic acid complexes approach the values for the very weak inorganic ion pairs (e.g., $MgCO_3^\circ$, $\log K = 2.9$; $CaSO_4^\circ$, $\log K = 3.4$). Morgan[32] has pointed out that the values for the metal-fulvic acid complexes were measured at much lower pH values than the K values for other complexing agents such as NTA and EDTA. The stability constants for NTA complexes reported in Table 5-10 are for the totally deprotonated species,

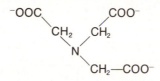

a species that is only predominant in solutions with pH of greater than 10.3. If the K values were measured where the species

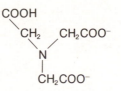

is predominant (e.g., at pH 7), the K values would be lower because the protonation reaction would have to be accounted for. Thus at pH > 10.3

$$Cu^{2+} + NTA^{3-} \rightleftharpoons Cu\text{-}NTA^-; \qquad K = 10^{13} \qquad (5\text{-}13)$$

(neglecting [$HNTA^{2-}$], [H_2NTA^-] and [H_3NTA]), but at 3 < pH < 10.3, the predominant NTA species is $HNTA^{2-}$ and the reaction

$$NTA^{3-} + H^+ \rightleftharpoons HNTA^{2-}; \qquad \frac{1}{K_{a,3}} = 10^{10.3} \qquad (5\text{-}14)$$

[31] M. Schnitzer and E. H. Hansen, *Soil Science,* 109: 333 (1970).

[32] J. J. Morgan, Discussion of paper by R. L. Malcolm, E. A. Jenne, and P. W. McKinley, "Conditional Stability Constants of a North Carolina Soil Fulvic Acid with Co^{2+} and Fe^{3+}" in *Organic Matter in Natural Waters,* D. W. Hood, ed., Inst. of Marine Science, University of Alaska, 1970, p. 479.

must be taken into account. Subtracting Eq. 5-14 from Eq. 5-13 yields

$$Cu^{2+} + HNTA^{2-} \rightleftharpoons Cu-NTA^- + H^+ \qquad (5-15)$$

$$K = 10^{13} \times K_{a,3} = 10^{13} \times 10^{-10.3} = 10^{2.7}$$

The measured values of the formation constant for Cu-NTA$^-$ at intermediate pH will then be $10^{2.7}$ which is significantly less than the value of 10^{13} measured at high pH. The same is true for the humic substance-metal ion complex equilibrium constants measured at low pH values. Morgan[33] estimates that at pH 2 to 3 the —COO$^-$ groups, which are important in forming fulvic acid-metal ion complexes, are virtually all in the —COOH or protonated form. Thus to obtain the value of the formation constant at intermediate pH for metal ion fulvic acid complexes, when the carboxyl groups are ionized, we should divide the equilibrium constant values presented in Table 5-13 by the dissociation constant of the —COOH group, that is,

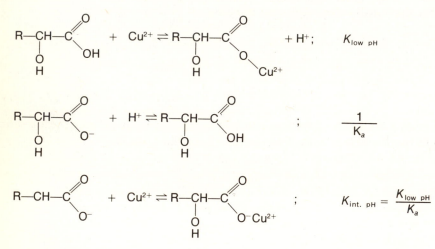

If we estimate the value of K_a to be $\sim 1 \times 10^{-5}$, then the true $K_{formation}$ values for metal ion-fulvic acid complexes would be in the range 10^7 to 10^{11}, which makes them moderately strong to strong complexes. They are, in general, weaker than the corresponding NTA and EDTA complexes, which accounts for the observed ability of EDTA to extract metals from humic materials.

Metal complexes of humic substances are important in natural waters because as we have already seen, they strongly influence the distribution of metals between the dissolved and particulate states. Because of this, the presence of complexing agents such as the humic substances may

[33] J. J. Morgan, Discussion of paper by R. L. Malcolm, E. A. Jenne, and P. W. McKenley, "Conditional Stability Constants of a North Carolina Soil Fulvic Acid with Co^{2+} and Fe^{3+}" in *Organic Matter in Natural Waters* D. W. Hood ed., Inst. of Marine Sciences, University of Alaska, 1970, p. 479.

TABLE 5-13   Formation Constants of Various Metal Ion Fulvic Acid Complexes (Ionic Strength $= 0.1 M$)

Metal Ion	pH of Measurement	log K for Metal Ion-Fulvic Acid complex[a]
Fe^{3+}	1.7	6.1
Al^{3+}	2.35	3.7
Cu^{2+}	3.0	3.3
Ni^{2+}	3.0	3.1
Co^{2+}	3.0	2.9
Pb^{2+}	3.0	2.6
Zn^{2+}	3.0	2.4
Mn^{2+}	3.0	2.1
Mg^{2+}	3.0	1.9

[a] Formation constants, K, are for the reaction M + fulvic acid $\rightarrow M \cdot$ fulvic acid.

influence the biological productivity or the amount of organisms such as algae and higher plants and animals that a natural water can support. For example, let us suppose that the growth-limiting nutrient for algae in a lake is iron, Fe(III). Let us assume that there are no humic substances in the lake water and that the Fe(III) concentration is controlled by its equilibrium with ferric hydroxide, $Fe(OH)_{3(s)}$. We shall see from calculations made in Chapter 6 that at pH 8, the maximum possible Fe(III) concentration is $10^{-11} M$, or 6×10^{-4} mg/liter. We also will learn in Chapter 6 that some precipitates like $Fe(OH)_{3(s)}$ are notoriously slow to reach equilibrium because they form slowly and dissolve slowly. Thus the water will contain only a minute amount of Fe(III) and as the growth of algae consumes the Fe(III), the dissolution of $Fe(OH)_{3(s)}$ to replace the depleted Fe(III) may be quite slow. Therefore, we might expect very little and very slow growth of algae under these circumstances.

Now consider the situation in the presence of humic substances that can complex Fe(III). The total amount of dissolved Fe(III) must be greater than in the absence of humic substances because, in addition to the Fe(III) in equilibrium with $Fe(OH)_{3(s)}$, we have in solution the Fe(III)-humic substance complexes. Let us also assume that the Fe(III)-humic substance complexes are labile—that is, they will dissociate rapidly to release Fe(III). In this situation we have not only seen an increase in the total dissolved iron but also an increase in the pool of readily available Fe(III). Observations made by Schnitzer[34] discussed previously have indicated

[34] M. Schnitzer, "Metal-Organic Matter Interactions in Soils and Waters," in *Organic Compounds in the Environment*, S. D. Faust and J. V. Hunter, eds., Dekker, 1971.

that the increase in total dissolved iron in the presence of humic substances can indeed be significant. In our example their presence would increase the amount and rate of algal growth; they might even increase the Fe(III) concentration to such a level that it would no longer be the nutrient in shortest supply that limits growth.

In this hypothetical example the presence of the complexing ability of humic substances was seen to result in an increase in dissolved iron species. The presence of complexing agents in a water may not always have this effect. For example, if the complexing agent (1) forms a stable and or inert complex with the metal of interest, or (2) is present at a sufficiently high concentration, it may have the effect of reducing the level and/or availability of metal ions below that found in the absence of any complexing agent. In such a situation we would expect a decrease in the productivity of the water over that of a water in which there was no complexing agent. As an example of this type of effect, Siegel[35] quotes unpublished work of Barber on the effect of several complexing agents on the productivity of water from the Gulf of Mexico. The complexing agents studied were DPTA (diethylenetriaminepentaacetic acid), EDTA, cysteine, and NTA. We can judge the relative "complexing ability" of these complexing agents by comparing the formation constant values of their complexes with Fe(III). These are DPTA; $K = 10^{28.6}$, EDTA, $K = 10^{25}$; cysteine, $K = 10^{10}$ and NTA, $K = 10^8$. At concentrations of $10^{-6} M$ the four complexing agents stimulated the productivity of the water in the order expected, that is, the strongest complexing agent (DPTA) stimulated the productivity the most, and the weakest complexing agent (NTA) stimulated the productivity the least. When the concentration of the weakest complexing agent was increased to $10^{-5} M$, it stimulated productivity more than did cysteine. However, when the strongest complexing agent (DPTA) concentration was increased to $10^{-5} M$, the productivity of the water was reduced below the level of the control containing no added complexing agents. Increasing the concentration of the next strongest complexing agent (EDTA) did not produce the same reduction in productivity. The productivity of the solutions was therefore in the following order:

$10^{-6} M$ EDTA
$10^{-5} M$ EDTA $> 10^{-5} M$ NTA $> 10^{-6} M$ cysteine $>$
$10^{-6} M$ DPTA $10^{-6} M$ NTA $>$ control $> 10^{-5} M$ DPTA

Apparently, DPTA is such a strong complexing agent that $10^{-5} M$ of it will reduce the metals level to a value below that required for as much growth as the control; increased levels of the weakest complexing agent, NTA, increase productivity while a tenfold difference in concentration of a

[35] A. Siegel, "Metal-Organic Reactions in the Marine Environment," in *Organic Compounds in Aquatic Environments*, S. D. Faust and J. V. Hunter, eds., Dekker, 1971, p. 265.

complexing agent of intermediate strength, EDTA, does not change its effect on productivity.

5.8.   PROBLEMS

1. Name the following complexes: $PtCl_4^{2-}$, $Hg(CN)_4^{2-}$, $Ni(NH_3)_6^{2+}$, $Fe(PO_4)_2^{3-}$, $Co(NH_3)_5(H_2O)^{3+}$, $CaHCO_3^{+}$.

2. In order to determine the hardness of a solution, a small amount of Eriochrome Black T is added to 100 ml of sample. After addition of 11.2 ml of 0.01 M EDTA solution, the color of the solution changes from red to blue. What is the hardness concentration in moles/liter and mg/liter as $CaCO_3$?

3. Show that $[HgOH^+]$ is negligible in Example 5-2.

4. A water, pH 5.5, is dosed with 80 mg/liter of alum. $C_{T,CO_3} = 1 \times 10^{-4}$.
    (a) How much hydrated lime, $Ca(OH)_2$, is needed so that the pH does not change during coagulation?
    (b) How much hydrated lime must be added to produce a water with a pH of 7.5 after coagulation?

5. The calculation in Section 5-6 showed that the only complexes of significance in a typical natural water are the sulfato calcium and magnesium complexes. Given a $C_{T,Ca} = 1 \times 10^{-3}$ M, $C_{T,Mg} = 5 \times 10^{-4}$ M and a very high sulfate concentration, $C_{T,SO_4} = 2 \times 10^{-3}$ M, how much of the Ca and Mg is complexed with sulfate?

6. Given the conditions listed for development of Fig. 5-6, and a pH of 8.5, what is the fraction of $C_{T,Ni}$ in the form of $Ni(CN)_4^{2-}$?

7. A solution contains 2×10^{-3} moles Ca^{2+}/liter and 3×10^{-4} moles Mg^{2+}/liter. Given the formation constants for $CaEDTA^{2-}$ and $MgEDTA^{2-}$ of $10^{10.6}$ and $10^{8.7}$, respectively, calculate
    (1) The concentration of $MgEDTA^{2-}$ after the addition of 1.5×10^{-3} moles $EDTA^{4-}$/liter of sample.
    (b) The concentration of $MgEDTA^{2-}$ after the addition of 2×10^{-3} moles $EDTA^{4-}$/liter of sample.
    Does the presence of Mg^{2+} significantly interfere with the determination of Ca^{2+} by the EDTA titration procedure?

8. What is the pH of a 10^{-5} M $Cu(ClO_4)_2$ solution? Check your answer to be certain that $Cu(OH)_{2(s)}$ will not precipitate.

9. Calculate the $Pb_4(OH)_4^{4+}$ concentration at pH 6 for the conditions given in Fig. 5-2. (*Note:* You may wish to use Fig. 5-2 to obtain a good initial estimate of the species concentrations of interest.)

10. A natural water contains equal concentrations in the micromolar range of Fe(III), Mg(II), Cu(II), and Pb(II), as well as 5 mg/liter of fulvic acid. Arrange

the metals in order of the extent of complexation with fulvic acid and give your reason for this order.

11. How can one explain the ability of EDTA in solution to extract metals from dissolved metal-humic acid complexes?

12. Under what conditions will the addition of a trace metal chelating agent aid the biological productivity of a water? Under what conditions will it decrease biological productivity?[36]

5.9. ADDITIONAL READING

COMPLEX FORMATION

Basolo, F., and R. G. Pearson, *Mechanisms of Inorganic Reactions, A Study of Metal Complexes in Solution*, 2nd ed. John Wiley, New York, 1967.

Beck, M., *Complex Formation*. Reinhold, New York, 1970.

Benson, D., *Mechanisms of Inorganic Reactions in Solution*. McGraw-Hill, London, 1968.

Stumm, W., "Metal Ions in Aqueous Solutions," in *Principles and Applications of Water Chemistry*, S. D. Faust and J. V. Hunter, eds. John Wiley, New York, 1967.

Stumm, W., and J. J. Morgan, *Aquatic Chemistry*. Wiley-Interscience, New York, 1970.

Wilkens, R. G., *The Study of Kinetics and Mechanisms of Reactions of Transition Metal Complexes*. Allyn & Bacon, Boston, Mass., 1974.

HUMIC SUBSTANCES

Chapters by R. F. Christman and R. A. Minear, A. Siegel, and M. Schnitzer in *Organic Compounds in Aquatic Environments*, S. D. Faust and J. V. Hunter, eds., Dekker, New York, 1971.

Gjessing, E. T., *Physical and Chemical Characteristics of Aquatic Humus*, Ann Arbor Science, Ann Arbor, Mich., 1975.

Schnitzer, M., and S. U. Khan, *Humic Substances in the Environment*, Dekker, New York, 1972.

---

[36] *Note:* Additional problems on the effect of complexation on solubility appear at the end of Chapter 6.

CHAPTER
6
PRECIPITATION AND DISSOLUTION

6.1. INTRODUCTION

Precipitation and dissolution phenomena are extremely important in both natural waters and water treatment processes. Dissolution of minerals is a prime factor in determining the chemical composition of natural waters. Natural water chemical composition can be altered by precipitation of minerals and the subsequent sedimentation of these solids from supersaturated solutions. Water and wastewater treatment processes such as lime-soda softening, iron removal, coagulation with hydrolyzing metal salts, and phosphate precipitation are based on precipitation phenomena.

Both equilibrium considerations and the rates of reactions are important. Knowledge of equilibrium relationships permits the calculation of concentrations at equilibrium and thus the amount of precipitate or amount of the material that dissolves can be calculated. In many instances rate controls the extent of reaction because insufficient time is available for equilibrium to be achieved.

We must be aware that equilibrium calculations and log-concentration diagrams of systems involving heterogeneous equilibria may only provide us with the boundary conditions of the system rather than the situation that truly exists. Thus equilibrium calculations involving solids in waters and wastewaters are generally less representative of the true situation than acid-base equilibrium calculations for the following reasons.

1. Some heterogeneous equilibria are slowly established.

2. The thermodynamically predicted stable solid phase for a given set of conditions may not indeed be the phase that is formed.

3. Solubility depends both on the degree of crystallinity and the particle size of the solids (which may vary from case to case).

4. Supersaturation may exist; that is, solution concentrations in excess of those predicted by equilibrium calculations may prevail.

5. The ions produced by the dissolution of solids may undergo further reaction in solution.

6. There is a wide variation in the reported values of heterogeneous equilibrium constants.

Because of the last reason it is imperative that one be sure of the conditions under which equilibrium constants were measured prior to using a particular value.

In this chapter we will deal with the kinetics and equilibrium calculations of heterogeneous systems. Graphical and computational methods for solving equilibrium problems will be presented. The precipitation of calcium carbonate will be discussed in some detail, and the chemistry of phosphorus will be used as a detailed example of several heterogeneous equilibria that are of relevance to natural waters and treatment processes.

6.2. PRECIPITATION AND DISSOLUTION KINETICS

Precipitation has generally been observed to occur in three steps: (1) nucleation, (2) crystal growth, and (3) agglomeration and the ripening of the solids.[1] It is possible for a solution that is only slightly supersaturated with respect to a solid phase to be stable indefinitely. When the degree of supersaturation is increased or when fine particles of a substance are mixed with this solution, precipitation occurs.

For example, Trussell[2] reports a simple experiment in which three solutions containing Ca^{2+} and carbonate species were titrated with NaOH. The turbidity of the solution was monitored after various increments of NaOH addition to detect the formation of a $CaCO_{3(s)}$ precipitate. Figure 6-1 shows that it was not until the value of the equilibrium constant for the precipitation of $CaCO_{3(s)}$ had been exceeded by 40 to 50 times that evidence of precipitate formation was shown.

6.2.1. Nucleation

A nucleus is a fine particle on which the spontaneous formation or precipitation of a solid phase can take place. Nuclei can be formed from clusters of a few molecules or ion pairs of component ions of the precipitate, or they may be fine particles unrelated chemically to the precipitate but with some similarity of crystal lattice structure. Precipitation from homogeneous solution (i.e., a solution with no solid phase present) requires that nuclei be formed from ions in solution. If the nuclei are formed from the component ions of the precipitate, the initial phase of precipitation is referred to as *homogeneous nucleation;* if the foreign particles are the nuclei, the nucleation is said to be *heterogeneous.* Because virtually all aqueous solutions contain fine particles of various types, most nucleation is heterogeneous.

---

[1] References such as A. G. Walton, *The Formation and Properties of Precipitates,* Wiley-Interscience, New York, 1967; W. Stumm and J. J. Morgan, *Aquatic Chemistry,* Wiley-Interscience, 1970; and A. E. Nielson, "Kinetics of Precipitation," Macmillan, 1964, should be consulted in addition to references given in this section.

[2] R. R. Trussell, "Systematic Aqueous Chemistry for Engineers," Ph.D. Thesis, University of California, Berkeley, California, 1972.

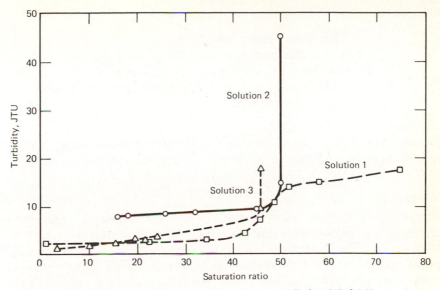

Fig. 6-1. The influence of the saturation ratio ($= [Ca^{2+}][CO_3^{2-}]/K_{so,CaCO_3}$) on the initiation of crystal formation for $CaCO_3$.

The formation of nuclei from precipitate ions is an energy-consuming process because one creates an organized structure with defined surfaces from a random arrangement of solution constituents. Because of this energy requirement it is necessary for solutions to be supersaturated, or have concentrations greater than that predicted by equilibrium with the precipitate, before the precipitate will form from a homogeneous solution. This driving force, or necessary degree of supersaturation, tends to be larger for homogeneous nucleation than for heterogeneous nucleation because the precipitate ions can collect on the surface of the foreign particle, and because the free energy to form a crystallite on a similar surface is less. In essence the foreign surface serves the same function as the surface of the precipitate particles; the only requirement is that the foreign surface be somewhat similar to the precipitate in lattice structure and distance between adjacent ions.[3]

6.2.2. Crystal Growth

Crystals form by the deposition of the precipitate constituent ions onto nuclei. Since water and wastewater treatment processes involving precipitation often do not reach equilibrium, the rate of crystal growth is of critical importance. Crystal growth rate can be expressed as

$$\frac{dC}{dt} = -kS(C - C^*)^n \qquad (6-1)$$

[3] J. H. Christiansen and A. E. Nielsen, *Acta Chem. Scand.* 5: 673 (1951).

where

C^* = saturation concentration, mole/liter
C = actual concentration of limiting ion, mole/liter
k = rate constant, liter^n time^{-1} mg^{-1}, mole$^{(1-n)}$
S = surface area available for precipitation, mg/liter of
    a given particle size
n = constant

For example, Nancollas and Reddy[4] found that the precipitation of calcium carbonate from a particular solution corresponded to the rate law:

$$\frac{d[Ca^{2+}]}{dt} = kS \, ([Ca^{2+}] - [Ca^{2+}]^*)^2 \tag{6-2}$$

where $k = 5.76$ liter^2 mg^{-1} mole^{-1} min^{-1}. Of course, a different rate equation is expected for solutions with other properties.

When the diffusion rate of ions to the surface of the crystal controls the crystal growth rate, the exponent n has a value of unity; when other processes such as the reaction rate at the crystal surface are rate limiting, the exponent may have a value other than unity. The value of k depends on solution conditions and the nature of the solid being precipitated.

6.2.3.   Agglomeration and Ripening

The initial solid formed by precipitation may not be the most stable solid (the thermodynamically stable phase) for the reaction conditions. If this is the situation, then over a period of time the crystal structure of the precipitate may change to that of the stable phase. This change may be accompanied by additional precipitation and, consequently, a reduced solution concentration because the more stable phase usually has a lower solubility than the initially formed phase. The changes in crystal structure that take place over time are often called aging. A phenomenon called ripening may also take place whereby the crystal size of the precipitate increases. Since very small particles have a higher surface energy than larger particles, the solution concentration in equilibrium with small particles is higher than that in equilibrium with larger particles. Consequently, in a mixture of particle sizes, the large particles will continue to grow because the solution is still supersaturated with respect to them. As the concentration in the solution is lowered through the growth of the larger particles, the smaller particles dissolve because the solution concentration is now below their saturation value. Conversion of small particles into larger particles is also enhanced by the agglomeration of particles to form larger particles.

This discussion should indicate that selection of an equilibrium constant for a precipitating solution is extremely difficult. Equilibrium constants

[4] G. H. Nancollas and M. M. Reddy, Chapter in *Aqueous Environmental Chemistry of Metals*, A. J. Rubin, ed., Ann Arbor-Science, 1975.

reported for individual solids may vary widely because beside the effects of the particle size and aging, one must account for factors such as complex formation, adsorption of impurities at the crystal interface, and formation of solid mixtures.

6.2.4. Dissolution

According to Walton,[5] dissolution is nearly always controlled by the rate of diffusion of the species away from the solids. Accordingly, the rate law,

$$\frac{dC}{dt} = kS\,(C^* - C) \tag{6-3}$$

is followed. On occasion the rate of dissolution of the substance may be affected by the formation of surface complexes.

6.3. EQUILIBRIUM CALCULATIONS

At the outset of any discussion on heterogeneous equilibria it is important to realize that heterogeneous equilibrium calculations only apply when there is an excess of the solid of interest present. If solid is absent, the equilibrium calculations are not valid for predicting solution composition; they may only be used to test whether saturation with a particular solid exists.

The situations commonly encountered are

1. *A solid phase or precipitate in a water containing no other constituents than those derived from the dissolution of the precipitate.* The solutes produced by the dissolution of the precipitate do not react (form complexes, hydrolyze, etc.) in solution at moderate concentrations and at an approximately neutral pH. Very few examples of this situation exist in water chemistry. Solids such as $CaF_{2(s)}$ and $BaSO_{4(s)}$ generally fit into this category under natural water conditions.

2. *Precipitates that dissolve to produce reactive solutes.* The reaction may be hydrolysis (protolysis), or complex formation. Examples are $CaCO_{3(s)}$ and $Al(OH)_{3(s)}$ dissolving in water.

3. *Solids that dissolve in a solution containing a species that is the same as one of the species that results from dissolution of the precipitate.* This situation results in the so-called *common ion effect* in which the presence of a common ion modifies the solubility of the solid. It may be further complicated by the production or presence of solutes that react further with solution constituents.

[5] A. G. Walton, *The Formation and Properties of Precipitates*, Wiley-Interscience, New York, 1967.

Examples are $CaF_{2(s)}$ dissolving in water containing Ca^{2+} ion and $CaCO_{3(s)}$ dissolving in a solution containing CO_3^{2-} ion.

4. *More complex systems where a variety of processes such as gas transfer and redox reactions are superimposed on the dissolution reactions and reactions of the dissolved solute.* These are the types of situation usually encountered in nature and in water and wastewater treatment processes; a typical one is the dissolution of $CaCO_{3(s)}$ from the soil during the passage of sewage effluent in groundwater recharge. Gas exchange of CO_2 with the soil atmosphere, oxidation of NH_4^+ to NO_3^- (nitrification), and reduction of NO_3^- to $N_{2(g)}$ (denitrification) are superimposed on the dissolution of $CaCO_{3(s)}$ and the reaction of CO_3^{2-} after $CaCO_{3(s)}$ dissolution in the soil water (Fig. 6-2).

These four situations are influenced by solution ionic strength. In general, the solubility of salts increases with increasing ionic strength. The magnitude of this effect can be assessed at most ionic strengths of interest to us using the DeBye-Hückel, or related, laws.

6.3.1. The Solubility Product

Solubility product is the name given to the equilibrium constant that describes the reaction by which a precipitate dissolves in pure water to form its constituent ions,

$$A_zB_{y(s)} \rightleftharpoons zA^{y+} + yB^{z-}$$

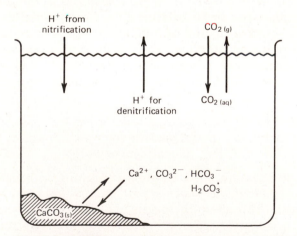

Fig. 6-2. $CaCO_{3(s)}$ equilibria in the presence of alkalinity. Strong acid addition occurs from nitrification; strong acid removal takes place with denitrification; and $CO_{2(g)}$ addition and removal occurs through gas transfer.

TABLE 6-1 Solubility Product Constants at 25°C[a]

Solid	pK_{so}	Solid	pK_{so}
$Fe(OH)_3$ (amorph)	38	$BaSO_4$	10
$FePO_4$	17.9	$Cu(OH)_2$	19.3
$Fe_3(PO_4)_2$	33	$PbCl_2$	4.8
$Fe(OH)_2$	14.5	$Pb(OH)_2$	14.3
FeS	17.3	$PbSO_4$	7.8
Fe_2S_3	88	PbS	27.0
$Al(OH)_3$ (amorph)	33	$MgNH_4PO_4$	12.6
$AlPO_4$	21.0	$MgCO_3$	5.0
$CaCO_3$ (calcite)	8.34	$Mg(OH)_2$	10.7
$CaCO_3$ (aragonite)	8.22	$Mn(OH)_2$	12.8
$CaMg(CO_3)_2$ (dolomite)	16.7	$AgCl$	10.0
CaF_2	10.3	Ag_2CrO_4	11.6
$Ca(OH)_2$	5.3	Ag_2SO_4	4.8
$Ca_3(PO_4)_2$	26.0	$Zn(OH)_2$	17.2
$CaSO_4$	4.59	ZnS	21.5
SiO_2 (amorph)	2.7		

[a] Equilibrium constants for the reaction $A_zB_{y(s)} \rightleftharpoons zA^{y+} + yB^{z-}$.

Generally stated, the equilibrium constant, K_{so}, is

$$K_{so} = \frac{\{A^{y+}\}^z \{B^{z-}\}^y}{\{A_zB_{y(s)}\}} \tag{6-4}$$

The activity of the solid phase can be taken as unity (see Section 3-3). The concentration product, $^cK_{so}$, has the same form as the equilibrium constant except that the concentrations of the species in solution are used instead of their activities. Accordingly,

$$^cK_{so} = [A^{y+}]^z [B^{z-}]^y = \frac{K_{so}}{(\gamma A^{y+})^z (\gamma B^{z-})^y} \tag{6-5}$$

Assuming that the solution components have activity coefficients of unity and that the solid has an activity of one, $^cK_{so} = K_{so}$. If the activity coefficients are not equal to unity, $^cK_{so} \neq K_{so}$ and $^cK_{so}$ is a function of ionic strength. Table 6-1 gives many K_{so} values of interest to us.

It is important to differentiate between the terms *solubility product* and *solubility*. Solubility product is the colloquial term for the equilibrium constant. Solubility is the amount of a substance in moles/liter or mg/liter that can dissolve in a solution under a given set of conditions. Solubility does not equal the solubility product, but the two quantities are interrelated.

Example 6-1

Calculate the solubility in mg/liter of CaF_2 in pure water at 25°C, neglecting ionic strength effects.

$$CaF_2 \rightleftharpoons Ca^{2+} + 2F^-$$

Solution

(1) $K_{so} = 5 \times 10^{-11} = {}^cK_{so} = [Ca^{2+}][F^-]^2$

Each mole of CaF_2 that dissolves in pure water produces 1 mole of Ca^{2+} and 2 moles of F^-. Therefore, if we set S = solubility in moles/liter, we can write

(2) $[Ca^{2+}] = S$

and

(3) $[F^-] = 2S$

From (1), (2), and (3)

$$S(2S)^2 = 4S^3 = 5 \times 10^{-11}$$

Therefore, $S = 2.32 \times 10^{-4}$ moles/liter.

The molecular weight of CaF_2 is 78; therefore,

$$S = 78 \text{ g/mole} \times 2.32 \times 10^{-4} \text{ moles/liter} \times 10^3 \text{ mg/g} = 18.1 \text{ mg/liter.}$$

It can be difficult to judge the solubility of a substance from its solubility product. For example, silver chloride, $AgCl$, has a $K_{so} = 1 \times 10^{-10}$, and silver chromate has a $K_{so} = 2.5 \times 10^{-12}$. Which silver salt will allow the greater solution concentration of silver ion, Ag^+, at equilibrium? This is equivalent to asking the question, which salt has the greater solubility?

For $AgCl$ at 25°C, neglecting ionic strength, we have $S = [Ag^+]$ and $S = [Cl^-]$, and

$$(S)(S) = K_{so} = 10^{-10}$$

Therefore,

$$S = 10^{-5} \text{ mole/liter}$$

For Ag_2CrO_4, under conditions where CrO_4^{2-} does not hydrolyze, $[CrO_4^{2-}] = S$ and $[Ag^+] = 2S$,

$$(S)(2S)^2 = K_{so} = 2.5 \times 10^{-12}$$

Therefore,

$$S = 8.6 \times 10^{-5} \text{ moles/liter}$$

Thus, although the solubility product of Ag_2CrO_4 is more than 2 orders of magnitude smaller than the solubility product of $AgCl$, it happens that Ag_2CrO_4 is about 8 times more soluble than $AgCl$ and $[Ag^+]$ is about 17 times greater in the Ag_2CrO_4 solution. This fact is used in the argentimetric method for determining chloride.[6] In this method a solution containing an unknown amount of chloride is titrated with silver nitrate solution in the presence of a small amount of the yellow-colored potassium chromate (used as an indicator). Silver chloride (white) precipitates first, since AgCl is less soluble than Ag_2CrO_4. When sufficient Ag^+ has been added to lower the Cl^- concentration to less than 10^{-5} moles/liter, the red Ag_2CrO_4 starts to precipitate, indicating the endpoint of the titration.

6.3.2. Effect of Temperature on Solubility

Temperature affects both the equilibrium position of the precipitation reaction and the reaction rate. In general, solubility increases with increasing temperature with a few notable exceptions such as $CaCO_3$, $Ca_3(PO_4)_2$, $CaSO_4$, and $FePO_4$, which are of importance in water chemistry. The effect of temperature on equilibrium is presented in Section 3-4 and the following example illustrates the effect on a typical precipitation reaction.

Example 6-2

Find K_{so} for $FePO_4$ at 50°C given the following information at 25°C for the reaction,

$$FePO_{4(s)} \rightleftharpoons Fe^{3+} + PO_4^{3-}$$

$$\Delta H° = -18.7 \text{ kcal}$$

$$\Delta G° = 24.4 \text{ kcal}$$

Solution

Knowing $\Delta G° = -RT \ln K$, we find

$$K_{so,25°C} = \exp\left[\frac{-\Delta G°}{RT}\right] = 1.19 \times 10^{-18}$$

We can then use the relationship derived by Van't Hoff from Chapter 3,

$$\ln\frac{K_1}{K_2} = \frac{\Delta H°}{R}\left[\frac{1}{T_2} - \frac{1}{T_1}\right] \tag{3-27}$$

Setting

$$K_1 = K_{so,50°C} \quad \text{and} \quad K_2 = K_{so,25°C}$$

Then

$$T_1 = 323°K \quad \text{and} \quad T_2 = 298°K$$

$$\ln\frac{K_{so,50°C}}{K_{so,25°C}} = \left[\frac{-18.7}{1.987 \times 10^{-3}}\left[\frac{1}{298} - \frac{1}{323}\right]\right] = -2.444$$

$$K_{so,50°C} = 1.03 \times 10^{-19}$$

Thus the $FePO_{4(s)}$ solubility decreases with increasing temperature.

---

[6] *Standard Methods for the Examination of Water and Wastewater*, Amer. Publ. Health Assoc., 14th ed., 1976, p. 303.

6.3.3.   The Common Ion Effect

When a solution contains an ion that is the same as one of the ions which result from the dissolution of the solid, the solubility of the solid will be less than that when the solid dissolves in pure water. The modification of solubility is called the *common ion effect*. When calculating the solubility of the solid in such a solution, we need to remember that the solubility is based on ions or species that come only from the dissolution of the solid.

Example 6-3

The solubility of stannous fluoride, $SnF_{2(s)}$, in water at 20°C is 0.012 g/100 ml. What is the solubility of $SnF_{2(s)}$ in a 0.08 M NaF solution neglecting ionic strength effects?

Solution

Let us first calculate the concentration product for the reaction,

$$SnF_{2(s)} \rightleftharpoons Sn^{2+} + 2F^-$$

$$K_{so} = {}^cK_{so} = [Sn^{2+}][F^-]^2$$

The molecular weight of SnF_2 is 125.6

$$S = 0.012 \frac{g}{100\ ml} \times \frac{1000\ ml}{1\ liter} \times \frac{1}{125.6} \frac{mole}{g}$$

$$= 9.6 \times 10^{-4}\ moles/liter$$

Therefore,

$$
\begin{aligned}
{}^cK_{so} &= (9.6 \times 10^{-4})(9.6 \times 10^{-4} \times 2)^2 \\
&= 3.5 \times 10^{-9}
\end{aligned}
$$

Now let $S' =$ moles/liter of SnF_2, which will dissolve. If 8×10^{-2} mole of F^-/liter is originally present,

$$[F^-] = (8 \times 10^{-2} + 2S')$$
$$[Sn^{2+}] = S'$$

and

$${}^cK_{so} = (8 \times 10^{-2} + 2S')^2(S') = 3.5 \times 10^{-9}$$

For this example let us assume that $2S'$ is negligible compared to 8×10^{-2}. Then

$${}^cK_{so} = [S'][0.08]^2 = 3.5 \times 10^{-9}$$

$$S' = \frac{3.5 \times 10^{-9}}{(0.08)^2} = 5.5 \times 10^{-7}\ moles/liter$$

Checking our assumption that $8 \times 10^{-2} > 2S'$, we see that $2S' = 11.0 \times 10^{-7} \ll 8 \times 10^{-2}$. The assumption is valid.

We see from this example that the solubility of SnF_2 is reduced from 9.6

$\times\ 10^{-4}$ mole/liter to 5.5×10^{-7} mole/liter by the presence of 0.08 M F$^-$. A similar calculation shown in Example 6-4 for a solution containing 10^{-4} mole NaF/liter shows essentially no decrease in solubility caused by the common ion.

Example 6-4

What is the solubility of SnF$_2$ in a 10^{-4} mole/liter NaF solution? Neglect ionic strength effects.

Solution

From Example 6-3,

$$^cK_{so} = 3.5 \times 10^{-9}$$

Let [Sn^{2+}] = S'. Then

$$[F^-] = 10^{-4} + 2S'$$

and

$$^cK_{so} = (S')(10^{-4} + 2S')^2$$

Assuming that $2\,S' \ll 10^{-4}$, we obtain

$$S' = \frac{3.5 \times 10^{-9}}{(10^{-4})^2} = 0.35$$

Checking the assumption that $2S' \ll 10^{-4}$, we find $2S' = 7 \times 10^{-1}$, which is not $\ll 10^{-4}$. The assumption is not valid. We will therefore proceed as we did in the solution of acid-base equilibrium problems and make the opposite assumption, that is, assume that $2S' \gg 10^{-4}$. Then

$$S' = 9.6 \times 10^{-4}$$

Checking the assumption, we find $2S' = 19.2 \times 10^{-4} \gg 10^{-4}$ so that the assumption is acceptable and the solubility is 9.6×10^{-4} mole/liter.

6.3.4. Conditional Solubility Product

As the name implies, a conditional solubility product is one whose value depends on solution conditions other than the presence of a common ion. Conditional solubility products are useful when determining the solubility of precipitates whose ions interact with solution constituents through reactions such as complexation or hydrolysis. This type of behavior is indeed the rule rather than the exception for the systems encountered in natural waters and treatment processes. Also, in many analytical measurements the total concentration of species is determined rather than the free ion concentration.

The conditional solubility product has the form

$$P_s = (C_{T,M})(C_{T,A}) \tag{6-6}$$

where

P_s = conditional solubility product

$C_{T,M}$ = total concentration of the metal ion M in all of its complex forms and

$C_{T,A}$ = total concentration of the anion A in all of its forms

The value of P_s may change as a function of solution properties such as pH and concentration of complex-forming species because the fractions of $C_{T,M}$ and $C_{T,A}$ present as the free cation and anion may vary as a function of these properties.

The conditional solubility product and K_{so}, the solubility product, are related. For any set of conditions, if α_M = fraction of $C_{T,M}$ present as the free metal ion, $[M^+] = \alpha_M (C_{T,M})$, and if α_A = fraction of $C_{T,A}$ present as free anion, $[A^-] = \alpha_A (C_{T,A})$. Then

$$K_{so} = [M^+][A^-]$$

or

$$K_{so} = \alpha_M \alpha_A (C_{T,M})(C_{T,A})$$

Therefore,

$$K_{so} = \alpha_M \alpha_A P_s$$

Example 6-5

Will $CaCO_{3(s)}$ precipitate at pH 9.1 from a solution containing $C_{T,Ca} = 10^{-2}$ M, $C_{T,CO_3} = 10^{-3} M$? The ionic strength is 10^{-2} and $P_s = 10^{-6.7}$ at pH 9.1.

Solution

$$(C_{T,Ca})(C_{T,CO_3}) = 10^{-2} \times 10^{-3} = 10^{-5}$$

Since $P_s = 10^{-6.7} < 10^{-5}$, the solution is supersaturated and $CaCO_{3(s)}$ will precipitate.

6.3.5. Log Concentration Diagrams

The presentation of precipitation equilibria on log concentration diagrams greatly aids the solution of problems and the understanding of these equilibria. For example, consider $Ca(OH)_{2(s)}$ and sparingly soluble $Mg(OH)_{2(s)}$, both of which are important in water treatment; $Ca(OH)_{2(s)}$ or hydrated (slaked) lime is commonly used to bring about the precipitation of Ca^{2+} as $CaCO_{3(s)}$ and Mg^{2+} as $Mg(OH)_{2(s)}$. The equilibrium relationships at 25°C are

$$Ca(OH)_{2(s)} \rightleftharpoons Ca^{2+} + 2OH^-; \qquad \log K_{so} = -5.3$$
$$Mg(OH)_{2(s)} \rightleftharpoons Mg^{2+} + 2OH^-; \qquad \log K_{so} = -10.74$$

Assuming that the activity effects are negligible, that is, activity coefficients are unity:[7]

$$[Ca^{2+}][OH^-]^2 = 10^{-5.3}$$

or

$$\log [Ca^{2+}] + 2 \log [OH^-] = -5.3$$

or

$$-\log [Ca^{2+}] + 2pOH = 5.3 \qquad (6\text{-}7)$$

Similarly, for $Mg(OH)_{2(s)}$,

$$-\log [Mg^{2+}] + 2pOH = 10.74 \qquad (6\text{-}8)$$

Equations 6-7 and 6-8 can be plotted as a log concentration diagram (Fig. 6-3). This diagram can be used handily to solve a variety of problems.

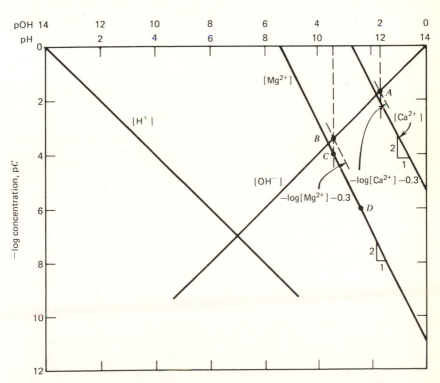

Fig. 6-3. The pC-pH diagram for Ca^{2+} in eqsilibrium with $Ca(OH)_{2(s)}$ and Mg^{2+} in equilibrium with $Mg(OH)_{2(s)}$ at 25°C.

[7] If the activity coefficients were not equal to unity, we would use $^cK_{so} = K_{so}/(\gamma_{Ca}^{2+})(\gamma_{OH^-})^2$ instead of K_{so}.

Example 6-6

Find the solubility of (1) $Ca(OH)_{2(s)}$ and (2) $Mg(OH)_{2(s)}$ in distilled water.

Solution

(1) $Ca(OH)_{2(s)}$

In addition to the equilibrium relationship for $Ca(OH)_{2(s)}$, Eq. 6-7, and K_w for water, which are presented in Fig. 6-3, the charge balance equation must be satisfied, that is,

$$2[Ca^{2+}] + [H^+] = [OH^-]$$

Assuming that $[H^+] \ll 2[Ca^{2+}]$,

$$2[Ca^{2+}] = [OH^-]$$

$$0.3 + \log [Ca^{2+}] = \log [OH^-]$$

or

$$pOH = -\log [Ca^{2+}] - 0.3$$

Plotting the line $(-\log [Ca^{2+}] - 0.3)$, we find this equation is satisfied at point A in Fig. 6-3 (pOH = 1.7). The assumption that $[H^+] \ll 2[Ca^{2+}]$ is valid at this point. When pOH = 1.7, from Fig. 6-3 we find $-\log [Ca^{2+}] = 2$. Since $S = [Ca^{2+}]$,

$$S = 10^{-2} M$$

(2) $Mg(OH)_{2(s)}$

Similarly, the charge balance for dissolution of $Mg(OH)_{2(s)}$ in distilled water is

$$2[Mg^{2+}] + [H^+] = [OH^-]$$

which is satisfied at point B in Fig. 6-3. At this point, pOH = 3.5. When pOH = 3.5, $-\log [Mg^{2+}] = 3.8$. Thus

$$S = [Mg^{2+}] = 10^{-3.8}$$

or

$$S = 1.6 \times 10^{-4} M$$

Example 6-7

Sodium hydroxide is added to a groundwater containing $10^{-4} M$ Mg^{2+} and $10^{-3} M$ Ca^{2+}. At what pH will $Mg(OH)_{2(s)}$ begin to precipitate? Will $Ca(OH)_{2(s)}$ precipitate at this pH? At what pH will $[Mg^{2+}]$ be reduced to $10^{-6} M$? The temperature is 25°C.

Solution

From Fig. 6-3, we find that the $10^{-4} M$ Mg^{2+} solution is in equilibrium with $Mg(OH)_{2(s)}$ only at pH 10.65 (point C). Thus $Mg(OH)_{2(s)}$ will precipitate when the pH exceeds 10.65. At this pH, the solution is very much undersaturated with $Ca(OH)_{2(s)}$. (From the diagram we see that more than $1 M$ Ca^{2+} is required for a saturated solution.)

Again from the diagram in Fig. 6-3, when $[Mg^{2+}] = 10^{-6} M$ and the solution is in equilibrium with $Mg(OH)_{2(s)}$, the pH is 11.65 (point D).

Master variables other than pH and pOH can be used in log concentration diagrams. Diagrams using the negative logarithm of concentrations of other ions are often useful in the solution of solubility problems. As an example consider the carbonate salts of Ca^{2+} and Mg^{2+}:

$$CaCO_{3(s)} \rightleftharpoons Ca^{2+} + CO_3^{2-}; \qquad -\log K_{so} = 8.34$$
$$MgCO_{3(s)} \rightleftharpoons Mg^{2+} + CO_3^{2-}; \qquad -\log K_{so} = 5$$

These equilibria can be represented in linear form as follows:

$$-\log [Ca^{2+}] - \log [CO_3^{2-}] = 8.34$$

or

$$-\log [Ca^{2+}] + pCO_3 = 8.34 \qquad (6\text{-}9)$$

and

$$-\log [Mg^{2+}] + pCO_3 = 5 \qquad (6\text{-}10)$$

where $pCO_3 = -\log [CO_3^{2-}]$. Since the carbonate species is common to each of the precipitates, it is useful as a master variable. Equations 6-9 and 6-10 are plotted in a log concentration diagram (Fig. 6-4), which can be used to solve problems such as those in Examples 6-8 and 6-9.

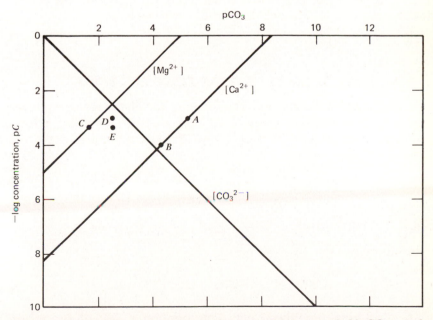

Fig. 6-4. The pC-pH diagram for Mg^{2+} in equilibrium with $MgCO_{3(s)}$ and Ca^{2+} in equilibrium with $CaCO_{3(s)}$.

Example 6-8

A surface water contains $[Ca^{2+}] = 10^{-3}$ M and $[Mg^{2+}] = 5 \times 10^{-4}$ M. At what concentration of CO_3^{2-} will (1) $[Ca^{2+}]$, and (2) $[Mg^{2+}]$ be reduced by one order of magnitude at equilibrium?

Solution

1. 10^{-3} M Ca^{2+} is in equilibrium with $CaCO_{3(s)}$ when $pCO_3 = 5.3$ (point A, Fig. 6-4). To reduce the Ca^{2+} concentration to 10^{-4}, pCO_3 must be reduced to 4.3, that is, the CO_3^{2-} concentration must be increased to $10^{-4.3}$ M at equilibrium (point B, Fig. 6-4).
2. Similarly for Mg^{2+}, if $[Mg^{2+}] = 5 \times 10^{-4}$ at equilibrium, $pCO_3 = 1.7$ ($[CO_3^{2-}] = 10^{-1.7}$) M, (point C, Fig. 6-4).

Example 6-9

The pH of the initial solution in Example 6-8 is 8.3 and the total alkalinity is 3×10^{-3} eq/liter. All of the alkalinity is converted to CO_3^{2-} by adding OH^-. After this conversion, is there sufficient CO_3^{2-} present to begin to precipitate the Ca^{2+} and the Mg^{2+}?

Solution

After conversion of 3×10^{-3} eq/liter of alkalinity (all present as HCO_3^-) to CO_3^{2-}, $pCO_3 = -\log (3 \times 10^{-3}) = 2.5$. From Fig. 6-4, when $[Ca^{2+}] = 10^{-3}$ and $pCO_3 = 2.5$, point D, the solution is supersaturated with $CaCO_3$ and it will precipitate. When $[Mg^{2+}] = 5 \times 10^{-4}$ ($-\log [Mg^{2+}] = 3.3$) and $pCO_3 = 2.5$, point E, the solution is undersaturated with $MgCO_3$ and thus it will not precipitate.

6.4. SOLUBILITY OF SALTS OF WEAK ACIDS AND BASES

Precipitation and dissolution phenomena become more complicated if the cation of the salt is a weak acid, such as NH_4^+, or if the anion is a weak base, such as CN^- or CO_3^{2-}. $AgCN_{(s)}$, for example, dissolves as follows:

$$AgCN_{(s)} \rightleftharpoons Ag^+ + CN^-; \qquad \log K_{so} = -13.8 \qquad (6\text{-}11)$$

and the CN^- ion reacts with water

$$CN^- + H_2O \rightleftharpoons HCN + OH^-; \qquad \log K_b = -4.85$$

or alternatively,

$$CN^- + H^+ \rightleftharpoons HCN; \qquad \log \frac{1}{K_a} = 9.15 \qquad (6\text{-}12)$$

Example 6-10 illustrates the effect of protonation of CN^- on the solubility of AgCN.

Example 6-10

Find the solubility of $AgCN_{(s)}$ in distilled water in a closed system (volatile HCN can not escape), neglecting ionic strength effects and Ag^+ complexes.

Solution

The unknowns are S, $[Ag^+]$, $[HCN]$, $[CN^-]$, $[H^+]$, and $[OH^-]$. Six independent equations are thus required.

The equilibrium constants for Eqs. 6-11 and 6-12 can be used together with the mass balances,

$$S = [Ag^+] = C_{T,Ag} \tag{6-13}$$

$$= [HCN] + [CN^-] = C_{T,CN} \tag{6-14}$$

the equilibrium constant,

$$K_w = 10^{-14} = [H^+][OH^-] \tag{6-15}$$

and the electroneutrality equation,

$$[H^+] + [Ag^+] = [CN^-] + [OH^-] \tag{6-16}$$

The most straightforward way to solve these equations is to substitute in the electroneutrality equation to obtain an equation with one unknown, say $[H^+]$, and then to solve it by trial and error. First, combine Eqs. 6-12 and 6-14 to yield

$$\alpha_1 = \frac{K_a}{[H^+] + K_a} \tag{6-17}$$

and

$$[CN^-] = \alpha_1 S \tag{6-18}$$

Also,

$$[OH^-] = \frac{K_w}{[H^+]}$$

and

$$K_{so} = [Ag^+][CN^-]$$
$$K_{so} = (S)(\alpha_1 S) = \alpha_1 S^2$$

Therefore,

$$S = \left(\frac{K_{so}}{\alpha_1}\right)^{1/2}$$

Substituting Eqs. 6-13 and 6-18 in Eq. 6-16 yields

$$[H^+] + S = \alpha_1 S + \frac{K_w}{[H^+]}$$

or

$$[H^+] + \left(\frac{K_{so}}{\alpha_1}\right)^{1/2} - (\alpha_1)\left(\frac{K_{so}}{\alpha_1}\right)^{1/2} = \frac{K_w}{[H^+]} = 0 \tag{6-19}$$

For an assumed pH, we can determine α_1 from Eq. 6-17, or from Appendix 1. Using these values, we then determine if Eq. 6-19 is satisfied. If not, a new value of pH is assumed and the process is repeated. The results of this process are given in Table 6-2. Plotting the value of Eq. 6-19 versus the assumed pH and extrapolating will enable good successive estimates of pH to be obtained.[8] We first plot the value of Eq. 6-19 (see Fig. 6-5a) at pH 7 and pH 8. The value changes from positive to negative at approximately pH 7.7. Further narrowing it down, we plot the value of Eq. 6-19 at pH 7.7 and pH 7.8, and from the graph (Fig. 6-5b) determine that the left-hand side of Eq. 6-19 has a value of zero at approximately pH 7.78. Rounding off to the nearest tenth of a pH unit, that is, pH 7.8, we find that

$$S = \left(\frac{K_{so}}{\alpha_1}\right)^{1/2} = 6.1 \times 10^{-7} M$$

TABLE 6-2   Solution to Eq. 6-19

pH	α_1	Value of Eq. 6-19
7	7.08×10^{-3}	1.58×10^{-6}
8	6.54×10^{-2}	-4.9×10^{-7}
7.7	3.39×10^{-2}	1.77×10^{-7}
7.8	4.24×10^{-2}	-2.9×10^{-8}

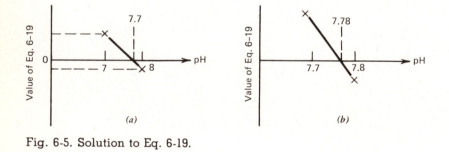

Fig. 6-5. Solution to Eq. 6-19.

It is interesting to examine the solubility of salts such as AgCN as a function of pH. For the purposes of this example let us assume that the solution pH is controlled by the addition of a strong acid or a strong base. Because we wish to obtain S as a function of pH, one less equation is needed than when the solubility of AgCN was determined in distilled water (Example 6-10). Usually, the electroneutrality equation is eliminated making it unnecessary to determine or know the amount of strong acid or strong base added.

Example 6-11

Determine the solubility of AgCN as a function of pH in a closed system.

[8] The use of a digital computer would further facilitate the solution.

The six unknowns are S, $[Ag^+]$, $[CN^-]$, $[HCN]$, $[H^+]$, and $[OH^-]$. The required number of equations is 5 (number of unknowns -1).

(1) $HCN \rightleftharpoons H^+ + CN^-$; $\log K_a = -9.15$

(2) $AgCN \rightleftharpoons Ag^+ + CN^-$; $\log K_{so} = -13.8$

(3) $S = [Ag^+] = C_{T,Ag}$

(4) $S = [HCN] + [CN^-] = C_{T,CN}$

(5) $K_w = [H^+][OH^-]$

Solution

From (4) and (1),

$$S = [HCN] + [CN^-]$$

$$= \frac{[H^+][CN^-]}{K_a} + [CN^-] = \frac{[CN^-]([H^+] + K_a)}{K_a}$$

or

$$S = \frac{[CN^-]}{\alpha_1}$$

or

(6) $[CN^-] = S\alpha_1$

From (2) and (3),

$$K_{so} = [Ag^+][CN^-]$$

$$[CN^-] = \frac{K_{so}}{[Ag^+]} = \frac{K_{so}}{S}$$

Substituting for $[CN^-]$ from (6), we obtain

$$S\,\alpha_1 = \frac{K_{so}}{S}$$

Therefore,

$$S = \left(\frac{K_{so}}{\alpha_1}\right)^{1/2} = \left[\left(\frac{K_{so}}{K_a}\right)([H^+] + K_a)\right]^{1/2}$$

When $[H^+] \ll K_a$,

$$S = [K_{so}]^{1/2}$$

and

$$\log S = \tfrac{1}{2}\log K_{so} = \tfrac{1}{2}(-13.8) = -6.9$$

Thus at pH > 9.15 the solubility is constant at $10^{-6.9}$ mole/liter. When $[H^+] \gg K_a$,

$$S = \left[\frac{K_{so}[H^+]}{K_a}\right]^{1/2}$$

and

$$\log S = \tfrac{1}{2}[\log K_{so} - \log K_a - pH]$$
$$\log S = \tfrac{1}{2}[-13.8 - (-9.15) - pH]$$
$$\log S = \tfrac{1}{2}[-4.65 - pH]$$

Thus at pH < 9.15 the solubility is a function of pH, increasing as pH decreases. The solubility variation is shown graphically in Fig. 6-6.

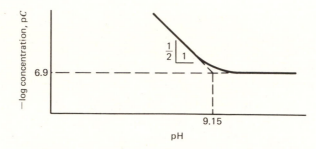

Fig. 6-6. Solubility of AgCN as a function of pH.

Both of the preceding examples were solved for closed systems so that HCN could not leave the liquid. If the system were open so that HCN could escape, the concentration at equilibrium would be a function of the HCN partial pressure; in an open system with no HCN$_{(g)}$ in the atmosphere above the liquid the equilibrium concentration in solution would be zero.

6.5. EFFECT OF COMPLEXATION ON SOLUBILITY

When any of the constituent ions of a solid participate in complex formation following dissolution, there will be an increase in the solubility of the solid. Consider the dissolution of cadmium hydroxide, Cd(OH)$_{2(s)}$, in water, the pH of which is controlled at 9.0 using strong acid or base. Cadmium is of particular importance because of the adverse health effects that it can cause if it is consumed in sufficient quantities. The following reactions are important.

(1) $Cd(OH)_{2(s)} \rightleftharpoons Cd^{2+} + 2OH^-$; $\log K_{so} = -13.65$

(2) $Cd(OH)_{2(s)} \rightleftharpoons CdOH^+ + OH^-$; $\log K_{s1} = -9.49$

(3) $Cd(OH)_{2(s)} \rightleftharpoons Cd(OH)_2{}^{\circ}$; $\log K_{s2} = -9.42$

(4) $Cd(OH)_{2(s)} + OH^- \rightleftharpoons HCdO_2{}^- + H_2O$; $\log K_{s3} = -12.97$

(5) $Cd(OH)_{2(s)} + 2OH^- \rightleftharpoons CdO_2{}^{2-} + 2H_2O$; $\log K_{s4} = -13.97$

Since the solution is equilibrated at pH 9, the solubility, $C_{T,Cd}$, can readily be calculated from the equilibrium equations. Since $[OH^-] = 10^{-5}$, from (1) we find

$$[Cd^{2+}] = \frac{K_{so}}{[OH^-]^2} = \frac{10^{-13.65}}{(10^{-5})^2}$$
$$= 10^{-3.65}$$

From (2),

$$[CdOH^+] = \frac{K_{s1}}{[OH^-]} = \frac{10^{-9.49}}{10^{-5}}$$
$$= 10^{-4.49}$$

From (3),

$$[Cd(OH)_2^0] = K_{s2} = 10^{-9.42}$$

From (4),

$$[HCdO_2^-] = K_{s3}[OH^-] = 10^{-12.97} \times 10^{-5} = 10^{-17.97}$$

From (5),

$$[CdO_2^{2-}] = K_{s4}[OH^-]^2 = 10^{-13.97} \times (10^{-5})^2 = 10^{-23.97}$$

From the mass balance on cadmium,

$$C_{T,Cd} = [Cd^{2+}] + [CdOH^+] + [Cd(OH)_2^0] + [HCdO_2^-] + [CdO_2^{2-}]$$
$$= 10^{-3.65} + 10^{-4.49} + 10^{-9.42} + 10^{-17.97} + 10^{-23.97}$$
$$C_{T,Cd} = 2.56 \times 10^{-4} M = S$$

This value can be compared to a solubility of $10^{-3.65}$ or 2.24×10^{-4}, the concentration of Cd^{2+} in equilibrium with $Cd(OH)_{2(s)}$, that would result if there were no complexes. Thus for cadmium hydroxide the formation of hydroxocadmium(II) complexes increases the solubility by approximately 14 percent. As the pH is increased, the various complex forms become more dominant; at lower pH values they are not present in significant concentrations.

Let us now assume that in addition to the hydroxide ion, there is sufficient Cl^- in solution so that the free Cl^- concentration is $10^{-3}\ M$. To calculate $C_{T,Cd}$ and $C_{T,Cl}$ the following reactions are important as well as those for the hydroxo complexes.

$$
\begin{array}{lll}
Cd^{2+} + Cl^- \rightleftharpoons CdCl^+; & \log K_1 = 1.32 \\
CdCl^+ + Cl^- \rightleftharpoons CdCl_2^0; & \log K_2 = 0.90 \\
CdCl_2^0 + Cl^- \rightleftharpoons CdCl_3^-; & \log K_3 = 0.09 \\
CdCl_3^- + Cl^- \rightleftharpoons CdCl_4^{2-}; & \log K_4 = -0.45
\end{array}
$$

The concentrations of the chlorocadmium(II) complexes are

$$[CdCl^+] = [Cd^{2+}][Cl^-]K_1$$
$$= 10^{-3.65} \times 10^{-3} \times 10^{1.32} = 10^{-5.33}\,M$$

$$[CdCl_2^0] = [CdCl^+][Cl^-]K_2$$
$$= 10^{-5.33} \times 10^{-3} \times 10^{0.9} = 10^{-7.43}\,M$$

$$[CdCl_3^-] = [CdCl_2^0][Cl^-]K_3$$
$$= 10^{-7.43} \times 10^{-3} \times 10^{0.09} = 10^{-10.34}\,M$$

$$[CdCl_4^{2-}] = [CdCl_3^-][Cl^-]K_4$$
$$= 10^{-10.34} \times 10^{-3} \times 10^{-0.45} = 10^{-13.79}$$

The total dissolved cadmium concentration is

$$C_{T,Cd} = [Cd^{2+}] + [CdOH^+] + [Cd(OH)_2^0] + [HCdO^-] + [CdO_2^{2-}]$$
$$+ [CdCl^+] + [CdCl_2^0] + [CdCl_3^-] + [Cd(Cl)_4^{2-}]$$
$$= 2.61 \times 10^{-4}\,M$$

Thus the presence of 10^{-3} M Cl^- has increased the total dissolved cadmium concentration even further. Although the increase is approximately 2 percent in this example because of the Cl^- concentration selected, quite significant increases in solubility are observed in solutions such as seawater, where the Cl^- concentration is approximately 20 g/liter or 0.56 M. In seawater of pH 8 the solubility of cadmium hydroxide (including all hydroxo and chlorocomplexes) is approximately $10^{+0.39}$ M compared with $10^{-1.65}$ M if no Cd(II) complexes were formed. This is an increase of 110 times in cadmium solubility. In seawater the major dissolved species is $CdCl_2^0$. The solubilization of metals such as cadmium and mercury by the formation of soluble chloro complexes has significance in relation to marine waste disposal. The discharge of freshwater streams containing these metals in suspension to a saline environment could well result in increased dissolved metal levels because of the formation of dissolved chloride complexes.

The hydroxo complexes, or hydrolysis products, of the trivalent metal ions and many other divalent metal ions have a dramatic effect on the solubility of these ions. Consider the following equilibria that relate to the behavior of Fe^{3+} in pure water:

$$Fe^{3+} + H_2O \rightleftharpoons FeOH^{2+} + H^+; \qquad \log K_1 = -2.16$$
$$Fe^{3+} + 2H_2O \rightleftharpoons Fe(OH)_2^+ + 2H^+; \qquad \log K = -6.74$$
$$Fe(OH)_{3(s)} \rightleftharpoons Fe^{3+} + 3OH^-; \qquad \log K_{so} = -38$$
$$Fe^{3+} + 4H_2O \rightleftharpoons Fe(OH)_4^- + 4H^+; \qquad \log K = -23$$
$$2Fe^{3+} + 2H_2O \rightleftharpoons Fe_2(OH)_2^{4+} + 2H^+; \qquad \log K = -2.85$$

The effect of these hydroxo complexes on solubility is most conveniently illustrated by using a pC-pH diagram. To plot these equations on a pC-pH diagram requires that they be expressed in terms of soluble Fe(III) species, H^+ and $Fe(OH)_{3(s)}$. We can rearrange these equations so that each Fe(III) species is in equilibrium with $Fe(OH)_{3(s)}$ and then assume that the activity of the solid $Fe(OH)_{3(s)}$ is unity so that the equilibrium constant is then in terms of soluble Fe(III) and H^+. For example, adding the equations,

$$Fe^{3+} + H_2O \rightleftharpoons Fe(OH)^{2+} + H^+; \quad \log K_1 = -2.16$$
$$Fe(OH)_{3(s)} \rightleftharpoons Fe^{3+} + 3OH^-; \quad \log K_{so} = -38$$
$$H^+ + OH^- \rightleftharpoons H_2O; \quad \log\left(\frac{1}{K_w}\right) = 14$$

we obtain

(1) $Fe(OH)_{3(s)} \rightleftharpoons Fe(OH)^{2+} + 2OH^-; \quad \log K_{s1} = -26.16$

Similarly, we can derive equations for $[Fe(OH)_2{}^+]$, $[Fe(OH)_4{}^-]$, and $[Fe_2(OH)_2{}^{4+}]$

(2) $Fe(OH)_{3(s)} \rightleftharpoons Fe(OH)_2{}^+ + OH^-; \quad \log K_{s2} = -16.75$

(3) $Fe(OH)_{3(s)} + OH^- \rightleftharpoons Fe(OH)_4{}^-; \quad \log K_{s4} = -5$

(4) $2Fe(OH)_{3(s)} \rightleftharpoons Fe_2(OH)_2{}^{4+} + 4OH^-; \quad \log K = -50.8$

Equations (1) through (4) can be plotted on a pC-pH diagram. For example, for equation (1), which describes the behavior of the species $Fe(OH)^{2+}$ with pH in a solution in equilibrium with $Fe(OH)_{3(s)}$,

$$K_{s1} = [Fe(OH)^{2+}][OH^-]^2$$

Taking logarithms, we obtain

$$\log K_{s1} = -26.16 = \log [Fe(OH)^{2+}] + 2 \log [OH^-]$$

or

$$\log [Fe(OH)^{2+}] = 2pOH - 26.16$$

and since pH + pOH = 14

$$\log [Fe(OH)^{2+}] = 1.84 - 2pH$$

This equation plots as line 1 in Fig. 6-7.

Similarly, for equation (2), describing the behavior of $Fe(OH)_2{}^+$ species, we have

$$\log [Fe(OH)_2{}^+] = \log K_{s2} + pOH$$

or

$$\log [Fe(OH)_2{}^+] = -16.74 + 14 - pH$$
$$\log [Fe(OH)_2{}^+] = -2.74 - pH$$

This is line 2 in Fig. 6-7. For $[Fe(OH)_4{}^-]$ we obtain

$$\log [Fe(OH)_4{}^-] = pH - 19$$

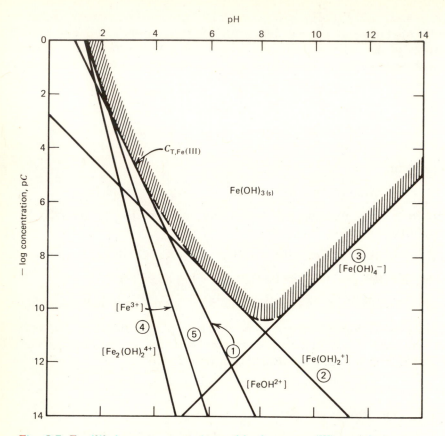

Fig. 6-7. Equilibrium concentrations of hydroxo iron(III) complexes in a solution in contact with freshly precipitated $Fe(OH)_{3(s)}$ at 25°C.

which is line 3 in Fig. 6-7. For $[Fe_2(OH)_2^{4+}]$ we find

$$\log [Fe_2(OH)_2^{4+}] = 5.2 - 4pH$$

which is line 4 in Fig. 6-7. The line for Fe^{3+} is obtained from the K_{so} equation as

$$\log [Fe^{3+}] = 4 - 3pH$$

and is drawn as line 5 in Fig. 6-7. This completes the pC-pH diagram for the species considered.

From an inspection of this diagram we can sketch in the region where $Fe(OH)_{3(s)}$ will precipitate. Since all of the lines represent the concentrations that may exist at various pH values when in equilibrium with solid $Fe(OH)_{3(s)}$, we can reason that any concentration of a particular species above the line in the diagram will be supersaturated with respect to $Fe(OH)_{3(s)}$ for that species. Moreover we can reason that the $Fe(OH)_{3(s)}$ boundary will be somewhere above the line that is at the highest concentration of iron species at any point in the diagram. For example, let us examine the situation at pH 4. As we move vertically up the diagram, we first encounter the line for $[Fe(OH)_4^-]$ at 10^{-15} M; next we encounter the line for $[Fe_2(OH)_2^{4+}]$ at $10^{-10.8}$ M; next is the $[Fe^{3+}]$ line at 10^{-8} M, then the $Fe(OH)_2^+$ line at $10^{-6.74}$ M and last the $FeOH^{2+}$ line at $10^{-6.16}$ M. The boundary of $Fe(OH)_{3(s)}$ at this pH is at the point represented by the sum of these concentrations, that is, $C_{T,Fe(III)} = [Fe^{3+}] + [FeOH^{2+}] + [Fe(OH)_2^+] + 2[Fe_2(OH)_2^{4+}] + [FeOH_4^-]$.

$$C_{T,Fe(III)} = 10^{-8} + 10^{-6.16} + 10^{-6.74} + 2 \times 10^{-10.8} + 10^{-15}$$
$$= 10^{-6.05}$$

By performing such an exercise at various pH values throughout the diagram, we can sketch in the $Fe(OH)_{3(s)}$ boundary. Intuition tells us that if one species predominates at a particular pH, then its concentration will coincide closely with the $Fe(OH)_{3(s)}$ boundary. We can see this happening in a couple of places in Fig. 6-7. Thus at pH values higher than 9, $Fe(OH)_4^-$ predominates and its concentration line is the $Fe(OH)_{3(s)}$ boundary in this region. Between pH 5–7 $Fe(OH)_2^+$ is the predominant species and likewise it forms the solution boundary for $Fe(OH)_{3(s)}$.

We can use pC-pH diagrams that include heterogeneous equilibria for the rapid evaluation of both the total concentration of all species present as well as to provide a graphic representation of the concentrations of individual species present at various pH values. From Fig. 6-7, for example, we can deduce that in the pH range from 4.5 to 8 (which covers virtually all natural waters), $Fe(OH)_2^+$ is the predominant soluble ferric iron species. The polymer (dimer) $Fe_2(OH)_2^{4+}$ does not predominate at any pH value, but it is a significant species at pH values below about 2.5. It is important to note that Fe^{3+}, ferric ion, controls the solubility of $Fe(OH)_{3(s)}$ only below about pH 2.5. Conversely stated, at typically encountered natural water pH values in the presence of $Fe(OH)_{3(s)}$, Fe^{3+} is a minor component of the ferric iron species.

A word of caution in the use of pC-pH diagrams that include heterogeneous equilibria is in order at this point. First, we must always realize that these diagrams are equilibrium diagrams. For them to be validly applied to the solution of natural water problems, we must assure ourselves that the solid of interest is present and that a heterogeneous equilibrium truly does exist. Second, for diagrams that involve solids containing an anion other than hydroxide, we must stipulate a certain solution concentration for that anion. For such precipitates (and we shall consider aluminum phosphate, $AlPO_{4(s)}$, as an example later) there is a unique pC-pH diagram for each concentration of the anion.

Example 6-12

An acid industrial waste discharge containing a high ferrous iron concentration is discharged into an oxygenated receiving water that has a neutral pH and is well buffered. What problems (if any) can be expected in the stream from increased dissolved iron? (Assume that only hydroxo complexes are of importance.)

Solution

First, ferrous iron will be rather rapidly oxidized to ferric iron at neutral pH in oxygenated water; thus only Fe(III) species are of interest. Second, since the water is well buffered and we are adding an acid waste, we should be interested in pH values below but near neutrality. Using Fig. 6-7 and knowing that $C_{T,Fe(III)}$ is the sum of all the concentrations of species present at a given pH, we find that total *soluble* Fe(III) concentration is 10^{-9} to $10^{-10}\ M$, or 0.06 to 0.006 μg Fe/liter, a minute amount. (The major problem may well be caused by the $Fe(OH)_{3(s)}$ that precipitates and either settles to the stream bottom or is carried in suspension.)

Aluminum ion, Al^{3+}, behaves very much like Fe^{3+} in solution except that it has a greater tendency to form polynuclear species. For the equilibria that exist in the presence of freshly precipitated aluminum hydroxide, $Al(OH)_{3(s)}$, we have

(1) $Al^{3+} + H_2O \rightleftharpoons Al(OH)^{2+} + H^+$; $\log K_1 = -5$

(2) $7Al^{3+} + 17H_2O \rightleftharpoons Al_7(OH)_{17}^{4+} + 17H^+$; $\log K = -48.8$

(3) $13Al^{3+} + 34H_2O \rightleftharpoons Al_{13}(OH)_{34}^{5+} + 34H^+$; $\log K = -97.4$

(4) $Al(OH)_{3(s)} \rightleftharpoons Al^{3+} + 3OH^-$; $\log K_{so} = -33$
     (fresh precipitate)

(5) $Al(OH)_{3(s)} + OH^- \rightleftharpoons Al(OH)_4^-$; $\log K_{s4} = 1.3$

(6) $2Al^{3+} + 2H_2O \rightleftharpoons Al_2(OH)_2^{4+} + 2H^+$; $\log K = -6.3$

These equations can be rearranged so that the concentration of each complex is expressed as an equilibrium with solid $Al(OH)_{3(s)}$ as was done for $Fe(OH)_{3(s)}$. The pC-pH diagram plotted from these equations (Fig. 6-8a) shows that the large, highly charged polymeric species such as $Al_{13}(OH)_{34}^{5+}$ and $Al_7(OH)_{17}^{4+}$ control $Al(OH)_{3(s)}$ solubility below a pH value of approximately 6.5.

Example 6-13

Compare the concentration of Al(III) in solution with the amount that precipitates [as $Al(OH)_{3(s)}$] per liter if 150 mg/liter filter alum ($Al_2(SO_4)_3 \cdot 14H_2O$) is used to coagulate a water (1) at pH 6.5 and (2) at pH 9.

Solution

The molecular weight of $Al_2(SO_4)_3 \cdot 14H_2O$ is 594. Thus 150 mg/liter $Al_2(SO_4)_3 \cdot 14H_2O$ = $(150/594) \times 2 = 5.05 \times 10^{-1}$ mmoles Al/liter, or $10^{-3.30}\ M$.

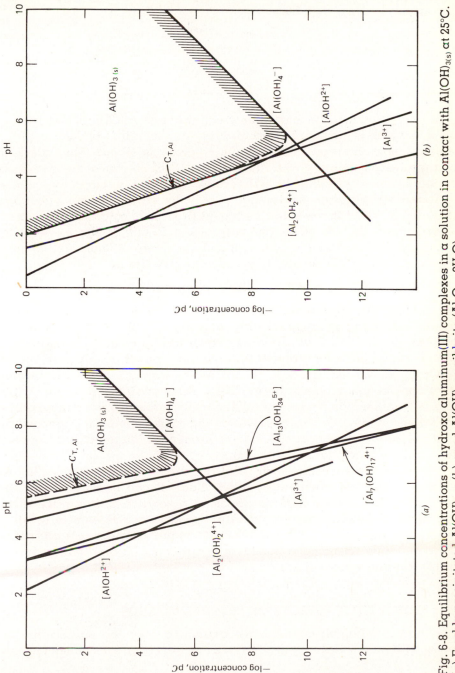

Fig. 6-8. Equilibrium concentrations of hydroxo aluminum(III) complexes in a solution in contact with $Al(OH)_{3(s)}$ at 25°C. (a) Freshly precipitated $Al(OH)_{3(s)}$; (b) aged $Al(OH)_{3(s)}$, gibbsite ($Al_2O_3 \cdot 3H_2O$).

From Fig. 6-8a at pH 6.5 and 9 the values of the total aluminum species in solution can be derived from the boundary of the $Al(OH)_{3(s)}$ region, which corresponds to $C_{T,Al(III)}$. At pH 6.5

$$C_{T,Al(III)} = 10^{-5.6} M$$

At pH 9.0

$$C_{T,Al(III)} = 10^{-3.5} M = 3.16 \times 10^{-4} M$$

At both pH values the filter alum precipitates as $Al(OH)_{3(s)}$. However, at pH 6.5 (which is close to the pH of minimum solubility of $Al(OH)_{3(s)}$) the solution concentration of aluminum species is about two orders of magnitude less than at pH 9.0. Further, at pH 9, 62 percent of the aluminum added $[(3.16 \times 10^{-4}/5.05 \times 10^{-4})(100)]$ does not precipitate.

Aluminum hydroxide precipitated freshly, for example, when filter alum is added to water in the water treatment process of coagulation, is more soluble than the thermodynamically-stable $Al(OH)_{3(s)}$ phase, gibbsite. This material has the same composition but a different, more well-defined crystal structure. The solubility product for gibbsite is

(7) $Al(OH)_{3(s),gibbsite} \rightleftharpoons Al^{3+} + 3OH^-$; $\log K_{so} = -36.3$

which is slightly smaller than the value of 10^{-33} for the freshly precipitated solid. Combining equation (7) with equations (1), (2), (3), (5), and (6) and plotting yields the pC-pH diagram shown in Fig. 6-8b. A comparison of the pC-pH diagram for gibbsite with the diagram for freshly precipitated $Al(OH)_{3(s)}$ shows that gibbsite predominates over a much larger region than does freshly precipitated $Al(OH)_{3(s)}$. Also, in equilibrium with gibbsite, Al^{3+} rather than the polymeric Al-species tends to predominate below pH values of about 5. This observation is consistent with the proposition that polymeric aluminum species are formed as intermediates between Al^{3+} and the precipitation of $Al(OH)_{3(s)}$.

Other species can compete with OH^- for coordination sites on Al(III). As evidence that such competition takes place, it can be shown[9] that the amount of OH^- which must be added to a solution to bring about the formation of $Al(OH)_{3(s)}$ is less when phosphates are present in solution than when they are not. Competitive effects probably cause a difference in coagulation phenomena based on a difference in the ligand concentrations in the water being treated.

6.6.   COMPETITIVE EFFECTS OF SEVERAL LIGANDS

In the preceding discussions we examined only the equilibrium of solids with pure water or with solutions containing an ion common to one

[9] W. Stumm and J. J. Morgan, "Chemical Aspects of Coagulation," *J. Am. Water Works Assoc.*, 54: 971 (1962).

of the constituent ions of the solid. This type of situation is rare in natural waters and wastewaters where there are a wide variety of ligands that can influence both the solubility of a solid and, indeed, govern the type of solid that forms. Two examples of this type of situation will be presented: the precipitation of ferrous hydroxide, $Fe(OH)_{2(s)}$, and ferrous carbonate, $FeCO_{3(s)}$, as may occur during the lime-soda softening of an iron-bearing groundwater; and the precipitation of aluminum hydroxide, $Al(OH)_{3(s)}$, and aluminum phosphate, $AlPO_{4(s)}$, as may occur when alum is added to wastewater to remove phosphate.

6.6.1. Precipitation of $Fe(OH)_{2(s)}$ and $FeCO_{3(s)}$—pC-pH and Predominance Area Diagrams

1. pC-pH Diagrams. We approach this problem by determining which soluble Fe(II) species exist in our system which might influence the solubility of the two solids. The following equilibria are pertinent to this case:

$$Fe(OH)_{2(s)} \rightleftharpoons Fe^{2+} + 2OH^-; \qquad \log K_{so} = -14.5 \qquad (6\text{-}20)$$

$$Fe(OH)_{2(s)} \rightleftharpoons FeOH^+ + OH^-; \qquad \log K_{s1} = -9.4 \qquad (6\text{-}21)$$

$$Fe(OH)_{2(s)} + OH^- \rightleftharpoons Fe(OH)_3^-; \qquad \log K_{s2} = -5.1 \qquad (6\text{-}22)$$

$$FeCO_{3(s)} \rightleftharpoons Fe^{2+} + CO_3^{2-}; \qquad \log K_{so} = -10.7 \qquad (6\text{-}23)$$

$$FeCO_{3(s)} + OH^- \rightleftharpoons FeOH^+ + CO_3^{2-}; \qquad \log K = -5.6 \qquad (6\text{-}24)$$

$$FeCO_{3(s)} + 3OH^- \rightleftharpoons Fe(OH)_3^- + CO_3^{2-}; \qquad \log K = -1.3 \qquad (6\text{-}25)$$

Using Eqs. 6-20 through 6-25, it is possible to construct two types diagrams to show the region of pH and total Fe(II) concentration over which each solid phase can exist. The first diagram is the pC-pH diagram as previously constructed for $Al(OH)_{3(s)}$; the second type is a predominance area diagram. The pC-pH diagram is constructed as follows:

1. Sketch a diagram for a solution containing all soluble Fe(II) species in equilibrium with $Fe(OH)_{2(s)}$. (If carbonato complexes were of importance, they would be included.)

2. Sketch a diagram showing all soluble Fe(II) species in equilibrium with $FeCO_3(s)$ for a constant total soluble carbonate concentration, $C_{T,CO_3} = [H_2CO_3^*] + [HCO_3^-] + [CO_3^{2-}]$. Note that a different diagram results if we use a different concentration of C_{T,CO_3}. In the example we will use $C_{T,CO_3} = 10^{-3}M$ and assume that we are dealing with a closed, constant C_{T,CO_3} system.

3. Combine the above two diagrams to show (i) the variation of soluble Fe(II) concentration with pH and (ii) the regions where the $FeCO_{3(s)}$ and $Fe(OH)_{2(s)}$ precipitates predominate.

Equations 6-20, 6-21, and 6-22 can be used to construct pC-pH diagrams in the same manner as for the $Fe(OH)_{3(s)}$ and $Al(OH)_{3(s)}$ diagrams constructed previously (Figs. 6-7 and 6-8). The equations plotted in Fig. 6-9 and their corresponding line numbers are

$$\log [Fe^{2+}] = 13.5 - 2pH; \quad \text{line 1}$$
$$\log [FeOH^+] = 4.6 - pH; \quad \text{line 2}$$
$$\log [Fe(OH)_3^-] = pH - 19.1; \quad \text{line 3}$$

The boundary line of $C_{T,Fe(II)} = [Fe^{2+}] + [FeOH^+] + [Fe(OH)_3^-]$ is shown in Fig. 6-9 and defines the cross-hatched area of predominance of $Fe(OH)_{2(s)}$.

The next step in construction of the diagram is to draw the area of predominance for $FeCO_{3(s)}$. To do this, we must first determine the variation of $[CO_3^{2-}]$ with pH for $C_{T,CO_3} = 10^{-3}$. From Chapter 4, Fig. 4-17, the variation of $[CO_3^{2-}]$ with pH is as plotted in Fig. 6-10, line 1. The $FeCO_{3(s)}$ solubility product equation, Eq. 6-23 must be satisfied and, therefore, in a pC-pH diagram it will be a mirror image of the $[CO_3^{2-}]$ line such that

$$\log [Fe^{2+}] + \log [CO_3^{2-}] = -10.7$$

Knowing the variation of $[CO_3^{2-}]$ with pH, we can plot the log $[Fe^{2+}]$ line in Fig. 6-10 (line 2). If we had a hypothetical system with no other ligands

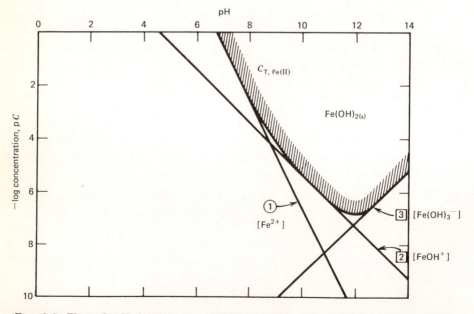

Fig. 6-9. The pC-pH diagram for soluble Fe(II) in equilibrium with $Fe(OH)_{2(s)}$, $T = 25°C$.

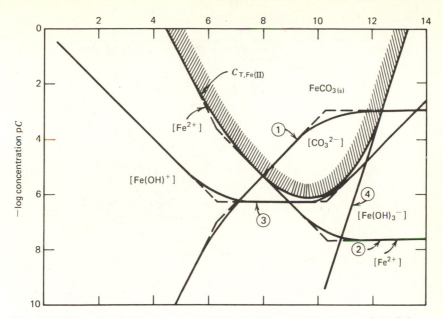

Fig. 6-10. The pC-pH diagram for soluble Fe(II) in equilibrium with $FeCO_{3(s)}$ $C_{T,CO_3} = 10^{-3}\ M$, $T = 25°C$.

save CO_3^{2-}, line 2 would represent the boundary of the predominance region for $FeCO_{3(s)}$. However, the ligand OH^- is present in any system in equilibrium with solid $FeCO_{3(s)}$. Assuming for now that $Fe(OH)_{2(s)}$ is not present, the additional hydroxo complexes that need to be considered are $Fe(OH)^+$ and $Fe(OH)_3^-$. Thus, using Eqs. 6-24 and 6-25, we obtain the equations for $[Fe(OH)^+]$ and $[Fe(OH)_3^-]$ in the presence of $FeCO_{3(s)}$,

$$\log\ [Fe(OH)^+] + \log\ [CO_3^{2-}] = -19.6 + pH$$

and

$$\log\ [Fe(OH)_3^-] + \log\ [CO_3^{2-}] = -43.3 + 3pH$$

Obtaining $[CO_3^{2-}]$ at each pH from Fig. 6-10, we can plot the $[Fe(OH)^+]$ and $[Fe(OH)_3^-]$ lines (lines 3 and 4, respectively, Fig. 6-10). The boundary of the $FeCO_{3(s)}$ predominance region (cross-hatched) is the line for $C_{T,Fe(II)}$.

Combining the predominance regions developed in Figs. 6-9 and 6-10, we obtain Fig. 6-11. From it we can determine the regions for a Fe(II) – OH^-– CO_3^{2-} system in which either $FeCO_{3(s)}$ or $Fe(OH)_{2(s)}$ predominates, and the condition under which both solid phases are stable. From this diagram it can be seen that the only solid below pH 10.5 is $FeCO_{3(s)}$ and that $Fe(OH)_{2(s)}$ is the only solid at pH above 10.5. Below pH 10.5, this figure shows that $FeCO_{3(s)}$ is less soluble than $Fe(OH)_{2(s)}$ (i.e., $C_{T,Fe(II)}$ in equilibrium with $FeCO_{3(s)}$ is less than $C_{T,Fe(II)}$ in equilibrium with $Fe(OH)_{2(s)}$) and $FeCO_{3(s)}$ is therefore the solid phase that is stable. Solid $Fe(OH)_2$ in

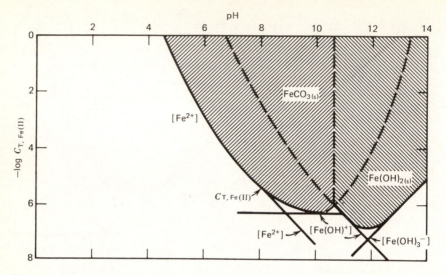

Fig. 6-11. The pC-pH diagram for soluble Fe(II) in equilibrium with $FeCO_{3(s)}$ or $Fe(OH)_{2(s)}$.

contact with solutions in this pH range will dissolve and be reprecipitated as $FeCO_{3(s)}$. Above pH 10.5, $Fe(OH)_{2(s)}$ becomes the least soluble phase so that here Fe(II) is precipitated as $Fe(OH)_{2(s)}$. That pH 10.5 is the dividing line between the two predominance areas, and that both phases are stable at pH 10.5 can be shown by combining Eqs. 6-21 and 6-24, thus

$$FeOH^+ + OH^- \rightleftharpoons Fe(OH)_{2(s)}; \qquad \log \frac{1}{K_{s1}} = 9.4$$

$$FeCO_{3(s)} + OH^- \rightleftharpoons Fe(OH)^+ + CO_3^{2-}; \qquad \log K = -5.6$$

$$FeCO_{3(s)} + 2OH^- \rightleftharpoons Fe(OH)_{2(s)} + CO_3^{2-}; \qquad \log K = 3.8$$

$$K = 10^{3.8} = \frac{[CO_3^{2-}]}{[OH^-]^2}$$

Taking logarithms and setting $[CO_3^{2-}] = 10^{-3.2} M$, the CO_3^{2-} concentration at pH 10.5 (see Fig. 6-10), we can show that this equation represents a vertical line at pH 10.5. Thus only at one point [pH = 10.5, $pC_{T,Fe(II)} = 5.8$] is the solution in equilibrium with *both* solid phases.

2. *Predominance Area Diagrams.* To present the equilibria in Eqs. 6-20 to 6-25 on a pC-pH diagram, it was necessary to fix the total carbonate concentration. In the second type of diagram—the predominance area diagram—we can show the effect of C_{T,CO_3} and pH on the predominant Fe(II) species. This diagram only shows the predominant species at any particular pH, however—not the concentration of each soluble species. To sketch such a diagram, it is necessary to fix the maximum total

concentration of Fe(II). We will assume that the maximum [Fe(II)] is 10^{-5} M; this concentration will only be reached if all precipitate dissolves. There are 5 different species, and it is necessary to develop equations for the boundaries between the regions where each of these species predominate.

There will be 10 possible equations interrelating all these species as follows:

Fe^{2+} and $FeOH^+$

Fe^{2+} and $Fe(OH)_{2(s)}$

Fe^{2+} and $Fe(OH)_3^-$

Fe^{2+} and $FeCO_{3(s)}$

$FeOH^+$ and $Fe(OH)_{2(s)}$

$FeOH^+$ and $Fe(OH)_3^-$

$FeOH^+$ and $FeCO_{3(s)}$

$Fe(OH)_{2(s)}$ and $Fe(OH)_3^-$

$Fe(OH)_{2(s)}$ and $FeCO_{3(s)}$

$FeCO_{3(s)}$ and $Fe(OH)_3^-$

We will state these equations and draw the lines they represent on a $pC_{T,CO_3} - pH$ predominance area diagram; then by deduction we will select the areas of predominance for each species. Our approach will be to first examine the equilibria between Fe^{2+} and the various hydroxo complexes, since these are independent of C_{T,CO_3} and will allow the establishment of pH regions of the predominance of the various hydroxo species. The task of deciding where to draw the lines that describe boundaries between hydroxo species and $FeCO_{3(s)}$ will then be made easier. Selection of the first relationship to plot is arbitrary; in this case we begin with $Fe(OH)_{2(s)}$ $- Fe(OH)_3^-$.

1. $Fe(OH)_{2(s)}$ and $Fe(OH)_3^-$.
    From Eq. 6-22,

$$10^{-5.1} = \frac{[Fe(OH)_3^-]}{\{Fe(OH)_{2(s)}\}[OH^-]}$$

$$-5.1 = \log \frac{[Fe(OH)_3^-]}{\{Fe(OH)_{2(s)}\}} + pOH$$

When precipitate is present in very small amounts and when $[Fe(OH)_3^-] = 10^{-5}\,M$ (the maximum soluble concentration),

$$pOH = -5.1 + 5.0 = -0.1 \quad \text{and} \quad pH = 14.1$$

This line is plotted in Fig. 6-12 (line 1). It is independent of C_{T,CO_3} and therefore plots as a vertical line at pH 14.1. When the pH is below 14.1, $Fe(OH)_{2(s)}$ precipitate predominates with respect to $Fe(OH)_3^-$ and the concentration of $Fe(OH)_3^-$ is less than $10^{-5}\,M$. We can therefore disregard any equilibrium relationships with $Fe(OH)_3^-$ below pH 14.1 because they represent relationships with minor or insignificant quantities relative to $Fe(OH)_{2(s)}$. Likewise above pH 14.1 equations containing $Fe(OH)_{2(s)}$ are not significant. We must keep these facts in mind because it will allow us to greatly simplify construction of the diagram.

2. $FeOH^+$ and $Fe(OH)_{2(s)}$.
   From Eq. 6-21,

$$10^{-9.4} = \frac{[Fe(OH)^+][OH^-]}{\{Fe(OH)_{2(s)}\}}$$

$$-9.4 = \log[OH^-] + \log\frac{[Fe(OH)^+]}{\{Fe(OH)_{2(s)}\}}$$

$$pOH = \log\frac{[Fe(OH)^+]}{\{Fe(OH)_{2(s)}\}} + 9.4$$

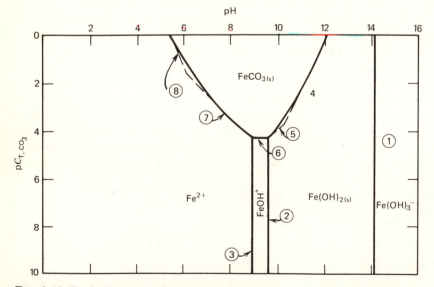

Fig. 6-12. Predominance area diagram for $FeCO_{3(s)}/Fe(OH)_{2(s)}/H_2O$ system at 25°C, $C_{T,Fe(II)} = 10^{-5}\,M$.

When the precipitate is present and $[Fe(OH)^+] = 10^{-5} M$, pOH = 4.4 and the pH = 9.6.

This is drawn as a vertical line (line 2) at pH 9.6 in Fig. 6-12. At pH values below 9.6, $FeOH^+$ predominates. The region of $Fe(OH)_{2(s)}$ predominance can be deduced to be between pH 9.6 and 14.1.

3. Fe^{2+} and $FeOH^+$.
Combining Eqs. 6-20 and 6-21 yields

$$Fe^{2+} + OH^- \rightleftharpoons FeOH^+; \qquad \log K = 5.1$$

$$5.1 = \log \frac{[FeOH^+]}{[Fe^{2+}]} + pOH$$

When $[Fe^{2+}] = [Fe(OH)^+]$, pOH = 5.1 and pH = 8.9.

This is a vertical line at pH 8.9 (line 3, Fig. 6-12). At pH < 8.9, Fe^{2+} predominates. The region of predominance for $FeOH^+$ is defined between pH 8.9 and 9.6.

There is now no reason to determine equations relating the species $Fe^{2+} - Fe(OH)_{2(s)}$, $Fe^{2+} - Fe(OH)_3^-$ and $FeOH^+ - Fe(OH)_3^-$. The boundary between Fe^{2+} and $Fe(OH)_{2(s)}$ would fall within the $FeOH^+$ predominance region, for example, and because we are interested in only the predominant species, this boundary is not of interest to us. Similar reasoning can be applied to the other two pairs of species.

If no carbonates were present, the diagram would be complete and we would not have gained much more useful information from it than would be obtainable from the simpler-to-construct pC-pH diagram. However, since $FeCO_{3(s)}$ also can form when carbonates are present the boundaries between $FeCO_{3(s)}$ and Fe^{2+} and the various hydroxo iron(II) species must be determined. Since the $FeCO_{3(s)}$ equilibria involve carbonate, we must express the variation of $[CO_3^{2-}]$ in terms of C_{T,CO_3} and pH.

From Chapter 4, we recall that

$$C_{T,CO_3} = [H_2CO_3^*] + [HCO_3^-] + [CO_3^{2-}]$$
$$K_{a,1} = 10^{-6.3}$$

and

$$K_{a,2} = 10^{-10.3}$$

so that when pH > 10.3,

$$C_{T,CO_3} \cong [CO_3^{2-}] \tag{6-26}$$

When 6.3 < pH < 10.3,

$$[CO_3^{2-}] \cong \frac{C_{T,CO_3} K_{a,2}}{[H^+] + K_{a,2}} \tag{6-27}$$

And when pH < 6.3,

$$[CO_3^{2-}] = \frac{C_{T,CO_3}K_{a,1}K_{a,2}}{[H^+]^2 + [H^+]K_{a,1} + K_{a,1}K_{a,2}} \tag{6-28}$$

Let us now determine the relationship between $FeCO_{3(s)}$ and each of the Fe^{2+} and hydroxo Fe(II) species, starting with $Fe(OH)_3^-$ at very high pH values and working our way down to lower pH values.

4. $FeCO_{3(s)}$ and $Fe(OH)_3^-$.
    From Eq. 6-25 we obtain

$$FeCO_{3(s)} + 3OH^- \rightleftharpoons CO_3^{2-} + Fe(OH)_3^-; \qquad \log K = -1.3$$

Therefore,

$$-1.3 = \log \frac{[Fe(OH)_3^-]}{\{FeCO_{3(s)}\}} + \log [CO_3^{2-}] - 3\log [OH^-]$$

At pH > 14, where $Fe(OH)_3^-$ is predominant, $[CO_3^{2-}] = C_{T,CO_3}$,

$$-1.3 = \log \frac{[Fe(OH)_3^-]}{\{FeCO_{3(s)}\}} + \log C_{T,CO_3} + 3pOH$$

When $[Fe(OH)_3^-] = C_{T,Fe(II)} = 10^{-5}\ M$ and when $FeCO_{3(s)}$ is present,

$$\log C_{T,CO_3} = 3.7 - 3pOH$$

This boundary exists only at extremely high C_{T,CO_3} concentrations and very high pH values. It is not within the limits of our diagram so we cannot plot it. The region it defines is beyond the limits of consideration for natural aquatic systems.

5. $FeCO_{3(s)}$ and $Fe(OH)_{2(s)}$.
    Combining Eqs. 6-21 and 6-24 yields

$$CO_3^{2-} + Fe(OH)_{2(s)} \rightleftharpoons FeCO_{3(s)} + 2OH^-$$

$$\log K = -3.8 = \log \frac{\{FeCO_{3(s)}\}}{\{Fe(OH)_{2(s)}\}} - 2pOH - \log [CO_3^{2-}]$$

When both solids are present at equilibrium,

$$2pOH = 3.8 - \log [CO_3^{2-}]$$

For pH > 10.3, $[CO_3^{2-}] = C_{T,CO_3}$ and the equation plots as line 4. For $6.3 < pH < 10.3$, from Eq. 6-27,

$$2pOH = 3.8 - \log \left[\frac{C_{T,CO_3}K_{a,2}}{([H^+] + K_{a,2})} \right]$$

which plots as shown—line 5. We only plot this line as far as the intersection with the boundary line between $Fe(OH)_{2(s)}$ and $FeOH^+$.

Beyond this point $Fe(OH)_{2(s)}$ is not a major species so that boundaries with it are not considered for this diagram. Examining Fig. 6-12, we see that we must seek a boundary between $FeCO_{3(s)}$ and $FeOH^+$ for the pH range from 9.6 to 8.9.

6. $FeCO_{3(s)}$ and $FeOH^+$.
   From Eq. 6-24,

$$-5.6 = \log \frac{[FeOH^+]}{\{FeCO_{3(s)}\}} + \log [CO_3^{2-}] - \log [OH^-]$$

When $FeCO_{3(s)}$ is present, $[FeOH^+] = 10^{-5}\ M$, and for pH in the region between 6.3 and 10.3 (from Eqs. 6-27),

$$pOH = -0.6 - \log \left[\frac{C_{T,CO_3} K_{a,2}}{([H^+] + K_{a,2})} \right]$$

This expression plots as line 6 in Fig. 6-12.

Below pH 8.9 the significant boundary will be between $FeCO_{3(s)}$ and Fe^{2+}.

7. $FeCO_{3(s)}$ and Fe^{2+}.
   From Eq. 6-23,

$$-10.7 = \log \frac{[Fe^{2+}]}{\{FeCO_{3(s)}\}} + \log [CO_3^{2-}]$$

When $[Fe^{2+}] = 10^{-5}$, $FeCO_{3(s)}$ is present and the solution is between pH 6.3 and 10.3, using Eq. 6-27, we obtain

$$-5.7 = +\log \left[\frac{C_{T,CO_3} K_{a,2}}{([H^+] + K_{a,2})} \right]$$

This plots as line 7 in Fig. 6-12. Below pH 6.3, this equation becomes

$$-5.7 = \log \left[\frac{C_{T,CO_3} K_{a,1} K_{a,2}}{([H^+]^2 + [H^+] K_{a,1} + K_{a,1} K_{a,2})} \right]$$

which plots as line 8 in Fig. 6-12 completing the diagram.

The diagram in Fig. 6-12 shows the pC_{T,CO_3} and pH regions in which $FeCO_{3(s)}$ and $Fe(OH)_{2(s)}$ will precipitate, and the regions in which Fe(II) is soluble. The diagram is a function of $C_{T,Fe(II)}$, but the approach to developing diagrams for other $C_{T,Fe(II)}$ values is identical to that used here.

6.6.2. Aluminum Phosphate Precipitation

Aluminum ion interacts with water to form hydroxo-complexes and solid $Al(OH)_{3(s)}$, and with orthophosphate to form the solid aluminum phosphate, $AlPO_{4(s)}$. The concentrations of PO_4^{3-} and OH^-, of course, are a function of pH. We would like to examine the amount of total phosphate that could be present in a solution in equilibrium with $AlPO_{4(s)}$ at various pH values. This knowledge is of importance in certain phosphate removal processes, where aluminum sulfate $(Al_2(SO_4)_3)$ is added to wastewater to remove phosphate as $AlPO_{4(s)}$.

We have already examined the precipitation of $Al(OH)_{3(s)}$ from solutions that do not contain phosphate (see Fig. 6-8). Our approach to this problem now will be to develop a pC-pH diagram for $AlPO_{4(s)}$ precipitation assuming no $Al(OH)_{3(s)}$ precipitates, and then to combine this diagram with Fig. 6-8 to obtain a diagram that will give us much useful information about the nature of $AlPO_{4(s)}$ precipitation. We will greatly simplify the problem by assuming that (1) we need be concerned only with freshly precipitated $Al(OH)_{3(s)}$, (2) there are no polymeric hydroxo-aluminum(III) complexes formed, and (3) the only aluminum phosphate solid formed is $AlPO_4$ (i.e., that no hydroxoaluminum phosphate will form). We will examine the importance of these assumptions later.

First, let us consider the interaction of aluminum ion with water, with the only important species in solution being Al^{3+}, $AlOH^{2+}$, and $Al(OH)_4^-$:

$$Al^{3+} + H_2O \rightleftharpoons AlOH^{2+} + H^+; \qquad \log K_1 = -5.0 \qquad (6\text{-}29)$$

$$Al^{3+} + 4H_2O \rightleftharpoons Al(OH)_4^- + 4H^+; \qquad \log K_4 = -21.7 \qquad (6\text{-}30)$$

Defining $C_{T,Al} = [Al^{3+}] + [AlOH^{2+}] + [Al(OH)_4^-]$, we can express $[Al^{3+}]$ in terms of $C_{T,Al}$, $[H^+]$ and constants as follows:

$$C_{T,Al} = [Al^{3+}] + \frac{K_1}{[H^+]}[Al^{3+}] + \frac{K_4}{[H^+]^4}[Al^{3+}]$$

$$= [Al^{3+}]\left(1 + \frac{K_1}{[H^+]} + \frac{K_4}{[H^+]^4}\right) \qquad (6\text{-}31)$$

$$[Al^{3+}] = C_{T,Al}\left(1 + \frac{K_1}{[H^+]} + \frac{K_4}{[H^+]^4}\right)^{-1}$$

We will now proceed in an identical fashion for the second component ion of the $AlPO_{4(s)}$ precipitate, PO_4^{3-}:

$$H_3PO_4 \rightleftharpoons H^+ + H_2PO_4^-; \qquad \log K_{a,1} = -2.1 \qquad (6\text{-}32)$$

$$H_2PO_4^- \rightleftharpoons H^+ + HPO_4^{2-}; \qquad \log K_{a,2} = -7.2 \qquad (6\text{-}33)$$

$$HPO_4^{2-} \rightleftharpoons H^+ + PO_4^{3-}; \qquad \log K_{a,3} = -12.3 \qquad (6\text{-}34)$$

Defining $C_{T,PO_4} = [H_3PO_4] + [H_2PO_4^-] + [HPO_4^{2-}] + [PO_4^{3-}]$, we obtain

$$C_{T,PO_4} = \frac{[PO_4^{3-}][H^+]^3}{K_{a,1}K_{a,2}K_{a,3}} + \frac{[PO_4^{3-}][H^+]^2}{K_{a,2}K_{a,3}} + \frac{[PO_4^{3-}][H^+]}{K_{a,1}} + [PO_4^{3-}]$$

$$[PO_4^{3-}] = C_{T,PO_4}\left(\frac{[H^+]^3}{K_{a,1}K_{a,2}K_{a,3}} + \frac{[H^+]^2}{K_{a,2}K_{a,3}} + \frac{[H^+]}{K_{a,1}} + 1\right)^{-1} \qquad (6\text{-}35)$$

The solubility product equation

$$AlPO_{4(s)} \rightleftharpoons Al^{3+} + PO_4{}^{3-}; \qquad \log K_{so} = -21 \qquad (6\text{-}36)$$

must also be used because the solution is either saturated or supersaturated. Assuming for this case that $C_{T,Al} = C_{T,PO_4}$, that is, that a stoichiometric amount of aluminum salt is added to the phosphate solution, and substituting from Eqs. 6-31 and 6-35 into Eq. 6-36, we obtain

$$K_{so} = C_{T,Al}\left(1 + \frac{K_1}{[H^+]} + \frac{K_4}{[H^+]^4}\right)^{-1}$$
$$\times C_{T,PO_4}\left(\frac{[H^+]^3}{K_{a,1}K_{a,2}K_{a,3}} + \frac{[H^+]^2}{K_{a,2}K_{a,3}} + \frac{[H^+]}{K_{a,1}} + 1\right)^{-1} \qquad (6\text{-}37)$$

Using Eq. 6-37, we can determine C_{T,PO_4} $(= C_{T,Al})$ as a function of pH. This relationship is plotted in Fig. 6-13. It shows that $AlPO_{4(s)}$ has a minimum solubility in the neighborhood of pH 5.5. This information is useful to us because it indicates that it would be inappropriate to attempt $AlPO_{4(s)}$ precipitation for phosphate removal from solutions containing in the range of 10^{-4} to 10^{-3} mole PO_4/liter (which is typical of wastewaters) at

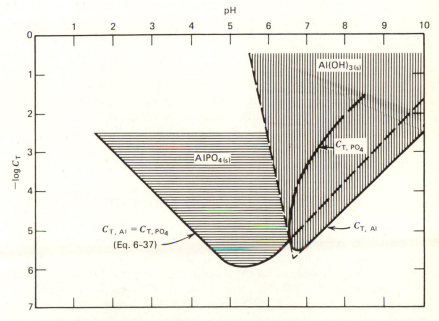

Fig. 6-13. Concentration of phosphate and aluminum that can exist at various pH values when $AlPO_{4(s)}$ is precipitated from or dissolved in pure water. Concentrations of phosphate and aluminum are controlled by $AlPO_{4(s)}$ solubility below pH 6.6. Above pH 6.6 $C_{T,Al}$ is controlled by $Al(OH)_{3(s)}$ solubility, and thus C_{T,PO_4} concentrations in this range are greater than $C_{T,Al}$.

pH values below 3 to 4 or above 6 to 7. We might deduce from Fig. 6-13 that the best place to remove phosphate from wastewater by $AlPO_{4(s)}$ precipitation would be at pH 5.5. This pH is just below the pH ($= 6.6$) at which we can expect $Al(OH)_{3(s)}$ to precipitate if the solid phase is a fresh precipitate, but it is somewhat above the pH at which $Al(OH)_{3(s)}$ will precipitate if the solid phase is aged $Al(OH)_{3(s)}$ (see Fig. 6-8b). Figure 6-13 shows the curve for C_{T,PO_4} in equilibrium with Al^{3+} in the region where $Al(OH)_{3(s)}$ (fresh precipitate) controls aluminum solubility. This curve is calculated from Eq. 6-37. If $Al(OH)_{3(s)}$ precipitates in addition to $AlPO_{4(s)}$, for example, before equilibrium is achieved, additional aluminum salt will have to be added to achieve good phosphate removal. Also, removal by phosphate adsorption onto or incorporation into the $Al(OH)_{3(s)}$ precipitate is possible. Such removals are not accounted for in Fig. 6-13.

Results of $AlPO_{4(s)}$ precipitation from wastewater agree in general with the above observations.[10] An optimum pH of about 6 was observed and it was necessary to use twice as much aluminum salt as required for phosphate precipitation because $Al(OH)_{3(s)}$ also precipitated. Also, the $Al(OH)_{3(s)}$ was of importance because it aided the removal of the fine, rather difficult to settle, $AlPO_{4(s)}$.

According to Fig. 6-13, we do not expect $Al(OH)_{3(s)}$ to form at pH 5 when $C_{T,Al(III)} = 10^{-3}$ to 10^{-4} M. Recht and Ghassemi[10] observed $Al(OH)_{3(s)}$ precipitate at pH above 5, however, indicating that our assumption of only pure, freshly precipitated $AlPO_{4(s)}$ to be the only solid phase present was incorrect. The $Al(OH)_{3(s)}$ may have been aged somewhat, and it did contain some phosphate; thus the solid phase was a hydroxoaluminum phosphate. Undoubtedly, there were also polymeric forms of hydroxoaluminum complexes present, but in spite of these differences, our calculations allowed us to make a reasonable assessment of what to expect when aluminum is used to precipitate phosphate. When making similar calculations, all conclusions made should be verified experimentally because of the limiting assumptions that must be made.

6.7.   CALCIUM CARBONATE SOLUBILITY AND WATER STABILITY

6.7.1.   Calcium Carbonate Solubility in Open and Closed Systems

Many of the concepts presented in Chapter 5 and in the previous sections of this chapter can be incorporated into calculations concerning calcium carbonate, $CaCO_{3(s)}$. Calcium ion forms weak soluble complexes with carbonates, such as $CaCO_3^{0}$ and $CaHCO_3^{+}$, and with hydroxides, such as $CaOH^{+}$ as well as forming the sparingly soluble solid, $CaCO_{3(s)}$.

---

[10] H. L. Recht and M. Ghassemi, "Kinetics and Mechanism of Precipitation and Nature of the Precipitate Obtained in Phosphate Removal from Wastewater Using Aluminum(III) and Iron(III) Salts," Report No. 17010EKI 04/70, U.S. Dept. of Interior, Federal Water Quality Adm., Cincinnati, 1970.

Pure calcium carbonate exists in two distinct crystalline forms. These are the trigonal solid calcite which in its pure forms has a solubility product K_{so} of $10^{-8.34}$ and the orthorhombic solid, aragonite, which in pure state has $K_{so} = 10^{-8.22}$, both at 25°C. When $CaCO_{3(s)}$ precipitation takes place from solutions containing high magnesium concentrations, for example, seawater with a $[Ca^{2+}]/[Mg^{2+}]$ ratio of $= 0.9$ or from the supernatants of some anaerobic sludge digesters, a more soluble solid—a so-called magnesium calcium carbonate—is formed. Chave et al.[11] found that the solubility product of calcium carbonate solid increased from 10^{-8} to $10^{-6.3}$ as the magnesium carbonate content of the solid increased from 3 to 20 percent.

In this section we will consider various situations that involve equilibria with $CaCO_{3(s)}$. The following three cases can serve as models for the behavior of natural systems in which $CaCO_{3(s)}$ participates.

CASE 1

$CaCO_{3(s)}$ is dissolving in pure water that is closed to the atmosphere, as shown in the following figure. An example in nature of this model system is $CaCO_{3(s)}$ dissolving from a sediment into the bottom of a stratified lake.

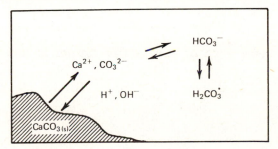

CASE 2

$CaCO_3$ dissolving or precipitating in water open to the atmosphere, as shown in the following figure. In nature this might apply to $CaCO_{3(s)}$ dissolving in the high P_{CO_2} atmosphere of the soil air. When the water rises to the surface ($P_{CO_2} = 10^{-3.5}$ atm) in a spring, $CaCO_{3(s)}$ may precipitate in the form of the mineral travertine.

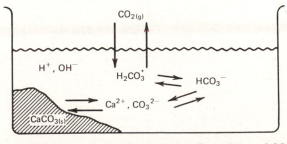

[11] K. E. Chave, K. S. Deffeyes, P. K. Weyl, R. M. Garrels, and M. E. Thompson, "Observation on the Solubility of Skeletal Carbonates in Aqueous Solutions," *Science, 137*: 33 (1962).

CASE 3

CaCO$_{3(s)}$ is in equilibrium with water, open to the atmosphere and to which various quantities of strong acid or base have been added, as shown in the following figure. This could model the dissolution of CaCO$_{3(s)}$ from soil by H$^+$, (strong acid) produced by the oxidation of ammonia present in sewage spread on the soil in, for example, a groundwater recharge operation.

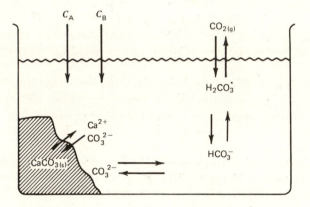

We will consider each of these cases further in the following three examples.

Example 6-14

Case 1

Find the concentrations of all species, the alkalinity and the hardness of the solution that results when CaCO$_{3(s)}$ is equilibrated with distilled water closed to the atmosphere. Neglect ionic strength effects and calcium complexation; the temperature is 25°C.

Solution

The unknowns are C_{T,CO_3}, S, [Ca^{2+}], [H$_2$CO$_3^*$], [HCO$_3^-$], [CO$_3^{2-}$], [H$^+$], and [OH$^-$]. Thus eight equations are required to define the system.

Equilibria

(1) $K_w = 10^{-14} = [\text{H}^+][\text{OH}^-]$

(2) $K_{a,1} = 10^{-6.3} = \dfrac{[\text{H}^+][\text{HCO}_3^-]}{[\text{H}_2\text{CO}_3^*]}$

(3) $K_{a,2} = 10^{-10.3} = \dfrac{[\text{H}^+][\text{CO}_3^{2-}]}{[\text{HCO}_3^-]}$

(4) $K_{so} = 10^{-8.3} = [\text{Ca}^{2+}][\text{CO}_3^{2-}]$

Mass Balances

(5) $S = [Ca^{2+}]$

(6) $S = [H_2CO_3^*] + [HCO_3^-] + [CO_3^{2-}]$

(7) $S = C_{T,CO_3}$

Charge Balance

(8) $2[Ca^{2+}] + [H^+] = [OH^-] + [HCO_3^-] + 2[CO_3^{2-}]$

Our approach to solving these problems will be to express the concentration of each constituent appearing in the electroneutrality equation in terms of $[H^+]$ and constants and then substitute into this equation. The resulting expression can then be solved for $[H^+]$ by trial-and-error. Then using the determined value for $[H^+]$, the values of each component can be determined. In making these manipulations, the α values given in Appendix 1 are helpful.

$$\alpha_0 = \frac{[H_2CO_3^*]}{C_{T,CO_3}}, \qquad \alpha_1 = \frac{[HCO_3^-]}{C_{T,CO_3}}$$

$$\alpha_2 = \frac{[CO_3^{2-}]}{C_{T,CO_3}}$$

Substituting equations (1), (5), and the α values into (—(, WE OBTAIN

$$2S + [H^+] = \frac{K_w}{[H^+]} + \alpha_1 C_{T,CO_3} + 2\alpha_2 C_{T,CO_3}$$

From equations (4), (5), and (7),

$$K_{so} = (S)(\alpha_2 S)$$

$$S = \left(\frac{K_{so}}{\alpha_2}\right)^{1/2}$$

Also from equations (5) and (7), and since $CaCO_{3(s)}$ is dissolving in pure water,

$$[Ca^{2+}] = C_{T,CO_3}$$

Substituting these values in (8), we find

(9) $\quad 2\left(\frac{K_{so}}{\alpha_2}\right)^{1/2} + [H^+] = \frac{K_w}{[H^+]} + \alpha_1\left(\frac{K_{so}}{\alpha_2}\right)^{1/2} + 2\alpha_2\left(\frac{K_{so}}{\alpha_2}\right)^{1/2}$

The only unknown in this equation is $[H^+]$. A trial-and-error solution yields: pH $= 9.95$, $[Ca^{2+}] = 1.27 \times 10^{-4}$, $[HCO_3^-] = 8.9 \times 10^{-5}$, $[CO_3^{2-}] = 3.8 \times 10^{-5}$, $[H_2CO_3^*]$ $= 2.5 \times 10^{-8}$, total alkalinity $= 2.5 \times 10^{-4}$ eq/liter, total hardness $= 1.27 \times 10^{-4}$ mole/liter, $S = 1.27 \times 10^{-4}$ mole/liter.

Example 6-15

Case 2

Find the solubility of $CaCO_{3(s)}$ in a distilled water system open to an atmosphere with, $P_{CO_2} = 10^{-3.5}$ atm. Determine the concentration of each species, the alkalinity, and the hardness. Neglect calcium complexation and ionic strength effects; the temperature is 25°C.

Solution

The unknowns are S, C_{T,CO_3}, $[H^+]$, $[OH^-]$, $[H_2CO_3^*]$, $[HCO_3^-]$, $[CO_3^{2-}]$, and $[Ca^{2+}]$. Thus eight equations are required.

Equilibria

(1) $K_w = 10^{-14} = [H^+][OH^-]$

(2) $K_{a,1} = 10^{-6.3} = \dfrac{[H^+][HCO_3^-]}{[H_2CO_3^*]}$

(3) $K_{a,2} = 10^{-10.3} = \dfrac{[H^+][CO_3^{2-}]}{[HCO_3^-]}$

(4) $K_{so} = 10^{-8.3} = [Ca^{2+}][CO_3^{2-}]$

(5) $[H_2CO_3^*] = K_H P_{CO_2} = (10^{-1.5}\text{mole/liter-atm})(10^{-3.5}\text{atm}) = 10^{-5}$

Mass Balances

(6) $S = [Ca^{2+}]$

(7) $C_{T,CO_3} = [H_2CO_3^*] + [HCO_3^-] + [CO_3^{2-}]$

Note that $S \neq C_{T,CO_3}$ because $[H_2CO_3^*]$ is controlled not by the dissolution of $CaCO_{3(s)}$ but rather by equilibration with the atmospheric P_{CO_2}.

Charge Balance

(8) $2[Ca^{2+}] + [H^+] = [OH^-] + [HCO_3^-] + 2[CO_3^{2-}]$

We will use the same technique as in Example 6-14, namely, finding concentrations in terms of $[H^+]$ and constants and then substituting them in the charge balance equation. From (5) and using α_0, we obtain

$$[H_2CO_3^*] = 10^{-5} = \alpha_0 C_{T,CO_3}$$

or

$$C_{T,CO_3} = \frac{10^{-5}}{\alpha_0}$$

From (4) and α_2,

$$[Ca^{2+}] = \frac{K_{so}}{[CO_3^{2-}]} = \frac{K_{so}}{(\alpha_2 C_{T,CO_3})}$$

Substituting $10^{-5}/\alpha_0$ for C_{T,CO_3}

$$[Ca^{2+}] = \frac{K_{so}\alpha_0}{(\alpha_2 10^{-5})}$$

and substituting into (8) yields

$$\frac{2\alpha_0 K_{so}}{10^{-5}\alpha_2} + [H^+] = \frac{K_w}{[H^+]} + \alpha_1\left(\frac{10^{-5}}{\alpha_0}\right) + 2\alpha_2\left(\frac{10^{-5}}{\alpha_0}\right)$$

The only unknown in this equation is $[H^+]$; accordingly, it can be solved by trial and error yielding pH = 8.3, $[H_2CO_3^*] = [CO_3^{2-}] = 10^{-5}\,M$, $[HCO_3^-] = 10^{-3}\,M$, $[Ca^{2+}]$ = 5 × $10^{-4}\,M$, total alkalinity = 10^{-3} eq/liter, total hardness = 5 × $10^{-4}\,M$, and S = 5 × $10^{-4}\,M$.

Note the effect of exposing the solution equilibrated with $CaCO_{3(s)}$ to an atmosphere containing only $10^{-3.5}$ atmospheres of CO_2. The pH of the solution has been depressed from 9.95 to 8.3 while the $CaCO_{3(s)}$ solubility has increased from 1.27 × $10^{-4}\,M$ to 5 × $10^{-4}\,M$ (or from 12.7 mg/liter as $CaCO_3$ to 50 mg/liter as $CaCO_3$)—an almost fourfold increase. This solubility increase is reflected in increases in alkalinity and total hardness. Comparing these two examples illustrates the solvent effect of CO_2 on carbonate minerals and provides an explanation for the way in which groundwaters pick up hardness and alkalinity. Consider the situation in a limestone strata that is in contact with an atmosphere containing 10^{-2} atm of CO_2—a reasonable amount of CO_2 for soil air. The water will have a pH of approximately 7.3, and a total hardness and alkalinity of approximately 125 to 150 mg/liter as $CaCO_3$ at equilibrium.

Example 6-16

Case 3

This case is identical to the second case, except that strong acid or strong base is added to the system. We must determine the same concentrations. In this example we will add 10^{-3} mole of the strong acid H_2SO_4 per liter of solution although a similar approach to the problem could be taken if a strong base were added.

Equations 1-7 in Example 6-15 are applicable, together with

(8) $C_{T,SO_4} = [SO_4^{2-}] = 1 \times 10^{-3}$

and a modified charge balance,

$$2[Ca^{2+}] + [H^+] = [OH^-] + [HCO_3^-] + 2[CO_3^{2-}] + 2[SO_4^{2-}]$$

Substituting in the same way as in Example 6-15, we obtain

$$\frac{2\alpha_0 K_{so}}{10^{-5}\alpha_2} + [H^+] = \frac{K_w}{[H^+]} + \frac{10^{-5}\alpha_1}{\alpha_0} + 2 \times \frac{10^{-5}\alpha_2}{\alpha_0} + (2 \times 10^{-3})$$

Solving by trial and error yields pH = 8.0, $[H_2CO_3^*] = 10^{-5}$, $[HCO_3^-] = 5 \times 10^{-4}$, $[CO_3^{2-}] = 3.6 \times 10^{-6}$, $[Ca^{2+}] = 2.5 \times 10^{-3}$, total alkalinity = 5 × 10^{-4} eq/liter, and total hardness = 2.5 × 10^{-3} mole/liter.

Two types of problems that often arise in conjunction with water treatment processes such as softening and water conditioning are: Is this solution saturated with respect to $CaCO_{3(s)}$? If not, is it undersaturated or oversaturated and what doses of which chemicals must be added to a water to produce a treated water of specified composition (alkalinity, pH, hardness, and degree of saturation with respect to $CaCO_{3(s)}$)?

The latter question is only addressed in an introductory fashion in this text, and the reader is referred to books on water treatment for a more detailed discussion of water conditioning. The use of diagrams such as the Arbatsky[12] and Caldwell–Lawrence[13] diagrams is common. More modern approaches utilize the computer to solve the equilibria involved in these problems. In water conditioning the water is treated as a closed system with $CaCO_{3(s)}$ present, but the calculations may be somewhat in error because the system does not entirely behave as a closed system. The composition of the system is computed prior to chemical dosing, chemicals (e.g., acid, base, CO_2, and alkalinity) are added, and the composition of the system is computed following chemical addition—again assuming a closed system.

The determination of whether a solution is in equilibrium with $CaCO_{3(s)}$ can be accomplished by calculating the free energy of the solution as described in Chapter 3.

Example 6-17

Water of the following composition is obtained following a softening/recarbonation process: $[Ca^{2+}] = 1 \times 10^{-3}$, $[HCO_3^-] = 2 \times 10^{-3}$, pH = 8.7, temperature = 10°C, $\mu = 5 \times 10^{-3}$. Is the water at equilibrium with $CaCO_{3(s)}$?

Solution

Combine the equations[14] using Table 4-4, Chapter 4 to obtain the necessary constants.

(1) $CaCO_{3(s)} \rightleftharpoons Ca^{2+} + CO_3^{2-}$; $\log K_{so} = -8.15$

(2) $H^+ + CO_3^{2-} \rightleftharpoons HCO_3^-$; $\log \dfrac{1}{K_{a,2}} = 10.49$

(3) $CaCO_{3(s)} + H^+ \rightleftharpoons Ca^{2+} + HCO_3^-$; $\log K = 2.34$

From Eqs. 3-20 and 3-22,

$$\Delta G = \Delta G^\circ + RT \ln Q$$
$$\Delta G^\circ = -RT \ln K$$

---

[12] J. W. Arbatsky, "Zeichnerische Ermittlung der Enhärtungs Verhaltnisse von Wassern. Nach dem Nomogramm von F. W. Staffeldt und dem Kalk-Soda-Wasserbild" *Gas und Wasserfach*, 83: 90–92 (1940).

[13] D. H. Caldwell and W. B. Lawrence, "Water Softening and Conditioning Problems," *Ind. Eng. Chem.*, 45: 535 (1953).

[14] Equation (1) could have been used, but the combination equation (3), is more appropriate, since HCO_3^- is the major carbonate species at the pH values typically encountered and it can be more accurately determined.

Let us first calculate ΔG° and then Q.

$$\Delta G^\circ = -(1.987 \times 10^{-3})(283) \ln 10^{2.34}$$

$$= -3.02 \, \text{kcal}$$

$$Q = \frac{\{Ca^{2+}\}\{HCO_3^-\}}{\{H^+\}}$$

$$pH = -\log \{H^+\} = 8.7$$

so that

$$\{H^+\} = 10^{-8.7}$$

Also,

$$\{Ca^{2+}\} = (\gamma_{Ca^{2+}})[Ca^{2+}]; \quad \{HCO_3^-\} = (\gamma_{HCO_3^-})[HCO_3^-]$$

From Fig. 3-4[15] for $\mu = 5 \times 10^{-3}$,

$$\gamma_{Ca^{2+}} = 0.75, \qquad \gamma_{HCO_3^-} = 0.93$$

$$Q = \frac{(0.75)(1 \times 10^{-3})(0.93)(2 \times 10^{-3})}{10^{-8.7}} = 6.99 \times 10^2$$

and using the values of ΔG° and Q we obtain

$$\Delta G = -3.02 + (1.987 \times 10^{-3})(283) \ln 6.99 \times 10^2$$
$$= 0.66 \, \text{kcal}$$

Because $\Delta G > 0$, the reverse of reaction (3) as written is proceeding spontaneously and precipitation of $CaCO_{3(s)}$ is taking place.

The Langelier Index   The problem of determining whether or not a water is in equilibrium with $CaCO_{3(s)}$ can be approached by determining whether the so-called Langelier Index (L.I.) (or Saturation Index, S.I.) is positive or negative. The L.I. is defined as the difference between the actual (or measured) pH of a water and the hypothetical pH the water would have if it were in equilibrium with $CaCO_{3(s)}$.[16]

$$\text{L.I.} = pH_a - pH_s \qquad (6\text{-}38)$$

where

   pH_a = actual pH of water
   pH_s = pH of water if it were in equilibrium with $CaCO_{3(s)}$ at
   the existing solution concentrations[17] of HCO_3^- and Ca^{2+}

[15] Note that the DeBye–Hückel limiting law could also have been used to calculate the activity coefficients.

[16] W. F. Langelier. "The Analytical Control of Anticorrosion Water Treatment," *J. Am. Water Works Assoc.*, 28: 1500 (1939).

[17] Some determine pH_s by using the total alkalinity in place of the $[HCO_3^-]$ concentration in the existing water. However, this leads to a reversal in sign for L.I. at pH > $pK_{a,2}$, i.e., precipitation occurs if the L.I. < 0, instead of precipitation occurring if pH > 0 as is true at pH < $pK_{a,2}$. (See R. E. Lowenthal and C. V. R. Marais, *Carbonate Chemistry of Aquatic Systems: Theory and Application*, Ann Arbor Science Publishers, 1976.) Defining pH_s as presented above avoids this problem. In the pH range normally encountered, the two approaches are essentially the same, since total alkalinity (eq/liter) = $[HCO_3^-]$.

If a water has a L.I. of zero, it is in equilibrium with $CaCO_{3(s)}$; if the L.I. is a positive value, the water is oversaturated with respect to $CaCO_{3(s)}$ and will tend to precipitate $CaCO_{3(s)}$; and if the L.I. is a negative value, the water is undersaturated with respect to $CaCO_{3(s)}$ and will tend to dissolve $CaCO_{3(s)}$.

The expression for the value of pH_s is derived from the equilibrium constant of the equation,

$$CaCO_{3(s)} + H^+ \rightleftharpoons Ca^{2+} + HCO_3^-; \qquad K = \frac{K_{so}}{K_{a,2}}$$

$$\frac{K_{so}}{K_{a,2}} = \frac{\{Ca^{2+}\}\{HCO_3^-\}}{\{H^+\}} = \frac{\gamma_{Ca^{2+}}[Ca^{2+}]\,\gamma_{HCO_3^-}[HCO_3^-]}{\{H^+\}}$$

$$\log\{H^+\} = \log K_{a,2} - \log K_{so} + \log[Ca^{2+}]$$
$$+ \log[HCO_3^-] + \log \gamma_{Ca^{2+}} + \log \gamma_{HCO_3^-}$$

or

$$pH_s = pK_{a,2} - pK_{so} + p[Ca^{2+}] + p[HCO_3^-] - \log \gamma_{Ca^{2+}} - \log \gamma_{HCO_3^-} \qquad (6\text{-}39)$$

The total alkalinity in eq/liter can be used interchangeably with $[HCO_3^-]$ for waters with pH < 9. When calcium complexes make up a significant portion of the total calcium concentration, the concentration of these complexes must be substracted from the total calcium to yield $[Ca^{2+}]$. Also, if species such as aqueous silica contribute to the total alkalinity (see Section 4.13.5) the portion of the total alkalinity attributable to these species must be subtracted from the total alkalinity; the difference is then used to calculate the HCO_3^- concentration.

Using the constants, the measured values of $[Ca^{2+}]$ and $[HCO_3^-]$ and the values of the activity coefficients as given in Example 6-17, we can solve for pH_s thus:

$$pH_s = 10.49 - 8.15 + 3 + 2.7 + 0.04 + 0.01 = 8.09$$

Therefore,

$$\text{L.I.} = pH_a - pH_s = 8.7 - 8.09 = +0.61$$

The L.I. is positive, so the water is "encrustive" and will tend to precipitate $CaCO_{3(s)}$.

The two procedures for assessing the state of a water with respect to $CaCO_{3(s)}$ are identical. This is shown as follows: Consider the reaction

$$CaCO_{3(s)} + H^+ \rightleftharpoons Ca^{2+} + HCO_3^-$$

$$K = \frac{\{Ca^{2+}\}\{HCO_3^-\}}{\{H^+\}} = \frac{\gamma_{Ca^{2+}}\gamma_{HCO_3^-}[Ca^{2+}][HCO_3^-]}{\{H^+\}}$$

$$Q = \frac{\gamma_{Ca^{2+}}\gamma_{HCO_3^-}[Ca^{2+}]_a[HCO_3^-]_a}{\{H^+\}_a}$$

where [ ] and { } denote equilibrium concentration and activity and [ ]$_a$ and { }$_a$ denote actual or measured concentrations or activity.

Now

$$\frac{Q}{K} = \frac{\gamma_{Ca^{2+}}\gamma_{HCO_3^-}[Ca^{2+}]_a[HCO_3^-]_a\{H^+\}}{\gamma_{Ca^{2+}}\gamma_{HCO_3^-}[Ca^{2+}][HCO_3^-]\{H^+\}_a}$$

From the definition of pH$_s$ given above,

$$[Ca^{2+}]_a = [Ca^{2+}]$$

and

$$[HCO_3^-]_a = [HCO_3^-]$$

then

$$\frac{Q}{K} = \frac{\{H^+\}}{\{H^+\}_a}$$

Since

$$pH_s = -\log\{H^+\}$$
$$pH_a = -\log\{H^+\}_a$$

We have

$$\log\frac{Q}{K} = pH_a - pH_s = L.I.$$

From Eq. 3-23,

$$\Delta G = 2.3\,RT\log\frac{Q}{K} = 2.3\,RT\,(L.I.) \qquad (6\text{-}40)$$

Thus a positive ΔG corresponds to a positive L.I.

The question of whether a water is oversaturated, undersaturated, or in equilibrium with calcium carbonate is important in the water industry, where processes called water conditioning are used to adjust the state of $CaCO_{3(s)}$ saturation. The importance of the $CaCO_{3(s)}$ saturation condition of a water relates to its behavior during transport, for example, in distribution systems and conduits and to some extent to its behavior during granular media filtration. Waters grossly oversaturated with $CaCO_3$ will tend to lay down precipitates of $CaCO_{3(s)}$ on the interior of the pipe, reducing its carrying capacity and in extreme cases blocking the pipe altogether (Fig. 6-14). Waters grossly undersaturated with respect to $CaCO_3$ tend to dissolve the protective $CaCO_{3(s)}$ coatings of transmission lines and are often classified as being "corrosive." Generally, an attempt is made in practice to maintain the L.I. slightly positive, although in some areas of the United States and Europe a slightly negative or zero value is used. If the water is supersaturated, it is desirable to know how

Fig. 6-14. Water distribution conduit virtually blocked with precipitated $CaCO_3$. Courtesy of R. R. Trussell, J. M. Montgomery Engineers, Pasadena, Calif.

much $CaCO_{3(s)}$ will precipitate or dissolve. The L.I. has limited application because it tells us only whether precipitation or dissolution, if any, will tend to take place. Indeed, a water with a L.I. of 0.3 and a high buffer intensity may precipitate more $CaCO_{3(s)}$ than one that has L.I. of 0.5 and a low buffer intensity.

Precipitation Potential   We can calculate the amount of $CaCO_{3(s)}$ that will precipitate or dissolve as the water equilibrates by applying the two following principles to calculate equilibrium concentrations:

1.   Total acidity does not change as $CaCO_{3(s)}$ precipitates or dissolves.

2.   Total alkalinity (eq/liter) − calcium (eq/liter) = constant, as precipitation or dissolution takes place.

The total acidity does not change because CO_3^{2-} does not enter into the equation for it, that is,

$$\text{Total acidity} = 2[H_2CO_3^*] + [HCO_3^-] + [H^+] - [OH^-]$$

This is analogous to the fact that the addition or elimination of CO_2 does not alter the total alkalinity if it does not cause $CaCO_{3(s)}$ to precipitate or dissolve. The total alkalinity, defined as

$$\text{Total alkalinity} = 2[CO_3^{2-}] + [HCO_3^-] + [OH^-] - [H^+]$$

changes as $CaCO_{3(s)}$ precipitates or dissolves but total alkalinity (eq/liter) $- Ca^{2+}$ (eq/liter) does not change because the number of eq/liter of Ca^{2+} removed or added via $CaCO_{3(s)}$ precipitation or dissolution is equal to the eq/liter of total alkalinity removed or added. The calculation procedure is illustrated in the following example.

Example 6-18

Calculate the amount of $CaCO_{3(s)}$, in eq/liter and mg/liter as $CaCO_3$, which will precipitate as the solution in Example 6-17 is equilibrated. Neglect ionic strength effects in this calculation.

Solution

The system is assumed to be closed so no CO_2 can enter or leave.

The unknown concentrations after equilibration are $[H_2CO_3^*]$, $[HCO_3^-]$, $[CO_3^{2-}]$, $[H^+]$, $[OH^-]$, $[Ca^{2+}]$, and C_{T,CO_3}. Thus seven equations are needed to determine the equilibrium concentrations.

From Example 6-17, before precipitation, we know $[Ca^{2+}]_{orig} = 1 \times 10^{-3}$, pH = 8.7, and $[HCO_3^-] = 2 \times 10^{-3}$. At 10°C, $K_{a,1} = 10^{-6.46}$, $K_{a,2} = 10^{-10.49}$, $K_{so} = 10^{-8.15}$ (from Table 4-4) and $K_w = 10^{-14.5}$ (interpolating from Table 4-2). Using $K_{a,1}$ and $K_{a,2}$, we calculate $[H_2CO_3^*] = 1.1 \times 10^{-5}$ and $[CO_3^{2-}] = 3.2 \times 10^{-5}$. Using these values in the equation for total acidity given above, we find total acidity $= 2.02 \times 10^{-3}$ eq/liter. Also, total alkalinity ($= 2.07 \times 10^{-3}$ eq/liter) $- Ca^{2+}$ ($= 2 \times 10^{-3}$ eq/liter) $= 0.07 \times 10^{-3}$ eq/liter.

After the solution equilibrates we can calculate the equilibrium concentrations knowing that total acidity and [total alkalinity $- Ca^{2+}$ (eq/liter)] does not change. The equations we need are

(1) Total acidity $= 2[H_2CO_3^*] + [HCO_3^-] + [H^+] - [OH^-] = 2.02 \times 10^{-3}$ eq/liter

(2) Total alkalinity $- Ca^{2+}$ (eq/liter) $= 2[CO_3^{2-}] + [HCO_3^-] + [OH^-] - [H^+] - 2[Ca^{2+}] = 7 \times 10^{-5}$

(3) $K_{a,1} = \dfrac{[H^+][HCO_3^-]}{[H_2CO_3^*]}$

(4) $K_{a,2} = \dfrac{[H^+][CO_3^{2-}]}{[HCO_3^-]}$

(5) $C_{T,CO_3} = [H_2CO_3^*] + [HCO_3^-] + [CO_3^{2-}]$

(6) $K_w = [H^+][OH^-]$

(7) $K_{so} = [Ca^{2+}][CO_3^{2-}]$

To solve, substitute into equation (2) to obtain one equation with one unknown, $[H^+]$, and solve by trial and error, just as was done in Examples 6-14, 6-15, and

6-16.   Remembering $\alpha_0 = [H_2CO_3^*]/C_{T,CO_3}$, $\alpha_1 = [HCO_3^-]/C_{T,CO_3}$, and $\alpha_2 = [CO_3^{2-}]/C_{T,CO_3}$. The result from combining equations (3), (4) and (5) and using equation (7) to obtain $[Ca^{2+}] = K_{so}/(\alpha_2 C_{T,CO_3})$ and substituting into equation (2) is

$$(8) \quad 7 \times 10^{-5} = C_{T,CO_3}(2\alpha_2 + \alpha_1) + \frac{K_w}{[H^+]} - \frac{2K_{so}}{\alpha_2 C_{T,CO_3}} - [H^+]$$

and from equation (1)

$$(9) \quad C_{T,CO_3} = \frac{2.02 \times 10^{-3} - [H^+] + (K_w/[H^+])}{2\alpha_0 + \alpha_1}$$

We can solve equations (8) and (9) by assuming $[H^+]$, obtaining α values from Appendix 1, calculating C_{T,CO_3} from equation (9), and then determining if equation (8) is satisfied. If not, a new value of $[H^+]$ is assumed. We can simplify the procedure somewhat in this case by assuming $[H^+]$ and $K_w/[H^+]$ to be small relative to the other terms in equations (8) and (9). Using this approach, we find $[H^+] = 10^{-8.08}$. From equation (7), $[Ca^{2+}] = 9.48 \times 10^{-4} M$.

Subtracting the $[Ca^{2+}]$ at equilibrium from the $[Ca^{2+}]$ originally present yields the Ca^{2+}, and thus the $CaCO_{3(s)}$, which precipitates per liter of solution.

$$[Ca^{2+}]_{orig} - [Ca^{2+}]_{equil} = 1 \times 10^{-3} - 9.48 \times 10^{-4}$$
$$= 0.52 \times 10^{-4} \, mole/liter$$
$$0.52 \times 10^{-4} \, mole/liter \times 2 \, eq/mole = 1.04 \times 10^{-4} eq/liter$$
$$1.04 \times 10^{-4} \, eq/liter \times 50,000 \, mg \, CaCO_3/eq = 5.2 \, mg/liter \, as \, CaCO_3$$

The theoretical precipitation potential of $CaCO_{3(s)}$ found in Example 6-18 was 5.2 mg/liter, which is within the range of 4 to 10 mg/liter suggested for a well-conditioned water by Merrill and Sanks.[18] A similar approach to that given in Example 6-18 can be used to determine the concentrations of conditioning chemicals to add if they are needed. Such calculations, as well as calculation of the precipitation potential, are facilitated by the use of diagrams such as the Caldwell–Lawrence diagrams; use of these diagrams is discussed by Merrill and Sanks. Water with the desired precipitation potential is thought to provide a thin protective coating of $CaCO_{3(s)}$ on the inside of distribution lines and so aid in preventing their deterioration by corrosion. The most effective $CaCO_{3(s)}$ layers appear to be those associated with iron hydroxide and iron carbonate precipitates.[19] A slight supersaturation with $CaCO_{3(s)}$ and high alkalinity also provides the water with pH buffering capacity against CO_2 generated in the distribution systems, by slime growths for example, and prevents corrosion by preventing low pH values from occurring (see Section 7-6 for a discussion of the effect of pH on corrosion). Waters that have a low calcium concentration and a low alkalinity, and thus a low buffer

---

[18] D. T. Merrill and R. L. Sanks, "Corrosion Control by Deposition of $CaCO_3$ Films," J. Am. Water Works Assoc., 69: 592 (1977); 69: 634 (1977); and 70: 12 (1978).
[19] C. Feigenbaum, L. Gal-or, and J. Yahalom, Corrosion, 34: 65 (1978); and 34: 133 (1978).

intensity, may require the use of L.I. values as high as 1.0 to prevent corrosion.[20] However, good protective films usually do not form under these conditions; Stumm found that when a high pH was necessary to create a positive L.I., scale formation was not uniform.[21]

6.7.2. The Effect of Complex Formation on the Solubility of $CaCO_{3(s)}$

The following carbonato and hydroxo complexes and their dissociation constants at 25°C have been reported for Ca^{2+}:

$$CaCO_3^0 \rightleftharpoons Ca^{2+} + CO_3^{2-}; \qquad \log K_d = -3.22 \qquad (6\text{-}41)$$

$$CaHCO_3^+ \rightleftharpoons Ca^{2+} + HCO_3^-; \qquad \log K_d = -1.26 \qquad (6\text{-}42)$$

$$CaOH^+ \rightleftharpoons Ca^{2+} + OH^-; \qquad \log K_d = -1.49 \qquad (6\text{-}43)$$

It should be noted that the reported values of constants for these equilibria vary widely. The values used here are representative. We can combine these equilibria with the equation for $CaCO_{3(s)}$ precipitation and then plot the resulting equations to determine the effect of these complexes on total soluble calcium. For $CaCO_3^0$,

$$CaCO_{3(s)} \rightleftharpoons Ca^{2+} + CO_3^{2-}; \qquad \log K_{so} = -8.3$$

$$Ca^{2+} + CO_3^{2-} \rightleftharpoons CaCO_3^0; \qquad \log \frac{1}{K_d} = 3.2$$

$$CaCO_{3(s)} \rightleftharpoons CaCO_3^0; \qquad \log K = -5.1$$

Thus

$$10^{-5.1} = [CaCO_3^0] = 8.3 \times 10^{-6}\,M \qquad (0.8 \text{ mg/liter as } CaCO_3) \qquad (6\text{-}44)$$

whenever $CaCO_{3(s)}$ is in equilibrium with solution. For $CaHCO_3^+$,

$$H^+ + CaCO_{3(s)} \rightleftharpoons CaHCO_3^+; \qquad \log K = 3.26$$

or

$$[CaHCO_3^+] = 1.8 \times 10^3\,[H^+] \qquad (6\text{-}45)$$

and for $CaOH^+$

$$[CaOH^+] = 31[Ca^{2+}][OH^-] \qquad (6\text{-}46)$$

Equations 6-44 to 6-46 are plotted in Fig. 6-15 together with the equation

$$C_{T,Ca} = [Ca^{2+}] + [CaOH^+] + [CaHCO_3^+] + [CaCO_3^0]$$

and the equation for $[Ca^{2+}]$ determined from K_{so} for $CaCO_{3(s)}$. The procedure used to develop the plot is illustrated in the following example.

[20] T. E. Larson, "Corrosion by Domestic Waters," Illinois State Water Survey Bulletin 59, Urbana, Ill., 1975.
[21] W. Stumm, "Investigations on the Corrosive Behavior of Water," Am. Soc. Civil Engrs., J. Sanit. Eng. Div., 86: 27 (1965).

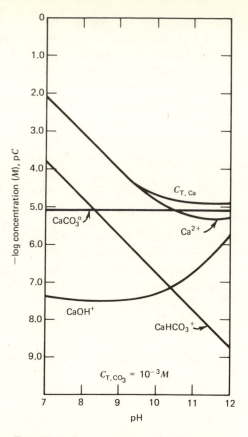

Fig. 6-15. Concentration of calcium species in equilibrium with $CaCO_{3(s)}$ at 25°C.

Example 6-19

The pH of a 25°C solution in equilibrium with $CaCO_{3(s)}$ is 12. The C_{T,CO_3} is measured and found to be 10^{-3} M. Assume a closed system and that activities are equal to concentrations. What is the concentration of dissolved calcium?

The unknowns are $C_{T,Ca}$, $[Ca^{2+}]$, $[CaOH^+]$, $[CaHCO_3^+]$, $[CaCO_3^0]$, $[H_2CO_3^*]$, $[HCO_3^-]$, and $[CO_3^{2-}]$. There are eight equations.

(1) $C_{T,CO_3} = 10^{-3} = [H_2CO_3^*] + [HCO_3^-] + [CaHCO_3^+] + [CaCO_3^0] + [CO_3^{2-}]$

(2) $C_{T,Ca} = [Ca^{2+}] + [CaHCO_3^+] + [CaCO_3^0] + [CaOH^+]$

(3) $K_{so} = 10^{-8.3}$

(5) $K_{a,1} = 10^{-6.3}$

(5) $K_{a,2} = 10^{-10.3}$

(6) $[CaCO_3^0] = 10^{-5.1}$

(7) $[CaHCO_3^+] = 10^{3.26} [H^+]$

(8) $[CaOH^+] = 31[Ca^{2+}][OH^-]$

Determine $[CO_3^{2-}]$ from equation (1). At pH 12,

$$[CO_3^{2-}] \gg [HCO_3^-] \gg [H_2CO_3^*]$$

Therefore, neglect $[HCO_3^-]$ and $[H_2CO_3^*]$. From equation (7)

$$[CaHCO_3^+] = 10^{3.26} \times 10^{-12} = 10^{-8.74}$$

$$[CaCO_3^0] = 10^{-5.1}$$

$$10^{-3} = [CO_3^{2-}] + 10^{-8.74} + 10^{-5.1}$$

Therefore,

$$[CO_3^{2-}] = 10^{-3}$$

Obtain $[Ca^{2+}]$ from equation (3),

$$[Ca^{2+}] = \frac{K_{so}}{[CO_3^{2-}]} = \frac{10^{-8.3}}{10^{-3}}$$

$$[Ca^{2+}] = 10^{-5.3}$$

From equation (8),

$$[CaOH^+] = 31[Ca^{2+}][OH^-]$$
$$= 31 \times 10^{-5.3} \times 10^{-2} = 10^{-5.8}$$

From the calcium mass balance, equation (2),

$$C_{T,Ca} = [Ca^{2+}] + [CaHCO_3^+] + [CaCO_3^0] + [CaOH^+]$$
$$= 10^{-5.3} + 10^{-8.74} + 10^{-5.1} + 10^{-5.8}$$
$$= 10^{-4.84} = 1.4 \times 10^{-5}$$

If the complexes had been neglected,

$$C_{T,Ca} = [Ca^{2+}] = \frac{K_{so}}{[CO_3^{2-}]}$$

$$C_{T,CO_3} \cong [CO_3^{2-}] = 10^{-3}$$

$$[Ca^{2+}] = \frac{10^{-8.3}}{10^{-3}} = 10^{-5.3} = 5 \times 10^{-6} \, M$$

From this calculation it can be deduced that if we had neglected the calcium complexes in our computation, we would have underestimated the dissolved calcium by

$$1.3 \times 10^{-5} - 5 \times 10^{-6} = 8 \times 10^{-6} \, M$$

or

$$8 \times 10^{-6} \times 10^5 \text{ mg/mole} = 0.8 \text{ mg/liter as } CaCO_3$$

This example provides a further illustration of the way in which complexation increases the solubility of sparingly soluble solids. The curves in Fig. 6-15 show that the consideration of calcium complexes becomes significant in determining calcium solubility at pH values above 9.

The solubility of calcium is affected by the SO_4^{2-} complex, and magnesium solubility is affected by OH^-, CO_3^{2-}, HCO_3^-, and SO_4^{2-} complex formation:

$$CaSO_4^0 \rightleftharpoons Ca^{2+} + SO_4^{2-}; \qquad \log K_d = -2.31$$

$$MgCO_3^0 \rightleftharpoons Mg^{2+} + CO_3^{2-}; \qquad \log K_d = -3.4$$

$$MgOH^+ \rightleftharpoons Mg^{2+} + OH^-; \qquad \log K_d = -2.6$$

$$MgHCO_3^- \rightleftharpoons Mg^{2+} + HCO_3^-; \qquad \log K_d = -1.2$$

$$MgSO_4^0 \rightleftharpoons Mg^{2+} + SO_4^{2-}; \qquad \log K_d = -2.36$$

The consideration of complex formation by both Ca^{2+} and Mg^{2+} is of significance in determining the lowest achievable residual hardness by the lime-soda water-softening process. Complexes of Ca^{2+} and Mg^{2+} are weak enough to be titrated by EDTA, that is, EDTA can break these complexes, in the commonly used, complexometric, titration analytical procedure and so will register as residual hardness. The presence of complexes may require that a higher value of L.I. be used to produce a noncorrosive water, assuming that the concentrations of complexes are not taken into account when the L.I. is calculated. For example, at pH 9, the presence of 200 mg SO_4^{2-}/liter will increase the solubility of $CaCO_{3(s)}$ from $1 \times 10^{-4} M$ to $1.4 \times 10^{-4} M$ when $C_{T,CO_3} = 1 \times 10^{-3} M$. When calculating the L.I., the concentrations of calcium complexes should be subtracted from the total calcium concentration to yield the Ca^{2+} concentration. This value can then be used directly in the L.I. equation, Eq. 6-39.

6.8. PHOSPHATE CHEMISTRY

6.8.1. Reactions of Phosphates

The chemistry of phosphates as relevant to aquatic systems is discussed as an example of the importance of heterogeneous equilibria in waters. Phosphorus, present as various forms of phosphate, is of central concern to a wide variety of biological and chemical processes in natural waters, wastewater, and water treatment. Phosphate is a nutrient required for the growth of all living protoplasm that contains approximately 2 percent phosphorus on a dry weight basis. As such, phosphorus can be the element in short supply that limits the growths of photosynthetic aquatic plants. Phosphate is used as a nutrient by organisms in biological waste treatment processes. Phosphates are used in industrial water-softening processes, where their ability to form sparingly soluble calcium salts is capitalized upon. Also, condensed phosphates are employed as agents for complexing Ca^{2+} and Fe^{2+} in boiler waters, as "threshold treatment agents" (e.g., the use of *sodium tripolyphosphate* which adsorbs on the surface of calcite crystallites and thus prevents $CaCO_{3(s)}$ precipitation) and as builders in synthetic detergent formulations (where tripolyphosphate stabilizes dirt particles, and complexes Ca^{2+} and Mg^{2+} to prevent them from combining with the detergent molecule).

Several of the more common classes of phosphorus-containing compounds are presented in Table 6-3. In the orthophosphate anion, the P atom is centrally bonded to the oxygen atoms, which are located at the corners of a tetrahedron. The condensed phosphates—the polyphosphates and metaphosphates—are formed by the condensation of two or more orthophosphate groups and have the characteristic P—O—P linkage. While polyphosphates are linear molecules, the metaphosphates are cyclic.

Typical concentrations of phosphates found in various waters are given in Table 6-4. In fresh raw domestic wastewater the phosphate is distributed approximately as follows: orthophosphate, 5 mg/liter as phosphorus, tripolyphosphate, 3 mg/liter as phosphorus, pyrophosphate, 1 mg/liter as phosphorus and organic phosphates < 1 mg/liter as phosphorus. Secondary biological treatment, and indeed prolonged contact with the microorganisms in raw sewage, ensures the hydrolysis, ("reversion") of condensed phosphates to orthophosphate. Tripolyphosphate, for example, hydrolyzes as follows:

$$HP_3O_{10}^{4-} + 2H_2O \rightleftharpoons 3HPO_4^{2-} + 2H^+$$

Hydrolysis is catalyzed by H^+. Examination of Fig. 2-8 shows that at 10°C the time for 5 percent hydrolysis of a pyrophosphate solution at pH 4 is about 1 year; at pH 7 is many years and at pH 10 is over a century! Of course, catalysis of the hydrolysis reaction by enzymes is important in nonsterile natural waters. Wastewater effluent and natural waters contain significant amounts of organically bound phosphate. Indeed, some estimates of the phosphate content of natural waters assign between 30 to 60 percent of the total phosphate to the organically bound category.

Phosphate solubility equilibrium constants and complexation equilibrium constants are given in Table 6-5.

The behavior of calcium phosphate precipitates in dilute aqueous solutions serves to illustrate some of the difficulties that one may face when attempting to apply equilibrium calculations to predict solution concentrations of ions in contact with a solid. The calcium phosphate system by itself is very complex. From Table 6-5 it can be learned that

TABLE 6-3   Classes of Phosphorus-Containing Compounds of Importance in Aquatic Systems

Group	Structural Representation (Typical)	Species of Importance	Acid Ionization Constants (25°C)
Orthophosphate		H_3PO_4, $H_2PO_4^-$, HPO_4^{2-}, PO_4^{3-}, HPO_4^{2-} complexes	$pK_{a,1} = 2.1$, $pK_{a,2} = 7.2$, $pK_{a,3} = 12.3$
Polyphosphates	pyrophosphate	$H_4P_2O_7$, $H_3P_2O_7^-$, $H_2P_2O_7^{2-}$, $HP_2O_7^{3-}$, $P_2O_7^{4-}$, $HP_2O_7^{3-}$ complexes	$pK_{a,1} = 1.52$, $pK_{a,2} = 2.4$, $pK_{a,3} = 6.6$, $pK_{a,4} = 9.3$
	tripolyphosphate	$H_3P_3O_{10}^{2-}$, $H_2P_3O_{10}^{3-}$, $HP_3O_{10}^{4-}$, $P_3O_{10}^{5-}$, $HP_3O_{10}^{4-}$ complexes	$pK_{a,3} = 2.3$, $pK_{a,4} = 6.5$, $pK_{a,5} = 9.2$
Metaphosphates	trimetaphosphate	$HP_3O_9^{2-}$, $P_3O_9^{3-}$	$pK_{a,3} = 2.1$
Organic phosphates	glucose 6-phosphate	Very many types, including phospholipids, sugar phosphates, nucleotides, phosphoamides, etc.	

there are a variety of solids that may form (and many more than listed may form); there are the acid-base equilibria of phosphoric acid to consider as well as complex formation of species such as $CaHPO_4^\circ$ and

TABLE 6-4   Typical Concentrations of Total Phosphorus in Water

Domestic waste water	3–15 mg/liter as P	$1–5 \times 10^{-4} M$
Agricultural drainage	0.05–1 mg/liter as P	$2–30 \times 10^{-6} M$
Lake surface water	0.01–0.04 mg/liter as P	$3–13 \times 10^{-7} M$

TABLE 6-5   Representative Heterogeneous and Complexation Equilibria of Phosphates (25°C)

Heterogeneous Equilibria		pK_{so}
Calcium hydrogen phosphate	$CaHPO_{4(s)} \rightleftharpoons Ca^{2+} + HPO_4^{2-}$	+ 6.66
Calcium hydrogen phosphate	$CaHPO_{4(s)} \rightleftharpoons Ca^{2+} + HPO_4^{2-}$	+ 6.66
Calcium dihydrogen phosphate	$Ca(H_2PO_4)_{2(s)} \rightleftharpoons Ca^{2+} + 2H_2PO_4^{-}$	+ 1.14
Hydroxyapatite	$Ca_5(PO_4)_3OH_{(s)} \rightleftharpoons 5Ca^{2+} + 3PO_4^{3-} + OH^{-}$	+55.9
β-Tricalcium phosphate	$\beta\text{-}Ca_3(PO_4)_{2(s)} \rightleftharpoons 3Ca^{2+} + 2PO_4^{3-}$	+24.0
Ferric phosphate	$FePO_{4(s)} \rightleftharpoons Fe^{3+} + PO_4^{3-}$	+21.9
Aluminum phosphate	$AlPO_{4(s)} \rightleftharpoons Al^{3+} + PO_4^{3-}$	+21.0
Complexation Equilibria		pK
With orthophosphate	$NaHPO_4^{-} \rightleftharpoons Na^{+} + HPO_4^{2-}$	+ 0.6
	$MgHPO_4^{\circ} \rightleftharpoons Mg^{2+} + HPO_4^{2-}$	+ 2.5
	$CaHPO_4^{\circ} \rightleftharpoons Ca^{2+} + HPO_4^{2-}$	+ 2.2
	$MnHPO_4^{\circ} \rightleftharpoons Mn^{2+} + HPO_4^{2-}$	+ 2.6
	$FeHPO_4^{+} \rightleftharpoons Fe^{3+} + HPO_4^{2-}$	+ 9.75
	$CaH_2PO_4^{+} \rightleftharpoons Ca^{2+} + HPO_4^{2-} + H^{+}$	− 5.6
With pyrophosphate	$CaP_2O_7^{2-} \rightleftharpoons Ca^{2+} + P_2O_7^{4-}$	+ 5.6
	$CaHP_2O_7^{-} \rightleftharpoons Ca^{2+} + HP_2O_7^{3-}$	+ 2.0
	$Fe(HP_2O_7)_2^{3-} \rightleftharpoons Fe^{3+} + 2HP_2O_7^{3-}$	+22
With tripolyphosphate	$CaP_3O_{10}^{3-} \rightleftharpoons Ca^{2+} + P_3O_{10}^{5-}$	+ 8.1

$CaH_2PO_4^{+}$. However, let us consider the magnitude of the phosphate concentration that we would predict to be in equilibrium with hydroxyapatite—the thermodynamically stable solid under typical natural water conditions of solution concentration, pH, and temperature.

From Table 6-5,

$$Ca_5(PO_4)_3(OH)_{(s)} \rightleftharpoons 5Ca^{2+} + 3PO_4^{3-} + OH^{-}; \quad K_{so} = 10^{-55.9}$$

$$[Ca^{2+}]^5 [PO_4^{3-}]^3 [OH^{-}] = 10^{-55.9}$$

For typical conditions of pH 8 and $[Ca^{2+}] = 1.5 \times 10^{-3} M$ (150 mg/liter as $CaCO_3$)

$$(1.5 \times 10^{-3})^5 [PO_4^{3-}]^3 (10^{-6}) = 10^{-55.9}$$

$$[PO_4^{3-}] = 1.2 \times 10^{-12} M$$

or

$$3.7 \times 10^{-5} \ \mu g/\text{liter as phosphorus}$$

Therefore, we might deduce that if natural waters were in equilibrium with hydroxyapatite, or stated another way "if the orthophosphate concentration of natural waters was controlled by hydroxyapatite," we would not be much concerned with problems of algal growth associated with phosphorus. We would also encounter remarkable success if we were to precipitate phosphate from wastewater by adding calcium salts such as lime. Because we know that phosphate levels in wastewater treated with calcium salts and in natural waters far exceed the calculated equilibrium value, we might deduce that hydroxyapatite does not control the orthophosphate level. What then is happening? While the entire answer to this question is beyond the scope of this text, some facets of the system can be explored to illustrate the practical consequences of the previous discussion in this chapter on precipitate formation by nucleation, phase transformation, and crystal growth. A general picture of the nature of calcium phosphate precipitation is provided by Fig. 6-16.

In a supersaturated solution there is at first a period of time before any decrease in dissolved phosphate takes place. This so-called induction period represents the time for nuclei to form. Ferguson et al.[10] showed

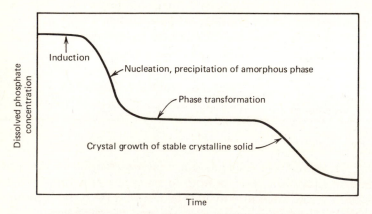

Fig. 6-16. Idealized scheme for calcium phosphate precipitation kinetics.

[10] J. F. Ferguson, D. Jenkins, and A. B. Menar, "Chemical Processes for Phosphate Removal," *Water Research*, 5: 369–381 (1971).

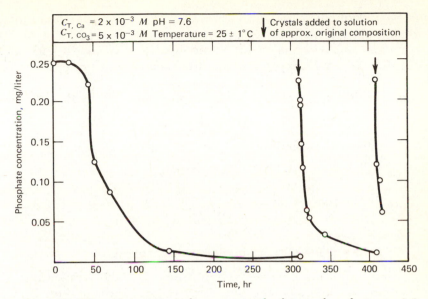

Fig. 6-17. Effect of crystal seeding on rate of calcium phosphate precipitation. J. F. Ferguson et al., *Water Research,* 5: 369–381 (1971). Reprinted by permission of Pergamon Press.

that if a crystal seed of precipitated calcium phosphate were added to a solution supersaturated with respect to hydroxyapatite, there was a virtual elimination of the induction period as the precipitate acted as a nucleus for further precipitation as shown in Fig. 6-17.

Following nucleation the formation of an amorphous precipitate occurs. The term *amorphous* refers to a solid that does not have a well-organized crystal structure and is almost always more soluble than the corresponding crystalline solid. The amorphous nature of calcium phosphate precipitates may indeed be an important reason why phosphate solution concentrations are present in natural waters at higher levels than predicted by equilibrium with the crystalline solid. For the example of calcium phosphate precipitation given, we can see that precipitation not only results in an amorphous solid but also a precipitate whose solubility is greater than the thermodynamically stable phase. Such a solid is not stable and will transform, albeit slowly, to the stable solid phase. However, the first-formed solid may be stable enough to control solution phosphate concentrations in a pseudo-equilibrium or steady state situation. An example of such a system is provided by the work of Menar and Jenkins,[11] who found that, in the presence of magnesium ion, the residual phosphate concentration in wastewater after treatment with calcium salts

[11] A. B. Menar and D. Jenkins, "Calcium Phosphate Precipitation in Wastewater Treatment," SERL Report 72-6, Sanitary Engineering Research Laboratory, University of California, Berkeley, 1972.

could be described by an equilibrium between it and the solid phase β-tricalcium phosphate, β-$Ca_3(PO_4)_2$, with a solubility product of 10^{-24}. Let us compare two systems at pH 8 and $[Ca^{2+}] = 1.5 \times 10^{-3}$ M, one in equilibrium with hydroxyapatite and the second with β-tricalcium phosphate.

From previously, for phosphate in equilibrium with hydroxyapatite,

$$PO_4^{3-} = 3.7 \times 10^{-5} \ \mu g/liter \text{ as phosphorus}$$

For β-tricalcium phosphate,

$$[Ca^{2+}]^3[PO_4^{3-}]^2 = 10^{-24}$$

$$[PO_4^{3-}] = \left(\frac{10^{-24}}{(1.5 \times 10^{-3})^3}\right)^{1/2} = 1.6 \times 10^{-8} M$$

or

$$PO_4^{3-} = 0.5 \ \mu g/liter \text{ as phosphorus}$$

This shows that the system in equilibrium with the less soluble solid can contain four orders of magnitude less orthophosphate than the water in equilibrium with β-tricalcium phosphate.

A further factor important in the control of phosphate levels by calcium phosphate precipitation is the rate of crystal growth. Ferguson et al.[12] showed that a rate law of the form

$$\frac{dC}{dt} = -kS(C - C^*)^n$$

was applicable for postinduction periods, or where an induction period was absent,

where

n = order of reaction
S = available crystal surface area
C = concentration of phosphate, the limiting reactant
C^* = equilibrium concentration
t = time from the end of the induction period
k = a constant

Since for hydroxyapatite, C^* is very low, we can assume it to be zero. Also, we can assume that S is a constant if a constant amount (excess) of seed particles of hydroxyapatite are present. Thus

$$\frac{dC}{dt} = K'C^n$$

where K' is a constant.

[12] J. F. Ferguson, D. Jenkins, and J. Eastman, "Calcium Phosphate Precipitation at Slightly Alkaline pH values," *J. Water Pollut. Control Fed.*, 45, 620–631 (1973).

By carrying out calcium phosphate precipitation in solutions containing various levels of carbonate, C_{T,CO_3}, which was shown to affect precipitation, an empirical rate expression was determined as follows:

$$\frac{dC}{dt} = -\left(\frac{4.1}{C_{T,CO_3}}\right) C^{2.7}$$

where C and C_{T,CO_3} are in units of millimoles/liter and t is in hours. The form of this rate equation has important practical implications for phosphate removal by calcium phosphate precipitation at slightly alkaline pH values. In addition to being an inverse function of the alkalinity (C_{T,CO_3}), the rate and extent of calcium phosphate precipitation are strongly dependent on reactor configuration. Reactors that feature a recycle stream are far more efficient than equivalent reactors that do not because of the presence of much larger surface area of preformed precipitate contributed by the recycled stream. The concentration exponent of 2.7 tells us that the rate is extremely sensitive to concentration. Thus reactors that have regions of high concentration will have precipitation rates greater than equivalent reactors that are mixed throughout. Plug-flow reactors are therefore more efficient than CSTR (completely stirred tank reactors). These features are illustrated by Fig. 6-18, which shows the predicted phosphate removal by flow types of reactors at various average hydraulic detention times.

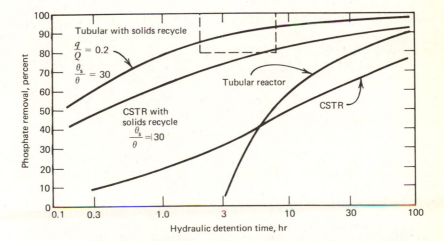

Fig. 6-18. Predicted phosphate removal versus hydraulic detention time for four reactor types. The dashed line indicates the limits of practical application, $2 \text{ hr} \leqslant \theta \leqslant 8 \text{ hr}$, 80 to 100 percent removal. Assumed conditions include $C_{T,Ca} = 2.0 \text{ m}M$, $C_{T,PO_4} = 0.25 \text{ m}M$, $C_{T,CO_3} = 1.3 \text{ m}M$, pH = 8.0, temperature = 25°C, recycle flow/throughput flow, $q/Q = 0.2$; solid residence time/liquid residence time, $\theta_x/\theta = 3.0$, $K' = 4.1$, and $n = 2.7$. From J. F. Ferguson et al., *J. Water Pollution Control Fed.*, 45: 620–631 (1973). @ 1973 Water Pollution Control Federation, reprinted with permission.

6.8.2. Magnesium Ammonium Phosphate Precipitation

The conditional solubility product of magnesium ammonium phosphate, struvite ($MgNH_4PO_{4(s)}$) illustrates a situation where more than one of the dissolving species is affected by solution pH. These species are the ammonium ion (NH_4^+) and the phosphate ion (PO_4^{3-}). Because an increase in pH will decrease the ammonium ion concentration and increase the phosphate ion concentration, it follows that there should be a pH value where the solubility of $MgNH_4PO_{4(s)}$ is a minimum, that is, the pH where the product $[Mg^{2+}][NH_4^+][PO_4^{3-}]$ is a minimum. Where is this point?

We can attack this problem by writing equations for the pH-dependent reactions involving each of the species in the precipitate. Thus

$$NH_4^+ \rightleftharpoons NH_{3(aq)} + H^+; \qquad \log K_a = -9.3$$

$$PO_4^{3-} + H^+ \rightleftharpoons HPO_4^{2-}; \qquad \log\left(\frac{1}{K_{a,3}}\right) = 12.3$$

$$HPO_4^{2-} + H^+ \rightleftharpoons H_2PO_4^-; \qquad \log\left(\frac{1}{K_{a,2}}\right) = 7.2$$

$$H_2PO_4^- + H^+ \rightleftharpoons H_3PO_4; \qquad \log\left(\frac{1}{K_{a,1}}\right) = 2.1$$

$$Mg^{2+} + OH^- \rightleftharpoons MgOH^+; \qquad \log\left(\frac{1}{K_{d,Mg}}\right) = 2.1$$

Defining the ionization fractions for Mg^{2+}, NH_4^+ and PO_4^{3-} as $\alpha_{Mg^{2+}} = [Mg^{2+}]/C_{T,Mg}$, $\alpha_{NH_4^+} = [NH_4^+]/C_{T,NH_3}$ and $\alpha_{PO_4^{3-}} = [PO_4^{3-}]/C_{T,PO_4}$, we can write the solubility product of $MgNH_4PO_{4(s)}$ as

$$K_{so} = \{Mg^{2+}\}\{NH_4^+\}\{PO_4^{3-}\}$$

$$= \gamma_{Mg^{2+}}[Mg^{2+}]\gamma_{NH_4^+}[NH_4^+]\gamma_{PO_4^{3-}}[PO_4^{3-}]$$

$$= \gamma_{Mg^{2+}}C_{T,Mg}\alpha_{Mg^{2+}}\gamma_{NH_4^+}C_{T,NH_3}\alpha_{NH_4^+}\gamma_{PO_4^{3-}}C_{T,PO_4}\alpha_{PO_4^{3-}}$$

where $C_{T,Mg}$, C_{T,NH_3}, and C_{T,PO_4} are the total analytical concentrations of magnesium, ammonia nitrogen, and orthophosphate, respectively. We can define the conditional solubility product as

$$P_s = C_{T,Mg}C_{T,NH_3}C_{T,PO_4} = \frac{K_{so}}{\alpha_{Mg^{2+}}\alpha_{NH_4^+}\alpha_{PO_4^{3-}}\gamma_{Mg^{2+}}\gamma_{NH_4^+}\gamma_{PO_4^{3-}}}$$

Then P_s is a function of pH and its minimum value will occur when the product $(\alpha_{Mg^{2+}}) \times (\alpha_{NH_4^+}) \times (\alpha_{PO_4^{3-}})$ is a maximum. The values of these ionization fractions are presented in Appendix 1; P_s is plotted as a function of pH in Fig. 6-19, assuming in one case that $\mu = 0$ and in the other that $\mu = 0.1$. Activity coefficients were obtained from Fig. 3-4. We can see from this figure that the minimum solubility occurs at about pH 10.7.

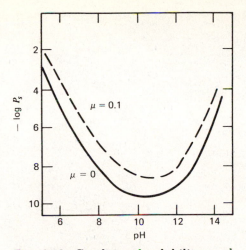

Fig. 6-19. Conditional solubility product, $P_s = C_{T,Mg} C_{T,NH_3} C_{T,PO_4}$, for magnesium ammonium phosphate at 25°C.

Magnesium ammonium phosphate precipitation is a recognized problem in anaerobic sludge digesters, where it precipitates in digester supernatant recycle lines, especially at elbows and on the suction side of pumps. This happens at these locations because they are regions of reduced pressure and CO_2 is released from solution. This causes a rise in pH of the digester supernatant with an accompanying increase in the product $[Mg^{2+}][NH_4^+][PO_4^{3-}]$ to values where K_{so} is exceeded. Figure 6-20 is an illustration of a severe case of magnesium ammonium phosphate (struvite) deposition. This photograph was taken from a digestor recycle line at the City of Los Angeles Hyperion plant and shows that the cross section of a 14-in. pipe has been reduced to less than 4 in. by struvite precipitation. Similar accumulation of struvite blocked the screens through which digested sludge was passed prior to discharge to the ocean. Because the Hyperion plant discharges both digested sludge and sewage effluent to the ocean, the solution to this problem was to dilute the digested sludge with effluent in a volume ratio of about 3:1. Let us examine the basis for this solution to the problem. Figure 6-21 shows the flow sheet of the wastewater treatment plant. Table 6-6 presents the concentrations of ammonia, orthophosphate, and magnesium and the pH of the raw sludge, digested sludge, and diluted digested sludge.

With the concentrations indicated in Table 6-6 we can compute the product $C_{T,Mg} \times C_{T,NH_3} \times C_{T,PO_4}$ and if this significantly exceeds the conditional solubility product, P_s, then struvite should precipitate.

For raw sludge,

$$C_{T,Mg} C_{T,NH_3} C_{T,PO_4} = (5 \times 10^{-3})(5 \times 10^{-3})(4 \times 10^{-2}) = 10^{-6}$$

Fig. 6-20. Magnesium ammonium phosphate deposition in digester pipeline.

From Fig. 6-19, at pH = 5.5 and $\mu = 0$, $P_s = 10^{-4}$. Because $10^{-4} \gg 10^{-6}$, the raw sludge is highly undersaturated with respect to struvite and it will not tend to precipitate.

For digested sludge,

$$C_{T,Mg}C_{T,NH_3}C_{T,PO_4} = 5 \times 10^{-3} \times 10^{-1} \times 7 \times 10^{-2}$$
$$= 3.5 \times 10^{-5} = 10^{-4.5}$$

From Fig. 6-19, at pH = 7.5 and $\mu = 0$, $P_s = 10^{-7.5}$. Because $10^{-4.5} \gg 10^{-7.5}$, the digested sludge is highly oversaturated with respect to struvite and should tend to precipitate.

For diluted digested sludge,

$$C_{T,Mg}C_{T,NH_3}C_{T,PO_4} = 10^{-3} \times 2.5 \times 10^{-2} \times 2 \times 10^{-2}$$
$$= 5 \times 10^{-7} = 10^{-6.3}$$

From Fig. 6-19, at pH = 7.5 and $\mu = 0$, $P_s = 10^{-7.5}$. Because $10^{-6.3} > 10^{-7.5}$, the diluted digested sludge is supersaturated with respect to struvite but not the same degree as the digested sludge. If P_s is corrected for activity,

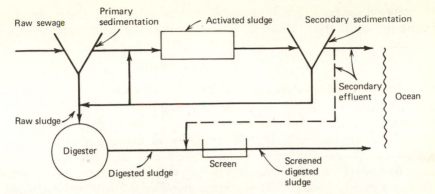

Fig. 6-21. Flowsheet for digested sludge treatment at the Hyperion plant.

assuming $\mu = 0.1$, $P_s = 10^{-6.1}$ so that the solution is actually undersaturated.

Thus the dilution provided appears to be adequate to prevent struvite precipitation. The solution to the problem is soundly based. The solution to this problem at the Hyperion plant represents something of a luxury because there are not many plants that have the ability to dilute digester supernatant with secondary effluent and then discharge the mixture. One such example is the Oakland, California, treatment plant of the East Bay Municipal Utility District. At this plant a $MgNH_4PO_{4(s)}$ scaling problem developed. Dilution to below P_s was not feasible; handling digester supernatant to keep CO_2 in solution (and therefore keep pH depressed) was not practicable. The solution to the scaling problem was to add a material that inhibited crystal formation (a polyacrylamide crystal inhibitor (*Cyanamer P-70*, American Cyanamid Corporation) at doses of 10 to 20 mg/liter). Such solutions to scaling problems are not as "safe and sure" as reducing the value of an ion product to below the solubility product of a scale-forming precipitate; however, they must be used when that

TABLE 6-6   Values of Constituents in Hyperion Sludges

Constituent	Raw Sludge		Digested Sludge		Diluted Digested Sludge	
	mg/liter	moles/liter	mg/liter	moles/liter	mg/liter	moles/liter
$C_{T,Mg}$	250	5×10^{-3}	200	5×10^{-3}	—	$\sim 10^{-3}$
Ammonia Nitrogen (total)	75	5×10^{-3}	~1400	10^{-1}	—	$\sim 2.5 \times 10^{-2}$
Phosphate Phosphorus (total)	1200	4×10^{-2}	2000	7×10^{-2}	—	$\sim 2 \times 10^{-2}$
pH	5.5		7.5		7.5	

alternative does not exist. It is useful to remember that thermodynamics is more reliable than kinetics in the solution of precipitation/scaling problems.

6.8.3. Iron Phosphate Precipitation from Wastewater

The addition of aluminum salts (see Section 6.6.2) or iron salts to wastewater for phosphate removal results in the precipitation of aluminum hydroxyphosphate $(Al_x(OH)_y(PO_4)_z)$ and ferric hydroxyphosphate $(Fe_x(OH)_y(PO_4)_z)$, respectively. These solids are incorporated into the sludge or underflow solids streams that in a typical municipal treatment plant are treated by anaerobic digestion (see Fig. 6-21). When Fe(III) salts, for example, $FeCl_3$ and $Fe_2(SO_4)_3$, are used as phosphate precipitants with the objective of removing phosphate, a question arises concerning the fate of this precipitate in the highly reducing conditions of the anaerobic digestion process. Under such conditions Fe(III) will be reduced to Fe(II)—what then is the fate of the phosphate that has been precipitated with the Fe(III)? Does it remain in the solid phase or is it released to solution to be recycled back through the treatment plant in the digester supernatant stream?

The early literature on the effect of iron and aluminum salt addition on phosphate concentrations in digester supernatant seemed to indicate that phosphate release was not occurring during digestion; rather, after metal salt addition phosphate levels in digester supernatants decreased. Singer[25] proposed that the reason for this was that the sparingly soluble ferrous phosphate, vivianite, was formed under the conditions of anaerobic sludge digestion.

$$2PO_4^{3-}$$
$$\downarrow$$
$$3Fe(III) \rightarrow 3Fe(II) \longrightarrow Fe_3(PO_4)_{2(s)}$$

The method used by Singer to determine the K_{so} value of $Fe_3(PO_4)_{2(s)}$ is a useful illustration of typical methods employed for equilibrium constant measurement. The experimental system is illustrated in Fig. 6-22.

A constant ionic strength medium (0.1 M $NaClO_4$) was prepared to eliminate the need for activity corrections. Ferrous perchlorate $Fe(ClO_4)_2$ and NaH_2PO_4 were added to water through which nitrogen gas was bubbled to remove all traces of dissolved oxygen and so to produce reducing conditions. The pH of individual bottles was then adjusted with Na_2CO_3 to several values in the pH range 4 to 8. After sealing, the bottles were stored for 1 month during which time a blue-white precipitate

[25] P. C. Singer, "Anaerobic Control of Phosphate by Ferrous Iron," *J. Water Pollut. Control Fed.*, **44**: 663 (1972).

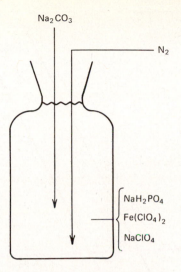

Fig. 6-22. Sample precipitation technique used to determine K_{so} for $Fe_3(PO_4)_{2(s)}$. From P. C. Singer, *J. Water Pollution Control Fed.*, 44: 663 (1972). © 1972 Water Pollution Control Federation, reprinted with permission.

formed. A final pH measurement was made, the samples were filtered, and Fe^{2+} and C_{T,PO_4} were measured. Now

$$Fe_3(PO_4)_{2(s)} \rightleftharpoons 3Fe^{2+} + 2PO_4^{3-}$$

$$^cK_{so} = [Fe^{2+}]^3[PO_4^{3-}]^2, \qquad \mu = 0.1$$

We can write the reaction as a dissolution of $Fe_3(PO_4)_{2(s)}$ by acid. Thus

$$Fe_3(PO_4)_{3(s)} + 4H^+ \rightleftharpoons 3Fe^{2+} + 2H_2PO_4^-$$

$$^cK' = \frac{^cK_{so}}{(^cK_{a,2}{}^cK_{a,3})^2} = \frac{[H_2PO_4^-]^2[Fe^{2+}]^3}{[H^+]^4} \qquad (6\text{-}47)$$

where $^cK_{a,2}$ and $^cK_{a,3}$ are the second and third dissociation constants of phosphoric acid. From Eq. 6-47,

$$p^cK_{so} - 2p^cK_{a,2} - 2p^cK_{a,3} = -3 \log [Fe^{2+}] - 2 \log [H_2PO_4^-] + 4 \log [H^+]$$

$$p^cK_{so} - 2p^cK_{a,2} - 2p^cK_{a,3} + 4pH = -\log [[Fe^{2+}]^3(\alpha_{2,PO_4}C_{T,PO_4})^2] \qquad (6\text{-}48)$$

All quantities in Eq. 6-48 are either known ($p^cK_{a,2}$, $p^cK_{a,3}$, α_{2,PO_4}) or

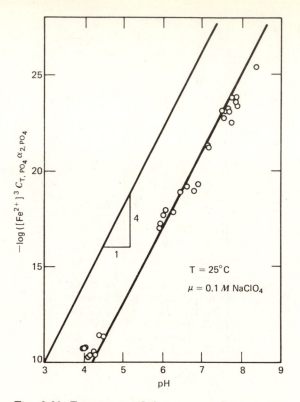

Fig. 6-23. Experimental data on solubility product of ferrous phosphate. From P. C. Singer, *J. Water Pollution Control Fed.*, **44**: 663 (1972). © 1972 Water Pollution Control Federation, reprinted with permission.

analytically measured, (Fe^{2+}, C_{T,PO_4}, $[H^+]$) except p^cK_{so}. A plot of $-\log([Fe^{2+}]^3(C_{T,PO_4}\alpha_{2,PO_4})^2)$ versus pH should give a line with a slope of $+4$ and an intercept of $p^cK_{so} - 2p^cK_{a,2} - 2p^cK_{a,3}$ when pH = 0. Figure 6-23 shows Singer's data plotted in this fashion.

The intercept at pH = 0 is

$$p^cK_{so} - 2p^cK_{a,2} - 2p^cK_{a,3} = -7.1$$

since $p^cK_{a,2} = 6.78$ and $p^cK_{a,3} = 11.7$

$$p^cK_{so} = 29.9; \qquad {}^cK_{so} = 1.3 \times 10^{-30} \qquad \text{at } \mu = 0.1$$

The determination of K_{so} using the activity coefficients in Fig. 3-4 yields a value of 8×10^{-34}. Examination of the solid by X-ray crystallography provided evidence that it was indeed vivianite. Similar study of solids from digesters that had received sludges from plants employing iron addition for phosphate removal also revealed the presence of vivianite.

6.9. PROBLEMS

1. Strontium, Sr^{2+}, is a heavy metal that frequently causes concern when it is present in water. Some preliminary calculations are necessary to determine whether it is feasible to precipitate Sr^{2+} as the sulfate. Given that 10^{-4} mole/liter of Sr^{2+} and 10^{-4} mole/liter of SO_4^{2-} are originally present and that $pK_{so} = 7.8$ for $SrSO_{4(s)}$,

   (a) How much Sr^{2+} will precipitate (moles/liter) if 1×10^{-3} mole Na_2SO_4 is added per liter of solution?

   (b) How much SO_4^{2-} and Sr^{2+} remain after precipitation?

   (c) Assuming that the residuals calculated in part (b) are satisfactory, what factors should be studied experimentally to determine the feasibility of the process?

2. A 25°C water sample has the following species present in addition to other ions:

$$[H^+] = 10^{-8.3} \text{ mole/liter}$$

$$[HCO_3^-] = 10^{-2} \text{ mole/liter}$$

$$[Mg^{2+}] = 10^{-4} \text{ mole/liter}$$

   (a) To what pH must the water be adjusted to precipitate $Mg(OH)_2$?

   (b) How many moles of OH^- must be added to reach the pH for precipitation of $Mg(OH)_2$? (Assume negligible Ca^{2+} in the solution and that any other ions or molecules that are present will not react with H^+, OH^- or the carbonates).

3. The pH of a 25°C groundwater containing 10^{-4} mole/liter of Mg^{2+} is raised to pH 12. No further base is added. What will the equilibrium pH and Mg^{2+} concentration be? (Assume that the concentration is equal to the activity, and that no other ions are present which will precipitate.)

4. Sketch the $pC-pCO_3^{2-}$ diagram for $CaCO_3$ and $MgCO_3$ at 25°C. Using this diagram, determine the $[CO_3^{2-}]$ in equilibrium with 5×10^{-5} M Ca^{2+}. What concentration of CO_3^{2-} is necessary to begin precipitating Mg^{2+} as the carbonate in a solution containing 3×10^{-4} M Mg^{2+}?

5. Sketch the $pC-pOH$ diagram for $Mg(OH)_2$ and $Ca(OH)_2$ at 25°C. To what level must the pH be raised to precipitate Mg^{2+} from a 3×10^{-3} M Mg^{2+} solution? To what level must the pH be raised to reduce $[Mg^{2+}]$ to 5×10^{-5} M at equilibrium? At what pH will Ca^{2+} just begin to precipitate if $[Ca^{2+}] = 1 \times 10^{-3}$ M?

6. Find the solubility of $Ag_2CrO_{4(s)}$ in distilled water at 25°C. Show how the formation of $HCrO_4^-$ affects the calculations. K_a for $HCrO_4^-$ is $10^{-6.5}$.

7. Sketch a $pC-pH$ diagram showing the concentration of Fe^{2+} versus pH in a 25°C solution that is in equilibrium with solid FeS. Make the simplifying assumption that the total sulfide concentration is 10^{-3} M at each pH, (i.e., $C_{T,S(II)} = [H_2S] + [HS^-] + [S^{2-}] = 10^{-3}$ at each pH), and that there are no Fe(II) complexes present at significant concentrations.

8. Calculate the conditional solubility product for $CaCO_{3(s)}$, $P_s = (C_{T,Ca})(C_{T,CO_3})$, where $C_{T,Ca}$ and C_{T,CO_3} represent the total concentrations of these species in a solution with $\mu = 10^{-2}$, pH = 8.7, the temperature = 25°C. Include ionic strength effects. Assume that the only soluble calcium species is Ca^{2+}.

9. When $C_{T,CO_3} = 10^{-4} M$ and $Fe^{2+} = 2 \times 10^{-4}$, at what pH will $FeCO_{3(s)}$ begin to precipitate as pH is increased? At what pH will $Fe(OH)_{2(s)}$ begin to precipitate? The temperature = 25°C.

10. Water is in contact with solid $CaCO_3$. Equilibrium conditions exist. What is the effect of small additions of the following on the calcium concentration and the total alkalinity of the solution? Neglect ionic strength effects.
    (a) KOH.   (b) $Ca(NO_3)_2$.
    (c) KCl.   (d) Na_2CO_3.
    (e) CO_2.

    Note: Consider that each addition is independent of the other additions. Indicate only whether the concentrations will increase or decrease.

11. Set up the equations necessary to calculate the solubility of calcium hydroxyapatite, $Ca_5(PO_4)_3OH_{(s)}$, in aqueous solution at 25°C. The complexes that are of importance are $CaHPO_4^{\circ}$ and $CaOH^+$.

12. Calculate the solubility of $Fe_3(PO_4)_{2(s)}$ as a function of pH and plot the diagram of $pC_{T,Fe(II)}$ versus pH at 25°C. Assume that $Fe_3(PO_4)_{2(s)}$ is the only solid phase present and that Fe^{2+}, $FeOH^+$, $Fe(OH)_3^-$, H_3PO_4, $H_2PO_4^-$, HPO_4^{2-}, and PO_4^{3-} are important. For this problem, $2C_{T,Fe(II)} = 3C_{T,PO_4}$. Compare the diagram obtained with Fig. 6-11 to determine for the applicable conditions whether $Fe_3(PO_4)_{2(s)}$ can be precipitated alone or whether $Fe(OH)_{2(s)}$ and/or $FeCO_{3(s)}$ are also likely to precipitate.

13. Groundwater 25°C equilibrates with calcareous ($CaCO_3$) rock in an atmosphere that has a CO_2 partial pressure of $10^{-1.5}$ atm. What is its pH, $[Ca^{2+}]$, total alkalinity, $[HCO_3^-]$, and total hardness (assume no Mg^{2+} is present)? A sample of this water is equilibrated with a $P_{CO_2} = 10^{-3.5}$ atm (atmospheric conditions) and $CaCO_{3(so)}$. How much $CaCO_3$ precipitate (dissolves)? What is the new pH, $[Ca^{2+}]$, total alkalinity, and $[HCO_3^-]$. Discuss the implications of this change with respect to water treatment and the pumping of water. (Neglect ionic strength effects. Note that the calculated concentrations change very much with small changes in α_2. Therefore an accurate value of α_2 should be used.)

14. Calculate the Langelier Index of a water with the following characteristics: total alkalinity = 8×10^{-4} eq/liter, $[Ca^{2+}] = 3 \times 10^{-4} M$, pH = 9.6, and total dissolved solids = 250 mg/liter. Include ionic strength effects; the temperature = 25°C.

15. Calculate the precipitation potential for the water described in Example 6-18, taking ionic strength effects into account.

16. List (but do not calculate) the equations required to determine the amount of $Ca(OH)_2$ required to produce a saturated water from one that has a negative Langelier Index.

17. List the equations required to determine the amount of $Ca(OH)_2$ required to produce a water with a theoretical precipitation potential of 5 mg/liter as $CaCO_3$ from one which has a negative Langelier Index.

6.10. ADDITIONAL READING

J. N. Butler, *Ionic Equilibrium*. Addison-Wesley, Reading, Mass., 1964.

A. E. Nielson, *Kinetics of Precipitation*. Macmillan, New York, 1964.

W. Stumm and J. J. Morgan, *Aquatic Chemistry*. Wiley-Interscience, New York, 1970.

A. G. Walton, *The Formation and Properties of Precipitates*. Wiley-Interscience, New York, 1967.

CHAPTER 7

OXIDATION-REDUCTION REACTIONS

7.1. INTRODUCTION

Oxidation-reduction or "redox" reactions play a central role in many of the reactions occurring in natural waters and in water and wastewater treatment processes. The behavior of compounds containing carbon, nitrogen, sulfur, iron, and manganese in natural waters and treatment processes is largely influenced by redox reactions. Redox reactions are encountered in many of the analyses conducted on water and wastewater; for example, the biochemical oxygen demand (BOD) and the chemical oxygen demand (COD) tests and the analysis of dissolved oxygen are based on them. Many of the chemicals employed in water and wastewater treatment processes are used to alter the chemical nature of water and wastewater constituents by oxidation-reduction processes, such as chlorine, chlorine dioxide, permanganate, hydrogen peroxide, oxygen, ozone, and sulfur dioxide. Oxidation-reduction reactions catalyzed by enzymes (microorganisms) form the basis for waste treatment processes like activated sludge, biological filtration, and anaerobic digestion. Such microbially mediated redox reactions are also significant in transformations of nutrients, metals, and other chemical species in natural waters.

In this chapter we will examine oxidation-reduction stoichiometry, equilibria, and the graphical representation of simple and complex equilibria, and the rate of oxidation-reduction reactions. The applications of redox reactions to natural waters will be presented in the context of a discussion of iron chemistry; the subject of corrosion will provide a vehicle for a discussion of the application of electrochemical processes; a presentation of chlorine chemistry will include a discussion of the kinetics of redox reactions and the reactions of chlorine with organic matter; finally, the application of redox reactions to various measurement methods will be discussed using electrochemical instruments as examples.

7.2. REDOX STOICHIOMETRY

The first step in solving any equilibrium problem is to determine the identity of the reactants and products and then to present them in a balanced chemical equation. In previous chapters we have seen how to do this for acid-base, complex, and heterogeneous equilibria using the principles of stoichiometry. In this section we will learn how to write a balanced chemical equation for an oxidation-reduction reaction. At the beginning we will not inquire whether such a reaction is possible (this question will be answered later in our discussion of the thermodynamics of redox systems).

A redox reaction consists of two parts or half-reactions. These are the *oxidation* reaction in which a substance *loses* or *donates* electrons and the reduction reaction in which a substance *gains* or *accepts* electrons. An oxidation reaction and a reduction reaction must always be coupled because "free" electrons cannot exist in solution and electrons must be conserved. The coupling between the two half-reactions is by the electrons that are either generated (by oxidation) or consumed (by reduction). We will use this fact in our technique for balancing redox reactions, which basically is a stepwise stoichiometric (mass) balancing of each constituent followed by a balancing of charge (electroneutrality).

Working with each half-reaction, we use the following procedure for balancing.

1. Identify the principal reactants and products, that is, species other than H^+, OH^-, and H_2O, in the oxidation half-reaction and the reduction half-reaction and write each half-reaction in crude form.

2. Then to obtain balanced half-reactions, balance the atoms other than hydrogen and oxygen by multiplying the reactants or products by appropriate integers.

3. Balance the oxygen using H_2O.

4. Balance the hydrogen with H^+.

5. Balance the charge with electrons.

6. Multiply each half-reaction by an appropriate integer so that both contain the same number of electrons.

7. Add the two balanced half-reactions.

8. Steps 1 to 7 will sometimes produce an equation that has H^+ as a reactant or a product. If it is known that the reaction takes place in alkaline solution, add the reaction for dissociation of water to the balanced equation to eliminate H^+ and form H_2O.

Example 7-1

Balance the reaction in which ferrous iron (Fe^{2+}) is oxidized to ferric iron (Fe^{3+}) by permanganate (MnO_4^-), which itself is reduced to manganese dioxide ($MnO_{2(s)}$). The reaction takes place in alkaline solution.

Solution

1. The reactants and products are

$$Fe^{2+} \rightleftharpoons Fe^{3+} \quad \text{(oxidation)}$$
$$MnO_4^- \rightleftharpoons MnO_{2(s)} \quad \text{(reduction)}$$

2. The atoms other than H and O are already balanced.

3. Balance the oxygen with water.

$$Fe^{2+} \rightleftharpoons Fe^{3+}$$
$$MnO_4^- \rightleftharpoons MnO_{2(s)} + 2H_2O$$

4. Balance the hydrogen with H^+.

$$Fe^{2+} \rightleftharpoons Fe^{3+}$$
$$4H^+ + MnO_4^- \rightleftharpoons MnO_{2(s)} + 2H_2O$$

5. Balance the charge with electrons, e^-.

$$Fe^{2+} \rightleftharpoons Fe^{3+} + e^-$$
$$3e^- + 4H^+ + MnO_4^- \rightleftharpoons MnO_{2(s)} + 2H_2O$$

6. Multiply the Fe half-reaction by 3, then add the two half-reactions, thus eliminating electrons

$$3 \times (Fe^{2+} \rightleftharpoons Fe^{3+} + e^-)$$
$$\underline{3e^- + 4H^+ + MnO_4^- \rightleftharpoons MnO_{2(s)} + 2H_2O}$$
$$4H^+ + 3Fe^{2+} + MnO_4^- \rightarrow 3Fe^{3+} + MnO_{2(s)} + 2H_2O$$

The equation is now balanced. In some instances it may be desirable to modify it to take into account other reactions and to make it more useful, as shown in steps (7) and (8).

7. The reaction takes place in alkaline solution. Add the water dissociation equation to eliminate H^+ as a reactant,

$$4H^+ + 3Fe^{2+} + MnO_4^- \rightleftharpoons 3Fe^{3+} + MnO_{2(s)} + 2H_2O$$
$$\underline{4H_2O \rightleftharpoons 4H^+ + 4OH^-}$$
$$2H_2O + 3Fe^{2+} + MnO_4^- \rightleftharpoons 3Fe^{3+} + MnO_{2(s)} + 4OH^-$$

8. Further, we know that in alkaline solution, Fe^{3+} and OH^- will combine to form $Fe(OH)_{3(s)}$. Adding this reaction, we obtain the final equation.

$$2H_2O + 3Fe^{2+} + MnO_4^- \rightleftharpoons 3Fe^{3+} + MnO_{2(s)} + 4OH^-$$
$$\underline{3 \times (Fe^{3+} + 3OH^- \rightleftharpoons Fe(OH)_{3(s)})}$$
$$2H_2O + 3Fe^{2+} + MnO_4^- + 5OH^- \rightleftharpoons 3Fe(OH)_{3(s)} + MnO_{2(s)}$$

This last step would not have been necessary if in step (1) we had taken $Fe(OH)_{3(s)}$ as the product rather than Fe^{3+}.

From this example we can see how a redox reaction is composed of the reduction half-reaction and the oxidation half-reaction. We note that ferrous iron is oxidized (it loses electrons) and permanganate is reduced (it gains electrons). Since the ferrous iron is donating electrons for the reduction of permanganate to $MnO_{2(s)}$, it is called a *reducing agent*. Conversely, the permanganate, which accepts the electrons from the oxidation of ferrous iron, is called an *oxidizing agent*. We encounter these terms widely in water chemistry. For example, if you thumb quickly through *Standard Methods for the Examination of Water and Wastewater*, you will discover statements such as "The chemical oxygen demand (COD) determination is a measure . . . of the organic matter . . . susceptible to oxidation by a *strong chemical oxidant*." "D.O. *(dissolved oxygen)* rapidly oxidizes an equivalent amount of the dispersed divalent manganese solution. . . ." "Ozone, a potent germicide, also is used as an oxidizing agent for the destruction of organic compounds . . . and for the oxidation of reduced iron or manganese salts to insoluble oxides. . . ."

Before we proceed further we should address one more aspect of stoichiometry and the balancing of reactions; that is, the question of equivalent weight (and normality) for oxidizing and reducing agents. *One equivalent weight of an oxidizing or reducing agent is the formula weight divided by the number of electrons taking part in the half-reaction.* For this reason, a knowledge of the specific half-reaction in which an oxidant or a reductant participates is needed to determine its equivalent weight and to make up a solution of desired normality. This is especially important for some oxidants and reductants that participate in different half-reactions depending on solution conditions such as pH. It should also be pointed out that the equivalent weight of a substance from an acid-base, charge, or precipitation standpoint bears no relationship to its equivalent weight in a redox system.

Example 7-2

A COD analysis is to be performed using 0.250 N potassium dichromate. How many grams of $K_2Cr_2O_7$ must be added to 2 liters of water to prepare this solution? The product of oxidation is chromic ion (Cr^{3+}) and the reaction takes place in acid solution.

Solution

First, we state the relevant reactant and product:

$$Cr_2O_7{}^{2-} \rightleftharpoons Cr^{3+}$$

Next, using the previously described technique, we balance the half-reaction, remembering that we are dealing with an acid solution.

$$6e^- + 14H^+ + Cr_2O_7{}^{2-} \rightleftharpoons 2Cr^{3+} + 7H_2O$$

Since there are $6e^-$ in the half-reaction, the equivalent weight of $K_2Cr_2O_7$ is its formula weight divided by 6. The formula weight of $K_2Cr_2O_7$ is 294.189. Thus

$$\text{Equivalent weight} = \frac{294.189}{6} = 49.0315$$

For 2 liters of a 0.250 N solution we require

$$0.250 \text{ eq/liter} \times 2 \text{ liter} \times 19.0315 \text{ g/eq} = 24.516 \text{ g}$$

This checks nicely with the instructions given in *Standard Methods* for preparing 0.250 N $K_2Cr_2O_7$—the addition of 12.259 g of $K_2Cr_2O_7$ to 1 liter of water is stipulated.

Example 7-3

The United Kingdom Department of the Environment[1] proposes a test method for "permanganate value" in which a diluted waste is oxidized for 4 hours at 27°C by a dilute acidic solution of potassium permanganate. The reagents for this test include 0.125 N potassium permanganate, $KMnO_4$. The instructions for preparing this reagent call for the addition of 4.0 g $KMnO_4$ to 1 liter distilled water. What is the rationale behind the recipe for this solution?

Solution

We must find what fraction of the formula weight of $KMnO_4$ corresponds to the situation that 4 g/liter $KMnO_4$ = 0.125 eq/liter (i.e., 0.125 N).

$$\text{Formula weight of } KMnO_4 \cong 158$$

One eighth of the formula weight of $KMnO_4$ = 19.75.

Thus, if there were one electron involved in the half-reaction, we would need to add 19.75 g (or approximately 20 g $KMnO_4$) to 1 liter to make a 0.125 N solution. However, the instructions only call for 4 g/liter. If we let x be the number of electrons involved in the half-reaction, then

$$\frac{20}{x} = 4$$

and $x = 5$.

Searching the literature we find the reaction where permanganate forms manganous ion,

$$MnO_4^- + 8H^+ + 5e^- \rightleftharpoons Mn^{2+} + 4H_2O$$

This fits the bill, since it has $5e^-$ and it takes place in acid solution.

Using the result of this computation, we can see that if the reaction from Example 7-1,

$$MnO_4^- + 4H^+ + 3e^- \rightleftharpoons MnO_{2(s)} + 2H_2O$$

in which MnO_4^- has an equivalent weight of $\frac{1}{3}$, instead of $\frac{1}{5}$, of the formula weight, had been used to compute the normality of the permanganate solution, we would obtain a solution with an entirely different composition. This raises the question of expressing concentrations of standard solutions (especially those used in oxidation-reduction reactions) in terms of normality. Imagine the mental anguish experienced by the chemist confronted with a bottle labeled 1 N $KMnO_4$. How does he know which reaction the normality is based upon? He does not, unless he made it up himself and can remember how much $KMnO_4$ he put in the bottle or unless he knows for what reaction the $KMnO_4$ is intended. It is therefore poor

[1] Analysis of Raw, Potable and Wastewater, London, H.M.S.O., Dept. of the Environment, 1972, p. 125.

practice to label bottles with "normality." If you use molar concentrations or weight concentrations, there will be no confusion.

The equivalent weights of oxidizing agents play an important practical role in water chemistry in the computation of the relative oxidizing capacity of various chlorine compounds. The strength of chlorine-containing disinfectants is often expressed in terms of "percent available chlorine," and incredible though it may seem, some of these agents have more than 100 percent available chlorine. The derivation of "percent available chlorine" is based on the half-reaction

$$2e^- + Cl_{2(g)} \rightleftharpoons 2Cl^-$$

in which chlorine gas is reduced to chlorine ions with the consumption of two electrons. In this equation the equivalent weight of chlorine is 2 × 35.5/2 = 35.5 g. Stated another way, we can say that each unit of oxidizing power (i.e., each mole of electrons consumed) is associated with 35.5 g of Cl_2.

Now let us consider another chlorine-containing oxidizing agent, for example, sodium hypochlorite, NaOCl. The half-reaction for NaOCl is

$$2e^- + 2H^+ + NaOCl \rightleftharpoons Cl^- + Na^+ + H_2O$$

In this reaction, each mole of electrons is associated with the formula weight of NaOCl/2 = 37.25 g of Cl_2.

For $Cl_{2(g)}$ we needed 35.5 g oxidizing agent to accept 1 mole of electrons; for NaOCl we need 37.25 g to do the job. Therefore, on an "electrons consumed per unit weight" basis, NaOCl is worse at consuming electrons than $Cl_{2(g)}$. It is precisely, (35.5/37.25) × 100 = 95 percent as good as $Cl_{2(g)}$. We say that NaOCl has "95 percent available chlorine" meaning, on a weight-to-weight basis, NaOCl has 95 percent of the oxidizing power of $Cl_{2(g)}$.

Example 7-4

What is the percent available chlorine in (1) monochloramine, NH_2Cl, and (2) high test hypochlorite, HTH, which is 70 percent by weight calcium hypochlorite, $Ca(OCl)_2$, given that the Cl in the NH_2Cl and HTH is converted to Cl^-, and that the N in NH_2Cl is converted to NH_4^+?

Solution

1. *Monochloramine, NH_2Cl*
   Half-reaction

$$2e^- + 2H^+ + NH_2Cl \rightleftharpoons Cl^- + NH_4^+$$

   Equivalent weight = formula weight of $NH_2Cl/2$ = 51.5/2 = 25.75

Percent available chlorine = 35.5/25.75 × 100 = 138 percent.

2. 70 percent $Ca(OCl)_2$,
    Half-reaction

$$4e^- + 4H^+ + Ca(OCl)_2 \rightleftharpoons 2Cl^- + Ca^{2+} + 2H_2O$$

Equivalent weight = formula weight of $Ca(OCl)_2/4$ = $143/4$ = 35.75

Percent available chlorine for $Ca(OCl)_2$ = $35.5/35.75 \times 100$ = 99.3 percent.
For 70 percent $Ca(OCl)_2$ = 99.3 percent $\times 70/100$ = 69.5 percent.

The unit mg/liter as Cl_2 is commonly used in the water treatment field. We use the concept of equivalents to convert from molar or mass concentration of a given chlorine species, such as $HOCl$, OCl^-, NH_2Cl, $NHCl_2$, to mass concentration as Cl_2. One mole each of $HOCl$, OCl^-, and NH_2Cl can accept 2 moles of electrons, just as Cl_2 can. The balanced half-reaction for $NHCl_2$ shows that it can accept 4 moles of electrons. The reaction products for the NH_2Cl and $NHCl_2$ reactions are NH_3 and Cl^-. Thus 1 mole of $HOCl$, OCl^-, or NH_2Cl is equivalent to 1 mole (71 g) of Cl_2 while $NHCl_2$ is equivalent to 2 moles (142 g) of Cl_2. Thus knowing the balanced half-reaction for a species allows us to convert to "mg/liter as Cl_2."

7.3. REDOX EQUILIBRIA

7.3.1. Direction of Reaction

In the previous section we learned how to balance redox reactions. However, once we had balanced the half-reactions and written a complete redox reaction we still did not know whether the reaction could proceed as written. In this section we will explore methods for making such a determination and in doing so will use many of the concepts and techniques developed in Chapter 3.

To illustrate the situation, we will examine the redox reaction between potassium dichromate, $K_2Cr_2O_7$, and ferrous iron, Fe^{2+}. This reaction is utilized in the titrimetric finish of the chemical oxygen demand (COD) test. The test consists of boiling a sample with a known amount of $K_2Cr_2O_7$ for 2 hours during which time some of the $K_2Cr_2O_7$ oxidizes the organic matter to CO_2 and water. Following this digestion, the excess dichromate must be measured. This is done by titrating with a standardized solution of ferrous ammonium sulfate (a source of ferrous ions) upon which the following reactions take place:

$$6e^- + 14H^+ + Cr_2O_7^{2-} \rightleftharpoons 2Cr^{3+} + 7H_2O$$

$$6 \times (Fe^{2+} \rightarrow Fe^{3+} + e^-)$$

$$14H^+ + Cr_2O_7^{2-} + 6Fe^{2+} \rightleftharpoons 2Cr^{3+} + 6Fe^{3+} + 7H_2O \qquad (7\text{-}1)$$

The dichromate is reduced to chromic ion while the ferrous ion is oxidized to ferric ion. When all of the $Cr_2O_7^{2-}$ has been reduced, it is possible for Fe^{2+} to exist in solution. An indicator called orthophenanthroline is added, which forms a red complex with Fe^{2+} and thereby detects the presence of free Fe^{2+}. (Incidentally, for purely aesthetic reasons this is one of the more delightful titrations to observe. The orange color of $Cr_2O_7^{2-}$ changes through an olive intermediate hue to the deep green color of the chromic ion and then suddenly, just like a stoplight changing, the green turns to the red of the orthophenanthroline-Fe^{2+} complex.)

If we were proposing, for the first time, to use Fe^{2+} as a reducing agent to determine the amount of $Cr_2O_7^{2-}$ in a solution, how could we be sure that the reaction would proceed as written in Eq. 7-1? The first approach would be to determine the standard free energy change, $\Delta G°$ of the reaction as written and determine if it was negative. Remember that the more negative the $\Delta G°$ value, the larger the equilibrium constant, K.

From Table 3-1,

Species	$\Delta \overline{G}_f°$, kcal/mole
H^+	0
$Cr_2O_7^{2-}$	-315.4
Fe^{2+}	-20.30
Cr^{3+}	-51.5
Fe^{3+}	-2.52
$H_2O_{(\ell)}$	-56.69

From Eq. 3-13,

$$\Delta G° = 7\Delta \overline{G}_{f,H_2O}° + 6\Delta \overline{G}_{f,Fe^{3+}}° + 2\Delta \overline{G}_{f,Cr^{3+}}° - 6\Delta \overline{G}_{f,Fe^{2+}}° - \Delta \overline{G}_{f,Cr_2O_7^{2-}}° - 14\Delta \overline{G}_{f,H^+}°$$
$$\Delta G° = 7(-56.69) + 6(-2.52) + 2(-51.5) - 6(-20.30) - (-315.4) - 14(0)$$
$$\Delta G° = -77.8$$

and since

$$\Delta G° = -RT \ln K$$
$$K = 10^{57.2}$$

K is certainly large and the reaction should be displaced far to the right at equilibrium.

Neglecting ionic strength effects, we assume that the titration has proceeded a little way and that the reactant concentrations are $[Fe^{3+}] = 10^{-3}$, $[Fe^{2+}] = 10^{-4}$, $[Cr^{3+}] = 10^{-2.4}$, $[Cr_2O_7^{2-}] = 10^{-2.7}$ and $[H^+] = 10^{+1.3}$. (These are realistic concentrations for the situation that exists in an actual COD titration.) Calculating the reaction quotient, Q, as follows,

$$Q = \frac{[Fe^{3+}]^6[Cr^{3+}]^2}{[Fe^{2+}]^6[Cr_2O_7^{2-}][H^+]^{14}}$$
$$= \frac{(10^{-3})^6(10^{-2.4})^2}{(10^{-4})^6(10^{-2.7})(10^{1.3})^{14}} = 10^{-14.3}$$

we can use Eq. 3-10 to calculate ΔG at 25°C,

$$\Delta G = \Delta G° + RT \ln Q$$

$$\Delta G = -77.8 + RT \ln 10^{-14.3}$$

$$= -97.2$$

Thus the reaction is spontaneous to the right.

7.3.2. Free Energy and Potential of Half-Reactions

Our calculation has involved the overall redox reaction and thus far we have done nothing different from our previous thermodynamic treatment of other types of equilibria. For the sake of convenience it would be profitable to be able to examine any two half-reactions and judge from them independently, in which direction a reaction could proceed spontaneously if they were combined.

One way to do this is to determine the $\Delta G°$ values for each half-reaction and then add them to determine the value of $\Delta G°$ for the complete redox reaction. Let us examine one of the steps of the Winkler method for determining dissolved oxygen (DO) as an example of this technique. The last step of the Winkler method is the titration of iodine ($I_{2(aq)}$) to iodide (I^-) using thiosulfate, $S_2O_3^{2-}$, which is itself reduced to tetrathionate, $S_4O_6^{2-}$. The reaction takes place in acid solution. The two half-reactions are

$$2e^- + I_{2(aq)} \rightleftharpoons 2I^- \tag{7-2}$$

$$2S_2O_3^{2-} \rightleftharpoons S_4O_6^{2-} + 2e^- \tag{7-3}$$

For Eq. 7-2,

$$\Delta G° = 2\Delta \overline{G}°_{f,I^-} - \Delta \overline{G}°_{f,I_{2(aq)}} - 2\Delta \overline{G}°_{f,e^-}$$

and for Eq. 7-3,

$$\Delta G° = \Delta \overline{G}°_{f,S_4O_6^{2-}} + 2\Delta \overline{G}°_{f,e^-} - 2\Delta \overline{G}°_{f,S_2O_3^{2-}}$$

In both of these equations we need to know the value of the free energy of formation of the electron $\Delta \overline{G}°_{f,e^-}$. Thinking one step ahead to the addition of Eqs. 7-2 and 7-3 to produce overall redox equation, we can see that $\Delta \overline{G}°_{f,e^-}$ will cancel out so that it does not really matter what value $\Delta G°_{f,e^-}$ has. By convention, however, $\Delta \overline{G}°_{f,e^-}$ is assigned a value of zero and this assignment permits us to examine half-reactions independent of the overall reaction.[2]

[2] With e^- we have a situation quite similar to the proton, H^+. Free electrons, similarly to free protons, do not exist in solution for a significant period of time. The proton is always attached to some molecule or ion in solution as is the electron. However, it is often useful to consider e^- as a separate entity such as in a half-reaction just as we did for the proton in acid-base equilibria.

From Table 3-1,

Species	ΔG_f°, kcal/mole
e^-	0
$I_{2(aq)}$	$+3.93$
I^-	-12.35
$S_2O_3^{2-}$	-127.2
$S_4O_6^{2-}$	-246.3

Thus for Eq. 7-2,

$$\Delta G^\circ = 2(-12.35) - (+3.93) - 2(0) = -28.63$$

and for Eq. 7-3,

$$\Delta G^\circ = -24.63 + 2(0) - 2(-127.2) = +8.1$$

Therefore, for the overall reaction,

$$\Delta G^\circ = -28.63 + 8.1 = -20.53 \text{ kcal}$$

Thus the reaction,

$$I_{2(aq)} + 2S_2O_3^{2-} \rightleftharpoons S_4O_6^{2-} + 2I^-$$

is spontaneous when all species are at unit activity, that is, when $Q = 1$. The electrons generated in the oxidation of $2S_2O_3^{2-}$ to $S_4O_6^{2-}$ drive the reduction of $I_{2(aq)}$ to $2I^-$.

In a well-mixed solution such as exists during the titration of $I_{2(aq)}$ with $S_2O_3^{2-}$, all of the participating species are in close contact with each other. The electron transfer takes place between species in intimate contact all throughout the solution. Imagine, however, if we were able to isolate the oxidation half-reaction and the reduction half-reaction into two separate containers. We then introduce a piece of platinum wire into both containers and connect the platinum wires externally from the solutions through a high-resistance voltmeter. We also ensure that there is connection between the two solutions so that ions formed in one container can eventually move to the other container and so maintain electroneutrality in the solutions. Without ion movement between the compartments, a transfer of e^- through the platinum wire could be accompanied by a buildup of negative charges in one compartment and positive charge in the other. Because this would violate the electroneutrality requirements, the reaction would not proceed. The system is illustrated in Fig. 7-1.

As soon as we connect the platinum wires, the voltmeter will register a voltage reading showing (1) that a difference in potential exists between the two electrodes, and (2) that electrons are flowing in the external circuit between the two compartments, assuming that the reactions in question can take place on the platinum surface. We have formed a *galvanic cell*

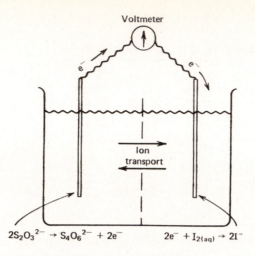

Fig. 7-1. Iodine-thiosulfate cell.

in which the chemical energy of the redox reaction is converted into electrical energy. We call the potential that develops the electromotive force (emf) of the cell; if all species are present at unit activity, the temperature is 25°C, and the pressure is 1 atm, we assign it a symbol of E°_{cell}. The cell emf for these conditions is the electrochemical version of ΔG° for the reaction. Since it is not necesary for the reaction to take place in a cell, the symbol E° is often used in place of E°_{cell} and is identical to it. Like ΔG°, the redox potential can be split up into two parts, one for the oxidation half-reaction and one for the reduction half-reaction:

$$E^\circ_{cell} = E^\circ_{ox} + E^\circ_{red}$$

where E°_{ox} and E°_{red} are, respectively, the half-reaction potentials for the oxidation half-reaction and the reduction half-reaction.

The E° values for various half-cells, like the $\Delta \overline{G}_f{}^\circ$ values for various species, cannot be determined absolutely. Like $\Delta \overline{G}_f{}^\circ$ values they are measured with reference to a particular half-reaction that is assigned an E° value of zero at 25°C and 1 atm. This reaction is the reduction of hydrogen ion to hydrogen gas:

$$H^+_{(aq)} + e^- \rightleftharpoons \tfrac{1}{2}H_{2(g)} \quad ; \quad E^\circ = 0 \tag{7-4}$$

This convention is consistent with the assignment of $\Delta \overline{G}_f{}^\circ = 0$ for $H_{2(g)}$ and H^+, since we have previously stated that $\Delta \overline{G}_{f,e^-}^\circ = 0$. E° is related to ΔG°, and E is related to ΔG, by

$$\Delta G^\circ = -nFE^\circ$$

or

$$\Delta G = -nFE \tag{7-5}$$

where

n = number of electrons involved in the reaction

F = the Faraday[3] = 23.06 kcal/volt-equivalent (or 96,500 coulombs/equivalent)

For Eq. 7-4,

$$\Delta G^\circ = \tfrac{1}{2}\Delta \overline{G}^\circ_{f,H_{2(g)}} - \Delta \overline{G}^\circ_{f,H^+_{(aq)}} - \Delta \overline{G}^\circ_{f,e^-}$$
$$\Delta G^\circ = 0 - 0 - 0$$

Since $\Delta G^\circ = 0$, from Eq. 7-5, $E^\circ = 0$.

The physical setup that defines E° for a half-reaction is a hydrogen gas electrode (Standard Hydrogen Electrode) connected to an electrode where the half-reaction of interest takes place (let us say the $I_{2(aq)} + 2e^- \rightleftharpoons 2I^-$ half-reaction; see Fig. 7-2). The cell reaction for standard conditions, 25°, unit activities, and 1 atm $H_{2(g)}$ pressure, is

$$\tfrac{1}{2}I_{2(aq)} + e^- \rightarrow I^-; \qquad E^\circ_{I_2,I^-}$$

$$\tfrac{1}{2}H_{2(g)} \rightarrow H^+ + e^-; \qquad E^\circ_{H_2,H^+}$$

$$\overline{\tfrac{1}{2}I_{2(aq)} + \tfrac{1}{2}H_{2(g)} \rightarrow H^+ + I^-; \qquad E^\circ_{cell}} \tag{7-6}$$

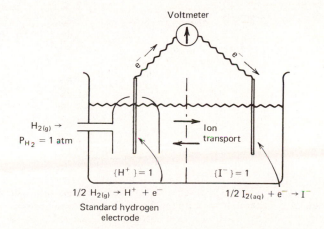

Fig. 7-2. Cell for definition of E° for a half-reaction.

[3] The Faraday, or Faraday's constant, is the charge per mole of electrons, or per equivalent. In this equation, nF represents the quantity of charge which is transported through a potential difference of E volts.

and

$$E^\circ_{\text{cell}} = E^\circ_{I_2,I^-} + E^\circ_{H_2,H^+} = E^\circ_{I_2,I^-} + 0$$

Therefore,

$$E^\circ_{\text{cell}} = E^\circ_{I_2,I^-}$$

Note that the half-reaction of interest is written as a reduction reaction [$I_{2(aq)}$ accepts electrons] and that the reaction at the hydrogen electrode is an oxidation reaction. Under these conditions the value of $E^\circ_{I_2,I^-}$ is called the *standard electrode potential* and has the same sign as the emf of the cell of which it is a component. If E° for the reaction in Eq. 7-6 is positive and all activities are unity, the reaction will proceed spontaneously as written because ΔG°, and thus ΔG, for the reaction is negative. If E° is negative, the reverse reaction will occur spontaneously. The convention for defining the standard electrode potential value as E° for the reduction half-reaction was adopted by the International Union of Pure and Applied Chemistry (IUPAC) in 1953, and we will use it throughout this text.[4] Table 7-1 presents a compilation of some E° values relevant to water chemistry.

TABLE 7-1   Standard Electrode Potentials at 25°C

Reaction	$\dfrac{E^\circ}{\text{Volt}}$	$p\epsilon^\circ\left(=\dfrac{1}{n}\log K\right)$
$H^+ + e^- \rightleftharpoons \frac{1}{2}H_{2(g)}$	0	0
$Na^+ + e^- \rightleftharpoons Na_{(s)}$	-2.72	-46.0
$Mg^{2+} + 2e^- \rightleftharpoons Mg_{(s)}$	-2.37	-40.0
$Cr_2O_7^{2-} + 14H^+ + 6e^- \rightleftharpoons 2Cr^{3+} + 7H_2O$	$+1.33$	$+22.5$
$Cr^{3+} + e^- \rightleftharpoons Cr^{2+}$	-0.41	-6.9
$MnO_4^- + 2H_2O + 3e^- \rightleftharpoons MnO_{2(s)} + 4OH^-$	$+0.59$	$+10.0$
$MnO_4^- + 8H^+ + 5e^- \rightleftharpoons Mn^{2+} + 4H_2O$	$+1.51$	$+25.5$
$Mn^{4+} + e^- \rightleftharpoons Mn^{3+}$	$+1.65$	$+27.9$
$MnO_{2(s)} + 4H^+ + 2e^- \rightleftharpoons Mn^{2+} + 2H_2O$	$+1.23$	$+20.8$
$Fe^{3+} + e^- \rightleftharpoons Fe^{2+}$	$+0.77$	$+13.0$
$Fe^{2+} + 2e^- \rightleftharpoons Fe_{(s)}$	-0.44	-7.4
$Fe(OH)_{3(s)} + 3H^+ + e^- \rightleftharpoons Fe^{2+} + 3H_2O$	$+1.06$	$+17.9$
$Cu^{2+} + e^- \rightleftharpoons Cu^+$	$+0.16$	$+2.7$

[4] Not all texts use the IUPAC convention so that caution must be exercised in using values of E° taken from tables. Notably, the extensive presentation of half-reaction potentials in *Oxidation Potentials*, 2nd ed., by W. M. Latimer (Prentice-Hall, Inc., Englewood Cliffs, N.J., 1952) employs a convention that is exactly opposite to that used here; i.e., the half-reaction is written as an oxidation rather than a reduction, and the E° value has a sign opposite to the standard reduction potential. In other textbooks, the E° value for the reduction half-reaction has a sign opposite to the standard electrode potential so that $\Delta G = nFE$.

TABLE 7-1 Standard Electrode Potentials at 25°C

Reaction	$E°$ Volt	$p\epsilon° \left(= \dfrac{1}{n} \log K \right)$
$Cu^{2+} + 2e^- \rightleftharpoons Cu_{(s)}$	+0.34	+5.7
$Ag^{2+} + e^- \rightleftharpoons Ag^+$	+2.0	+33.8
$Ag^+ + e^- \rightleftharpoons Ag_{(s)}$	+0.8	+13.5
$AgCl_{(s)} + e^- \rightleftharpoons Ag_{(s)} + Cl^-$	+0.22	+3.72
$Au^{3+} + 3e^- \rightleftharpoons Au_{(s)}$	+1.5	+25.3
$Zn^{2+} + 2e^- \rightleftharpoons Zn_{(s)}$	−0.76	−12.8
$Cd^{2+} + 2e^- \rightleftharpoons Cd_{(s)}$	−0.40	−6.8
$Hg_2Cl_{2(s)} + 2e^- \rightleftharpoons 2Hg_{(\ell)} + 2Cl^-$	+0.27	+4.56
$2Hg^{2+} + 2e^- \rightleftharpoons Hg_2^{2+}$	+0.91	+15.4
$Al^{3+} + 3e^- \rightleftharpoons Al_{(s)}$	−1.68	−28.4
$Sn^{2+} + 2e^- \rightleftharpoons Sn_{(s)}$	−0.14	−2.37
$PbO_{2(s)} + 4H^+ + SO_4^{2-} + 2e^- \rightleftharpoons PbSO_{4(s)} + 2H_2O$	+1.68	+28.4
$Pb^{2+} + 2e^- \rightleftharpoons Pb_{(s)}$	−0.13	−2.2
$NO_3^- + 2H^+ + 2e^- \rightleftharpoons NO_2^- + H_2O$	+0.84	+14.2
$NO_3^- + 10H^+ + 8e^- \rightleftharpoons NH_4^+ + 3H_2O$	+0.88	+14.9
$N_{2(g)} + 8H^+ + 6e^- \rightleftharpoons 2NH_4^+$	+0.28	+4.68
$NO_2^- + 8H^+ + 6e^- \rightleftharpoons NH_4^+ + 2H_2O$	+0.89	+15.0
$2NO_3^- + 12H^+ + 10e^- \rightleftharpoons N_{2(g)} + 6H_2O$	+1.24	+21.0
$O_{3(g)} + 2H^+ + 2e^- \rightleftharpoons O_{2(g)} + H_2O$	+2.07	+35.0
$O_{2(g)} + 4H^+ + 4e^- \rightleftharpoons 2H_2O$	+1.23	+20.8
$O_{2(aq)} + 4H^+ + 4e^- \rightleftharpoons 2H_2O$	+1.27	+21.5
$SO_4^{2-} + 2H^+ + 2e^- \rightleftharpoons SO_3^{2-} + H_2O$	−0.04	−0.68
$S_4O_6^{2-} + 2e^- \rightleftharpoons 2S_2O_3^{2-}$	+0.18	+3.0
$S_{(s)} + 2H^+ + 2e^- \rightleftharpoons H_2S_{(g)}$	+0.17	+2.9
$SO_4^{2-} + 8H^+ + 6e^- \rightleftharpoons S_{(s)} + 4H_2O$	+0.35	+6.0
$SO_4^{2-} + 10H^+ + 8e^- \rightleftharpoons H_2S_{(g)} + 4H_2O$	+0.34	+5.75
$SO_4^{2-} + 9H^+ + 8e^- \rightleftharpoons HS^- + 4H_2O$	+0.24	+4.13
$2HOCl + 2H^+ + 2e^- \rightleftharpoons Cl_{2(aq)} + 2H_2O$	+1.60	+27.0
$Cl_{2(g)} + 2e^- \rightleftharpoons 2Cl^-$	+1.36	+23.0
$Cl_{2(aq)} + 2e^- \rightleftharpoons 2Cl^-$	+1.39	+23.5
$2HOBr + 2H^+ + 2e^- \rightleftharpoons Br_{2(\ell)} + 2H_2O$	+1.59	+26.9
$Br_2 + 2e^- \rightleftharpoons 2Br^-$	+1.09	+18.4
$2HOI + 2H^+ + 2e^- \rightleftharpoons I_{2(s)} + 2H_2O$	+1.45	+24.5
$I_{2(aq)} + 2e^- \rightleftharpoons 2I^-$	+0.62	+10.48
$I_3^- + 2e^- \rightleftharpoons 3I^-$	+0.54	+9.12
$ClO_2 + e^- \rightleftharpoons ClO_2^-$	+1.15	+19.44
$CO_{2(g)} + 8H^+ + 8e^- \rightleftharpoons CH_{4(g)} + 2H_2O$	+0.17	+2.87
$6CO_{2(g)} + 24H^+ + 24e^- \rightleftharpoons C_6H_{12}O_6(glucose) + 6H_2O$	−0.01	−0.20
$CO_{2(g)} + H^+ + 2e^- \rightleftharpoons HCOO^-(formate)$	−0.31	−5.23

Source: L. G. Sillen and A. E. Martell, "Stability Constants of Metal Ion Complexes," The Chemical Society, London, Special Publication No. 16, 1964. W. Stumm and J. J. Morgan, Aquatic Chemistry, Wiley-Interscience, New York 1970.

For the $I_{2(aq)}/I^-$ half-reaction the standard electrode potential, $E°$, is +0.62 volts. This means that the reaction in Eq. 7-6 will proceed spontaneously as written if all reactants and products are at unit activity, and that $I_{2(aq)}$ has a greater affinity for electrons than H^+, that is, $I_{2(aq)}$ is a better oxidizing agent than H^+. In the terminology of the corrosion chemist we would say that $I_{2(aq)}$ is more "noble" than H^+.

We can make a similar series of statements about I^-. The positive $E°$ value tells us that I^- has less tendency to donate electrons than $H_{2(g)}$, that is, it is a weaker reducing agent than $H_{2(g)}$, or it is less noble than $H_{2(g)}$.

The standard electrode potential for the $S_4O_6^{2-}/S_2O_3^{2-}$ half-reaction is $E° = +0.18$ volts. Because $E°$ is positive, the overall reaction

$$S_4O_{6(aq)}^{2-} + H_{2(g)} \rightleftharpoons 2S_2O_{3(aq)}^{2-} + H_{(aq)}^+$$

can proceed spontaneously as written when all reactants and products are at unit activity.

7.3.3.   Combination of Half-Reactions

We can now combine the two half-reactions (Eqs. 7-2 and 7-3) and determine the value of $nFE°_{cell}$ for the cell in Fig. 7-1 from the two $nFE°$ values for the half-reactions. This is exactly equivalent to adding $\Delta G°$ values for the half-reactions. For example, using the values of $\Delta G°$ previously determined,

	$\Delta G°(kcal)$	$-nFE°$
$I_{2(aq)} + 2e^- \rightarrow 2I^-$	-28.63	$-2F(+0.62)$
$2S_2O_3^{2-} \rightarrow S_4O_6^{2-} + 2e^-$	$+8.1$	$-2F(-0.18)$
$I_{2(aq)} + 2S_2O_3^{2-} \rightarrow 2I^- + S_4O_6^{2-}$	-20.53	$-2F(+0.44)$

Since $-2FE°_{cell} = -2F(0.44)$, $E°_{cell} = 0.44$ volts.

Note that the value of $E°$ is independent of reaction stoichiometry, that is, the number of electrons involved in the reaction. Thus if the reaction had been written $\frac{1}{2}I_{2(aq)} + S_2O_3^{2-} \rightarrow I^- + \frac{1}{2}S_4O_6^{2-}$, the value of $E°$ would still be $+0.44$. However, the value of $\Delta G°$ would change, since $\Delta G° = -nFE°$ and $n = 1$ rather than $n = 2$.

In this example and in the previous computations with the cell involving the standard hydrogen electrode, we added the $E°$ values for the half-reactions to obtain the $E°_{cell}$ values. This is only possible because the nF component of the term $-nFE°$ was identical for both half-reactions; that is, the same number of electrons were involved in each half-reaction. This will always be the case when two half-reactions are added to give an overall reaction because the equations must be added so that the electrons cancel. The addition of $E°$ values for half-reactions is not always correct, however. For example, when we compose a new half-reaction by adding two half-reactions, the $E°$ value of the new half-reaction may only be computed by the summation of $\Delta G°\,(= -nFE°)$ values. In all instances $-nFE°$ values are additive.

Example 7-5

Find $E°$ for the half-reaction,

$$Fe^{3+} + 3e^- \rightleftharpoons Fe_{(s)}$$

knowing that

(1) $Fe^{2+} + 2e^- \rightleftharpoons Fe_{(s)};\qquad E° = -0.44$ volt

(2) $Fe^{3+} + e^- \rightleftharpoons Fe^{2+};\qquad E° = +0.77$ volt

Solution

Adding (1) and (2), we obtain

$$Fe^{3+} + 3e^- \rightleftharpoons Fe_{(s)}$$

$$-nFE°_{Fe^{3+},Fe_{(s)}} = -nFE°_{Fe^{2+},Fe_{(s)}} + (-nFE°_{Fe^{3+},Fe^{2+}})$$

$$3FE°_{Fe^{3+},Fe_{(s)}} = -2F(-0.44) - F(+0.77)$$

$$E°_{Fe^{3+},Fe_{(s)}} = 0.037 \text{ volt}$$

7.3.4. The Nernst Equation

Thus far we have confined our discussion to cells and half-reactions at standard state, that is, 25°C, 1 atm pressure, and unit activity for all species. To determine the effect of reactant and product concentration, we need to draw further on our analogy between free energy change and electrode potential or cell emf. Using the free energy equation as developed in Chapter 3,

$$\Delta G = \Delta G° + RT \ln Q \tag{7-7}$$

and dividing each term in the equation by $-nF$, we obtain

$$\frac{\Delta G}{-nF} = \frac{\Delta G°}{-nF} + \frac{RT}{-nF} \ln Q$$

Substituting from Eq. 7-5 yields

$$E = E° - \frac{RT}{nF} \ln Q \tag{7-8}$$

which is the *Nernst equation*. In simplified form it is,

$$E = E° - \frac{0.059}{n} \log Q$$

at 25°C, since the term $2.3\,RT/F$ has the value of 0.059. E is the cell potential and is a function of the reaction quotient, Q. From Eq. 7-5 we see that E has the opposite sign to ΔG; thus if ΔG is negative, E is positive and the reaction will proceed spontaneously to the right. Conversely, if ΔG is positive, E is negative and the reaction will proceed spontaneously to the left. If $\Delta G = 0$, $E = 0$, and from Eq. 7-8 we obtain

$$E° = \frac{RT}{nF} \ln K \qquad (7\text{-}9)$$

because, at equilibrium, $Q = K$.

The application of the Nernst equation to the reactions that take place in an electrochemical cell and in solution is illustrated in Examples 7-6 and 7-7, respectively.

Example 7-6

A Daniell cell consists of a zinc electrode in a zinc chloride solution connected to a copper electrode in a cupric chloride solution,

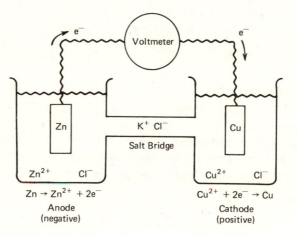

What is the equilibrium constant of the cell reaction at 25°C? From Table 7-1, $E°_{Zn^{2+},Zn_{(s)}} = -0.76$ volt and $E°_{Cu^{2+},Cu_{(s)}} = +0.34$ volt.

Solution

Assuming that Zn is oxidized and Cu^{2+} is reduced in the cell, the cell reaction is

	$-nFE°$
$Zn_{(s)} \rightleftharpoons Zn^{2+} + 2e^-$	$-2F(+0.76)$
$2e^- + Cu^{2+} \rightleftharpoons Cu_{(s)}$	$-2F(+0.34)$
$Zn_{(s)} + Cu^{2+} \rightleftharpoons Zn^{2+} + Cu_{(s)}$	$-2F(+1.10)$

Thus $E°_{cell} = +1.10$ volts. [Note that if we had assumed that $Cu_{(s)}$ was oxidized and Zn^{2+} was reduced, $E°_{cell}$ would have the same absolute magnitude but the opposite sign.] At equilibrium, when no current flows through the circuit, $E = 0$ and from Eq. 7-8 we have

$$E° = +1.1 = \left(\frac{0.059}{2}\right) \log K$$
$$K = 10^{+37.2}$$

Example 7-7

The Winkler determination of dissolved oxygen involves the formation of $MnO_{2(s)}$ by the oxidation of Mn^{2+} with oxygen. The $MnO_{2(s)}$ is then reacted with I^- to form $I_{2(aq)}$. Determine the E value of a solution that contained 8 mg DO/liter at a point when half of the $MnO_{2(s)}$ formed by reaction with DO has been reduced to Mn^{2+} by I^- (which itself is oxidized to $I_{2(aq)}$). The H^+ concentration is approximately 1 M. The following reagent concentrations and quantities are used in the determination:

1. $MnSO_4 \cdot 2H_2O$: 2 ml of a 400 g/liter solution per 300-ml sample volume.

2. KI : 2 ml of a 150 g/liter solution per 300-ml sample volume.

Solution

Neglect the volume changes caused by the addition of $MnSO_4$ and KI solutions.

$C_{T,Mn}$,

400 g/liter $MnSO_4 \cdot 2H_2O$ = 2.14 moles/liter
(2 ml/300 ml sample)(2.14 moles/liter) = 1.43×10^{-2} M

$C_{T,I}$,

150 g/liter KI = 0.90 mole/liter
(2 ml/300 ml sample)(0.90 mole/liter) = 6×10^{-3} M

Manganous ion reacts with DO to form manganese dioxide,

$$\frac{\begin{array}{l} 2 \times (Mn^{2+} + 2H_2O \rightleftharpoons MnO_{2(s)} + 4H^+ + 2e^-) \\ 4e^- + 4H^+ + O_{2(aq)} \rightleftharpoons 2H_2O \end{array}}{2Mn^{2+} + O_{2(aq)} + 2H_2O \rightleftharpoons 2MnO_{2(s)} + 4H^+}$$

Thus 8 mg/liter DO [= (8 mg/liter)/(32,000 mg/mole) = 2.5×10^{-4} moles/liter] will produce $2 \times 2.5 \times 10^{-4}$ = 5.0×10^{-4} mole $MnO_{2(s)}$/liter.

After reduction of half of the $MnO_{2(s)}$ to Mn^{2+} with I^-,

$$[Mn^{2+}] = C_{T,Mn} - \text{number of moles of } MnO_{2(s)}/\text{liter}$$
$$= 1.43 \times 10^{-2}M - 0.5(5.0 \times 10^{-4}M)$$
$$= 1.40 \times 10^{-2}M$$

The reaction of $MnO_{2(s)}$ with I^- yields

$$\frac{\begin{array}{l} 2I^- \rightleftharpoons I_{2(aq)} + 2e^- \\ MnO_{2(s)} + 4H^+ + 2e^- \rightleftharpoons Mn^{2+} + 2H_2O \end{array}}{MnO_{2(s)} + 2I^- + 4H^+ \rightleftharpoons I_{2(aq)} + Mn^{2+} + 2H_2O}$$

Then

$$E°_{cell} = E°_{I^-,I_{2(aq)}} + E°_{MnO_2,Mn^{2+}} = -0.62 + 1.23 = +0.61$$

1 mole of $MnO_{2(s)}$ produces 1 mole of $I_{2(aq)}$. Therefore, $2.5 \times 10^{-4} M\ MnO_{2(s)}$ produces $2.5 \times 10^{-4} M\ I_{2(aq)}$.

$$[I^-] = C_{T,I} - 2[I_{2(aq)}]$$
$$= 6 \times 10^{-3} - 2(2.5 \times 10^{-4})$$
$$= 5.5 \times 10^{-3} M$$

and

$$Q = \frac{[I_{2(aq)}][Mn^{2+}]}{\{MnO_{2(s)}\}[I^-]^2[H^+]^4}$$

$$= \frac{(2.5 \times 10^{-4})(1.4 \times 10^{-2})}{(1)(5.5 \times 10^{-3})^2(1)^4} = 0.116$$

Applying the Nernst equation, Eq. 7-8, to the overall reaction, we find that

$$E = E^\circ - \frac{0.059}{n} \log Q$$

$$= +0.61 - \frac{0.059}{2} \log 0.116$$

$$= +0.64 \text{ volt}$$

Thus the remaining $MnO_{2(s)}$ will continue to oxidize the I^-.

We can make Eq. 7-8 applicable to a half-reaction by writing the half-reaction in combination with the standard hydrogen electrode reaction. For example, for the reduction of cupric copper (Cu^{2+}) to cuprous copper (Cu^+),

$$Cu^{2+} + e^- \rightleftharpoons Cu^+; \qquad E^\circ = 0.16$$

$$\tfrac{1}{2}H_{2(g)} \rightleftharpoons H^+ + e^-; \qquad E^\circ = 0$$

$$\overline{Cu^{2+} + \tfrac{1}{2}H_{2(g)} \rightleftharpoons Cu^+ + H^+; \qquad E^\circ = 0.16}$$

Applying Eq. 7-8, we obtain

$$E = E^\circ - \frac{RT}{nF} \ln Q$$

$$= 0.16 - \frac{RT}{F} \ln \frac{\{H^+\}\{Cu^+\}}{\{Cu^{2+}\}(P_{H_2})^{1/2}}$$

Because $\{H^+\}$ and $P_{H_{2(g)}}$ are unity in the standard hydrogen electrode,

$$E = 0.16 - \frac{RT}{F} \ln \frac{\{Cu^+\}}{\{Cu^{2+}\}} \qquad (7\text{-}10)$$

and when activity effects are negligible,

$$E = 0.16 - \frac{RT}{F} \ln \frac{[Cu^+]}{[Cu^{2+}]}$$

Thus the Nernst equation for an overall reaction involving the standard hydrogen electrode is the same as the Nernst equation for the half-reaction and the potential, E, is a function of the activities of the reactants and products of the half-reaction other than e^-. In the following sections of this book, we will follow the practice of writing the Nernst equation for half-reactions without explicit reference to the standard hydrogen electrode.

Furthermore, we will follow the common practice of referring to the potential E as E_H (the redox potential) when the half-reaction is written as a reduction reaction. The subscript H is used to emphasize that the potential only has meaning in reference to the standard hydrogen electrode reaction. Using this nomenclature, Eq. 7-10 is written as

$$E_H = 0.16 - \frac{RT}{F} \ln \frac{\{Cu^+\}}{\{Cu^{2+}\}}$$

Example 7-8

The half-reaction for reduction of sulfate to sulfite is

(1) $SO_4^{2-} + 2H^+ + 2e^- \rightleftharpoons SO_3^{2-} + H_2O$

What is the redox potential of this half-reaction, E_H, at 25°C if it takes place in the presence of 10^{-3} moles SO_3^{2-}/liter and 10^{-4} moles SO_4^{2-}/liter at a pH of 8? Neglect ionic strength effects.

Solution

From Table 7-1, $E° = -0.04$ for the reduction half-reaction. Applying the Nernst equation, Eq. 7-8, we obtain

$$E_H = E° - \frac{0.059}{n} \log Q$$

$$= -0.04 - \frac{0.059}{2} \log \left(\frac{[SO_3^{2-}]}{[H^+]^2 [SO_4^{2-}]} \right)$$

$$= -0.04 - \frac{0.059}{2} \log \left[\frac{10^{-3}}{(10^{-8})^2 (10^{-4})} \right]$$

$$= -0.50 \text{ volt}$$

Note that determining the potential of a half-reaction is the same as determining the emf of a cell in which one electrode is the standard hydrogen electrode and at the other electrode the half-reaction in question proceeds. If the potential is to be called a redox potential, the latter reaction must be written as a reduction reaction.

We must be careful not to use E_H values for half-reactions in the same way as the E values for overall reactions. The value of E_H for a half-reaction simply tells us the potential relative to the $H^+/H_{2(g)}$ half-reaction, whereas the value of E for an overall reaction tells us whether or not that reaction is at equilibrium. Remembering this will eliminate much confusion in the use of E_H values.

7.3.5. Formal Potentials

The E and $E°$ values as used in the Nernst equation are defined in terms of the activities of individual species. In very dilute solutions in which there is no formation of complexes, concentration can be used in place of activity and there is no need to account for the portion of the total concentration of a species that might be complexed. In many solutions, however, ionic interactions, complex formation, and acid-base reactions must be taken into account. Formal potentials, E' and $E°'$ are often used for this purpose and apply only to a given set of solution conditions. For example, for the half-reaction

$$Fe^{3+} + e^- \rightleftharpoons Fe^{2+}$$

the Nernst equation, Eq. 7-8, is

$$E_H = E° - \frac{RT}{F} \ln \frac{\{Fe^{2+}\}}{\{Fe^{3+}\}}$$

whereas the Nernst equation using formal potentials is

$$E_H' = E°' - \frac{RT}{F} \ln \frac{C_{T,Fe(II)}}{C_{T,Fe(III)}}$$

where $C_{T,Fe(II)}$ and $C_{T,Fe(III)}$ represent the analytical concentrations of Fe(II) and Fe(III), respectively, regardless of the degree of complexation, ionic interactions, and so forth. Whereas $E°$ has a value of $+0.77$ volts, the value of $E°'$ changes as the solution composition changes. For example, $E°' = +0.68$ volts in a $C_{T,Fe} = 0.1\ M$, $H_2SO_4 = 1\ M$ solution.

7.3.6. The Electron Balance and Equilibrium Calculations

In making equilibrium calculations for redox reactions it is often necessary to make use of the *electron balance*. The equation is analogous to the proton balance and is based on the principle that electrons are conserved. For example, when Cl_2 is added to a solution, the following half-reactions take place:

$$Cl_2 + 2e^- \rightarrow 2Cl^-$$
$$Cl_2 + 2H_2O \rightarrow 2HOCl + 2H^+ + 2e^-$$

Given that these half-reactions are the only ones of importance and that no Cl^- or HOCl was present before the Cl_2 was added, we obtain the electron balance by establishing an *electron reference level* (ERL) as the species that are either oxidized or reduced and with which the solution was prepared. In this example, the solution was prepared with H_2O and

Cl_2, but only Cl_2 is oxidized or reduced so it alone is the ERL. Each Cl^- formed has one electron more than the Cl at the ERL; each Cl in HOCl has one electron less than the ERL. Thus the electron balance is

$$[HOCl] = [Cl^-]$$

In the oxidation of Fe^{2+} by $Cr_2O_7^{2-}$,

$$6Fe^{2+} \rightarrow 6Fe^{3+} + 6e^-$$
$$Cr_2O_7^{2-} + 14H^+ + 6e^- \rightarrow 2Cr^{3+} + 7H_2O$$

The ERL is Fe^{2+} and $Cr_2O_7^{2-}$. One electron is produced for each Fe^{3+} formed and three electrons are used in the formation of each Cr^{3+}. Since there are no other sources or sinks of electrons and given that no Fe^{3+} or Cr^{3+} was in the solution prior to adding $Cr_2O_7^{2-}$ and Fe^{2+}, the electron balance yields

$$[Fe^{3+}] = 3[Cr^{3+}]$$

Example 7-9

To 100 ml of 0.01 M $K_2Cr_2O_7$ in 1 M H_2SO_4, add 20 ml of 0.1 M Fe^{2+} (in the form of ferrous ammonium sulfate). Calculate $C_{T,Fe(II)}$, $C_{T,Fe(III)}$, C_{T,Cr_2O_7}, and $C_{T,Cr(III)}$. The formal potentials are

$$Fe^{3+} + e^- \rightleftharpoons Fe^{2+}; \qquad E^{\circ\prime} + 0.68 \text{ volts}$$

$$Cr_2O_7^{2-} + 14H^+ + 6e^- \rightleftharpoons 2Cr^{3+} + 7H_2O; \qquad E^{\circ\prime} = 1.33 \text{ volts}$$

Solution

There are five unknowns (E_H', $C_{T,Fe(II)}$, $C_{T,Fe(III)}$, $C_{T,Cr(III)}$, and C_{T,Cr_2O_7}) and thus five equations are needed. Because 1 M H_2SO_4 is used, $[H^+]$ can be considered constant and it is thus not an unknown.

For this problem we can write the following equations.

Mass Balances

(1) $C_{T,Cr} = C_{T,Cr(III)} + 2C_{T,Cr_2O_7} = \dfrac{\text{number of moles of Cr}}{\text{liter}}$

$\qquad = \dfrac{(2)(0.01 \text{ mole/liter})(0.1 \text{ liter})}{(0.1 + 0.02)\text{liter}}$

$\qquad = 0.0167 \, M$

(2) $C_{T,Fe} = C_{T,Fe(II)} + C_{T,Fe(III)} = \dfrac{(0.1 \text{ mole/liter})(0.02 \text{ liter})}{(0.1 + 0.02)\text{liter}}$

$\qquad = 0.0167 \, M$

Equilibrium Relationships

$$E_H' = E^{\circ\prime} - \frac{RT}{nF} \ln Q$$

(3) $E'_H = 0.68 - \dfrac{(1.987 \times 10^{-3})(298)}{(1)(23.06)} \ln \dfrac{C_{T,\text{Fe(II)}}}{C_{T,\text{Fe(III)}}}$

(4) $E'_H = 1.33 - \dfrac{(1.987 \times 10^{-3})(298)}{(6)(23.06)} \ln \dfrac{(C_{T,\text{Cr(III)}})^2}{C_{T,\text{Cr}_2\text{O}_7}[H^+]^{14}}$

Electron Balance

(5) $C_{T,\text{Fe(III)}} = 3C_{T,\text{Cr(III)}}$

Because $\text{Cr}_2\text{O}_7{}^{2-}$ is a very strong oxidizing agent, we can assume that the Fe^{2+} which is added reacts completely. Thus

$$C_{T,\text{Cr(III)}} = \tfrac{1}{3}C_{T,\text{Fe}} = \tfrac{1}{3}(0.0167) = 5.57 \times 10^{-3}\,M$$

Then from (1),

$$2C_{T,\text{Cr}_2\text{O}_7} = 0.0167 - 5.57 \times 10^{-3} = 1.11 \times 10^{-2}\,M$$

or

$$C_{T,\text{Cr}_2\text{O}_7} = 5.56 \times 10^{-3}\,M$$

and combining these values with equations (1) and (4) yields

$$E'_H = 1.33 - 0.00427 \ln \left[\frac{(5.57 \times 10^{-3})^2}{(2)^{14}(5.56 \times 10^{-3})} \right]$$

$$= 1.39\,\text{volts}$$

At equilibrium, E'_H for the $\text{Cr}_2\text{O}_7{}^{2-}/Cr^{3+}$ couple is equal to that for the Fe^{3+}/Fe^{2+} couple. Thus from (3)

$$1.39 = 0.68 - 0.0257 \ln \frac{C_{T,\text{Fe(II)}}}{C_{T,\text{Fe(III)}}}$$

$$\ln \frac{C_{T,\text{Fe(II)}}}{C_{T,\text{Fe(III)}}} = -28$$

$$\frac{C_{T,\text{Fe(II)}}}{C_{T,\text{Fe(III)}}} = 10^{-12.2}$$

and when $C_{T,\text{Fe(III)}} = 0.0167\,M$,

$$C_{T,\text{Fe(II)}} = 1.05 \times 10^{-14}$$

Substituting these values into the mass balances, we see that our assumption of complete reaction was satisfactory.

7.4.   ELECTRON ACTIVITY AND $p\epsilon$

The concept of electron activity is used in the description of oxidation-reduction equilibria and especially in the solution of problems that involve both redox and other equilibria such as acid-base, complexation, and so forth. In addition, this concept provides a useful basis for the graphical representation of complicated redox equilibria. The approach is extremely useful when working with redox equilibria in natural waters.

However, when problems are concerned with topics such as analytical methods and corrosion that involve electrochemical cells, it is often more convenient to work with potentials and the Nernst equation because this can be related directly to voltage measurements. Whatever the system or situation it is beneficial to know how to "do it both ways." We will instruct the reader how to readily use both methods interchangeably.

The negative logarithm of electron activity, $p\epsilon = -\log \{e^-\}$, is analogous to pH, the negative logarithm of the hydrogen ion activity. Whereas pH is a measure of the *availability* of protons in solution (even though no free protons exist in solution), $p\epsilon$ is a measure of the availability of electrons in solution (even though no free electrons exist in solution). The equations for the development of pH and $p\epsilon$ can be compared,

$$\text{pH}$$

$$\begin{array}{lll} (1) & HA \rightleftharpoons H^+ + A^-; & K_1 \\ (2) & H^+ + H_2O \rightleftharpoons H_3O^+; & K_2 \\ \hline (3) & HA + H_2O \rightleftharpoons H_3O^+ + A^-; & K_3 \end{array}$$

By convention, $K_2 = 1$. Thus $K_3 = K_1 K_2 = K_1$ and $K_1 = \{H^+\}\{A^-\}/\{HA\}$ or

$$pH = pK_1 + \log \frac{\{A^-\}}{\{HA\}}$$

where $pH = -\log \{H^+\}$, and $pK_1 = -\log K_1$, *where K_1 is for the acid dissociation half-reaction* (with the proton on the right-hand side of the equation).

$$p\epsilon$$

$$\begin{array}{lll} (1) & M^{2+} + e^- \rightleftharpoons M^+; & K_1 \\ (2) & \tfrac{1}{2}H_{2(g)} \rightleftharpoons H^+ + e^-; & K_2 \\ \hline (3) & M^{2+} + \tfrac{1}{2}H_{2(g)} \rightleftharpoons M^+ + H^+; & K_3 \end{array}$$

By convention, $K_2 = 1$. Thus $K_3 = K_1 K_2 = K_1$ and

$$K_1 = \frac{\{M^+\}}{\{e^-\}\{M^{2+}\}}$$

or

$$p\epsilon = p\epsilon^\circ - \log \frac{\{M^+\}}{\{M^{2+}\}}$$

where $p\epsilon = -\log \{e^-\}$ and $p\epsilon^\circ = \log K_1$, *where K_1 is for the reduction half-reaction* (*with the electron on the left-hand side* of the equation).

For acid-base equilibria the proton exchange reaction between H_3O^+ and H_2O at standard conditions is assigned a free energy change of zero and an equilibrium constant of unity. It provides the datum to which all other acid base reactions are referenced. For redox equilibria, the reduction of H^+ to $\frac{1}{2}H_{2(g)}$ at standard conditions is assigned a free energy change of zero and an equilibrium constant of unity. It provides the reference point for defining $p\epsilon$.

Taking the reduction of ferric ion, Fe^{3+}, to ferrous ion, Fe^{2+}, as an example, we can write

		$E°$, volts	$K = \exp\left(\dfrac{nFE°}{RT}\right)$
(1)	$Fe^{3+} + e^- \rightarrow Fe^{2+}$	$+0.77$	$K_1 = 10^{13}$
(2)	$\frac{1}{2}H_{2(g)} \rightarrow H^+ + e^-$	0	$K_2 = 1$
(3)	$Fe^{3+} + \frac{1}{2}H_{2(g)} \rightarrow H^+ + Fe^{2+}$	$+0.77$	$K_3 = 10^{13}$

Writing the equilibrium constant for (1), we have

$$K_1 = \frac{\{Fe^{2+}\}}{\{Fe^{3+}\}\{e^-\}}$$

Taking logarithms, we obtain

$$\log K_1 = \log\frac{\{Fe^{2+}\}}{\{Fe^{3+}\}} - \log\{e^-\}$$

Setting $p\epsilon = -\log\{e^-\}$ and $p\epsilon° = \log K_1$, we find that

$$p\epsilon = p\epsilon° - \log\frac{\{Fe^{2+}\}}{\{Fe^{3+}\}} \tag{7-11}$$

When $\{Fe^{2+}\} = \{Fe^{3+}\}$,

$$p\epsilon = p\epsilon° = \log K,$$

For comparison, knowing that $\{H^+\} = \{H_{2(g)}\} = 1$, we can write for reaction (3)

$$\Delta G = \Delta G° + RT\ln\left(\frac{\{Fe^{2+}\}}{\{Fe^{3+}\}}\right) \tag{7-12}$$

and

$$E_H = E° - \left(\frac{RT}{nF}\right)\ln\left(\frac{\{Fe^{2+}\}}{\{Fe^{3+}\}}\right) \tag{7-13}$$

Rearranging this equation and substituting $n = 1$ yields

$$\left(\frac{F}{2.3RT}\right)E_H = \left(\frac{F}{2.3RT}\right)E° - \log\frac{\{Fe^{2+}\}}{\{Fe^{3+}\}}$$

Comparing this equation with Eq. 7-11 for 25°C, we find that

$$p\epsilon = \frac{FE_H}{(2.3RT)} = 16.9E_H \tag{7-14}$$

$$p\epsilon° = \frac{FE°}{(2.3RT)} = 16.9E° \tag{7-15}$$

where $E°$ is the standard potential for the reduction half-reaction. Several useful pϵ relationships are summarized in Table 7-2.

For any half-reaction,

$$ox + ne^- \rightleftharpoons red$$

a similar development shows,

$$p\epsilon = p\epsilon° - \frac{1}{n}\log\left(\frac{\{red\}}{\{ox\}}\right) \tag{7-16}$$

where

$$p\epsilon° = \frac{1}{n}\log K \tag{7-17}$$

where K is the equilibrium constant for the reduction half-reaction. Again

$$p\epsilon = \frac{FE_H}{(2.3RT)}(= 16.9E_H \text{ at } 25°C) \tag{7-18}$$

and

$$p\epsilon° = \frac{FE°}{(2.3RT)}(= 16.9E° \text{ at } 25°C) \tag{7-19}$$

where E_H and $E°$ apply to the reduction half-reaction. Also, since $\Delta G = -nFE$,

$$p\epsilon = \frac{-\Delta G}{2.3nRT} \tag{7-20}$$

and

$$p\epsilon° = \frac{-\Delta G°}{2.3nRT} \tag{7-21}$$

From these interrelationships we can deduce that $p\epsilon°$ is proportional to the free energy change accompanying the transfer of 1 mole of electrons from a reducing agent at unit activity to H^+ at unit activity. Several pϵ relationships for use when more than one electron is exchanged are also summarized in Table 7-2.

Note that pϵ only applies to a half-reaction; thus its value cannot be used to indicate position of that half-reaction with respect to equilibrium, contrary to the value of E for an overall reaction. However, if two half-reactions in the same solution have equal values of pϵ, those half-reactions are in equilibrium.

The application of pϵ is illustrated in the following example.

Example 7-10

The pH of a stabilization pond effluent is measured in the field and found to be 7.8 at 25°C. A sample of the effluent is taken back to the laboratory by an

TABLE 7-2   pϵ Relationships

For a one electron exchange reaction written as a reduction reaction	For a multiple electron exchange reaction written as a reduction reaction
$M^{2+} + e^- \rightarrow M^+;\ E^\circ$	$ox + ne^- \rightarrow red;\ E^\circ$
$p\epsilon = p\epsilon^\circ - \log \dfrac{\{M^+\}}{\{M^{2+}\}}$	$p\epsilon = p\epsilon^\circ - \dfrac{1}{n}\log \dfrac{\{red\}}{\{ox\}}$
where	where
$p\epsilon = -\log\{e^-\}$ $p\epsilon^\circ = \log K$	$p\epsilon = -\log\{e^-\}$ $p\epsilon^\circ = \dfrac{1}{n}\log K$
and	and
$\log K = \dfrac{FE^\circ}{2.3RT}$	$\dfrac{1}{n}\log K = \dfrac{FE^\circ}{2.3RT}$
Also	Also
$p\epsilon = \dfrac{FE_H}{2.3RT}$ $\quad = 16.9E_H$ at 25°C $p\epsilon^\circ = \dfrac{FE^\circ}{2.3RT}$ $\quad = 16.9E^\circ$ at 25°C $p\epsilon = \dfrac{-\Delta G}{2.3RT}$ $p\epsilon^\circ = \dfrac{-\Delta G^\circ}{2.3RT}$	$p\epsilon = \dfrac{FE_H}{2.3RT}$ $\quad = 16.9E_H$ at 25°C $p\epsilon^\circ = \dfrac{FE^\circ}{2.3RT}$ $\quad = 16.9E^\circ$ at 25°C $p\epsilon = \dfrac{-\Delta G}{2.3nRT}$ $p\epsilon^\circ = \dfrac{-\Delta G^\circ}{2.3nRT}$

inexperienced graduate student who had the misfortune to place the sample container in full sunlight in the back of a pick-up truck. Photosynthesis occurred in the sample. On arrival at the laboratory the sample pH was found to be 10.2; the atmosphere above the sample was found to contain 40 percent oxygen and the temperature was 25°C. Assuming that the oxygen reduction half-reaction

$$4e^- + 4H^+ + O_{2(g)} \rightleftharpoons 2H_2O$$

governed the redox conditions in the sample, what was the change in (1) pϵ, and (2) redox potential (E_H, volts) of the sample?

Solution

We will solve this problem first using the pϵ method and then the Nernst equation.

1. The oxygen reduction half-reaction governs the redox conditions; from Table 7-1, $p\epsilon^\circ = +20.8$. From Eq. (7-16),

$$p\epsilon = p\epsilon^\circ - \frac{1}{n}\log \frac{1}{P_{O_{2(g)}}[H^+]^4}$$

Let us asume that the stabilization pond effluent is in equilibrium with the normal atmosphere that contains 21 percent oxygen. Knowing that the pH is 7.8, we find that

$$p\epsilon = 20.8 - \tfrac{1}{4}\log\left[\frac{1}{(0.21)(10^{-7.8})^4}\right] = 12.83$$

When the sample is brought back to the laboratory, $P_{O_2} = 0.4$ atm and pH $= 10.2$.

$$p\epsilon = 20.8 - \tfrac{1}{4}\log\frac{1}{P_{O_2}[H^+]^4}$$

$$= 20.8 - \tfrac{1}{4}\log\frac{1}{0.4 \times (10^{-10.2})^4}$$

$$= 10.5$$

The decrease in $p\epsilon = 12.83 - 10.5 = 2.33$.

2. For the oxygen reduction half-reaction, applying Eq. 7-8, we obtain

$$E_H = E^\circ - \left(\frac{0.059}{4}\right)\log\left\{\frac{1}{P_{O_2}[H^+]^4}\right\}$$

From Table 7-1, $E^\circ = +1.23$ volts.
  For the field sample, pH $= 7.8$ and $P_{O_2} = 0.21$ atm.

$$E_H = 1.23 - \left(\frac{0.059}{4}\right)\log\left(\frac{1}{0.21 \times (10^{-7.8})^4}\right)$$

$$= +0.76 \text{ volt}$$

For the sample in the laboratory, pH $= 10.2$ and $P_{O_2} = 0.4$ atm

$$E_H = 1.23 - \left(\frac{0.059}{4}\right)\log\frac{1}{0.4 \times (10^{-10.2})^4}$$

$$= +0.62 \text{ volt}$$

The decrease in $E_H = 0.76 - 0.62 = 0.14$ volt.

7.5. GRAPHICAL REPRESENTATION OF REDOX EQUILIBRIA

Graphical presentation of redox equilibria, like the graphical treatment of acid-base, complexation, and precipitation equilibria is helpful in understanding complicated problems and in obtaining approximate solutions to equilibrium questions. For redox systems in natural waters the equilibrium condition is truly a boundary condition. In many cases, natural systems are not at equilibrium from a redox standpoint. The diagrams usually present an idea of what is possible, not necessarily of the existing or imminent situation. The graphical presentations of redox equilibria are seldom simple because redox reactions usually involve changes in solution composition other than electron transfer. Usually, pH changes must be considered and often complexation and solubility equilibria must be superimposed on redox equilibrium diagrams to obtain a realistic picture of the situation.

In this section we will first present the development of a simple diagram (the so-called $p\epsilon$-pC or E_H-pC diagram) for a reaction that involves only electron transfer in a homogeneous system. Next we will consider the predominance area diagrams for presenting combined redox and acid-base equilibria. Finally, we will add a third dimension and consider systems with multiple phases using the $p\epsilon$-pH predominance area diagram.

7.5.1.   The $p\epsilon$-pC Diagrams

Let us first examine an aqueous solution at 25°C in which the half-reaction between Fe^{3+} and Fe^{2+} governs the redox conditions. The solution is maintained at pH 2 so that Fe^{3+} and Fe^{2+} are the major ferric and ferrous iron species, respectively, and the formation of hydrolysis products is negligible. There is a total analytical concentration of iron of $C_{T,Fe} = 10^{-4}$ M.

We proceed with the construction of this diagram using methods similar to those employed for preparing acid-base equilibrium diagrams. Although we use $p\epsilon$ in our development of the diagram (Fig. 7-3), we can show E_H versus pC too, since $p\epsilon$ has a fixed relationship to E, for example, at 25°C, $p\epsilon = 16.9 E_H$ (Eq. 7-19).

Equilibrium Relationship (from Table 7-1)

$$Fe^{3+} + e^- \rightleftharpoons Fe^{2+}; \quad E° = +0.77 \text{ volt}, p\epsilon° = 13.0$$

From Eq. 7-16, neglecting ionic strength effects,

$$p\epsilon = p\epsilon° - \log \frac{[Fe^{2+}]}{[Fe^{3+}]}$$

$$p\epsilon = 13 - \log \frac{[Fe^{2+}]}{[Fe^{3+}]} \tag{7-22}$$

Mass Balance

$$C_{T,Fe} = [Fe^{2+}] + [Fe^{3+}] = 10^{-4} \tag{7-23}$$

Like the pC-pH diagrams, the first equation we plot is the mass balance, which in logarithmic form is

$$-\log C_{T,Fe} = 4 \tag{7-24}$$

Equation 7-24 is not a function of $p\epsilon$ so it plots as a horizontal line at pC = 4 (line 1, Fig. 7-3).

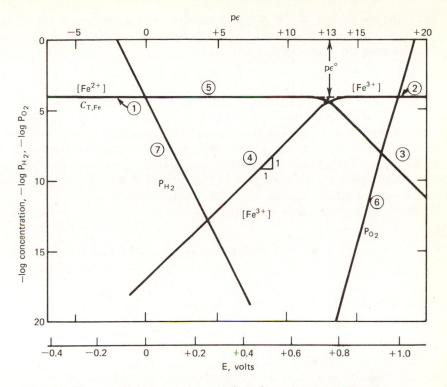

Fig. 7-3. The $p\epsilon$-pC diagram for a Fe^{2+}/Fe^{3+} system at 25°C, pH = 2.

Now we will develop the $-\log$ concentration versus $p\epsilon$ lines for Fe^{3+} and Fe^{2+}. In logarithmic form Eq. 7-22 becomes

$$p\epsilon - 13 = \log [Fe^{3+}] - \log [Fe^{2+}] \tag{7-25}$$

When $p\epsilon \gg p\epsilon^\circ\ (= 13)$, Eq. 7-25 indicates that $\log [Fe^{3+}] \gg \log [Fe^{2+}]$. For example, when $p\epsilon = +15$, Eq. 7-25 becomes

$$15 - 13 = \log [Fe^{3+}] - \log [Fe^{2+}] = 2$$

Thus at $p\epsilon = +15$, $[Fe^{3+}] = 100[Fe^{2+}]$ and therefore $[Fe^{2+}]$ is negligible compared to $[Fe^{3+}]$. From the mass balance equation,

$$C_{T,Fe} = [Fe^{2+}] + [Fe^{3+}]$$

with

$$[Fe^{3+}] \gg [Fe^{2+}]$$

$$C_{T,Fe} = [Fe^{3+}] = 10^{-4}$$

$$-\log [Fe^{3+}] = 4 \tag{7-26}$$

Thus in the region of the $p\epsilon$-pC diagram where $p\epsilon \gg p\epsilon°$, the $-\log$ $[Fe^{3+}]$ line is a horizontal line at $pC = 4$ (line 2, Fig. 7-3).

Substituting Eq. 7-26 in Eq. 7-25 yields

$$p\epsilon = -\log [Fe^{2+}] + 9 \qquad (7\text{-}27)$$

which again is valid for the region $p\epsilon \gg p\epsilon°$. Equation 7-27 has a slope of $+1$ and passes through the point $(-\log [Fe^{2+}] = 4, p\epsilon = 13)$. This line is plotted as line 3 in Fig. 7-3. It is dashed in the region where $p\epsilon$ is close to $p\epsilon°$ because here the assumption that $p\epsilon \gg p\epsilon°$ does not hold.

By a similar type of reasoning, a line for $-\log [Fe^{3+}]$ versus $p\epsilon$ and $-\log [Fe^{2+}]$ versus $p\epsilon$ can be developed in the region where $p\epsilon \ll p\epsilon°$. These are plotted in Fig. 7-3 as lines 4 and 5, respectively.

Equation 7-25 shows that when $p\epsilon = p\epsilon° = +13$, $\log [Fe^{2+}] = \log [Fe^{3+}]$, and when

$$C_{T,Fe} = [Fe^{2+}] + [Fe^{3+}] = 10^{-4}$$

we find

$$[Fe^{2+}] = [Fe^{3+}] = 5 \times 10^{-5}$$

or

$$-\log [Fe^{2+}] = -\log [Fe^{3+}] = 4.3$$

After plotting these points, we can complete the diagram in the vicinity of $p\epsilon = p\epsilon°$ as shown. The straight line portions of each line are connected with the curved lines in the region approximately one $p\epsilon$ unit from $p\epsilon°$, just as was done for the pC-pH diagrams for acid base systems. However, this latter step depends on the slopes of the concentration lines.

It is interesting to plot on the same diagram two equations that show the redox stability limits of water. Since the reactions of solutes we wish to examine take place in aqueous solution, we should know whether the redox reactions we are describing are possible while the solvent, water, is stable. At high $p\epsilon$, or under highly oxidizing conditions, H_2O can be converted to oxygen. The half-reaction,

$$4e^- + 4H^+ + O_{2(g)} \rightleftharpoons 2H_2O; \qquad E° = +1.23 \text{ volts}$$

Knowing $p\epsilon° = 16.9 \, E°$ at 25°C, Eq. 7-16,

$$p\epsilon = p\epsilon° - \frac{1}{n} \log \frac{1}{P_{O_2}[H^+]^4}$$

yields

$$p\epsilon = 20.8 + \tfrac{1}{4} \log P_{O_2} - pH \qquad (7\text{-}28)$$

For $pH = 2$,

$$p\epsilon = 18.8 + \tfrac{1}{4} \log P_{O_2} \qquad (7\text{-}29)$$

This equation plots as line 6 in Fig. 7-3 with $-\log P_{O_2}$ on the $-\log$ concentration scale. If the $p\epsilon$ is controlled by the $O_{2(g)}/H_2O$ half-reaction, this line shows that a solution in equilibrium with 0.21 atm of oxygen at pH 2 has a $p\epsilon = +18.6$. (*Note:* Two half-reactions taking place in the same solution may not be in equilibrium with each other. If not, the $p\epsilon$ calculated using one half-reaction will differ from the $p\epsilon$ calculated from the other half-reaction.)

Next the reduction of water to hydrogen gas must be considered as the lower bound of the stability field of water.

$$2H_2O + 2e^- \rightleftharpoons H_{2(g)} + 2OH^-; \qquad E° = -0.83 \text{ volt}$$

From Eq. 7-16,

$$p\epsilon = p\epsilon° - \tfrac{1}{2}\log P_{H_2} \times [OH^-]^2$$

Since $p\epsilon° = 16.9E° = -14$ at 25°C,

$$p\epsilon = -14 - \tfrac{1}{2}\log P_{H_2} + pOH \tag{7-30}$$

When pH = 2, pOH = 12 and

$$p\epsilon = -2 - \tfrac{1}{2}\log P_{H_2} \tag{7-31}$$

which plots as shown, line 7, Fig. 3.

The $p\epsilon$-pC diagram conveys the idea that both Fe^{2+} and Fe^{3+} have regions where they can be the predominant species in water at pH 2. However, if Fe^{2+} is to predominate and be in equilibrium with the O_2/H_2O couple, the solution must essentially be stripped of dissolved oxygen. For example, at $p\epsilon$ values of less than 13, Fe^{2+} will predominate when the oxygen partial pressure in equilibrium with the water couple is less than 10^{-22} atm.

Example 7-11

1. What is the predominant iron species in oxygenated water at pH 2 and 25°C? Assume a normal earth atmosphere and that the iron species are in equilibrium with the O_2/H_2O couple.

2. What are the predicted equilibrium concentrations of Fe^{2+} and Fe^{3+} under these conditions for a solution containing 10^{-4} M total iron?

3. Does $p\epsilon$ change significantly when the pH is changed to 6?

4. At pH 2 what partial pressure of oxygen would be in equilibrium with a solution containing 10^6 times as much Fe^{2+} as Fe^{3+}?

Neglect solids and hydroxo-Fe complexes.

Solution

1. From Fig. 7-3, a solution in equilibrium with $P_{O_2} = 0.21$ atm has $p\epsilon = +18.6$. At this $p\epsilon$, Fe^{3+} predominates.

2. At $p\epsilon = +18.6$, $[Fe^{3+}] = 10^{-4} M$ and $[Fe^{2+}] = 10^{-9.6} M$.

3. As the pH is changed to 6, the $p\epsilon$ value of the O_2/H_2O couple will change because $[H^+]$ is involved in this equilibrium. From Eq. 7-28

$$p\epsilon = p\epsilon° + \tfrac{1}{4} \log P_{O_2} - pH$$
$$= 20.8 - 0.17 - 6$$
$$= +14.6$$

From Fig. 7-3, at $p\epsilon = +14.6$, $[Fe^{3+}] = 10^{-4}$ and $[Fe^{2+}] = 10^{-5.6}$.

Ferric iron is still the major species and the equilibrium concentrations have not changed significantly from the standpoint of the iron couple. However, as we shall see later in our discussion of iron chemistry, the kinetics of ferrous iron oxidation by oxygen plays a highly significant role in determining the form of iron found in oxygenated solutions of different pH values. Just to whet the reader's appetite, we can state that although the equilibrium calculations show Fe^{3+} as the predominant species at both pH 2 and pH 6, experiments show that, at pH 2, Fe^{2+} is essentially not oxidized in oxygenated solutions while at pH 6 its oxidation to Fe^{3+} is rapid.

4. From Fig. 7-3, $[Fe^{2+}] = 10^6 [Fe^{3+}]$ when $p\epsilon = +7.0$. Since we assume that the oxygen couple is in equilibrium with the iron, this will also be the $p\epsilon$ value for the O_2/H_2O couple. Because the graph for P_{O_2} versus $p\epsilon$ does not extend to $p\epsilon = 7$, we use Eq. 7-29 to calculate P_{O_2}

$$p\epsilon = +7 = +18.8 + \tfrac{1}{4} \log P_{O_2}$$
$$P_{O_2} = 10^{-47} \text{ atm}$$

Thus Fe^{2+} is only stable at minutely small partial pressures of oxygen and, based on equilibrium calculations, should be oxidized to Fe^{3+} when there is any measurable quantity of dissolved oxygen in water.

7.5.2. The $p\epsilon$-pH Predominance Area Diagram

The construction of a $p\epsilon$-pH or E_H-pH diagram is the next step in complexity in the graphical representation of redox equilibria. In such diagrams we establish areas of predominance in a $p\epsilon$-pH coordinate system for various species involved in redox, acid-base, precipitation, and complexation equilibria.

We will illustrate the construction of a $p\epsilon$-pH diagram for the aqueous chlorine system. For the construction of such diagrams we must first state the ground rules for drawing boundaries between any two species. There are two basic rules:

1. Boundaries will only be drawn between the two major species under a given set of conditions.

2. A boundary will be drawn between two species at the place where the concentration of the two species is equal. Thus, on one side of the boundary, one of the species predominates in concentration

and, on the other side, the second species predominates. Combining these ground rules allows us to state that boundaries in a $p\epsilon$-pH diagram are drawn at places of significant equality.

The first thing to consider is the stability domain of water. For the half-reaction,

$$4e^- + 4H^+ + O_{2(g)} \rightleftharpoons 2H_2O$$

we have from Eq. 7-28

$$p\epsilon = +20.8 + \tfrac{1}{4}\log P_{O_2} - pH$$

At $P_{O_2} = 0.21$ atm

$$p\epsilon = 20.6 - pH \qquad (7\text{-}32)$$

In the $p\epsilon$-pH diagram this plots as a straight line, line 1 in Fig. 7-4. At a given pH, if $p\epsilon$ lies above line 1, H_2O is converted to $O_{2(g)}$. If $p\epsilon$ lies below the line, H_2O is stable.

For the reduction of water,

$$2H_2O + 2e^- \rightleftharpoons H_{2(g)} + 2OH^-$$

From Eq. 7-30,

$$p\epsilon = -14 - \tfrac{1}{2}\log P_{H_2} + pOH$$

and when $P_{H_2} = 1$,

$$p\epsilon = -14 + pOH \qquad (7\text{-}33)$$

This plots as a straight line in the $p\epsilon$-pH diagram (line 2, Fig. 7-4). At a given pH and a $H_{2(g)}$ pressure of 1 atm, if $p\epsilon$ is above this line H_2O is stable, while if it is below this line, H_2O is reduced to $H_{2(g)}$. As P_{H_2} decreases, the H_2O/H_2 line is raised thereby making smaller the area of H_2O stability.

Next we must list all the species involved in the reactions of interest. For the aqueous chlorine system these are

$$Cl_{2(aq)}, HOCl, OCl^-, \text{ and } Cl^-$$

Then we set up equations in terms of pH and $p\epsilon$ relating each species to each of the other species. The necessary equilibria can either be obtained from Table 7-1, or from combining the equilibria in this table. In writing the redox equilibria, we must remember to follow the golden rule of *always writing the reactions as reductions*. We have four basic equations to work with.

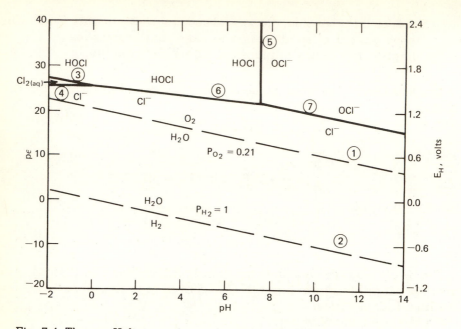

Fig. 7-4. The $p\epsilon$-pH diagram for aqueous chlorine; 25°C, $C_{T,Cl} = 1 \times 10^{-4} M$.

The formation of aqueous chlorine from hypochlorous acid:

(1) $e^- + H^+ + HOCl \rightleftharpoons \frac{1}{2}Cl_{2(aq)} + H_2O$; $p\epsilon° = +27.0$

The formation of chloride ion by the reduction of aqueous chlorine:

(2) $2e^- + Cl_{2(aq)} \rightleftharpoons 2Cl^-$; $p\epsilon° = +23.5$

The dissociation of hypochlorous acid:

(3) $HOCl \rightleftharpoons H^+ + OCl^-$; $pK_a = 7.5$

The value for the total concentration of chlorine-containing species, $C_{T,Cl}$, in this example:

(4) $C_{T,Cl} = 10^{-4} M = 2[Cl_{2(aq)}] + [HOCl] + [OCl^-] + [Cl^-]$

With these equations we must find relationships in terms of only $p\epsilon$ and pH between

$Cl_{2(aq)}$ and $HOCl$

$Cl_{2(aq)}$ and Cl^-

$HOCl$ and OCl^-

HOCl and Cl^-

OCl^- and Cl^-

$Cl_{2(aq)}$ and OCl^-

Let us first work with the species in the higher oxidation state (i.e., the better oxidants) and thus develop the portion of the diagram at high $p\epsilon$.

$Cl_{2(aq)}$ and HOCl
    From equation (1) we find

$$p\epsilon = p\epsilon^\circ - \log \frac{[Cl_{2(aq)}]^{\frac{1}{2}}}{[HOCl][H^+]}$$

or

$$p\epsilon = +27.0 - pH - \log \frac{[Cl_{2(aq)}]^{\frac{1}{2}}}{[HOCl]}$$

At the boundary, when $[HOCl] = [Cl_{2(aq)}]$, the ratio $[Cl_{2(aq)}]^{\frac{1}{2}}/[HOCl]$ is a function of $C_{T,Cl}$. Now, in this region,

$$2[Cl_{2(aq)}] + [HOCl] \gg [OCl^-] + [Cl^-]$$

Substituting in the mass balance, equation (4), yields

$$C_{T,Cl} = 2[Cl_{2(aq)}] + [HOCl] = 10^{-4}$$

Now at the boundary, $[HOCl] = [Cl_{2(aq)}]$; therefore,

$$3[Cl_{2(aq)}] = 10^{-4} \quad \text{and} \quad [Cl_{2(aq)}] = 3.3 \times 10^{-5}$$

Evaluating the logarithmic term, we find

$$\frac{[Cl_{2(aq)}]^{\frac{1}{2}}}{[HOCl]} = \frac{[Cl_{2(aq)}]^{\frac{1}{2}}}{C_{T,Cl} - 2[Cl_{2(aq)}]} = \frac{(3.3 \times 10^{-5})^{\frac{1}{2}}}{3.3 \times 10^{-5}} = 174$$

Substituting this value in the $p\epsilon$ equation, we find

$$p\epsilon = +27.0 - pH - \log 174$$
$$p\epsilon = 24.7 - pH \qquad\qquad (7\text{-}34)$$

This equation plots as line 3 in Fig. 7-4.

$Cl_{2(aq)}$ and Cl^-
    From equation (2) we obtain

$$p\epsilon = p\epsilon^\circ - \frac{1}{n} \log \frac{[Cl^-]^2}{[Cl_{2(aq)}]}$$

In this region of the diagram, $2[Cl_{2(aq)}] + [Cl^-] \gg [HOCl] + [OCl^-]$. Substituting in equation (4), we find that

$$C_{T,Cl} = 10^{-4} \cong 2[Cl_2] + [Cl^-],$$

and since

$$[Cl_{2(aq)}] = [Cl^-]$$
$$[Cl_{2(aq)}] = 3.3 \times 10^{-5}$$
$$p\epsilon = 23.5 - \tfrac{1}{2}\log 3.3 \times 10^{-5}$$
$$p\epsilon = 25.7 \tag{7-35}$$

This equation plots as a horizontal line (line 4) in Fig. 7-4.

HOCl and OCl$^-$

From equation (3), when $[HOCl] = [OCl^-]$,

$$\log \frac{[OCl^-]}{[HOCl]} = 0$$

$$pH = 7.5 \tag{7-36}$$

This line (line 5 in Fig. 7-4) is a vertical line at pH = 7.5.

To find the other chlorine species relationships we must use combinations of equations (1) through (3).

HOCl and Cl$^-$

Adding equations (1) and (2), we obtain

$$\Delta G^\circ = -nFE^\circ$$

(1) $HOCl + H^+ + e^- \to \tfrac{1}{2}Cl_{2(aq)} + H_2O$ $-F(+1.60)$
(2) $e^- + \tfrac{1}{2}Cl_{2(aq)} \to Cl^-$ $-F(+1.39)$

---

(5) $HOCl + H^+ + 2e^- \to Cl^- + H_2O$ $-2FE^\circ = -F(+1.60 + 1.39)$
      $E^\circ = +1.50$

and from Eq. 7-15, $p\epsilon^\circ = 16.9\,E^\circ = +25.4$. Thus

$$p\epsilon = p\epsilon^\circ - \tfrac{1}{2}\log \frac{[Cl^-]}{[HOCl][H^+]}$$

$$p\epsilon = +25.4 - \tfrac{1}{2}pH - \tfrac{1}{2}\log([Cl^-]/[HOCl])$$

When $[HOCl] = [Cl^-]$,

$$p\epsilon = +25.4 - \tfrac{1}{2}pH \tag{7-37}$$

This equation is plotted as line 6 in Fig. 7-4.

OCl$^-$ and Cl$^-$

To relate OCl$^-$ and Cl$^-$, we subtract equation (3) from equation (5),

$$\Delta G^\circ = -nFE^\circ$$

(5) $HOCl + H^+ + 2e^- \rightleftharpoons Cl^- + H_2O;$ $-2F\,(+1.50) = 69.2$

−(3) $OCl^- + H^+ \rightleftharpoons HOCl;$ $-RT\ln 10^{7.5} = 10.2$

(6) $OCl^- + 2H^+ + 2e^- \rightleftharpoons Cl^- + H_2O;$ -79.4

$$E^\circ = -79.4/2F = +1.72$$

$$p\epsilon^\circ = 16.9E^\circ = +29.1$$

For equation (6) we have

$$p\epsilon = p\epsilon^\circ - \tfrac{1}{2}\log\frac{[Cl^-]}{[OCl^-][H^+]^2}$$

$$= 29.1 - pH - \tfrac{1}{2}\log\frac{[Cl^-]}{[OCl^-]}$$

When $[Cl^-] = [OCl^-]$,

$$p\epsilon = 29.1 - pH \qquad\qquad (7\text{-}38)$$

This equation appears as line 7 in Fig. 7-4.

We can now use reasoning to finalize the drawing of the diagram. For example, we have not yet dealt with the $OCl^-/Cl_{2(aq)}$ relationship. This line will fall somewhere between the OCl^- and Cl_2 regions in Fig. 7-4. However, we have already shown that to the right of line 3, and below line 4, $Cl_{2(aq)}$ is not a *predominant* species. Thus we can save ourselves the time of establishing this line, since we are only interested in boundaries between predominant species. We can also ask why line 5 terminates at the intersection of lines 6 and 7. Below line 7, Cl^- predominates over OCl^-; below line 6, Cl^- predominates over $HOCl$. Again we are only interested in boundaries between predominant species and we must therefore terminate line 5 as shown.[5]

From the diagram we can draw the following conclusions for a $10^{-4}\,M$ $C_{T,Cl}$ solution.

1. $Cl_{2(aq)}$ as a predominant species only exists at low pH values (below pH 0). As $C_{T,Cl}$ increases, similar calculations show the pH value at which $Cl_{2(aq)}$ predominates increases.

2. At higher pH values, $Cl_{2(aq)}$ disproportionates into $HOCl$ and Cl^-.

3. Chloride ion is the stable chlorine-containing species in the $p\epsilon$-pH range of natural waters.

[5] Systems much more complex than aqueous chlorine are time-consuming to draw. Other advanced texts in this area (e.g., W. Stumm and J. J. Morgan, *Aquatic Chemistry*, Wiley-Interscience, 1970) illustrate procedures whereby such diagrams can be drawn.

4. $Cl_{2(aq)}$, HOCl, and OCl^- are stronger oxidants than O_2 in accordance with the reaction,

$$2HOCl \rightleftharpoons 2Cl^- + 2H^+ + O_{2(aq)}$$

which is the sum of the two half-reactions,

$$2H_2O \rightleftharpoons 4H^+ + 4e^- + O_{2(aq)}$$

$$2HOCl + 2H^+ + 4e^- \rightleftharpoons 2Cl^- + 2H_2O$$

This reaction is basically the oxidation of water to oxygen by chlorine, which itself is reduced to chloride ion. The reaction proceeds only when catalyzed, for example, by ultraviolet light. It is the reaction that accounts for the "decay" of chlorine solutions that are exposed to sunlight.

5. Only $C_{T,Cl} = 10^{-4}$ was used in these calculations, but the use of other values of $C_{T,Cl}$ shows that the boundaries in Fig. 7-4 vary little with $C_{T,Cl}$. Thus the diagram in Fig. 7-4 is generally applicable at other total concentrations as well.

7.5.3.   Other Predominance Area Diagrams

In the previous section we used pH and $p\epsilon$ as the coordinates of the diagram. Frequently, the concentration of one species of a system can significantly affect the concentrations of other species that are present; in this case a diagram with the concentration of that species as one variable, together with pH as the other variable, can be most useful. To develop such a diagram, it is necessary to combine half-reactions to form overall reactions, thus eliminating $p\epsilon$, or E_H, as a variable.

For example, let us consider such a predominance area diagram for aqueous bromine species at 25° in which the diagram axes are pH and the negative log of the bromide concentration.

The overall reactions relating the species of interest are

1. $Br_{2(aq)} + H_2O \rightleftharpoons HOBr + H^+ + Br^-; \quad K_h = 6 \times 10^{-9}.$

2. $HOBr \rightleftharpoons OBr^- + H^+; \quad pK_a = 8.4.$

3. $Br_{2(aq)} + Br^- \rightleftharpoons Br_3^-; \quad K = 15.9.$

We will use a similar approach to that used for constructing $p\epsilon$-pH diagrams. Lines representing equality of concentration between significant species are sought in the form of equations containing $[Br^-]$ and pH — the two axes of the diagram. The species are

$$Br_{2(aq)}, \; Br_3^-, \; HOBr, \; OBr^-$$

Possible relationships are

HOBr and OBr⁻

HOBr and $Br_{2(aq)}$

HOBr and Br_3^-

OBr⁻ and Br^{3-}

OBr⁻ and $Br_{2(aq)}$

$Br_{2(aq)}$ and Br_3^-

HOBr and OBr⁻ are related by the K_a equation,

$$K_a = 10^{-8.4} = \frac{[H^+][OBr^-]}{[HOBr]}$$

When [OBr⁻] = [HOBr], this equation simplifies to

$$pH = 8.4$$

which plots as line 1 in Fig. 7-5.

To derive the relationship for HOBr and Br_3^- it is necessary to combine equations (1) and (3). In logarithmic form we have

$$-9.4 = 2 \log [Br^-] - pH + \log \frac{[HOBr]}{[Br_3^-]}$$

When [HOBr] = [Br₃⁻],

$$\log [Br^-] = -4.7 + \tfrac{1}{2}pH \tag{7-39}$$

This is plotted as line 2 in Fig. 7-5.

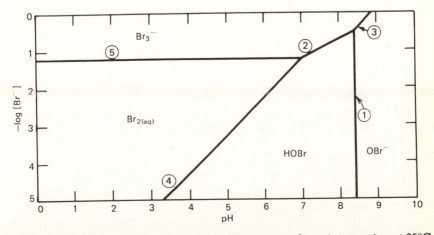

Fig. 7-5. Predominance area diagram for aqueous bromine species at 25°C.

Combining the relationship derived above for [HOBr] and [Br$_3^-$] with equation (2), we obtain

$$2 \log [Br^-] - 2pH + \log \frac{[OBr^-]}{[Br_3^-]} = -17.4$$

When [OBr$^-$] = [Br$_3^-$]

$$\log [Br^-] = pH - 8.9 \qquad (7\text{-}40)$$

This plots as line 3 in Fig. 7-5.

Combining equations (1) and (2), we obtain

$$\log [Br^-] - 2\,pH + \log \frac{[OBr^-]}{[Br_{2(aq)}]} = -16.6$$

When [OBr$^-$] = [Br$_{2(aq)}$]

$$\log [Br^-] = 2\,pH - 16.6 \qquad (7\text{-}41)$$

This equation plots in the region where [HOBr] predominates over [OBr$^-$] so that it does not represent a significant equality and is therefore not shown in Fig. 7-5.

From equation (1) we obtain the HOBr/Br$_2$ relationship,

$$\log (6 \times 10^{-9}) = \log [H^+] + \log [Br^-] + \log \frac{[HOBr]}{[Br_{2(aq)}]}$$

When [HOBr] = [Br$_{2(aq)}$]

$$\log [Br^-] = -8.22 + pH$$

which plots as line 4 in Fig. 7-5.

From equation (3) we obtain the relationship between Br$_{2(aq)}$ and Br$_3^-$,

$$1.2 = -\log [Br^-] + \log \frac{[Br_3^-]}{[Br_{2(aq)}]}$$

When [Br$_3^-$] = [Br$_{2(aq)}$]

$$\log [Br^-] = -1.2 \qquad (7\text{-}42)$$

This plots as a horizontal line at $-\log [Br^-] = 1.2$ (line 5 in Fig. 7-5).

Using the same type of reasoning that was employed in the construction of the pϵ-pH diagram in Fig. 7-4, we can map out the significant boundaries in Fig. 7-5.

The bromide concentration of most fresh natural waters is ≤ 1 mg/liter. Seawater contains about 70 mg Br$^-$/liter ($\sim 10^{-3}$ M). Therefore, from this diagram we can conclude that in most natural waters bromine will be present as the undissociated HOBr.

Comparing the diagram in Fig. 7-5 with those developed for the other halogens (Fig. 7-6) used as water disinfectants, chlorine and iodine, we

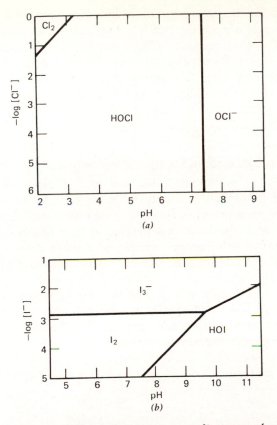

Fig. 7-6. Predominance area diagrams for aqueous chlorine and aqueous iodine at 25°C. (a) Aqueous chlorine; and (b) aqueous iodine. After D. G. Taylor and J. D. Johnson, chapter in *Chemistry of Water Supply, Treatment and Distribution*, A. J. Rubin, ed., Ann Arbor Science, 1974. Reprinted by permission of Ann Arbor Science Publishers, Inc., Mich.

see that bromine occupies an intermediate position as far as (1) the relative predominance of the trihalide anion (X_3^-), (2) the relative predominance of the diatomic aqueous halogen molecule [$X_{2(aq)}$], and (3) the relative predominance of the undissociated hypohalous acid (HOX).

The Cl_3^- species does not exist as a predominant species within the bounds of the coordinates given; Br_3^- is important at Br^- concentrations of greater than approximately 10^{-1} M and pH below about 7; I_3^- is important at lower [I^-] levels (> approximately 10^{-3} M) and up to higher pH levels. $Cl_{2(aq)}$ is only important in a limited pH range (below pH 3 and for Cl^- levels above 10^{-2} M) while $Br_{2(aq)}$ and $I_{2(aq)}$ progressively extend over a greater range of halide and pH values. HOCl and HOBr are the major halogen species in oxygenated water at natural water pH values. HOI may be a major species, but $I_{2(aq)}$ can also be present in significant quantities. OCl^- and possibly OBr^-, but never OI^-, can be important species under natural water conditions. These species distributions are extremely important from the standpoint of disinfection efficiency. For example, the ability of "chlorine" to kill coliform organisms is 80 to 100 times greater when the chlorine is present as HOCl than when it is present as OCl^-.

7.5.4.   The pε-pH Diagrams Incorporating Solids

The last step in complexity that we will explore in detail for the construction of diagrams to illustrate redox equilibria involves the addition of heterogeneous equilibria to redox and acid-base equilibrium diagrams. We will illustrate this system with a pε-pH diagram for iron species in aqueous solution containing no anions other than hydroxide. We will expand on this diagram later in this chapter during the discussion of iron chemistry.

First we must identify which solids and solution components are important. For an unknown system we would achieve this by searching the literature for relevant reactions and equilibrium data. In this example the following components are of interest,

SOLIDS	SOLUTION COMPONENTS
$Fe(OH)_{2(s)}$	Fe^{3+}
$Fe(OH)_{3(s)}$	$FeOH^{2+}$
	$Fe(OH)_2^+$
	Fe^{2+}
	$FeOH^+$

The task of diagram construction is broken down into four stages.

1. Boundaries are constructed between the solids of interest if more than one solid exists.

2. Boundaries are constructed between solids and individual solution components. A concentration of dissolved component is selected and the stipulation is made that the solid/solution boundary is valid for the selected concentration of the indicated solution component. For our example we will use a Fe concentration of $10^{-7} M$ (5.6 μg Fe/ liter) although in the later discussion of iron chemistry several solid-solution boundaries are entered on a single diagram to represent different solution concentrations of Fe species at the boundary.

3. The complete solids-solution boundary (with various solution com-

ponents in different parts of the diagram) is constructed. When significant boundaries are in doubt, they can be established by testing various pϵ-pH coordinates in the equations in Table 7-3 to determine the species of significance.

4. For the region of the diagram that describes solution (away from solids predominance areas), the equilibria between solution components are plotted for all combinations of solution components. Significant boundaries are deduced as lines of equal concentration of solution components.

We will not present many of the detailed considerations required for deciding significant boundaries between the various solids and species because this would needlessly complicate the presentation. For a full description of the procedure for developing a diagram such as this using other types of diagrams as aids, see Stumm and Morgan.[6]

To develop the diagram, we first enter in Fig. 7-7 the stability limits of water that were established for Fig. 7-4. Next we refer to Table 7-3 that presents equations that interrelate all species and solids. We would find, from a detailed study of the diagram construction, that only a few of these equations describe significant boundaries; moreover, these equations apply only in specific regions of the diagram.

Line 1 is the boundary between the two solids $Fe(OH)_{2(s)}$ and $Fe(OH)_{3(s)}$. Along this line $Fe(OH)_{3(s)}$ and $Fe(OH)_{2(s)}$ are in equilibrium with each other. At very low pϵ values, a third solid, $Fe_{(s)}$ (metallic iron) may be formed (e.g., below pϵ = −7.4 in the pH range 0–8). We will not consider this solid as part of the diagram.

We now develop the diagram for the equilibria of the solids $Fe(OH)_{2(s)}$ and $Fe(OH)_{3(s)}$ with soluble species. These lines describe the boundaries of the two solids. The region to the right of and above these lines is the area of predominance of the solids. To the left of and below these lines, dissolved components predominate. The dividing line between solids and solution is based on a maximum soluble iron concentration of 10^{-7} M. Of the 10 equations in Table 7-3 relating solids to soluble Fe-species, only 4 are employed in the diagram construction. The rest describe insignificant boundaries.

We next proceed to the solution equilibria, (equations 12 through 21 in Table 7-3). Of these 10 equations, 5 describe significant boundaries and are plotted in Fig. 7-7.

If we examine Fig. 7-7, we find that a $Fe(OH)_2{}^+$ predominance region does not appear. However an area occupied by $Fe(OH)_2{}^+$ does exist but because of the selection of a solids/solution equilibrium concentration, $C_{T,Fe}$, of 10^{-7} and because of our rounding off of equilibrium constants,

[6] W. Stumm and J. J. Morgan, *Aquatic Chemistry*, Wiley-Interscience, New York, 1970.

TABLE 7-3 Equilibria and Equations for Construction of Aqueous Fe(II)-Fe(III) pε-pH Diagram (Figure 7-7)

Boundary	Reaction	Log K	Equilibrium Statement	Line Number (Figure 7-7)
1. For Boundaries between Two Solids				
Fe(OH)$_{2(s)}$/Fe(OH)$_{3(s)}$	Fe(OH)$_{3(s)}$ + H$^+$ + e$^-$ → Fe(OH)$_{2(s)}$ + H$_2$O	+4.62	pH + pε = 4.62	1[a]
2. For Boundaries between Solids and Solution Components				
Fe(OH)$_{2(s)}$/Fe^{2+}	Fe(OH)$_{2(s)}$ → Fe^{2+} + 2OH$^-$	−15.1	pH = +6.5 − ½ log [Fe^{2+}]	2
Fe(OH)$_{2(s)}$/FeOH$^+$	Fe(OH)$_{2(s)}$ + H$^+$ → FeOH$^+$ + 2H$_2$O	+4.6	pH = +4.6 − log [FeOH$^+$]	3[a]
Fe(OH)$_{2(s)}$/Fe^{3+}	Fe(OH)$_{2(s)}$ → Fe^{3+} + 2OH$^-$ + e$^-$	−28.2	pε = +0.2 + 2 pH + log [Fe^{3+}]	4
Fe(OH)$_{2(s)}$/FeOH^{2+}	Fe(OH)$_{2(s)}$ + H$^+$ → FeOH^{2+} + H$_2$O + e$^-$	−2.2	pε = +2.2 + pH − log [FeOH^{2+}]	5
Fe(OH)$_{2(s)}$/Fe(OH)$_2^+$	Fe(OH)$_{2(s)}$ → Fe(OH)$_2^+$ + e$^-$	−6.9	pε = +6.9 + log [Fe(OH)$_2^+$]	6
Fe(OH)$_{3(s)}$/Fe^{3+}	Fe(OH)$_{3(s)}$ → Fe^{3+} + 3OH$^-$	−37.2	pH = +1.6 − ⅓ log [Fe^{3+}]	7
Fe(OH)$_{3(s)}$/FeOH^{2+}	Fe(OH)$_{3(s)}$ + 2H$^+$ → FeOH^{2+} + 2H$_2$O	+2.4	pH = +1.2 − ½ log [FeOH^{2+}]	8
Fe(OH)$_{3(s)}$/Fe(OH)$_2^+$	Fe(OH)$_{3(s)}$ + H$^+$ → Fe(OH)$_2^+$ + H$_2$O	−2.3	pH = +2.3 − log [Fe(OH)$_2^+$]	9[a]
Fe(OH)$_{3(s)}$/Fe^{2+}	Fe(OH)$_{3(s)}$ + 3H$^+$ + e$^-$ → Fe^{2+} + 3H$_2$O	+17.9	pε = +17.9 − 3pH − log [Fe^{3+}]	10[a]
Fe(OH)$_{3(s)}$/FeOH$^+$	Fe(OH)$_{3(s)}$ + 2H$^+$ + e$^-$ → FeOH$^+$ + 2H$_2$O	+9.25	pε = +9.25 − 2pH − log [FeOH$^+$]	11[a]

TABLE 7-3 (continued)

3. For Boundaries between Solution Components

Boundary	Reaction	Log K	Equilibrium Statement	Line Number (Figure 7-7)
Fe^{3+}/Fe^{2+}	$Fe^{3+} + e^- \rightarrow Fe^{2+}$	$+13.1$	$p\epsilon = +13.0 - \log \dfrac{[Fe^{2+}]}{[Fe^{3+}]}$	12[a]
$Fe^{3+}/FeOH^{2+}$	$Fe^{3+} + H_2O \rightarrow FeOH^{2+} + H^+$	-2.4	$pH = +2.4 + \log \dfrac{[FeOH^{2+}]}{[Fe^{3+}]}$	13[a]
$Fe^{3+}/FeOH^+$	$Fe^{3+} + H_2O + e^- \rightarrow FeOH^+ + H^+$	$+4.4$	$p\epsilon = +4.4 + pH - \log \dfrac{[FeOH^+]}{[Fe^{3+}]}$	14
$Fe^{3+}/Fe(OH)_2^+$	$Fe^{3+} + 2H_2O \rightarrow Fe(OH)_2^+ + H^+$	-7.1	$pH = +3.6 + \tfrac{1}{2} \log \dfrac{[Fe(OH)_2^+]}{[Fe^{3+}]}$	15
$Fe^{2+}/FeOH^+$	$Fe^{2+} + H_2O \rightarrow FeOH^+ + H^+$	-8.6	$pH = +8.6 + \log \dfrac{[FeOH^+]}{[Fe^{2+}]}$	16[a]
$Fe^{2+}/FeOH^{2+}$	$Fe^{2+} + H_2O \rightarrow FeOH^{2+} + H^+ + e^-$	-15.5	$p\epsilon = +15.5 - pH + \log \dfrac{[FeOH^{2+}]}{[Fe^{2+}]}$	17[a]
$Fe^{2+}/Fe(OH)_2^+$	$Fe^{2+} + 2H_2O \rightarrow Fe(OH)_2^+ + 2H^+ + e^-$	-20.2	$p\epsilon = +20.2 - 2pH + \log \dfrac{[Fe(OH)_2^+]}{[Fe^{2+}]}$	18
$FeOH^+/FeOH^{2+}$	$FeOH^+ \rightarrow FeOH^{2+} + e^-$	-6.9	$p\epsilon = +6.9 - \log \dfrac{[FeOH^{2+}]}{[FeOH^+]}$	19
$FeOH^+/Fe(OH)_2^+$	$FeOH^+ + H_2O \rightarrow Fe(OH)_2^+ + H^+ + e^-$	-11.6	$p\epsilon = +11.6 - pH + \log \dfrac{[Fe(OH)_2^+]}{[FeOH^+]}$	20
$FeOH^{2+}/Fe(OH)_2^+$	$Fe(OH)_2^+ + H^+ \rightarrow FeOH^{2+} + H_2O$	$+4.7$	$pH = +4.7 - \log \dfrac{[FeOH^{2+}]}{[Fe(OH)_2^+]}$	21[a]

[a] Only these lines describe significant equalities and thus appear in some region of Figure 7-7.

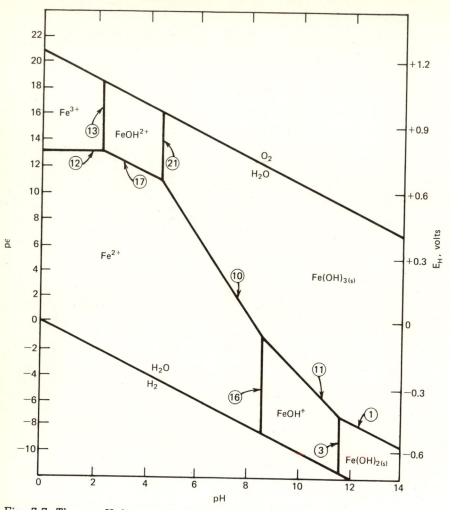

Fig. 7-7. The pϵ-pH diagram for a Fe(II)-Fe(III) system at 25°C.

the area becomes so small it is difficult to draw in on the scale used. Equation 21 (Table 7-3) states:

$$[FeOH^{2+}] = [Fe(OH)_2^+] \text{ at pH } 4.7$$

while equation 8 is

$$pH = 1.2 - \tfrac{1}{2} \log [FeOH^{2+}]$$

which, when $[FeOH^{2+}] = 10^{-7}$ M, reduces to pH $= 4.7$. Thus the solid phase boundary with $FeOH^{2+}$ is superimposed on the equality line between $[FeOH^{2+}]$ and $[Fe(OH)_2^+]$. A small region would exist for $Fe(OH)_2^+$ had our choice of solution $C_{T,Fe}$ been smaller than 10^{-7} M.

One striking feature of the diagram is the large area occupied by Fe^{2+}.

From this observation we might suspect that, in the absence of oxygen, iron in solution has a very good chance of being present as ferrous ion. In the later section on iron chemistry we will follow up on this observation.

This is as far as we will go in the details of diagram construction. However, we will continue to present diagrams (some of greater complexity), knowing that the reader will understand them and appreciate the time that it takes to construct them. The diagram for the iron system in Fig. 7-7 only considers the soluble species Fe^{3+}, Fe^{2+}, and hydroxo iron species. Such diagrams become more useful (as we shall see in the section on iron chemistry) when other iron-containing solids and species such as carbonates and sulfides are introduced. When this is done the diagram is more complex but the rules of the game are the same. An additional constraint is introduced in that the total concentration of anion (e.g., total carbonate carbon or total sulfur) must be specified. There is an individual diagram for each specified anion level. A further useful presentation is to draw more than one boundary between solid phases and solution species. Each boundary represents a different selected total solution concentration of iron (e.g., 10^{-7}, 10^{-6}, $10^{-5} M$, etc). Such diagrams help us to judge the solution concentration of species at various pϵ and pH values.

7.6. CORROSION

Corrosion is the deterioration of metallic structures, usually with the loss of metal to solution. It manifests itself in many ways including rusting, pitting, tuberculation, cracking (or embrittlement) etc. Corrosion is a costly and wasteful process; it has been estimated that some 25 percent of the annual production of steel in the United States is for the replacement of material lost to corrosion.[7]

All refined metals have a tendency to revert to a thermodynamically more stable form such as those in which they occur naturally on earth. Thus one of the corrosion products of iron is iron oxide ($Fe_2O_{3(s)}$) which is one form of iron ore. Almost all types of corrosion can be explained in terms of electrochemistry (oxidation-reduction reactions); for this reason we will consider corrosion as an example of the application of redox chemistry or electrochemistry to a practical situation. We will not present a detailed quantitative analysis of corrosion and the design of corrosion-control systems. Other texts should be consulted for this type of information.

7.6.1. The Corrosion Cell

For corrosion to occur, the presence of all the components of an electrochemical cell is required. These components are an *anode*, a

---

[7] R. O. Dean "Corrosion in Irrigation Systems," presented at the 39th Annual Rural Electric Conference, University of California, Davis, February 1964.

cathode, an *external circuit,* or a connection between the anode and cathode, and an *internal circuit* or conducting solution (electrolyte) between the anode and cathode. An anode and cathode are sites on a metal that have a difference in potential between them. If any one of these components is absent, a corrosion cell does not exist and no corrosion will occur.

That mystery often surrounds the process of corrosion is probably because of the hard-to-recognize forms that the electrochemical cell takes. Persons accustomed to the laboratory will visualize an electrochemical cell as a beaker containing electrolyte in which to pieces of metal are immersed and joined externally with a wire. It is difficult to make the translation between this situation and that of a water pipe running through alternate marshy and sandy patches of soil, yet both are electrochemical cells—and both will be subject to the reactions that go to make up corrosion.

Figure 7-8 illustrates the general features of a galvanic cell indicating the direction of current and electron flow. The oxidation (removal of electrons) or dissolution of metal occurs at the anode, which consequently is negatively charged. Reduction (consumption of electrons) or deposition of metal occurs at the cathode, which consequently is positively charged. We have previously discussed a galvanic cell of this kind in Example 7-6, where zinc metal was the anode and corroded to form zinc ions; copper was the cathode and copper ions in the copper chloride solution surrounding the cathode plated out as copper metal. We can visualize a similar situation existing when a copper water pipe is connected to a galvanized (zinc-coated steel) pipe (Fig. 7-9). The galvanized pipe is the anode and the copper pipe is the cathode. The electrolyte is the water flowing

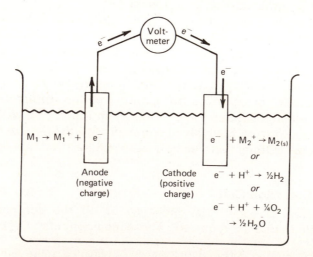

Fig. 7-8. Features of a galvanic cell.

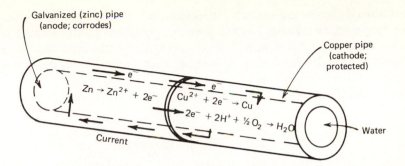

Fig. 7-9. Corrosion cell between galvanized and copper pipes joined together.

through the pipe. The external circuit is the pipe material, and contact between the zinc and copper is required.

Corrosion cells can be far more subtle. For example, electrical code states that a neutral wire must be bonded to a water pipe. If it is connected to a galvanized pipe in one residence and a copper pipe in a neighboring dwelling, we have the makings of a corrosion cell (Fig. 7-10). The galvanized pipe is the anode; the copper pipe is the cathode. The common neutral wire is the external circuit. The moist soil surrounding the pipes is the electrolyte.

Corrosion cells can be produced by the interaction of small, local, adjacent anodes and cathodes on the same piece of metal. These so-called "local-action cells" form because the surface of a piece of metal is not uniform. Small variations in composition, local environment, orientation of the grain structure, and differences in the amount of stress and surface imperfections all may contribute to the creation of tiny areas of

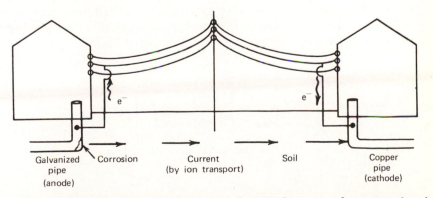

Fig. 7-10. Corrosion cell between galvanized pipe and copper pipe in separate residences.

metal with different potentials. Therefore, a piece of iron is covered with many tiny corrosion cells. The reader can prove this by immersing a piece of iron into a dilute solution of HCl. The gas that bubbles from the surface of the iron is hydrogen being produced at the small local cathodes (Fig. 7-11).

The anodic, cathodic, and net reactions that are taking place are

$$Fe_{(s)} \rightarrow Fe^{2+} + 2e^- \quad \text{(anode)}$$

$$\underline{2e^- + 2H^+ \rightarrow H_{2(g)} \quad \text{(cathode)}}$$

$$Fe_{(s)} + 2H^+ \rightarrow Fe^{2+} + H_{2(g)} \quad \text{(net)}$$

As we shall see later, differences in the external environment (such as the amount of moisture, temperature, and dissolved oxygen concentration) also can produce corrosion cells on the same piece of metal.

7.6.2. The Galvanic (or Electromotive) Series

The galvanic series (Table 7-4) is a listing of metals presented in the order of their tendency to corrode (or go into solution). The metals are listed in groups, and the metals within these groups can generally be used together safely (i.e., without either one corroding significantly). If metals from separate groups are connected in a situation where a corrosion cell forms, the metal that is highest on the list will corrode. In general, the farther apart the metals, the greater will be the potential for corrosion because the potential difference between them will be greater. Basically, this list is in the order of decreasing oxidation potentials, but not exactly so. This is because the behavior of some metals from a corrosion standpoint is not solely dictated by differences in the electrode potential. For example, chromium and stainless steel each appear at two widely separated places in the list. At the higher position where they occur they are listed as "active"; at the lower position they carry the designation "passive." These metals become "active" (more subject to corrosion) when they are present in an oxygen-starved environment (the higher listing); they are passive (less subject to corrosion) when they are in an environment with oxygen present (the lower listing).

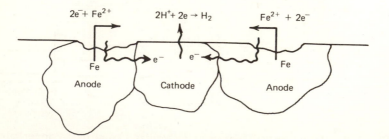

Fig. 7-11. Corrosion of iron in acid solution.

TABLE 7-4   The Galvanic Series

Anode	Anode Reaction	Potential $E°$ (volts)
Magnesium	$Mg_{(s)} \rightarrow Mg^{2+} + 2e^-$	+2.37
Aluminum	$Al_{(s)} \rightarrow Al^{3+} + 3e^-$	+1.68
Zinc	$Zn_{(s)} \rightarrow Zn^{2+} + 2e^-$	+0.76
Cadmium	$Cd_{(s)} \rightarrow Cd^{2+} + 2e^-$	+0.40
Steel or iron	$Fe_{(s)} \rightarrow Fe^{2+} + 2e^-$	+0.44
Cast iron	$Fe_{(s)} \rightarrow Fe^{2+} + 2e^-$	+0.44
Chromium (active)	$Cr_{(s)} \rightarrow Cr^{3+} + 3e^-$	+0.74
Stainless steel (active)	$Fe_{(s)} \rightarrow Fe^{2+} + 2e^-$	+0.44
Soft solder		
Tin	$Sn_{(s)} \rightarrow Sn^{2+} + 2e^-$	+0.136
Lead	$Pb_{(s)} \rightarrow Pb^{2+} + 2e^-$	+0.126
Nickel	$Ni_{(s)} \rightarrow Ni^{2+} + 2e^-$	+0.25
Brass		
Bronze		
Copper	$Cu_{(s)} \rightarrow Cu^{2+} + 2e^-$	−0.345
Silver solder		
Chromium (passive)	$Cr_{(s)} \rightarrow Cr^{3+} + 3e^-$	+0.74
Stainless steel (passive)	$Fe_{(s)} \rightarrow Fe^{2+} + 2e^-$	+0.44
Silver	$Ag_{(s)} \rightarrow Ag^+ + e^-$	−0.8
Gold	$Au_{(s)} \rightarrow Au^{3+} + 3e^-$	−1.5
Platinum	$Pt_{(s)} \rightarrow Pt^{2+} + 2e^-$	−1.42

Source: M. J. Orton and R. O. Dean, "Swimming Pool Corrosion", paper to San Francisco Bay Section of the National Association of Corrosion Engineers, October 1963.

The rate of corrosion (loss of metal) is proportional to the amount of current that flows in the corrosion cell. From Ohm's law we know that, for direct current,

$$E = IR$$

where

E = potential difference, volts
I = current, amperes
R = resistance, ohms

and

$$I = \frac{E}{R}$$

Thus the greater the potential difference (i.e., the greater the separation in the galvanic series) and the lower the resistance, the greater will be the amount of current carried in the corrosion cell. Faraday's law states that the amount of a material that can be produced electrochemically is proportional to the amount of charge in coulombs. A coulomb is the amount of charge transferred when 1 ampere flows for 1 second. One equivalent of a material on a charge basis (Avagadro's number of positive or negative charges) is produced by 1 Faraday or 96,500 coulombs. That is,

$$\text{Number of equivalents reacted} = \text{current (amp)}$$
$$\times \text{ time (sec)}/(96,500 \text{ coulombs/equivalent})$$
$$= \frac{It}{F}$$

where F = the Faraday constant.
and t = time in seconds

Table 7-5 translates Faraday's law into practical terms by indicating the pounds of various metals that would be lost from an anode of the metal in a corrosion cell with a current of 1 amp flowing for 1 year.

7.6.3.   Corrosion Reactions

Reactions occurring during corrosion are conveniently divided into those taking place at the anode and those taking place at the cathode. The reactions may be (and usually are) more extensive and complex than the primary oxidation reduction half-reactions. Let us consider the corrosion of a piece of iron pipe. The basic galvanic cell is illustrated in Fig. 7-12.

At anodic areas an oxidation reaction releases Fe^{2+} ions into solution, leaves electrons behind in the metal, and produces a negative charge on the metal.

$$Fe \rightleftharpoons Fe^{2+} + 2e^-$$

TABLE 7-5   Corrosion Loss by Various Metals

Metal	lb Lost/amp-year
Lead	75
Copper	46
Tin	43
Zinc	23
Iron	20
Magnesium	9
Aluminum	6
Carbon	2

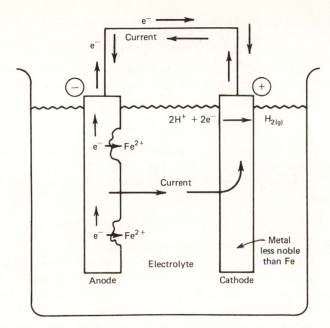

Fig. 7-12. Galvanic cell for corrosion of iron.

The electrons travel in the external circuit from the anode to the cathode; at the cathode they may react with H^+ ions in solution to form hydrogen gas, which bubbles out of solution,

$$2H^+ + 2e^- \rightleftharpoons H_{2(g)}$$

This is the reaction we mentioned previously in our discussion of the corrosion of iron in dilute HCl. It is referred to as "hydrogen evolution corrosion."

Alternatively, if there is oxygen in the solution surrounding the cathode, the cathode reaction may produce water thus:

$$4e^- + 4H^+ + O_{2(aq)} \rightleftharpoons 2H_2O$$

In both of these cathodic reactions, H^+ is consumed so that the pH rises and significant amounts of OH^- ions appear. These ions migrate through the electrolyte toward the anode, thus conducting current through the solution. At the anode, if no oxygen is present and pH becomes sufficiently high, ferrous hydroxide will precipitate,

$$Fe^{2+} + 2OH^- \rightleftharpoons Fe(OH)_{2(s)}$$

In oxygenated water Fe^{2+} is converted to Fe^{3+}

$$4Fe^{2+} + 4H^+ + O_{2(aq)} \rightleftharpoons 4Fe^{3+} + 2H_2O$$

which in turn may precipitate as ferric hydroxide at the anode:

$$Fe^{3+} + 3OH^- \rightleftharpoons Fe(OH)_{3(s)}$$

$Fe(OH)_{3(s)}$ is sparingly soluble in waters with pH values above 5 to 6 while $Fe(OH)_{2(s)}$ is much more soluble. The precipitate $Fe(OH)_{3(s)}$ dehydrates to form Fe_2O_3, ferric oxide, which has the familiar red color of rust.

$$2Fe(OH)_{3(s)} \rightleftharpoons Fe_2O_{3(s)} + 3H_2O$$

If the precipitate of $Fe(OH)_{3(s)}$, [or $Fe(OH)_{2(s)}$ which is then oxidized to $Fe(OH)_3$], sloughs off the pipe into the water, *red water* results. Also, depending on the CO_3^{2-} concentration and whether oxygen is absent, Fe^{2+} may precipitate as a carbonate solid.

$$Fe^{2+} + CO_3^{2-} \rightarrow FeCO_{3(s)} \text{ (siderite)}$$

Corrosion of an iron pipe under certain conditions can also manifest itself in the formation of "pits" and "tubercules" (Fig. 7-13). The OH^- produced by the cathodic reaction not only may precipitate $Fe(OH)_{2(s)}$ and $Fe(OH)_{3(s)}$ but can also react with bicarbonate ion (alkalinity) in the water to produce carbonate ion. In the locally high pH region around the cathode, both $FeCO_{3(s)}$ and calcium carbonate may precipitate

$$HCO_3^- + OH^- \rightleftharpoons H_2O + CO_3^{2-}$$
$$Ca^{2+} + CO_3^{2-} \rightleftharpoons CaCO_{3(s)}$$

The iron and calcium precipitates together form the familiar knarled tubercules that surround the anodic areas where Fe^{2+} dissolves to form pits. However, under other conditions $CaCO_{3(s)}$, and possibly iron precipitates, may form a thin protective coating over the metal surface.

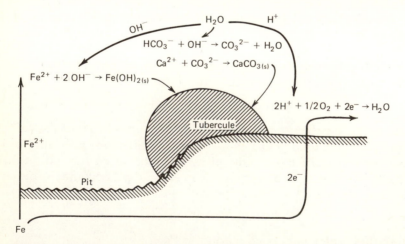

Fig. 7-13. Formation of pits and tubercules in the corrosion of iron pipe.

The exterior of iron pipes in marshy (anaerobic) soils are often coated with a black film of material that results from a mode of iron corrosion. The ferrous ions produced at the anode react with sulfide present in the anaerobic soil (bacteria may reduce sulfate to sulfide in the absence of oxygen). The black ferrous sulfide is produced by reactions,

At the anode,

$$Fe_{(s)} \rightleftharpoons Fe^{2+} + 2e^-$$

In the anaerobic soil,

$$8e^- + 9H^+ + SO_4^{2-} \rightleftharpoons HS^- + 4H_2O$$

At the interface between the pipe and the soil,

$$Fe^{2+} + HS^- \rightleftharpoons FeS_{(s)} + H^+$$

7.6.4. Concentration Cells

If adjacent portions of a surface are in contact with a solution having metal ions of different concentration, a potential will develop that can result in corrosion. The corrosion process in such a situation will always tend to equalize the concentrations of metal ion by dissolving metal from the metal surface. Therefore, we can deduce that the portion of metal in contact with the dilute solution will become the anode—it will corrode and contribute metal ions to solution. The part of the metal in the more concentrated metal ion solution will become the cathode and either metal ions will plate out on it or some oxidizing agent such as O_2 will be reduced at its surface.

To illustrate such a concentration cell let us consider the galvanic cell made up of a piece of zinc metal with two sites A and B (Fig. 7-14). Site

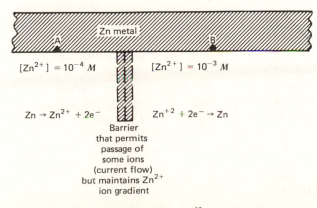

Fig. 7-14. Zinc concentration cell.

A is in contact with a solution containing $10^{-4} M$ zinc ion, and site B is in contact with a $10^{-3} M$ zinc ion solution. From the Nernst equation we can show that site A is the anode and site B is the cathode. To prove this, we will write the reaction assuming site A to be the anode and site B to be the cathode, and then examine the resulting potential to determine if the reaction will be spontaneous as written.

	$E°$ (volts)
at Site A: $Zn_A \rightarrow Zn_A{}^{2+} + 2e^-$;	$+0.76$
at Site B: $Zn_B{}^{2+} + 2e^- \rightarrow Zn_B$;	-0.76
$Zn_B{}^{2+} + Zn_A \rightarrow Zn_A{}^{2+} + Zn_B$;	0.0

From Eq. 7-8, neglecting ionic strength effects,

$$E = E° - \frac{2.3RT}{nF} \log Q$$

$$= 0 - \frac{0.059}{2} \log \frac{[Zn_A{}^{2+}]}{[Zn_B{}^{2+}]} = \frac{-0.059}{2} \log \frac{10^{-4}}{10^{-3}}$$

$$= +0.03$$

Since $\Delta G = -nFE$, ΔG is negative and the reaction is spontaneous as written. Site A, in contact with the lower metal ion concentration, is the anode and site B, in contact with the higher metal ion concentration, is the cathode.

One of the more common causes of corrosion due to concentration galvanic cells is the presence of different concentrations of dissolved oxygen or hydrogen ion at different sites on a metal surface. When caused by dissolved oxygen, this is often referred to as "differential oxygenation corrosion." Some examples of these situations are illustrated in Fig. 7-15. Common areas for differential oxygenation corrosion are between two metal surfaces, for example, under rivets, washers, or in crevices. Also common is the area beneath various types of surface adhesion, which may be things such as a barnacle, a bacterial slime, or a piece of mill scale. As with the concentration cell formed by two metal ion solutions of differing concentrations, the reaction in a differential aeration cell will proceed in a manner that tends to equalize the oxygen concentrations. Since oxygen is not produced by either the anode or cathode reaction, the corrosion will proceed in such a way that the higher oxygen concentration is reduced. Oxygen participates in the cathodic reaction

$$O_2 + 4H^+ + 4e^- \rightleftharpoons 2H_2O$$

Thus the part of metal in contact with the higher oxygen concentration

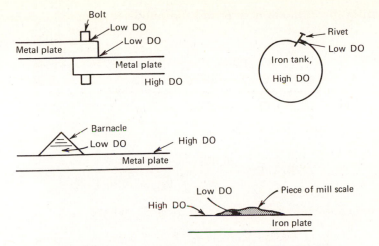

Fig. 7-15. Examples of differential oxygenation corrosion.

will be the cathode and the metal in contact with the lower oxygen concentration will be the anode and will corrode. This is consistent with the observation that corrosion takes place underneath adhesions on metal surfaces.

Because O_2 reduction depends on H^+ concentration, differences in pH have the same effect as O_2 concentration differences. Waters that have a high buffer intensity, β, also will have less tendency to develop regions of different pH, and thus will not be as corrosive.

Let us consider the example of a piece of iron with a barnacle growing on it (Fig. 7-15). Underneath the barnacle the oxygen concentration is lower than outside the confines of the barnacle shell. The ferrous iron and pH levels in solution at both places are assumed to be identical. We will examine the tendency of the corrosion reaction to proceed at both locations (A under the barnacle and B outside the barnacle).

If we assume that the cathode is at A and the anode is at B,

	$E°$ (volts)
$2Fe \rightarrow 2Fe_B^{2+} + 4e^-$	$+0.44$
$4e^- + 4H_A^+ + O_{2(aq)_A} \rightarrow 2H_2O$	$+1.27$
$2Fe + 4H_A^+ + O_{2(aq)_A} \rightarrow 2Fe_B^{2+} + 2H_2O$	$+1.71$

From Eq. 7-8,

$$E_A = E_A° - \frac{0.059}{4} \log \frac{[Fe_B^{2+}]^2}{[O_{2(aq)_A}][H_A^+]^4}$$

Assuming the cathode is at B and the anode is at A,

$$E° \text{ (volts)}$$

$$
\begin{array}{ll}
2Fe \rightarrow 2Fe_A{}^{2+} + 4e^- & +0.44 \\
4e^- + 4H_B{}^+ + O_{2(aq)_B} \rightarrow 2H_2O & +1.27 \\
\hline
2Fe + 4H_B{}^+ + O_{2(aq)_B} \rightarrow 2Fe_A{}^{2+} + 2H_2O & +1.71
\end{array}
$$

From Eq. 7-8,

$$E_B = E_B° - \frac{0.059}{4} \log \frac{[Fe_A{}^{2+}]^2}{[O_{2(aq)_B}][H_B{}^+]^4}$$

For the concentration cell, the potential is E_R, where

$$E_R = E_B - E_A = \frac{-0.059}{4} \log \frac{[Fe_B{}^{2+}]^2}{[O_{2(aq)_A}][H_A{}^+]^4} + \frac{0.059}{4} \log \frac{[Fe_A{}^{2+}]^2}{[O_{2(aq)_B}][H_B{}^+]^4}$$

$$= \frac{0.059}{4} \log \frac{[O_{2(aq)_A}][H_A{}^+]^4[Fe_A{}^{2+}]^2}{[O_{2(aq)_B}][H_B{}^+]^4[Fe_B{}^{2+}]^2}$$

Given that the pH and Fe^{2+} concentration are the same at both sites, $[H_A{}^+]$ $= [H_B{}^+]$, and $[Fe_A{}^{2+}] = [Fe_B{}^{2+}]$, and since $[O_{2(aq)_A}] < [O_{2(aq)_B}]$,

$$E_R = \frac{0.059}{4} \log \frac{[O_{2(aq)_B}]}{[O_{2(aq)_A}]} > 0$$

Since $\Delta G = -nFE$, the free energy change is negative and there is a greater tendency for the corrosion reaction to take place when B is the cathode and A, the area of low oxygen concentration, is the anode rather than vice versa. If E_R has been negative, B would have been the anode, and if $[O_{2(aq)_A}] = [O_{2(aq)_B}]$, $E_R = 0$ and no corrosion would take place.

Calculation of E_A and E_B for this problem using typical concentrations such as $[Fe^{2+}] = 10^{-5}$ M, $[O_{2(aq)}] = 10^{-4}$ M and pH = 7 show that both potentials are positive. Because there is oxygen at both sites, it is necessary to examine the value of the relative potential, E_R, rather than the individual values of E_A and E_B.

7.6.5. Corrosion Control

We have previously mentioned that all of the components of a galvanic cell must be present for corrosion to occur. Corrosion control functions by eliminating or reducing the effectiveness of one or more of these components. Thus we can control corrosion by eliminating anodes and cathodes, by eliminating or reducing differences in potential between metallic sites, and by breaking internal or external circuits. Some of these objectives are achieved by subtle methods but the secret in corrosion control, as in assessing a corrosion problem, is to reduce the situation to basic principles.

1. *Selection of Materials.* The most obvious corrosion-combating measure in this category is to select nonmetallic materials or materials low in the galvanic series so that they are more likely to become cathodic rather than anodic. The choice, however, cannot be based solely upon the position of a metal in the galvanic series. Under certain conditions some metals high in the galvanic series form corrosion products, for example, metal oxides, that cling tenaciously to the metallic surface and block further corrosive attack or significantly reduce the rate of corrosion. They effectively protect the base metal beneath an oxide layer and render it "passive." A good example of this type of behavior is aluminum whose oxide is always present at an aluminum surface. Aluminum saucepans are dull and greyish after use because of the accumulation of the protective oxide film. Corrosion is very slow beneath the film. Some ions such as Cl^- can penetrate the aluminum oxide film and promote corrosion of the underlying metal. The Cl^- ion can apparently aid in the release of aluminum ions, possibly through the formation of soluble chloroaluminum complexes, and also aid in the passage of the electrical current necessary to maintain corrosion. For this reason aluminum is a poor choice of metal to be in contact with saline water. The City of San Francisco does not use aluminum light poles to the west of Twin Peaks because of the chloride content of the atmosphere close to the Pacific Ocean. When aluminum is immersed in a dilute mineral acid, the oxide film is dissolved away and rapid corrosion of the aluminum, accompanied by hydrogen evolution, occurs.

    Other metals also can be "passivated." For example the presence of 12 percent chromium in steel renders the steel passive *in an oxygenated environment* because it promotes the formation of a thin but tightly-bound oxide layer. The presence of chromates in the electrolyte will encourage the formation of γ-Fe_2O_3 on iron surfaces, isolating the surfaces from corrosion.

2. *Coatings.* Paints, cements, bituminous materials, wrappings, and precipitates, such as calcium carbonate, can all act to combat corrosion by isolating either or both the anode or the cathode.

    Painting for corrosion protection usually involves a thorough cleaning of the metal surface to remove any corrosion products followed by application of a primer containing a corrosion inhibitor such as zinc chromate or calcium plumbate; next follows a heavily pigmented undercoat for the purpose of reducing the permeation of water to the metal surface (elimination of the internal circuit—the electrolyte), finally a third decorative coat that is resistant to the atmosphere.

Coating metallic structures to isolate them from the corrosive environment is often practiced, especially along buried pipelines that have anodic and cathodic regions. The cathodic areas are wrapped. The rationale is that if a small hole develops in the coating on a cathode, corrosion will proceed at a very low rate from a large anode to a small cathode (Fig. 7-16). The electrons generated over the large anodic area must be discharged to an oxidizing agent such as oxygen over a very small area. However, if the same small hole were to develop in a protective coating on an anode there would be a tiny anode area and a huge cathode. The electrons generated at the small anode could be rapidly discharged. All the current is generated from a small area and a pit, or hole in a pipe develops rapidly (Fig. 7-16).

The protection of metal surfaces with a thin layer of $CaCO_{3(s)}$ (or $CaCO_{3(s)}$ containing iron salts) is one of the objectives underlying the conditioning of municipal water supplies to have a slightly positive Langelier Index prior to distribution (see Sec 6-7). It should be kept in mind, however, that the pH immediately adjacent to a metal surface may be different from that in the bulk solution. Thus a water that has a tendency to precipitate, that is, a positive Langelier Index, may not actually precipitate on the metal surface because of localized conditions in the immediate vicinity of the surface. The coating of metal pipes with scales such as $CaCO_{3(s)}$ can lead to corrosion problems if the scale accumulation is excessive or uneven. Uneven deposition and poor adhesion of scale is a problem, for example, when alkalinity and calcium concentrations

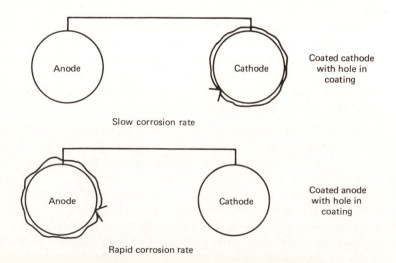

Fig. 7-16. Corrosion control by coating cathodes.

are low and pH is high.[8] Thus scale may be washed off by the flowing water in the pipe and leave exposed metal areas where corrosion can proceed.

3. *Insulation*. It is often necessary to connect two types of metal, for example, galvanized water service pipes to a hot water heater from which copper hot water lines emerge. The corrosion cell that would develop in this case (Fig. 7-9) can be eliminated by inserting a dielectric coupling between the galvanized pipe and the copper pipe which effectively breaks the external circuit and eliminates the corrosion cell. However, this is not common practice because brass fixtures are used in nearly all systems instead of galvanized material.

4. *Chemical Treatment*. A wide variety of chemicals and water treatments are used for corrosion control. Corrosion inhibitors usually act by forming some type of impervious layer on the metallic surface of either the anode or cathode that impedes the reaction at the electrode and thereby slows or inhibits the corrosion reaction. For example, various alkali metal hydroxides, carbonates, silicates, borates, phosphates, chromates, and nitrites promote the formation of a stable surface oxide on metals. The presence of these chemicals in the electrolyte allows any faults in the metal surface or its oxide film to be repaired. If they are used in too small a quantity as anodic inhibitors, they may promote intense local attack because they can leave a small unprotected area on the anode where the current density will be very high. This is particularly true of chromates and polyphosphates.

   Zinc sulfate can be used as a cathodic inhibitor. The Zn^{2+} ions in solution will react with OH^- produced by the cathodic reactions or with carbonates to form sparingly soluble zinc precipitates that coat the cathode.

   We have seen that the presence of dissolved oxygen is important in several corrosion reactions such as differential aeration corrosion and the cathodic reaction with H^+ to produce H_2O. The elimination of dissolved oxygen from the electrolyte will prevent these reactions from occurring. It is common industrial water treatment practice to deoxygenate waters for circulating heating and cooling systems and boilers. Typical methods are to use sulfur dioxide (SO_2) or sodium bisulfite ($NaHSO_3$) with cobalt as a catalyst, hydrazine (N_2H_4), or to use a degasifier. Steam degasification is among the more common physical methods.

5. *Cathodic Protection*. Cathodic protection is the technique of con-

---

[8] W. Stumm, "Investigations on the Corrosive Behavior of Water" *Am. Soc. Civil Engrs, J. Sanit. Engr. Div., 86*: 27 (1965).

verting all of the structural components of interest into cathodes. It can be achieved in two ways:

1. by using a so-called sacrificial anode electrically connected to the material to be protected. The anode material is anodic to the entire system (less noble than the material to be protected) and will therefore convert it into a cathode.

2. By impressing upon the system a d-c current of a magnitude that opposes the current generated by the galvanic corrosion cell. In this case a piece of metal (e.g., scrap iron or graphite) is used as the anode.

These two methods are illustrated in Fig. 7-17. To minimize the rate of degradation of the sacrificial anode, the external surface of the metal being protected is usually coated with a substance such as bitumin, thereby increasing resistance to the current flow.

Sacrificial anodes for galvanic cathodic protection are commonly made from magnesium, which from Table 7-4 one can see is the least noble (most anodic) of all the metals listed. Zinc is also utilized, but it has a lower oxidation potential. Good electrical connections (by soldering or brazing) must be made between the sacrificial electrode and the structure to be protected.

Galvanizing is another form of cathodic protection. A galvanized pipe is a steel pipe coated with a thin layer of zinc. Since zinc is anodic to steel (Table 7-4), the zinc will corrode preferentially to the iron and so protect it. Moreover, the corrosion products of zinc, the carbonates and hydroxides, adhere to the galvanized surface and render it passive.

7.7. IRON CHEMISTRY

The redox reactions of iron are involved in several important phenomena occurring in natural waters and water treatment systems. The oxidation of reduced iron minerals, such as pyrite ($FeS_{2(s)}$), produces acidic waters and the problem of acid mine drainage. The oxidation/reduction of iron in soil and groundwaters determines the iron content of these waters. Redox reactions are intimately involved in the removal of iron from waters. As we have already seen in Section 7-6, the oxidation of metallic iron is an important corrosion reaction.

The purpose of this section is to expand the discussion on the chemistry of iron and to use it as an example of redox reactions in natural waters. The reaction kinetics of redox reactions will be discussed using ferrous iron oxidation as an example.

7.7.1. Iron in Groundwaters

Iron equilibria in groundwaters can be nicely modeled with a pϵ-pH diagram that includes the interaction of iron species with sulfide and

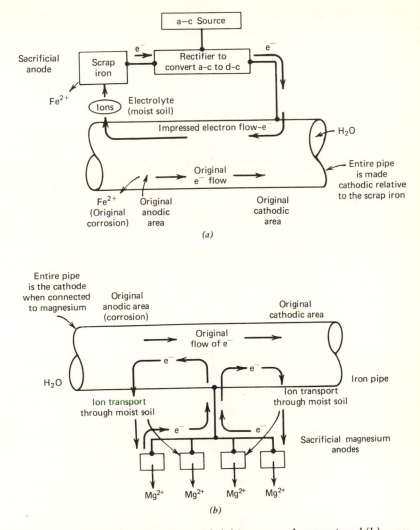

Fig. 7-17. Cathodic protection with (a) impressed current and (b) magnesium sacrificial anodes.

bicarbonate. The diagram presented in Fig. 7-18 bears some resemblance to the pϵ-pH diagram for iron species in pure water presented earlier (Fig. 7-7). Two more solids are evident: ferrous carbonate ($FeCO_{3(s)}$) and ferrous disulfide ($FeS_{2(s)}$). The stability region of ferrous hydroxide ($Fe(OH)_2$) has been reduced in size by $FeS_{2(s)}$ and $FeCO_{3(s)}$. It is important to note that the diagram in Fig. 7-18 has been drawn for a specified level of total inorganic carbon ($C_{T,CO_3} = 10^{-3}M$) and of total inorganic sulfur ($C_{T,s} = 10^{-4}$ M). Another feature of this particular pϵ-pH diagram is the presentation of several boundaries between solid and solution phases, each representing a different allowable solution concentration of iron. These boundaries range all the way from 10 M (570 g Fe/liter) to 10^{-5} M (0.57 mg Fe/liter).

When a well is drilled into the ground, one may visualize it as passing vertically down through the pϵ-pH diagram in Fig. 7-18. Such a "well," with three inlets labeled 1, 2, and 3, has been drawn on Fig. 7-18. For our purposes, constant pH with depth will be assumed.

Intake 1 would be for a shallow well that has its source of water in aerated (unsaturated) soil. The iron here is near equilibrium with atmospheric oxygen (oxygenated water), and the predominant iron-containing mineral is $Fe(OH)_{3(s)}$. The iron content of the water is therefore governed by the equilibrium:

$$Fe(OH)_{3(s)} + 3H^+ \rightleftharpoons Fe^{3+} + 3H_2O; \qquad K = 10^3$$

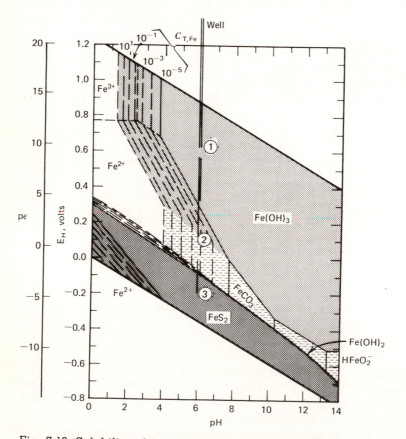

Fig. 7-18. Solubility of iron in relation to pH and pϵ at 25°C and 1 atm. $C_{T,S} = 10^{-4}$ M; $C_{T,CO_3} = 10^{-3}$ M. After J. D. Hem, "Some Chemical Relationships Among Sulfur Species and Dissolved Ferrous Iron," U.S. Geological Service Water Supply Paper, Washington, D.C., 1960.

The iron content of waters from this region is very low because at pH 6, for example, only a fraction of a μg Fe^{3+}/liter is in solution at equilibrium (see Fig. 6-7). They produce waters with a low pH because they are often in the biologically active zone of the soil where considerable CO_2 production takes place. Table 7-6 shows a partial analysis for this type of well water ($Fe(OH)_{3(s)}$ Zone). The iron concentration shown is more than predicted, but this may be due to iron contained in organic complexes.

Well inlet 3 (Fig. 7-18) would be for a deep well that draws water from soil that is in equilibrium with pyrite, $FeS_{2(s)}$. In this soil zone the iron is in the form of ferrous iron and the sulfur in the form of sulfide; iron disulfide has precipitated. The iron content of the well water will be controlled by the reaction

$$FeS_{2(s)} \rightleftharpoons Fe^{2+} + S_2^{2-}; \qquad K = 10^{-26}$$

This reaction will allow the solution of less than 1 μg/liter of iron assuming that S_2^{2-} is the only sulfide species.[9] Waters of this type have low iron contents and low sulfate levels, and they often contain traces of hydrogen sulfide. The analysis for this type of water shown in Table 7-6 shows only slightly more iron than predicted. Unfortunately, no hydrogen sulfide determination was made.

Well intake 2 (Fig. 7-18) is in an intermediate zone between the $Fe(OH)_{3(s)}$ and $FeS_{2(s)}$ regions. In this region the mineral that controls the solution concentration of iron is ferrous carbonate ($FeCO_{3(s)}$), usually the mineral siderite.

$$FeCO_{3(s)} + H^+ \rightleftharpoons Fe^{2+} + HCO_3^-; \qquad K = 10^{-4.4}$$

At pH 6 and $[HCO_3^-] = 10^{-3}$ M, the predicted iron concentration is approximately 20 mg/liter. Thus waters in this region can have significantly higher dissolved iron contents than waters in equilibrium with $Fe(OH)_{3(s)}$ and $FeS_{2(s)}$ at a comparable pH. In Table 7-6 is shown a partial analysis for this type of water. Note also the increase in sulfate concentration in the well water from the $FeCO_{3(s)}$ zone. This arises because we can treat the formation of an $FeCO_{3(s)}$-zone water as the oxidation of $FeS_{2(s)}$ to ferrous sulfate, thus:

$$2 \times (8H_2O + FeS_{2(s)} \rightleftharpoons Fe^{2+} + 2SO_4^{2-} + 16H^+ + 14e^-)$$

$$7 \times (4e^- + 4H^+ + O_2 \rightleftharpoons 2H_2O)$$

$$\overline{2H_2O + 2FeS_{2(s)} + 7O_2 \rightleftharpoons 2Fe^{2+} + 4SO_4^{2-} + 4H^+}$$

This equilibrium picture can be used to teach several lessons in well placement relative to water quality. Wells in the $Fe(OH)_{3(s)}$ zone have a high CO_2 content and a low pH value; they are potentially corrosive. For wells drilled into the $FeS_{2(s)}$ zone one must make certain that the surface

[9] Other polysulfide and sulfide species will exist in the same solution under certain conditions, but $C_{T,Fe}$ will not be greatly affected.

TABLE 7-6   Typical Waters from the Three Iron-Mineral Zones

Constituent as mg/liter of Indicated Species Except Where Noted	$Fe(OH)_{3(s)}$ Zone[a]	$FeCO_{3(s)}$ Zone[b]	$FeS_{2(s)}$ Zone[c]
TDS (calculated), mg/liter	163	378	329
pH, pH units	4.3	7.4	7.3
Ca^{2+}	6	42	3.5
Mg^{2+}	4.9	17	0.9
Na^+	} 11	50	121
K^+		7.6	2.0
Fe (total)	0.18	24	0.06
HCO_3^-	55	102	248
SO_4^{2-}	22	162	48
Cl^-	0.2	31	18
SiO_2	62	18	13
Specific conductance, μmho at 25°C	213	585	534

Source: (Data provided by Dr. John D. Hem, U.S. Geological Service, Menlo Park, Calif.)

[a] Center Point Community Center, 8.5 mi Southeast of Pittsburgh, 23 ft, July 27, 1961.
[b] Lloyd Justice, Route 2, Dangerfield, Tex., 700 ft, June 1, 1960.
[c] City of Pittsburgh, Well No. 3, 641 ft, July 16, 1963.

casing extends down to the $FeS_{2(s)}$ zone so that oxygen-containing water cannot enter this zone. In addition, the surface of the casing should be cemented to prevent corrosion of the casing in the $Fe(OH)_{3(s)}$ zone that it passes through. This will also prevent oxidizing water from the $Fe(OH)_{3(s)}$ zone from running down the well casing into the $FeS_{2(s)}$ zone. Screens for such wells are set only in the $FeS_{2(s)}$ zone, and pumping is at a rate that keeps water above the screens from being drawn down into the $FeS_{2(s)}$ zone.

Wells should not be located in the zone where iron solubility is governed by $FeCO_{3(s)}$, since they will produce a water that is unacceptably high in iron.

7.7.2. Acid Mine Drainage

Reaction kinetics are an interesting aspect of iron chemistry in addition to the redox and precipitation reactions discussed in the previous section. The consideration of kinetics is very important in problems of water-quality significance such as acid mine drainage and the removal of iron from water. At pH values of greater than 5.5 the rate law for the oxygenation of ferrous iron is[10]

$$\frac{-d[Fe(II)]}{dt} = k[Fe(II)][OH^-]^2 P_{O_2} \tag{7-43}$$

[10] W. Stumm and G. F. Lee, "Oxygenation of Ferrous Iron," *Ind. Eng. Chem. 53:* 143 (1961).

where

    k = rate constant
      = $8 (\pm 2.5) \times 10^{13}$ liter^2/(atm-min-mole^2) at 20°C[11]
    P_{O_2} = partial pressure of oxygen, atm

From this equation we can see that the rate of ferrous iron oxidation is first order with respect to oxygen and Fe^{2+} and second order with respect to OH^-.

The rate equation can be transformed to

$$\frac{-d([Fe(II)])}{[Fe(II)] \, dt} = k[OH]^2 P_{O_2} = \frac{-d \ln [Fe^{2+}]}{dt} \qquad (7\text{-}44)$$

For a constant partial pressure of oxygen we have

$$\frac{-d \ln [Fe^{2+}]}{dt} = k' \, [OH^-]^2 = \text{rate}$$

Taking logarithms of both sides and substituting $K_w/[H^+]$ for $[OH^-]$ we have

$$\log (\text{rate}) = \log k'' + 2pH \qquad (7\text{-}45)$$

A plot of log (rate) versus pH should give a straight line with a slope of 2. Figure 7-19 shows that this is indeed true for experimental data at pH 5.5 or above. At pH values below this the rate law appears to approach the form

$$\frac{-d \ln [Fe^{2+}]}{dt} = k_1$$

or, when P_{O_2} is not a constant,

$$\frac{-d \, [Fe^{2+}]}{dt} = k_2 \, [Fe^{2+}] \, P_{O_2} \qquad (7\text{-}46)$$

where $k_2 = 1 \times 10^{-25}$ (1/atm-min) at 25°C. This is an extremely slow rate of iron oxidation. Indeed, below pH 5.5 the rate is negligible (half-life of many years), so on this basis we conclude that, at this pH, Fe^{2+} is stable in oxygenated solutions. This conclusion is at odds with the known phenomenon of acid mine drainage in which pyrite ($FeS_{2(s)}$) in mine waters is oxidized very rapidly to Fe^{3+} at pH values of 2 to 3. Such acid mine drainage gives rise to waters of low pH, high iron, high hardness, high TDS and with a brown-yellow color (Table 7-7).

[11] W. Stumm and J. J. Morgan, *Aquatic Chemistry*, Wiley-Interscience, New York, 1970, p. 534.

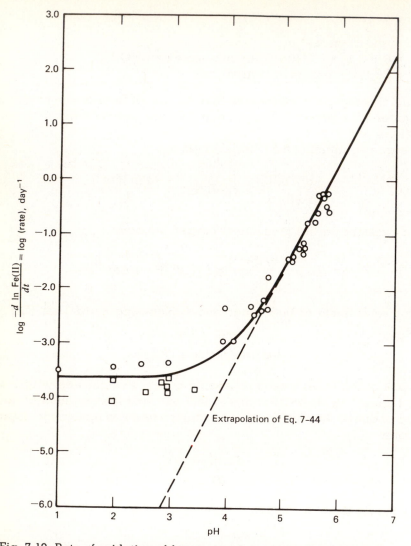

Fig. 7-19. Rate of oxidation of ferrous iron by oxygen. Experimental points obtained in this study: ○, exposed to light; □ in darkness. $k'' = -d \ln [Fe(II)]/dt$, $P_{O2} = 0.20$ atm, and temperature = 25°C. After P. C. Singer and W. Stumm, "Oxygenation of Ferrous Iron," U.S. Dept. of Interior, Fed. Water Quality Adm., Water Polln. Cont. Res. Series Rept. 14010-06/69, June 1970.

The classical stoichiometric picture of acid mine drainage is

$$4FeS_{2(s)} + 14O_2 + 4H_2O \rightleftharpoons 4Fe^{2+} + 8H^+ + 8SO_4^{2-} \tag{7-47}$$

$$4Fe^{2+} + 8H^+ + O_2 \rightleftharpoons 4Fe^{3+} + 2H_2O \tag{7-48}$$

$$4Fe^{3+} + 12H_2O \rightleftharpoons 4Fe(OH)_{3(s)} + 12H^+ \tag{7-49}$$

TABLE 7-7   Water Quality from Roaring Creek-Grassy Run
Watershed of Elkins, W.Va.

Parameter	Concentration Range
pH	2.4–3.3
Mineral acidity, mg/liter as $CaCO_3$	204–980
Fe, mg/liter	35–260
SO_4, mg/liter	340–1650
Total hardness, mg/liter as $CaCO_3$	190–740

The net result of these reactions is that 4 moles of pyrite are oxidized to produce 4 moles of $Fe(OH)_{3(s)}$, which causes the yellow-brown discoloration of the water; 12 moles of H^+ (strong acid) are produced which then react with calcareous minerals in the soil to give rise to waters that have a high hardness and high total dissolved solids content. Another reaction of importance in acid mine drainage is the oxidation of $FeS_{2(s)}$ by ferric iron (Fe^{3+}). Stoichiometrically, this reaction is

$$FeS_{2(s)} + 14Fe^{3+} + 8H_2O \rightleftharpoons 15Fe^{2+} + 2SO_4^- + 16H^+ \tag{7-50}$$

These classical stoichiometric statements do not give a true picture of what is happening in acid mine drainage. A more accurate picture[12] is given by the scheme shown in Fig. 7-20. Some pyrite is oxidized by oxygen to produce Fe^{2+} and this in turn is oxidized by oxygen to Fe^{3+}. The ferric ion may either precipitate as $Fe(OH)_{3(s)}$ or be available to oxidize more $FeS_{2(s)}$ to Fe^{2+} by the reaction in Eq. 7-50. From our previous discussion it would appear that the rate-limiting step in this cyclical oxidation of pyrite would be the slow oxidation of Fe^{2+} at low pH. However, it has been shown that various microorganisms such as *Thiobacillus thiooxidans*, *Thiobacillus ferrooxidans*, and *Ferrobacillus ferrooxidans*, are able to catalyze the oxygenation of ferrous iron. *Thiobacillus ferrooxidans* catalyzes or mediates the redox reactions of Fe^{2+}

$$Fe^{2+} \rightleftharpoons Fe^{3+} + e^-$$

as well as

$$4H_2O + S_{(s)} \rightleftharpoons SO_4^{2-} + 8H^+ + 8e^-$$

while *Ferrobacillus ferroxidans* mediates the couple,

$$Fe^{2+} \rightleftharpoons Fe^{3+} + e^-$$

[12] K. L. Temple and E. W. Delchamps, "Autotrophic Bacteria and the Formation of Acid in Bituminous Coal Mines," *Appl. Microbiol,* **1:** 255 (1953).

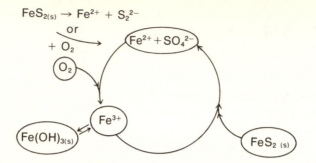

Fig. 7-20. Mechanism of pyrite oxidation.

The importance of microbial catalysis can be demonstrated by comparing the rate of Fe^{2+} oxidation in waters from a mine where acid mine drainage occurs with the rate in the same water that has been sterilized. The rate in the unsterilized water was shown to be about 10^6 times greater than in the sterilized water.[13] From Fig. 7-20 we can deduce that once the pyrite oxidation cycle has started [i.e., once a small amount of $FeS_{2(s)}$ has been oxidized by O_2] further oxygen is only needed for the microbially catalyzed oxidation of Fe^{2+} to Fe^{3+}. Further $FeS_{2(s)}$ is oxidized by Fe^{3+} according to Eq. 7-50. An ample supply of Fe^{3+} for this reaction is guaranteed because $Fe(OH)_{3(s)}$ precipitates generated by the reaction in Eq. 7-49 will be present right at the site of iron oxidation.

The observations made above have an important bearing on the methods for controlling acid mine drainage. Sealing mines to keep out oxygen will not necessarily be successful for two reasons. Oxygen is only needed initially for $FeS_{2(s)}$ oxidation and the microorganisms that catalyze Fe^{2+} oxidation are microaerophiles, that is, they need only low levels of oxygen to survive.

7.8. CHLORINE CHEMISTRY

Chlorine is used widely in water and waste treatment both as an oxidizing agent and a disinfectant. As an oxidizing agent it is used for taste and odor control and color removal in municipal water treatment (oxidation of organics); it is used for oxidation of Fe(II) and Mn(II) in groundwater supplies; in industrial waste treatment it is employed for cyanide oxidation; in domestic waste treatment its uses include odor control, sulfide oxidation, ammonia removal, and disinfection. As a disinfectant it is used in municipal drinking water treatment and for wastewater disinfection. Chlorine is employed for slime or biofouling

[13] P. C. Singer and W. Stumm, "Oxygenation of Ferrous Iron," U.S. Department of Interior Federal Water Quality Administration, Water Pollution Cont. Res. Series Rept. 14010-06/69, June 1970.

control in industrial water treatment applications such as cooling towers and condensers. Also in a role that may be regarded as selective disinfection, or selective killing, chlorine is utilized for the control of filamentous microorganisms (bulking) in the activated sludge treatment of wastewater. Chlorine is also widely used as a disinfectant in swimming pools. Some undesirable compounds may be formed under certain conditions when chlorine reacts with organic matter as we will see in Section 7.8.4.

7.8.1. Forms of Aqueous Chlorine

The term *aqueous chlorine* as it is used in water and wastewater treatment refers not only to the elemental chlorine species Cl_2, but also to a variety of other species including hypochlorous acid, HOCl, hypochlorite ion, OCl^-, and several chloramine species such as monochloramine, NH_2Cl, and dichloramine, $NHCl_2$. Chlorine is used (or applied) in several forms. At the present time it is most often used as a gas, $Cl_{2(g)}$, generated from the vaporization of liquid chlorine stored under pressure. A typical use is disinfection of potable water and wastewater. Salts of HOCl such as sodium hypochlorite or bleach, NaOCl, and calcium hypochlorite or HTH (high-test hypochlorite), $Ca(OCl)_2$, are also used. All these types of "chlorine" find use in swimming pool disinfection. In addition, organic chlorine-containing compounds such as chlorinated cyanuric acid,

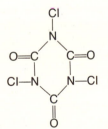

which is stable in the presence of sunlight, are used for swimming pool disinfection. The chlorinated cyanuric acid will hydrolyze in solution to yield free chlorine.[14]

At room temperature and atmospheric pressure, Cl_2 is a pale-green gas. It can be compressed at room temperatures to a yellow-green liquid. Both the gaseous and liquid chlorine react with water to become hydrated. Liquid chlorine forms the compound $Cl_2 \cdot 8H_2O$, "chlorine ice," below 49°F.

---

[14] J. E. O'Brien, J. C. Morris, and J. N. Butler, "Equilibria in Aqueous Solutions of Chlorinated Isocyanurate," chapter in *Chemistry of Water Supply, Treatment and Distribution*, A. J. Rubin, ed., Ann Arbor Science Publishers, Ann Arbor, Mich., 1974.

Chlorine gas dissolves in water thus

$$Cl_{2(g)} \rightleftharpoons Cl_{2(aq)}; \qquad K_H = 6.2 \times 10^{-2} \tag{7-51}$$

7.8.2. Chlorine Equilibria

As we have seen previously in the development of $p\epsilon$-pH diagrams, $Cl_{2(aq)}$ reacts with water, with one atom being oxidized to $Cl(+I)$, the other being reduced to $Cl(-I)$. This is often referred to as disproportionation.

$$Cl_{2(aq)} + H_2O \rightleftharpoons HOCl + H^+ + Cl^-; \qquad K_h = 4 \times 10^{-4} \tag{7-52}$$

Equation 7-52 is a combination of the two half-reactions,

$$e^- + \tfrac{1}{2}Cl_{2(aq)} \rightleftharpoons Cl^-$$

$$\underline{H_2O + \tfrac{1}{2}Cl_{2(aq)} \rightleftharpoons HOCl + H^+ + e^-}$$

$$Cl_{2(aq)} + H_2O \rightleftharpoons HOCl + H^+ + Cl^-$$

The HCl formed is completely dissociated under usual dilute aqueous solution conditions. Hypochlorous acid, on the other hand, is a relatively weak acid:

$$HOCl \rightleftharpoons H^+ + OCl^-; \qquad pK_a = 7.5 \tag{7-53}$$

From Eqs. 7-52 and 7-53 it is evident that the relative amount of the various oxidized chlorine species is a function of pH.

From the 25°C pK_a value it can be deduced that at pH 7.5 the activities $\{HOCl\}$ and $\{OCl^-\}$ are equal. At pH values below 7.5, HOCl predominates while above pH 7.5, OCl^- is the predominant species. This is of more than academic interest because the disinfecting ability of HOCl is generally regarded to be far greater than that of OCl^-; HOCl is about 80 to 100 times more effective at killing $E.\ coli$ than is OCl^-.

We can illustrate the distribution of the various chlorine species with pH using a distribution diagram in which the fraction of the total aqueous chlorine that is a particular species ($Cl_{2(aq)}$, HOCl, or OCl^-) is plotted versus pH for a fixed Cl^- concentration. Figure 7-21 is such a diagram for

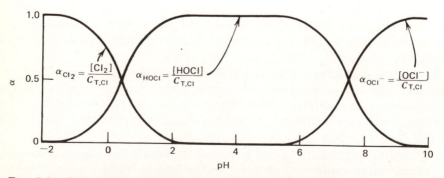

Fig. 7-21. Distribution diagram for chlorine species, 25°C, $[Cl^-] = 10^{-3}\ M$, $C_{T,Cl} = [Cl_2] + [HOCl] + [OCl^-]$.

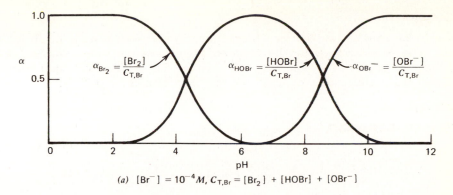

(a) $[Br^-] = 10^{-4}M,\ C_{T,Br} = [Br_2] + [HOBr] + [OBr^-]$

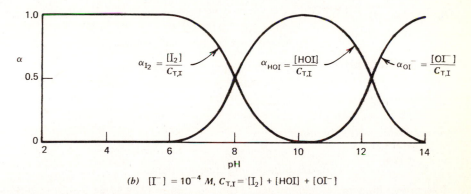

(b) $[I^-] = 10^{-4}\ M,\ C_{T,I} = [I_2] + [HOI] + [OI^-]$

Fig. 7-22. Distribution diagrams for (a) aqueous bromine and (b) iodine at 25°C.

a fixed $[Cl^-]$ of 10^{-3} M (35.5 mg Cl^-/liter). The diagram shows what previous diagrams in Figs. 7-4 and 7-6 (pϵ-pH and predominance area) have hinted: under typical natural water conditions, HOCl and OCl⁻ are the major chlorine species. $Cl_{2(aq)}$ is unimportant above pH 2. For comparison, Fig. 7-22 shows distribution diagrams for the other two halogens used as disinfectants (bromine and iodine). For bromine all three species $(Br_{2(aq)},$ HOBr and OBr⁻) could conceivably be important species under natural water conditions. With iodine only $I_{2(aq)}$ and HOI are important.

The diagram for chlorine shows that in the pH range of natural waters (pH 6 to 9), the relative amounts of HOCl and OCl⁻ are very sensitive to pH. As we have previously mentioned, HOCl and OCl⁻ differ greatly in their disinfecting ability so that pH control can be a critical factor in determining the degree of disinfection achieved by a certain level of chlorine.

Example 7-12

A chlorine dose of 1 mg/liter as Cl_2 satisfactorily disinfects a water at pH 7.0. What dose of chlorine would be required to achieve the same disinfection efficiency if the pH of the water was 8.5? Assume that HOCl is 100 times more effective as a disinfectant than OCl^- and that the temperature is 25°C.

Solution

From Fig. 7-21 at pH 7,

$$\alpha_{Cl_2} = 0$$
$$\alpha_{HOCl} = 0.76$$
$$\alpha_{OCl^-} = 0.24$$

Since $C_{T,Cl} = 1$ mg/liter as Cl_2, HOCl = 0.76 mg/liter as Cl_2, and $OCl^- = 0.24$ mg/liter.

The relative disinfection effectiveness, HOCl = 100, $OCl^- = 1.0$.

Disinfection effectiveness = concentration × relative disinfection effectiveness
$$= 0.76 \times 100 + 0.24 \times 1.0$$
$$= 76 \text{ units per mg/liter of chlorine}$$

From Fig. 7-21 at pH 8.5,

$$\alpha_{Cl_2} = 0$$
$$\alpha_{HOCl} = 0.09$$
$$\alpha_{OCl^-} = 0.91$$

The disinfection effectiveness is

$$= 0.09 \times 100 + 0.91 \times 1.0$$
$$= 9.9 \text{ units per mg/liter of chlorine}$$

To achieve the same effectiveness of disinfection at pH 8.5, we would therefore need 76/9.9 = 7.7 times the dose applied at pH 7.

7.8.3.   Chlorine Reactions with Inorganic Species

The form in which chlorine is added to a water affects some of the chemical properties of the water. The addition of chlorine gas to a water will lower its alkalinity because of the production of the strong acid and HOCl by the reaction in Eq. 7-52. However, if chlorine is dosed as a salt of hypochlorous acid:

$$NaOCl \rightarrow Na^+ + OCl^-$$

and

$$OCl^- + H_2O \rightleftharpoons HOCl + OH^-$$

there will be an increase in alkalinity to the extent that OCl^- reacts with H_2O. The use of calcium hypochlorite (HTH) increases both the alkalinity and the total hardness (Ca^{2+}) of a water

$$Ca(OCl)_2 \rightarrow Ca^{2+} + 2OCl^-$$
$$2OCl^- + 2H_2O \rightleftharpoons 2HOCl + 2OH^-$$

These considerations are important because they show that the form in which chlorine is added can affect the water chemistry in different ways. The addition of significant amounts of $Cl_{2(g)}$ to a low alkalinity water could depress pH to an unacceptably low value. Bleach addition would not cause such a pH depression. The use of HTH as a disinfectant in swimming pools is not to be recommended because the increase in hardness and alkalinity that accompanies its addition can aggravate scale ($CaCO_3$ precipitation) problems.

If we were to add (or dose) a water with a known amount of one or another of the forms of chlorine, and then after a time interval (contact time) were to analyze the water for chlorine (the chlorine residual), we would find less chlorine present than we added. We say that the water has a "chlorine demand," where after a certain contact time,

$$\text{chlorine demand} = \text{chlorine dose} - \text{chlorine residual}$$

The chlorine demand is the result of a variety of reactions in which chlorine is consumed by various constituents of the water and by decomposition. The reactions of chlorine with water constituents can be conveniently grouped into (1) those promoted by sunlight, (2) those with inorganic compounds, (3) those with ammonia, and (4) those with organic compounds. We will consider each of these types of reaction in this section as an illustration of the importance of kinetics and reaction mechanisms of redox reactions.

1. *Sunlight.*   Aqueous chlorine solutions are not stable when exposed to sunlight. Radiation in the ultraviolet region provides energy for the reaction of chlorine with water,

$$2HOCl + 2H^+ + 4e^- \rightleftharpoons 2Cl^- + 2H_2O$$

$$2H_2O \rightleftharpoons 4H^+ + O_2 + 4e^-$$

$$\overline{2HOCl \rightleftharpoons 2H^+ + 2Cl^- + O_2}$$

This reaction accounts for a major part of the chlorine consumption in outdoor swimming pools. It is the reason why bleach is sold in opaque plastic bottles and it should be the reason that you should never buy bleach in a clear bottle "on sale" from your neighborhood supermarket, that is, unless you are interested in purchasing a bottle of salt solution.

2. *Reactions with Inorganics.*   The reactions between chlorine and reduced inorganic compounds (e.g., Mn(II), Fe(II), NO_2^-, S($-$II)) are usually rapid. Thus when chlorine is used to oxidize ferrous iron to ferric iron, for example, in groundwater treatment, the reactions,

$$Cl_{2(aq)} + 2Fe^{2+} \rightleftharpoons 2Fe^{3+} + 2Cl^-$$

or

$$HOCl + H^+ + 2Fe^{2+} \rightleftharpoons 2Fe^{3+} + Cl^- + H_2O$$

proceed almost instantaneously at pH values near and above neutrality.

If we were to conduct an experiment in which a ferrous iron solution was dosed with increasing amounts of chlorine, we would obtain a plot of dosed chlorine versus residual chlorine like that in Fig. 7-23. No residual chlorine appears until all the Fe^{2+} has been oxidized to Fe^{3+}. When this has taken place, chlorine residuals appear and the further dosing of chlorine results in the appearance of chlorine equal to the additional dose. Reactions of chlorine with $S(-II)$, $Mn(II)$, and NO_2^- all follow this pattern except that the reaction with $Mn(II)$ only occurs at pH > 8.5. When present at high pH values, chlorine oxidation of sulfide tends to form polysulfides. It is a common complaint of operators of activated sludge plants that encounter partial nitrification that "It is impossible for me to maintain a chlorine residual." Almost without fail this is because of the presence of nitrite, NO_2^-, in the effluent that chlorine will oxidize to nitrate, NO_3^-.

$$H_2O + NO_2^- \rightleftharpoons NO_3^- + 2H^+ + 2e^-$$

$$\underline{2e^- + H^+ + HOCl \rightleftharpoons Cl^- + H_2O}$$

$$HOCl + NO_2^- \rightleftharpoons NO_3^- + Cl^- + H^+$$

Each mole of NO_2^- (46 g as NO_2^- or 14 g as N) oxidized requires 1 mole HOCl (52.5 g as HOCl or 71 g as Cl_2). Thus each mg NO_2^-–N/liter will consume approximately 5 mg/liter of HOCl as Cl_2.

3. Reactions with Ammonia. The reactions of chlorine with ammonia and organic nitrogen are quite different from those of chlorine with other inorganic and organic compounds. Let us illustrate the situation with the most studied of these reactions—that between chlorine and ammonia.

Simply stated, chlorine reacts with ammonia to produce a series of chlorinated ammonia compounds called chloramines (Table 7-8) and eventually oxidizes the ammonia to nitrogen gas ($N_{2(g)}$) or a variety of nitrogen-containing chlorine-free products (Table 7-9).

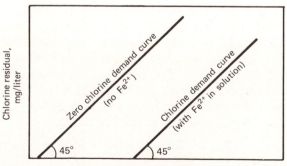

Fig. 7-23. Chlorine dose-residual curve for Fe^{2+} oxidation.

TABLE 7-8   Chlorinated Ammonia Compounds

Name	Formula
Monochloramine	NH_2Cl
Dichloramine	$NHCl_2$
Trichloramine or nitrogen trichloride	NCl_3

The reaction mechanism is complex, not completely understood, and the products vary with conditions such as the pH, ratio of Cl_2 added to ammonia present, and contact time.

Formation of the chloramines can be depicted as a stepwise process:

$$NH_{3(aq)} + HOCl \rightleftharpoons NH_2Cl + H_2O$$
$$NH_2Cl + HOCl \rightleftharpoons NHCl_2 + H_2O$$
$$NHCl_2 + HOCl \rightleftharpoons NCl_3 + H_2O$$

These reactions are written here in terms of HOCl, but Cl_2 may be used instead. Since

$$Cl_2 + H_2O \rightleftharpoons HOCl + H^+ + Cl^-$$

a mole of HOCl (or OCl^-) can be used interchangeably on a mole-for-mole basis with Cl_2 in order to determine the required dosage to achieve the reaction.

The oxidation of ammonia to the various end products listed in Table 7-9 requires different amounts of chlorine per unit amount of ammonia-nitrogen oxidized. These "Cl_2 reduced to NH_3–N oxidized" ratios are given in Table 7-9 for each possible end product on both a molar and a weight basis. These ratios are calculated from the pertinent redox reaction stoichiometry. For example, the two most common end products of

TABLE 7-9   Possible Products of Oxidation of Ammonia by Chlorine

Oxidation Products		Ratio of $\dfrac{Cl_2 \text{ reduced}}{NH_3\text{–N oxidized}}$	
Name	Formula	Mole Basis	Weight Basis
Hydrazine	N_2H_4	0.5	2.54
Hydroxylamine	NH_2OH	1.0	5.07
Nitrogen	N_2	1.5	7.61
Nitrous oxide	N_2O	2.0	10.1
Nitric oxide	NO	2.5	12.7
Nitrite	NO_2^-	3.0	15.2
Nitrogen tetroxide	N_2O_4	3.5	17.7
Nitrate	NO_3^-	4.0	20.3

ammonia oxidation are nitrogen gas and nitrate; for nitrogen gas we have

$$3Cl_2 + 2NH_3 \rightleftharpoons N_{2(g)} + 6HCl$$

The chlorine reduced to ammonia oxidized ratio, $[Cl_2]/[NH_3]$, is $\frac{3}{2} = 1.5$ on a mole basis or $1.5 \times (71$ g $Cl_2/mole)/(14$ g NH_3—N/mole$) = 7.6$ on a weight basis.

For oxidation to NO_3^- we have

$$3H_2O + NH_{3(aq)} \rightleftharpoons NO_3^- + 9H^+ + 8e^-$$

$$\underline{(2e^- + Cl_2 \rightleftharpoons 2Cl^-) \times 4}$$

$$4Cl_2 + NH_{3(aq)} + 3H_2O \rightleftharpoons 8Cl^- + NO_3^- + 9H^+$$

The chlorine reduced to ammonia-oxidized mole ratio is 4, while on a weight basis the ratio is $4 \times \frac{71}{14} = 20.3$.

Chloramine formation and oxidation of ammonia by chlorine combine to create a unique dose-residual curve for the addition of chlorine to ammonia-containing solutions (Fig. 7-24). As the chlorine dose increases, the chlorine residual at first rises to a maximum at a $[Cl_2]$ dose to $[NH_3]$ molar ratio of about 1.0. As the chlorine dose is increased further, the chlorine residual falls to a value close to zero. The chlorine dose corresponding to this minimum is called the "breakpoint" dose, and it occurs at a molar ratio of 1.5:1 to 2:1, depending upon solution conditions. The primary reaction that causes the residual chlorine concentration to decrease and thus to form the breakpoint is the breakpoint reaction, which can be represented as

$$2NH_2Cl + HOCl \rightleftharpoons N_{2(g)} + 3H^+ + 3Cl^- + H_2O \tag{7-54}$$

or as

$$3HOCl + 2NH_{3(aq)} \rightleftharpoons 3H^+ + 3Cl^- + 3H_2O + N_{2(g)} \tag{7-55}$$

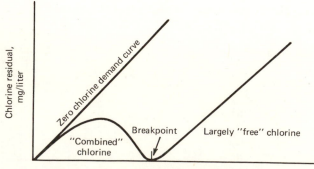

Fig. 7-24. Dose-demand curve for chlorine-ammonia reaction after approximately 1 hr at pH 7 to 8.

These are not mechanistic representations but they do account for the observed stoichiometry of 3 moles chlorine per 2 moles of NH_3. If the observed stoichiometry is greater than 3:2 (or 1.5), the increase in dose can generally be attributed to the formation of some NO_3^- and/or NCl_3.

An examination of the nature of the chlorine residual present prior to the breakpoint will reveal that it is composed almost entirely of chloramines. In the jargon of water chemistry these are referred to as "combined chlorine residual" as opposed to Cl_2, OCl^-, and $HOCl$ which are known as "free chlorine residuals."

As the chlorine dose is increased in excess of that required to produce the breakpoint, the increase in residual chlorine will be approximately equal to the excess. If we were to examine the nature of the chlorine residual after the breakpoint, we would find that it is largely free chlorine with some combined chlorine.

A more detailed examination of the types of chlorine residual would produce the picture presented in Fig. 7-25. Prior to the breakpoint, monochloramine and dichloramine are produced (i.e., under the "hump" of chlorine residual). Following the breakpoint, free chlorine ($HOCl$ and OCl^-) predominate but some dichloramine and trichloramine may be in evidence. A similar plot for nitrogen-containing species (Fig. 7-26) shows that there is no free ammonia present at Cl_2 dose to initial NH_3 molar ratios of greater than 1.0 and that nitrogen is *not eliminated* until the ratio is equal to or greater than 1.0. The figures further indicate that the chloramines increase in concentration up to a ratio of 1.0 and thereafter decrease. A reaction mechanism must therefore describe the formation of chloramines and their subsequent disappearance, and the formation of the known common oxidation products, nitrogen gas ($N_{2(g)}$) and nitrate (NO_3^-). Prior to examination of such a mechanistic model for the breakpoint chlorination of ammonia, let us look at the rates of some of the important constituent reactions.

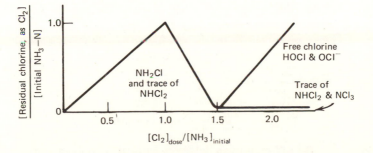

Fig. 7-25. Chlorine residual as a function of chlorine dose/initial ammonia ratio at near-neutral pH.

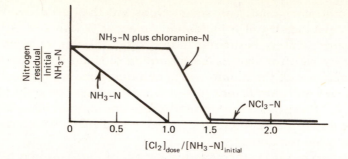

Fig. 7-26. Nitrogen residual as a function of the chlorine dose/initial ammonia ratio at near-neutral pH.

The formation of monochloramine has been shown to be an elementary reaction as follows.[15]

$$HOCl + NH_{3(aq)} \rightleftharpoons NH_2Cl + H_2O \qquad (7-56)$$

with a rate constant of 5.1×10^6 liter/mole-sec. The reaction is first order with respect to both HOCl and NH_3 and second order overall. Since the reaction occurs between HOCl and $NH_{3(aq)}$, pH can be expected to exert a dramatic effect on the reaction rate. Lowering the pH from 9 to 8 will increase [HOCl], ($pK_a = 7.5$), and decrease $[NH_{3(aq)}]$, ($pK_a = 9.3$); raising pH from the range for optimum reaction rate will decrease [HOCl] and increase $[NH_{3(aq)}]$. Because of this influence of pH on the speciation of reactants, the reaction rate can be written in terms of the dissociation constants of HOCl and $NH_{3(aq)}$. From the rate law, we find

$$-\frac{d[HOCl]}{dt} = \frac{d[NH_2Cl]}{dt} = k_1[HOCl][NH_3]$$

Let $C_{T,Cl(+I)} = [HOCl] + [OCl^-]$. Then

$$[HOCl] = \alpha_{HOCl}C_{T,Cl(+I)}$$

where $\alpha_{HOCl} = [HOCl]/C_{T,Cl(+I)}$.

Let $C_{T,NH_3} = $ total ammonia concentration, $[NH_{3(aq)}] + [NH_4^+]$. Then

$$[NH_3] = \alpha_{1,NH_3}C_{T,NH_3}$$

where $\alpha_{1,NH_3} = [NH_3]/C_{T,NH_3}$. Then

$$\frac{d[NH_2Cl]}{dt} = k_1\alpha_{HOCl}C_{T,Cl(+I)}\alpha_{1,NH_3}C_{T,NH_3}$$

[15] J. C. Morris, "Kinetics of Reactions Between Aqueous Chlorine and Nitrogenous Compounds," in *Principles and Applications of Water Chemistry*, S. D. Faust and J. V. Hunter, eds., John Wiley, New York, 1967.

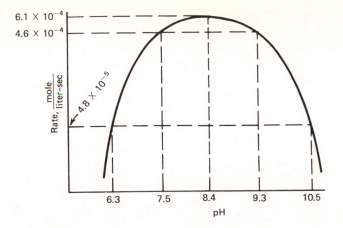

Fig. 7-27. Rate of monochloramine formation as a function of pH.

The α values at a particular pH can be determined from the mass balances and equilibrium constants.

Figure 7-27 shows the variation of the rate of monochloramine formation as a function of pH.

The back reaction in Eq. 7-56—decomposition of monochloramine to NH_3 and HOCl—is considerably slower than the forward reaction. It is an elementary reaction with a rate constant of 3×10^{-5} sec^{-1}.[16] The ratio of forward to back reaction rate constants yields an equilibrium constant for monochloramine formation of 1.7×10^{11}, Eq. 7-56.

The formation of dichloramine,

$$NH_2Cl + HOCl \rightleftharpoons NHCl_2 + H_2O \qquad (7\text{-}57)$$

is also elementary and thus is first order with respect to each reactant and second order overall. The reaction is catalyzed by H^+. The rate constant has the general form:

$$k_2 = 3.4 \times 10^2(1 + 5 \times 10^4\ [H^+])$$

The formation of NCl_3,

$$NH_2Cl + HOCl \rightleftharpoons NCl_3 + H_2O \qquad (7\text{-}58)$$

is first order with respect to each reactant and second order overall.[17] The rate constant for formation at pH 4.0 is 1.6 liter/mole-sec, a value that is

---

[16] J. C. Morris, "Kinetics of Reactions Between Aqueous Chlorine and Nitrogenous Compounds," in *Principles and Applications of Water Chemistry*, S. D. Faust and J. V. Hunter, eds., John Wiley, New York, 1967.

[17] J. L. S. Saguinsin and J. C. Morris, "The Chemistry of Aqueous Nitrogen Trichloride," chapter in *Disinfection*, J. D. Johnson, ed., Ann Arbor Science, Ann Arbor, Mich., 1975.

about two orders of magnitude less than for $NHCl_2$ and NH_2Cl. The rate of NCl_3 formation increases significantly as the pH decreases below pH 3 because of the presence of $Cl_{2(aq)}$, which is about 10^4 times more reactive toward $NHCl_2$ than is HOCl. NCl_3 decomposes by hydrolysis to $NHCl_2$ at pH values greater than 7, with the decomposition rate increasing rapidly as pH increases in the range 7 to 9. Sanguinsin and Morris[17] found that the decomposition reaction could be given as

$$NCl_3 + H_2O \rightarrow NHCl_2 + HOCl$$

with a rate law of

$$\frac{-d[NCl_3]}{dt} = k[NCl_3]$$

where $k = 3.2 \times 10^{-5}(1 + 5.88 \times 10^{-5} [OH^-])sec^{-1}$, thus indicating catalysis by OH^- or reaction of the NCl_3 with OH^-. They also found that NCl_3 would react with NH_3 and thereby be removed from solution.

Because of the effect of pH on the formation of chloramine, pH has a profound effect on the species of combined chlorine that make up the "hump" of the breakpoint curve. At pH 7 and above, essentially only NH_2Cl is present, while below 7, significant amounts of $NHCl_2$ are found, particularly at mole ratios between 1:1 and the breakpoint.[18] $NHCl_2$ at mole ratios less than 1:1 is favored only below a pH of approximately 5. NCl_3 will not appear in significant quantities under the "hump" unless the pH is of the order of 4 or lower.

The mechanism of the breakpoint reaction is, as yet, unresolved. The most comprehensive model is that of Wei and Morris[19] who proposed a scheme consisting of reactions that formed chloramines; reactions that converted chloramines into a hypothetical intermediate, NOH; and reactions by which NOH decomposed to form N_2 and NO_3^- (Table 7-10).

Saunier and Selleck[20] examined this model and experimentally determined the rate constants for each of the reactions in Table 7-10. They concluded that the model fits breakpoint chlorination results. They suggested, but did not prove, that hydroxylamine (NH_2OH) and possibly hydrazine (N_2H_4) could be intermediates in the breakpoint reaction—perhaps the hypothetical NOH of Wei and Morris.

[18] A. T. Palin, "Water Disinfection—Chemical Aspects and Analytical Control," in *Disinfection*, J. D. Johnson, ed., Ann Arbor Science, Ann Arbor, Mich., 1975.

[19] I. W. Wei and J. C. Morris, "Dynamics of Breakpoint Chlorination," in *Chemistry of Water Supply, Treatment and Distribution*, A. J. Rubin ed., Ann Arbor Science, Ann Arbor, Mich., 1974.

[20] B. Saunier and R. E. Selleck, "Kinetics of Breakpoint Chlorination and of Disinfection," Sanitary Engineering Research Laboratory Report 76-2, University of California, Berkeley, May 1976.

TABLE 7-10   Model Reactions for Breakpoint Chlorination

$HOCl + NH_3 \rightarrow NH_2Cl + H_2O$	(1)
$HOCl + NH_2Cl \rightarrow NHCl_2 + H_2O$	(2)
$HOCl + NHCl_2 \rightarrow NCl_3 + H_2O$	(3)
$NCl_3 + H_2O \rightarrow NHCl_2 + HOCl$	(4)
$NHCl_2 + H_2O \rightarrow NOH + 2H^+ + 2Cl^-$	(5)
$NOH + NH_2Cl \rightarrow N_2 + H_2O + H^+ + Cl^-$	(6)
$NOH + NHCl_2 \rightarrow N_2 + HOCl + H^+ + Cl$	(7)
$NOH + 2HOCl \rightarrow NO_3^- + 3H^+ + 2Cl^-$	(8)

7.8.4.   Chlorine Reactions with Organic Substances

Chlorine at concentrations used for water and wastewater treatment readily reacts with organic compounds in the water. In some reactions, such as those with organic nitrogen compounds and phenols, Cl is substituted for a hydrogen atom, thus producing the chlorinated compound. Chlorine can also be incorporated into a molecule by addition reactions, or it may react with a compound to oxidize it without chlorinating it. Although some of the chlorinated organics that form during the chlorination of natural and wastewaters are known, many still remain to be identified.

1. *Reactions with Organic Nitrogen.*   Chlorine reacts readily with many organic nitrogen compounds just as it does with ammonia. The organic amines, which have the group $-NH_2$, $-NH-$, or $-N=$ as part of their molecule, are very common. The elementary reaction with methylamine, CH_3NH_2, is typical

$$HOCl + CH_3NH_2 \rightarrow CH_3NHCl + H_2O$$

The rate constant for this reaction, $k = 10^{8.56}$ liter/mole-sec, is higher than that found for formation of NH_2Cl.[21] Also, the formation of dichloromethylamine,

$$HOCl + CH_3NHCl \rightarrow CH_3NCl_2 + H_2O$$

is acid catalyzed just as is the formation of dichloramine from monochloramine. Analysis of other similar reactions showed that as the base strength of the amine (indicated by the basicity constant K_b) increased, the rate of the reaction with HOCl increased. Amides, compounds in which the group $-OCNH_2$ or $-OCNH-$ is incorporated, behave in a similar fashion to the amines but do not react as rapidly because they are not as basic.

[21] J. C. Morris, "Kinetics of Reactions Between Aqueous Chlorine and Nitrogen Compounds" in *Principles and Applications of Water Chemistry*, S. D. Faust and J. V. Hunter, eds., John Wiley, New York, 1967.

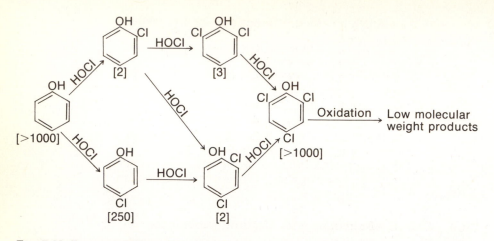

Fig. 7-28. Reaction scheme for the chlorination of phenol. [ ] indicates odor threshold concentration (μg/liter). From R. H. Burttschell, A. A. Rosen, F. M. Middleton, and M. B. Ettinger, "Chlorine Derivatives of Phenol Causing Taste and Odor," *J. Am. Water Works Assoc., 51:* 205–214 (1959). Reprinted by permission of the American Water Works Association, © 1959.

The reaction of chlorine with organic nitrogen is important because it exerts a demand, that is, it requires that more chlorine be added to achieve a given level of disinfection. The chlorine attached to the nitrogen has not in all cases lost its capacity to oxidize, but it is generally not as potent an oxidant as HOCl or NH_2Cl. Some chlorinated organic nitrogen compounds react as residual chlorine, along with HOCl, OCl^-, NH_2Cl, and so forth, in the various analytical procedures for residual chlorine.

2. Reactions with Phenols. Chlorine readily substitutes on to phenol, ⬡—OH, and compounds containing the phenolic group. Because these compounds can be present in water supplies resulting from industrial discharges or natural decay processes, and because several of the chlorinated phenols are very odorous, their formation has long been a concern of water treatment plant operators. A typical reaction between HOCl and phenol is shown in Fig. 7-28. The threshold odor numbers of each chlorinated species indicating the concentrations that will just produce a detectable odor are also shown. The amount of any species present at a given time depends upon pH, chlorine dose, phenol concentration, and temperature.[22] Predicted threshold odor versus time curves that show the effect of reactant concentration and pH are presented in Fig. 7-29. The curves give an indication of the parameters which can be changed to control a chlorophenol odor problem.

[22] G. F. Lee, "Kinetics of Reactions Between Chlorine and Phenolic Compounds," in *Principles and Applications of Water Chemistry,* S. D. Faust and J. V. Hunter, eds., John Wiley, New York, 1967.

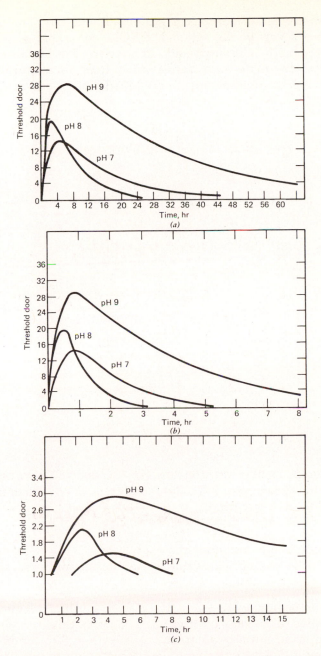

Fig. 7-29. Threshold odor from chlorination of phenol. (a) Initial chlorine 0.2 ppm, initial phenol 50 ppb; (b) initial chlorine 1.0 ppm, initial phenol 50 ppb; and (c) initial chlorine 0.2 ppm, initial phenol 5.0 ppb. All were 25°C. From G. F. Lee, chapter in *Principles and Applications of Water Chemistry*, S. D. Faust and J. V. Hunter, eds., John Wiley, 1967. Reprinted by permission of John Wiley & Sons, Inc.

3. Trihalomethane Formation.   Trihalomethanes have the general form CHX_3 where X can be Cl, Br, or I. $CHCl_3$, chloroform, is of particular interest because it is a suspected carcinogen. The health effects of the other species are unknown. A series of reactions demonstrating the basic steps by which chloroform may be produced during water treatment is as follows:

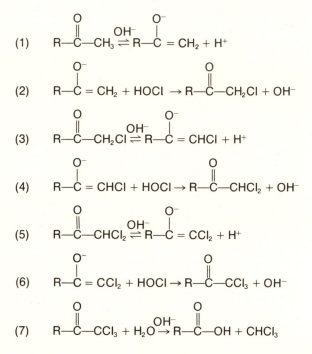

The slow steps in the reaction are (1), (3), and (5); because these are favored in the presence of OH^-, the reaction proceeds much more rapidly at high pH than at low pH.

Not all compounds that have the acetyl group, $-\overset{\overset{\displaystyle O}{\|}}{C}-CH_3$, react rapidly enough to pose a problem during water treatment.[23] For example, the reaction with acetone, $CH_3-\overset{\overset{\displaystyle O}{\|}}{C}-CH_3$, at concentrations found in polluted water supplies is too slow for it to be of concern. However, since there is organic matter in natural water and wastewater with functional groups that undergo attack by chlorine to form chloroform rapidly, its formation

[23] J. C. Morris and B. Baum, "Precursors and Mechanisms of Haloform Formation in the Chlorination of Water Supplies," chapter in *Water Chlorination*, Vol. 2, R. L. Jolley, H. Gorchev, and D. H. Hamilton, Jr., eds., Ann Arbor Science, Ann Arbor, Mich., 1978.

is of concern during water treatment. The major source of these groups appears to be in the so-called humic substances, which were discussed in Section 5-7.

Many treated waters contain not only chloroform but other chlorinated and brominated trihalomethanes such as $CHCl_2Br$, $CHClBr_2$, and $CHBr_3$. These brominated species form because aqueous chlorine converts Br^- ion in the water to HOBr. The bromine compounds then react with the organic matter in the same way as HOCl.[24] If I^- is also present in the water, presumably HOCl will also oxidize it and the product will react similarly to HOCl, producing additional trihalomethanes species. Dichloroiodomethane is one of the species commonly observed.

4. *Addition and Oxidation Reactions.*   Chlorine can undergo an addition reaction if the organic compound has a double bond,

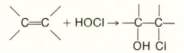

For many compounds with double bonds, this reaction is too slow to be of importance in water treatment.[25] However, more must be learned of the characteristics of some of the substances that are found in water before we can be certain that this will always be the case.

Chlorine can also oxidize organic compounds as follows:

$$R{-}CHO + HOCl \rightarrow R{-}\overset{\overset{\displaystyle O}{\displaystyle \|}}{C}{-}OH + H^+ + Cl^-$$

For example, continued reaction of 2,4,6-trichlorophenol (see Fig. 7-28) will break the aromatic ring of the phenol forming 2-carbon residues,

such as oxalic acid, $HO{-}\overset{\overset{\displaystyle O}{\displaystyle \|}}{C}{-}\overset{\overset{\displaystyle O}{\displaystyle \|}}{C}{-}OH$, and Cl^-, and continued reaction with chlorine results in conversion of the 2-carbon residues to CO_2 and H_2O.

The oxidation reaction with a carbohydrate (e.g., lactose) or with a fat or fatty acid such as oleic acid is generally quite slow. The organic compound is eventually converted to CO_2 and H_2O. The reactions are generally far too slow to be considered significant in the context of the reaction times involved in various chlorination practices. The dose-residual curves presented in Fig. 7-30 represent the slow progress of the

[24] J. C. Morris, "The Chemistry of Aqueous Chlorine in Relation to Water Chlorination," in *Water Chlorination—Environmental Impact and Health Effects*, Vol. 1, R. L. Jolley, ed., Ann Arbor Science, Ann Arbor, Mich., 1978.

[25] J. C. Morris, "The Chemistry of Aqueous Chlorine in Relation to Water Chlorination," in *Water Chlorination—Environmental Impact and Health Effects*, Vol. 1, R. L. Jolley, ed., Ann Arbor Science, Ann Arbor, Mich., 1978.

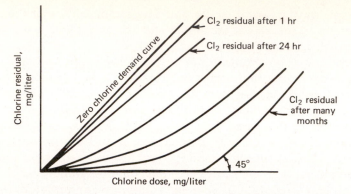

Fig. 7-30. Chlorine dose-residual curves for lactose oxidation.

reaction. The pertinent half-reactions and the net reaction for lactose $(C_{12}H_{22}O_{11})$ are

$$13H_2O + C_{12}H_{22}O_{11} \rightleftharpoons 12CO_2 + 48H^+ + 48e^-$$

$$48e^- + 24H^+ + 24HOCl \rightleftharpoons 24Cl^- + 24H_2O$$

$$C_{12}H_{22}O_{11} + 24HOCl \rightleftharpoons 12CO_2 + 11H_2O + 24Cl^- + 24H^+$$

7.9. BIOLOGICALLY IMPORTANT REDOX REACTIONS

Many of the phenomena encountered in natural waters and many of the processes that make up wastewater treatment are due to redox reactions that are catalyzed (or mediated) by biological systems, for example, bacteria, algae, and other microorganisms. It is important to realize that microorganisms, like inanimate objects, must obey the laws of thermodynamics. One might wonder about this statement in view of the well-known observation that photosynthetic organisms have the ability to synthesize organic matter from CO_2 and H_2O–a reaction that is not thermodynamically spontaneous. However, it must be realized that photosynthetic organisms must have an input of energy (sunlight) to drive the reaction. They are able to mediate a reaction that goes in the opposite direction to thermodynamic predictions only because they have devised a means of capturing energy from an external source and using it to reverse the internal energy gradient of the system. Because photosynthetic organisms are active in most natural waters and in some waste treatment systems we must be careful in some calculations, to treat them as steady-state open systems with an energy input from a source external to the system, rather than as closed systems.

Microorganisms (or any other organisms) do not *perform* chemical reactions, they *catalyze* them and *use* them for purposes such as deriving energy for metabolic processes or source materials for biosynthesis. Nevertheless the importance of the microbial catalysis of redox reactions cannot be stressed enough. We have already seen that the production of acid mine drainage is made possible by the microbial catalysis of an oxidation that in the absence of microorganisms would proceed very slowly if at all. In this section we will examine the reactions of the nitrogen cycle, the sequence of the use of various electron acceptors in biological systems, and the relationship between the amount of energy obtained from redox reactions by microorganisms and the amount of cell material (yield) that they can produce.

7.9.1. The Nitrogen Cycle

The way in which nitrogen circulates on the earth's surface and in its atmosphere is usually depicted as the nitrogen-cycle (Fig. 7-31). Many of the reactions in the nitrogen cycle are microbially catalyzed redox reactions. These are the oxidation of NH_4^+ to NO_2^- and then to NO_3^- (nitrification), the reduction of NO_3^- to NO_2^- and then to NH_4^+ (nitrate reduction), the reduction of NO_3^- to $N_{2(g)}$ (denitrification), and the reduction of $N_{2(g)}$ to NH_4^+ (nitrogen fixation). The incorporation of NH_4^+ into nitrogen-containing organic matter (amination) or its release (deamination or ammonification) is the only nonredox reaction involving a nitrogen transformation in the entire nitrogen cycle.

The various redox reactions of inorganic nitrogen containing species are presented in Table 7-1. These reactions can be used to construct a pC-

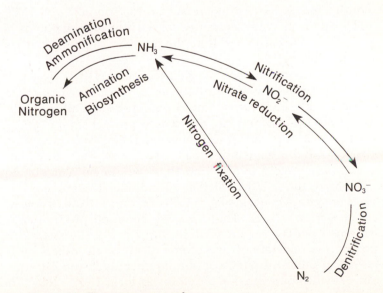

Fig. 7-31. The nitrogen cycle.

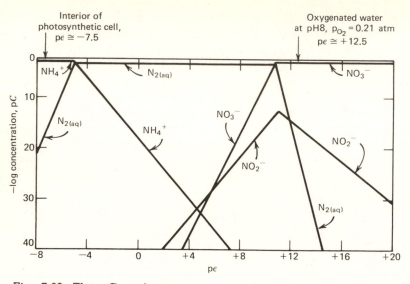

Fig. 7-32. The pC-pε diagram for nitrogen system, $C_{T,N} = 10^{-3}$, in equilibrium with $P_{N_2} = 0.77$ atm.

pε diagram for a model system at pH 8 containing a total dissolved nitrogen concentration of 10^{-3} M ($C_{T,N} = 10^{-3} = 2[N_{2(aq)}] + [NH_4^+] + [NO_2^-] + [NO_3^-]$). In the pε range where $N_{2(aq)}$ predominates, the solution is approximately in equilibrium with an atmosphere containing a partial pressure of 0.77 atm of $N_{2(g)}$—the typical $N_{2(g)}$ content of the normal surface earth's atmosphere (Fig. 7-32). Marked on the pε axis is the pε value of water in equilibrium with an atmosphere containing 0.21 atm of O_2 at pH 8 (pε = +12.5) and the pε value estimated for the interior of a photosynthetic cell (pε = −7.0 to −7.5, estimated from the pε of the reaction in which CO_2 is converted to glucose).

Since all of the major nitrogen-containing species, NH_4^+, $N_{2(aq)}$ and NO_3^- have regions of predominance within the pε range over which microbially catalyzed reactions can occur, it would be expected that their interconversion would be rapid. Moreover, one would predict from Fig. 7-32 that, in oxygenated water at pH 8, all dissolved nitrogen gas should be converted to NO_3^-. If this were indeed the case, then the earth's atmosphere would be virtually devoid of $N_{2(g)}$ and NO_3^- would be a major aqueous species. We know that neither of these predictions are accurate, since the earth's atmosphere contains $P_{N_2} = 0.77$ atm and nitrate is a very minor component of natural water. The disparity between prediction and observation stems from the slowness and complexity of reactions involving $N_{2(g)}$. Nitrogen gas consists of two atoms of nitrogen joined by a triple bond (N≡N). The other nitrogen-containing species in the nitrogen cycle all contain one nitrogen atom. Therefore, to form NO_3^-, NO_2^-, or NH_4^+ from $N_{2(aq)}$ or $N_{2(g)}$, the very strong N≡N bond must be broken. Because of this the rate of these reactions is quite slow. However, the conversion of N_2 to NH_4^+ does take place in organisms that fix $N_{2(g)}$:

$$N_{2(g)} + 8H^+ + 6e^- \rightarrow 2NH_4^+; \qquad p\epsilon^\circ = +4.68$$

Since this reaction, at pH 8, is only spontaneous below $p\epsilon \cong -6$, it is necessary for the microorganisms to generate a highly reducing local environment (such as that produced by photosynthesis) to drive the nitrogen fixation reaction. Such environments exist in the interior of photosynthetic cells including the blue-green algae that fix nitrogen (e.g., *Microcystis*, *Anabaena*, *Aphanizomenon*) and in the symbiotic association in root nodules between bacteria of the genus *Rhizobium* and certain leguminous plants.

Denitrification [the conversion of NO_3^- to $N_{2(aq)}$] is a common reaction catalyzed by many bacteria and proceeds rapidly. However, it appears that microbial catalysts are unable to catalyze the reverse of the denitrification reaction ($N_{2(aq)} \rightarrow NO_3^-$).

From this discussion it is evident that reactions involving the conversion of $N_{2(aq)}$ to NH_4^+ or NO_3^- proceed slowly. However, it is a well-supported observation that the interconversion of the other nitrogen-containing species of the nitrogen cycle proceeds relatively easily and quite rapidly. Because of these observations it is profitable to redraw the pC-pϵ diagram for $C_{T,N} = 10^{-3}$ and pH 8, omitting $N_{2(aq)}$ and $N_{2(g)}$. This is tantamount to converting the system from an open system in contact with an atmosphere containing $N_{2(g)}$ to a closed system in which $N_{2(g)}$ and therefore $N_{2(aq)}$ is absent. Such a diagram is drawn in Fig. 7-33. It shows that NH_4^+, NO_2^- and NO_3^- are all major species in the very narrow pϵ range from +4.5 to +6.2. Such a presentation is consistent with the observation that these species are easily and rapidly interconvertible (by nitrification and nitrate reduction). Also note that the pϵ region over which NO_2^- is the predominant species is extremely narrow. This is consistent with the observation that NO_2^- is a transitory species in natural waters and waste treatment processes—it only exists in significant quantities over a very narrow pϵ range and it is readily interconverted to NO_3^- and NH_4^+.

7.9.2.   Electron Acceptors in Microbial Systems

It is common in waste treatment practice and in water pollution control to classify environments or processes as either "aerobic" or "anaerobic." These gross classifications are usually made on the basis of whether or not dissolved oxygen is present in the water or wastewater. Aerobic systems are those in which oxygen is present, and anaerobic environments are those in which oxygen is absent. It is observed that the chemical reactions that take place and the types of organisms that predominate are different in aerobic and anaerobic systems. A closer examination of natural waters and waste treatment processes reveals that the dividing

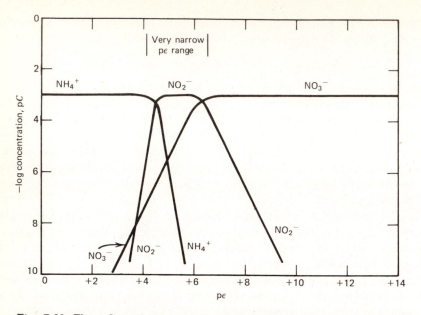

Fig. 7-33. The pC-pϵ diagram for nitrogen species ($C_{T,N} = 10^{-3} M$, pH 8) omitting nitrogen gas and aqueous N_2.

line between aerobic and anaerobic environments is not sharply drawn. To say that a water or process was aerobic would be tantamount to stating that in the aerobic system the redox conditions are controlled by the reaction

$$O_{2(g)} + 4H^+ + 4e^- \rightarrow 2H_2O; \qquad p\epsilon^\circ = 20.8$$

whereas in the anaerobic environment (oxygen absent) the redox conditions are controlled by another reaction. When we examine which "other" reactions are likely candidates for controlling the redox environment of anaerobic processes, we find that there are many. Table 7-11 lists these reactions in the order of descending $p\epsilon^\circ$ values.

Microorganisms use these reduction reactions to consume the electrons generated by the oxidation of their energy-yielding substrate. Although this substrate can be any of a variety of substances, let us take for an example the oxidation of formaldehyde (CH_2O) to CO_2 and H_2O by the reaction:

$$CO_{2(g)} + 4H^+ + 4e^- \rightarrow CH_2O + H_2O; \qquad p\epsilon^\circ = -1.20$$

This reaction can be coupled with any one of the electron accepting (reduction) reactions listed in Table 7-11. From the ΔG° values of these coupled reactions it can be deduced that the amount of free energy that a microorganism can obtain from the coupled redox reactions listed in Table 7-11 is in direct proportion to its $p\epsilon^\circ$ value.

TABLE 7-11   Electron Acceptance Reactions (Reduction Reactions) (Energy-Producing Reactions) in Aerobic and Anaerobic Systems

	$p\epsilon^{\circ}$	Name of Reaction
Aerobic		
$O_{2(g)} + 4H^+ + 4e^- \rightarrow 2H_2O$	+20.8	Aerobic respiration
Anaerobic		
$2NO_3^- + 12H^+ + 10e^- \rightarrow N_{2(g)} + 6H_2O$	+21.0	Denitrification
$NO_3^- + 10H^+ + 8e^- \rightarrow NH_4^+ + 3H_2O$	+14.9	Nitrate reduction
$CH_2O + 2H^+ + 2e^- \rightarrow CH_3OH$	+3.99	Fermentation
(Formaldehyde) (Methanol)		
$SO_4^{2-} + 9H^+ + 8e^- \rightarrow HS^- + 4H_2O$	+4.13	Sulfate reduction
$CO_{2(g)} + 8H^+ + 8e^- \rightarrow CH_{4(g)} + 2H_2O$	+2.87	Methane fermentation

This trend is supported by many observations in natural waters and waste treatment systems. It is generally supported by the observation that "anaerobic" communities cannot develop in oxic (or oxygen-containing) environments. Using the same substrate, the anaerobic organisms cannot get as much energy per mole by its oxidation as can the aerobic organism. Therefore, they cannot grow as fast as the aerobes and they become overgrown. Some anaerobes, such as the methane fermenters, are poisoned by oxygen. In this general sense, the redox conditions of a water determine the type of biological community that develops. The effect is not always as dramatic as indicated, since many organisms—the so-called facultative aerobes—have the ability to use both oxygen and nitrate as electron acceptors. They only use nitrate when the concentration of dissolved oxygen is very low, for the simple reason that they do not get as much energy per mole out of doing so. Greenwood[26] has found that, for soil bacteria, the concentration of dissolved oxygen in contact with the bacteria must fall to about 4×10^{-6} M (corresponding to an atmosphere containing 0.3 percent O_2) before the use of nitrate as an electron accepter starts. Horstkotte et al.[27] found that dentrification of nitrate ion in a secondary effluent would proceed in a biofilm when the solution concentration of dissolved oxygen was approximately 1 mg/liter ($\sim 3.5 \times 10^{-5}$ M).

When nitrate is utilized, the reaction to produce $N_{2(g)}$ (denitrification) is favored over the reduction of NO_3^- through NO_2^- to NH_4^+ (nitrate reduction) for the same reason: it yields more usable energy to the organism that catalyzes it. For example, in a nitrifying activated sludge plant, there is

[26] D. J. Greenwood, "Nitrification and Nitrate Dissimilation in Soil. Part 2. Effect of Oxygen Concentration," *Plant and Soil*, 17: 365–378 (1962).

[27] G. A. Horstkotte, D. G. Niles, D. S. Parker, and D. H. Caldwell, "Full Scale Testing of a Water Reclamation System," *J. Water Pollut. Control Fed.*, 46: 181–197 (1974).

usually dissolved oxygen and nitrate present in the aeration basin. When the activated sludge settles in the secondary sedimentation basin, a sludge blanket forms in which high concentrations of activated sludge continue to respire, using electron acceptors. If the sludge is maintained in this quiescent, unmixed and unaerated, blanket for an extended period of time, all of the dissolved oxygen will be consumed. Then some of the microorganisms in the activated sludge (those that have the enzyme, nitrate reductase) will turn to the nitrate for an electron acceptor. Denitrification (not nitrate reduction) will occur with the accompanying production of the sparingly soluble $N_{2(g)}$. Small bubbles of $N_{2(g)}$ are produced within the activated sludge blanket (possibly within the activated sludge floc itself), and they rise to the surface of the secondary sedimentation basin, carrying with them particles of activated sludge which are more often than not carried over into the secondary effluent. This is the vexing problem of "blanket rising" or "sludge rising." In this example the oxidation of ammonia to nitrate (nitrification) and denitrification occur in the same process with the only difference being environments where dissolved oxygen is present or absent.

There are other such examples. Many unsaturated soils are known to convert NH_4^+ to $N_{2(g)}$ via NO_3^- (i.e., nitrification then denitrification). They achieve this because there are local oxic and anoxic environments in the soil waters that, respectively, allow nitrification and denitrification to proceed. In stratified lakes, nitrification may occur in the oxygenated epilimnion (upper layer) and denitrification in the hypolimnion (bottom water) and in the sediment pore water where dissolved oxygen concentrations fall to zero. The nitrification and denitrification process is important in preserving the fishery in Indian Creek Reservoir in the Sierra Nevada mountains. This reservoir is fed by the tertiary effluent from the City of South Lake Tahoe sewage treatment plant. The effluent has at times contained 15 to 20 mg NH_4^+–N/liter. Levels of ammonia of this magnitude are toxic to fish, yet in the reservoir there is a thriving fishery. This is achieved because the top waters of the lake nitrify the ammonia to nitrate and this is reduced to $N_{2(g)}$ by the anoxic bottom waters. The summer concentrations of nitrogen species of the reservoir[28] are approximately 4 mg NO_3^-–N/liter and 4 mg NH_4–N/liter.

When domestic wastewater is treated by chemical precipitation followed by carbon adsorption, one of the problems encountered is that the carbon columns become highly odorous. The odor derives from the reduction of SO_4^{2-} to HS^- by microorganisms growing within the carbon column on biodegradable adsorbed organic matter. This is the sulfate-reduction reaction. Attempts to alleviate the situation by injecting air into the columns have met with little success because the oxidation of sulfide by oxygen is not always a rapid process, especially if the sulfides

[28] Eutrophication of Surface Water—Lake Tahoe's Indian Creek Reservoir, EPA Report No. 660/3-75-003, February 1975.

TABLE 7-12   Control Methods for Sulfide in Carbon Adsorption Columns

Sulfide Control Method	Mean Total Sulfide Content in Carbon Column Effluent, mg S/liter
(1) Surface and backwash	2.9
(2) Surface and backwash plus increase of column influent to 2–6 mg/liter DO	1.9
(2) plus 40 mg Cl_2/liter to column influent	1.1
(1) plus 5.3 mg NO_3–N/liter	0.02
(1) plus 5.4 mg NO_3–N/liter	0

Source: L. S. Directo, C. L. Chen and I. J. Kugelman, "Pilot-Plant Study of Physical-Chemical Treatment," J. Water Pollut. Control Fed, 49: 2081 (1977).

are present in a particulate form. One successful method of preventing the sulfide odor has been to feed sodium nitrate to the carbon column influent. In the presence of NO_3^-, sulfide generation ceases because the microorganisms would rather reduce NO_3^- to $N_{2(g)}$ than SO_4^{2-} to HS^-, since they get far more energy out of it (Table 7-12). The data in Table 7-12 are from work conducted by Directo et al.[29] in which both oxygen and chlorine were unsuccessfully used to eliminate sulfides from carbon column effluents. The successful use of nitrate for preventing sulfide production in the columns is clearly evident.

In Table 7-11 the electron acceptance reaction with the lowest energy yield for microorganisms is the reduction of $CO_{2(g)}$ to methane, the so-called methane fermentation reaction. The reaction occurs only under strictly anaerobic conditions and the organisms using the reaction grow very slowly compared with aerobic heterotrophic bacteria because they obtain so little energy per mole of substrate reduced. When the methane fermentation reaction occurs, we can predict from Table 7-11 that SO_4^{2-} will be absent having been completely reduced to sulfide. Using this observation and the knowledge that many toxic heavy metals form insoluble sulfides, Lawrence and McCarty[30] derived an ingenious technique for detecting the presence of potentially harmful concentrations of toxic metals in the sludge fed to anaerobic sludge digestion units at waste treatment plants. The series of redox, heterogeneous, and acid-base equilibria shown in Fig. 7-34 control the atmospheric H_2S concentration (P_{H_2S}) in the digester gas. With a normal (nontoxic level of metals in the digesting sludge, the P_{H_2S} value will be constant and can be

[29] L. S. Directo, C. L. Chen, and I. J. Kugelman, "Pilot-Plant Study of Physical-Chemical Treatment," J. Water Pollut. Control Fed., 49: 2081 (1977).

[30] A. W. Lawrence and P. L. McCarty, "The Role of Sulfide in Preventing Heavy Metal Toxicity in Anaerobic Digesters," J. Water Pollut. Control Fed., 37: 392 (1965).

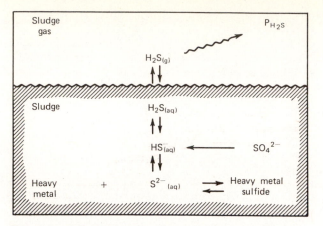

Fig. 7-34. Schematic of controls on P_{H_2S} in an anaerobic sewage sludge digester.

measured and recorded. If a high concentration of heavy metals is fed to
the digester, more metal sulfide solids will form and the concentrations
of S^{2-}, HS^- and $H_2S_{(aq)}$ will decrease. These will be accompanied by a fall
in P_{H_2S}, since its concentration is determined by $[H_2S_{(aq)}]$. By determining
the acceptable range of P_{H_2S} levels in the digester gas, therefore, an early
warning system for toxic levels of heavy metals is available.

7.9.3.   Microbial Yields

In the previous section we saw how microorganisms use or become
established to use electron-acceptance reactions strictly in the order of
the amount of energy per mole of electrons consumed that they can derive
from them. The amount of energy available to a microorganism is,
however, a function of both the electron-generating (oxidation) half-
reaction and the electron-acceptance (reduction) reaction. Organisms that
oxidize organic matter (CH_2O) all the way to CO_2 and H_2O (respiration)
obtain more energy per mole than do organisms that only take "CH_2O"
as far as, for example, formic acid (an example of a fermentation reaction).
The $p\epsilon$ values for the following equations illustrate this:

$$CO_2 + 4H^+ + 4e^- \rightarrow CH_2O + H_2O; \qquad p\epsilon° = -1.20$$
$$HCOO^- + 3H^+ + 2e^- \rightarrow CH_2O + H_2O; \qquad p\epsilon° = +2.82$$

Organisms that use substrates with lower energy content for energy
generation produce even less energy per mole of substrate oxidized.

The result of the various combinations of redox couples used by different
microorganisms for energy generation is that the amount of energy
available per mole of substrate for processes such as the biosynthesis of
new cell material varies widely from organism to organism. Expressed
in microbial terms, we say that the yield (g cell material per g substrate

consumed) varies widely. McCarty[31] has related cell yield for various microorganisms to the amount of free energy produced by coupled substrate oxidation and electron-accepting reduction reactions. Some of his data are plotted in Fig. 7-35. The plot shows that nitrifying organisms have cell yields on the order of 0.1 cell equivalents/equivalent of electrons transferred in the substrate oxidation/electron acceptance reaction. These microorganisms oxidize ammonia to nitrite and reduce carbon dioxide to their cell material. The free energy change for this reaction is −13.4 kcal/

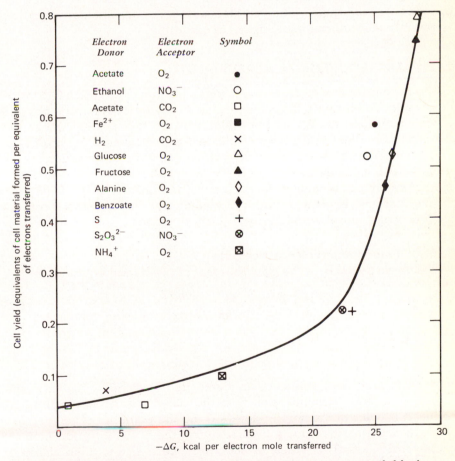

Fig. 7-35. Relationship of microbial cell yield to free energy available from substrate oxidation. After P. L. McCarty, *Progress in Water Technology*, 7, 1: 157 (1975); and P. L. McCarty, chapter in *Organic Compounds in Aquatic Environments*, S. D. Faust and J. V. Hunter, eds., Dekker, New York, 1971.

[31] P. L. McCarty, "Energetics and Bacterial Growth," in *Organic Compounds in Aquatic Environments*, S. D. Faust and J. V. Hunter, eds., Dekker, New York, 1971.

mole of electrons. This cell yield can be compared with the value of 0.74 for a heterotrophic aerobic microorganism oxidizing fructose to CO_2 and H_2O and using O_2 as an electron acceptor. The free energy change for this redox couple is -28.7 kcal per electron mole.

Not only does the amount of cell material that a microorganism can produce per unit amount of substrate processed vary with the energy available from the redox reaction that the microorganism catalyzes, but also the maximum rate at which various microorganisms can grow is related to the available energy. Table 7-13 shows typical growth rates of microorganisms growing under various redox conditions. The maximum growth rates are quite clearly related to the amount of energy available to the microorganisms from substrate oxidation.

7.10. ELECTROCHEMICAL MEASUREMENTS

The properties of electrochemical cells allow us to use them in a variety of ways for determining the concentrations of individual ions on physical-chemical properties. We will not deal with all types of electrochemical measurements here; rather we will select examples to illustrate the use of electrochemical techniques in water chemistry. Our examples will include the measurement of activity (concentration) by potentiometric methods using galvanic cells and specific ion electrodes, and the measurement of activity (concentration) by electrolytic cells using techniques such as polarography and amperometric titration.

7.10.1. Potentiometric Measurements

1. Electrode Systems.   The general form of the Nernst equation is

$$E = E° - \frac{RT}{nF} \ln \frac{\{red\}}{\{ox\}} \tag{7-59}$$

for the reaction

$$ox + ne^- \rightleftharpoons red$$

where "ox" and "red" indicate "oxidized" and "reduced" species, respectively. This equation indicates that the activities of the oxidized and reduced species are factors in determining the potential of an electrode. We can turn this fact to use by measuring the potential of an electrode at several known activities of reacting species, and then using the potential/activity relationship to determine the activities in unknown solutions from potential measurements of these solutions. If we wish to achieve this objective for the measurement of a single species or a single ion activity, we have several hurdles to overcome.

First, the measuring instrument, or *indicator electrode*, must only respond to the activity of one component of the reaction taking place at the electrode; this means that the other components of the reaction must

TABLE 7-13   Relationship of Maximum Microorganism Growth Rate Under Various, Practically Defined Redox Conditions

Redox Condition	Maximum Growth Rate, day^{-1}	Typical Free Energy Change of Redox Reaction $\Delta G^\circ/n$, kcal/electron mole
Aerobic treatment of domestic sewage[a]	3.7	-28
Denitrification using methanol[b]	1.8	-25
Nitrite oxidation in estuary water[c]	0.65	-18
Ammonia oxidation in activated sludge[d]	0.33	-10
Methane fermentation of acetic acid[e]	0.13	< -1

[a] P. Benedek and I. Horath, "A Practical Approach to Activated Sludge Kinetics," *Water Res.*, *1*: 663 (1963).
[b] H. D. Stensel, R. C. Loehr, and A. W. Lawrence, "Biological Kinetics of Suspended-Growth Denitrification," *J. Water Pollut. Control Fed.*, *45*: 249 (1973).
[c] G. Knowles, A. L. Downing, and M. J. Barnett, "Determination of Kinetic Constants for Nitrifying Bacteria in Mixed Culture with the Aid of an Electronic Computer," *J. Gen. Microbiol.*, *38*: 263 (1965).
[d] A. L. Downing, H. A. Painter, and G. Knowles, "Nitrification in the Activated Sludge Process," *J. Inst. Sew. Purif.*, Part 2: 130 (1964).
[e] A. W. Lawrence and P. L. McCarty, "Kinetics of Methane Fermentation in Anaerobic Treatment," *J. Water Pollut. Control Fed.*, *41*: RI (1969).

be kept constant. Second, the response of the electrode must be specific to the activity of the ion or species being measured; that is, the electrode must be highly selective. Third, the so-called indicator electrode must be made part of an electrochemical cell with another electrode—*the reference electrode*—whose potential remains constant over the range of conditions in which the cell is used.

The indicator electrode and the reference electrode are joined externally through a voltmeter (potentiometer) that is of the type which draws very little current because it has near-infinite internal resistance. Because there is little current the reaction at the indicator electrode does not shift perceptibly from equilibrium and there is no significant consumption of the species of interest at the electrode. The internal contact between the indicator and reference electrodes is through a salt bridge or liquid junction that allows the passage of ions but does not permit significant mixing of solutions. The potential of the cell is

$$E_{cell} = E_{\text{reference electrode}} + E_{\text{indicator electrode}} + E_{\text{junction}}$$

where E_{junction} is the potential across the liquid junction. The $E_{\text{reference electrode}}$ and E_{junction} are designed to be virtually constant so that ideally

$$E_{cell} = \text{constant} + E^\circ - \frac{RT}{nF} \ln \frac{\{red\}}{\{ox\}} \tag{7-60}$$

where

$$\text{constant} = E_{\text{reference electrode}} + E_{\text{junction}}$$

and

$$E_{\text{indicator electrode}} = E° - \frac{RT}{nF} \ln \frac{\{\text{red}\}}{\{\text{ox}\}}$$

If we can keep one of the reaction component activities constant at the indicator electrode, we can use the cell to measure the activity of the other component. For example, consider the electrode reaction

$$Ag^+ + e^- \rightleftharpoons Ag_{(s)}; \qquad E° = 0.799 \text{ volt}$$

With a silver metal electrode, $\{Ag_{(s)}\}$ is virtually constant and is taken as unity.

Substituting into Eq. 7-60, we find

$$E_{\text{cell}} = \text{constant} + 0.799 + \frac{RT}{nF} \ln \{Ag^+\}$$

If the activity is approximately equal to the concentration, then

$$E_{\text{cell}} = \text{constant} + 0.799 + 0.059 \log [Ag^+]$$

and the potential of the cell is a logarithmic function of $[Ag^+]$. For every tenfold change (decade) in $[Ag^+]$ concentration, the potential will change 0.059 volt (V) or 59 millivolts (mV). This type of response, that is, the change in mV reading per concentration decade can be used to judge whether an indicator electrode is behaving in an ideal or a "Nernstian" fashion.

2. *Calomel Electrode.* The normal hydrogen electrode [a platinum wire in 1.288 N HCl solution ($\{H^+\} = 1$) with $H_{2(g)}$ at 1 atm pressure bubbling through it] was used to define the standard electrode potential scale (see Section 7.3.2). This electrode is not convenient to use on an everyday basis, so a series of secondary reference electrodes has been developed for this purpose. One of the most commonly used laboratory reference electrodes is the *saturated calomel electrode*. The electrode (Fig. 7-36) consists of a platinum wire set in a paste that is a mixture of mercury ($Hg_{(\ell)}$), mercurous chloride (calomel, $Hg_2Cl_{2(s)}$), and potassium chloride (KCl). The paste is in contact with a solution that is saturated with KCl and $Hg_2Cl_{2(s)}$. The electrode can be represented as

$$Hg/Hg_2Cl_{2,\text{sat}}, \, KCl_{\text{sat}}(4.2M)$$

The salt bridge or liquid junction between the saturated KCl/Hg_2Cl_2 solution and the solution being measured can be made through a capillary tube filled with asbestos fibers (Fig. 7-36). The potential of the saturated calomel electrode is determined by the reaction

$$Hg_2Cl_{2(s)} + 2e^- \rightarrow 2Hg_{(\ell)} + 2Cl^-$$

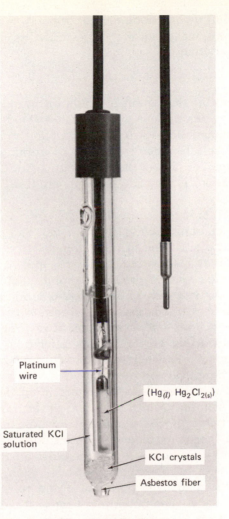

Platinum wire

$(Hg_{(l)}\ Hg_2Cl_{2(s)})$

Saturated KCl solution

KCl crystals

Asbestos fiber

Fig. 7-36. The calomel reference electrode. Courtesy of the Corning Glass Works, Corning, N.Y.

and is +0.244 volt at 25°C. Since both $Hg_2Cl_{2(s)}$ and $Hg_{(l)}$ have unit activity the potential of the electrode can be written

$$E = E° - \frac{RT}{2F} \ln \{Cl^-\}^2$$

and since $\{Cl^-\}$ is constant (the KCl solution is saturated), the potential of the electrode should be constant at a given temperature, that is,

$$E = E° - \text{constant} = +0.244 \text{ volt at } 25°C$$

3. Copper/Copper Sulfate Electrode.   The copper $(Cu_{(s)})$, saturated copper sulfate $(CuSO_4)$ reference electrode represented as $Cu_{(s)}/CuSO_{4,sat}$ and with an electrode reaction of

$$Cu_{sat}^{2+} + 2e^- \rightleftharpoons Cu_{(s)}; \qquad E° = +0.34$$

finds use in the field measurement of potential in applications such as corrosion control. The electrode is sturdily constructed with a wood outer case and a screw connection for a liquid junction made from a porous plug. To satisfy yourself that the electrode is in working order you merely shake it to hear if you can detect the rattle of the $CuSO_{4(s)}$ crystals in the internal $CuSO_4$ solution. If they are present, the $CuSO_4$ solution will be saturated.

4. Glass Electrode.   Three indicator electrodes are discussed to illustrate the various types of construction and modes of operation of these electrodes. The *glass electrode* (Fig. 7-37) consists of a bulb of thin glass (a glass membrane) that is sensitive to changes in H^+ activity. The inside of the glass membrane contains a reference solution of fixed $[H^+]$, usually a 0.1 M HCl solution into which is immersed an internal reference electrode (usually Ag/AgCl). The glass membrane separates the internal, constant $\{H^+\}$ solution from the test solution of variable $\{H^+\}$; see Fig. 7-37. The potential of the glass electrode appears to develop because of an ion exchange reaction that takes place at the glass surface. Hydrogen ions selectively exchange at the glass surface, allowing charges to be transported through the glass, thereby determining the potential difference between the constant internal $\{H^+\}$ and the variable external $\{H^+\}$. Ion exchange only takes place satisfactorily when the glass electrode surface is hydrated. Therefore, glass electrodes must be "conditioned" by soaking in dilute acid prior to use. When the glass electrode is made part of a galvanic cell with a reference electrode, such as the saturated calomel electrode, the measured potential of the cell is as indicated in Fig. 7-38 between A and E. The potential of interest is between C and D. The potentials between A and B and between B and C are constant because the internal reference solution concentration is constant. The potential between D and E is approximately constant. Its variation over a wide range of $\{H^+\}$ is minimized by "standardizing" the electrode system at $\{H^+\}$ levels that are close to those which are anticipated in the test solutions to be measured. This process of standardization leads to the operational definition of pH. We recall that pH is defined mathematically as

$$pH = -\log \{H^+\}$$

where $\{H^+\}$ is the activity of the H^+ ion.

It is not possible to measure this or any other single ion activity absolutely because single ions cannot be isolated in solution nor can potential measurements be obtained from one electrode in isolation. This

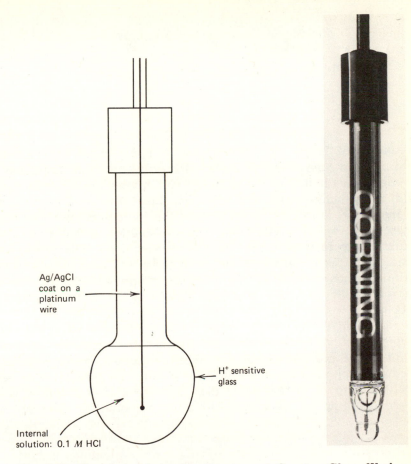

Ag/AgCl
coat on a
platinum
wire

H⁺ sensitive
glass

Internal
solution: 0.1 M HCl

Fig. 7-37. The glass electrode. Courtesy of Corning Glass Works, Corning, N.Y.

leads to the need for an operational definition of activity and concentration as determined from electrochemical measurements.

For the glass electrode,

$$E = E° + \frac{2.3RT}{nF} \log \{H^+\}$$

and since pH $= -\log \{H^+\}$,

$$E = E° - 0.059 \text{ pH}$$

The potential of the glass electrode/saturated calomel electrode cell is

$$E_{cell} = E_{calomel} + E_{glass} + E_{junction}$$

or

$$E_{cell} = \text{constant} - 0.059 \text{ pH}$$

where "constant" $= E^{\circ}_{glass} + E_{calomel} + E_{junction}$. Therefore,

$$\text{pH} = \frac{\text{constant}}{0.059} - \frac{E_{cell}}{0.059} \qquad (7\text{-}61)$$

Equation 7-61 is the operational definition of pH. In the process of standardization one selects a buffer of known $\{H^+\}$ and immerses the electrode system in it. The dial of the pH meter (which is basically a high-resistance voltmeter with a scale that has been converted to read in pH units instead of millivolts) is set to read the indicated pH of the buffer. The process is repeated with another buffer solution with a different pH. The first buffer is reread and the meter readjusted. The technique is repeated until the meter reads the indicated pH of both buffers. By this technique we have "dialed in" the value of constant/0.059 for the pH range of interest so that the value of $E_{cell}/0.059$ will be directly proportional to the pH.

5. *Divalent Cation Electrode.*   The divalent cation electrode for hardness is a selective ion electrode that detects Ca^{2+} and Mg^{2+} and other divalent ions in much the same way that the glass electrode detects H^+. Instead of developing a potential across a glass membrane, it develops a potential over an inert porous disc that is saturated with a water-immiscible, liquid ion exchanger which is selective for divalent ions,

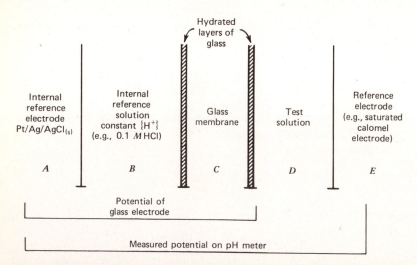

Fig. 7-38. Potentials in a glass electrode-calomel electrode system commonly used for pH measurement.

especially Ca^{2+} and Mg^{2+} (Figure 7-39). The internal solution is a calcium chloride solution saturated with the ion exchanger. There is an internal Ag/AgCl reference electrode. The liquid ion exchanger is, for example, the sparingly soluble calcium salt of dodecyl phosphoric acid dissolved in a di-n-octyl phenyl phosphonate and immobilized in a cellulose membrane. The calcium dodecyl phosphoric acid transports Ca^{2+} ions across the membrane and is responsible for establishing the potential of the electrode. Because the dodecyl phosphoric acid-Ca^{2+} salt is sparingly soluble, some Ca^{2+} ion is in solution in equilibrium with it so that the electrode potential reaches a plateau value at $[Ca^{2+}] = 10^{-6}$. This particular electrode responds to ions other than Ca^{2+} and Mg^{2+}, ions that may be regarded as interferences. The degree of interference is expressed by the selectivity constant K_x, where K_x is defined by the equation

$$E = E^\circ + \frac{0.059}{2} \log \left([Ca^{2+}] + [Mg^{2+}] + K_x[X^{2+}] \right)$$

as the relative response produced by 1 mole of $[X^{2+}]$ to that produced by 1 mole of either $[Ca^{2+}]$ or $[Mg^{2+}]$. For the divalent cation electrode some selectivity constants are 3.5 for $[Fe^{2+}]$, 3.1 for $[Cu^{2+}]$, and 0.54 for $[Sr^{2+}]$.

The pH affects the operating region of the electrode. The lower bound

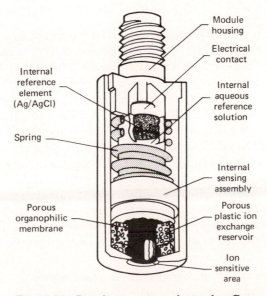

Fig. 7-39. Divalent cation electrode. Courtesy of Orion Research, Inc., Cambridge, Mass.

of pH depends on the $[Ca^{2+}] + [Mg^{2+}]$ concentrations and varies from about 6.5 at $10^{-5} M$ divalent cation to about 5 at $10^{-1} M$. The upper bound is about pH 10 and is set by the presence of the $MgOH^+$ ion, which is not detected by the electrode.

6. *Oxidation-Reduction Potential.* The measurement of the so-called "oxidation reduction potential" (ORP) of natural waters, wastewaters, and waste treatment systems has been advanced by some workers as a useful index of the state of the system (i.e., aerobic or anaerobic) or the degree of treatment (i.e., extent of biological oxidation of organics in wastewater). The measurement of ORP has found use in nonbiological situations such as the control of chlorination and dechlorination of sewage effluents. The oxidation-reduction potential is measured in a galvanic cell consisting of a reference electrode (e.g., calomel) and an indicating electrode of a highly noble metal (e.g., platinum or gold). The calomel electrode is the cathode and the inert platinum or gold electrode is the anode. The anode is made of a highly noble metal so that the potential for its oxidation is less than that of any oxidizable solution components. The anode thus is a site of the oxidation of solution constituents but ideally is not affected itself.

Oxidation-reduction potential measurements in natural waters and wastewaters are difficult to interpret. The only potentials that will register in the ORP cell are those from species that can react at the indicator electrode surface—we call these electroactive species. In natural waters only a few reactions proceed at the electrode surface, for example,

$$Fe^{3+} \rightleftharpoons Fe^{2+} + e^- \quad \text{and} \quad Mn(IV) \rightleftharpoons Mn^{2+} + 2e^-$$

All the important redox reactions involved in the nitrogen cycle, the sulfur cycle, and the carbon cycle are not completed at the indicator electrode in an ORP cell. At best, too, the voltage reading produced by an ORP cell is a reflection of many reactions—it is a "mixed potential" and its value is difficult if not impossible to interpret in any fundamental chemical terms. Moreover, when an ORP electrode combination is immersed in a water the voltage reading will vary with time, usually falling from the initial reading obtained. This behavior is due to the general process of polarization and of "poisoning" of the indicator electrode surface by the accumulation of oxidation products on the surface of the electrode.

Despite all of these limitations ORP measurements have been used widely in biological systems where, if they are treated as indices or "black box measurements" rather than fundamental indicators of a specific chemical environment, they can be of qualitative use.

7.10.2. Amperometric (Polarographic) Measurements

As the name implies, amperometric techniques involve the measurement of current. The term *polarography* derives from the fact that the electrode at which the reaction of interest occurs is in a polarized condition. The polarization of an electrode occurs when the products of a reaction that occurs at it accumulate to such an extent that they limit the rate of reaction at the electrode. When an electrode is in such a polarized condition, we can visualize that the concentration of the *product* of the electrode reaction is very high close to the electrode surface. Conversely, the concentration of the *reactants* at the electrode surface will be very low. In fact, for a completely polarized electrode we can safely assume that at the electrode surface the reactant concentration is zero. In such a condition the amount of current that the electrode will pass is directly related to the mass flux of reactant from the bulk of the electrolyte solution to the electrode surface across the concentration gradient established by polarization. In turn, the rate of mass flux or diffusion of reactant is (by Fick's law) a function of the bulk concentration of reactant.

$$\frac{dM'}{dt} = \frac{Ad}{\delta}(C - C_0) \tag{7-62}$$

M' = mass of reactant
t = time
A = area of electrode surface
D = diffusion coefficient of reactant
δ = thickness of layer around electrode through which diffusion takes place
C = bulk concentration of reactant
C_0 = electrode surface concentration of reactant

For a polarized electrode $C_0 = 0$. Therefore,

$$\frac{dM'}{dt} = \frac{ADC}{\delta} \tag{7-63}$$

Since the reactants at the electrode are the current-carrying vehicles in the internal circuit (i.e., in the electrolyte solution), the current generated in a cell with a polarized electrode can be made a function of the bulk solution concentration of reactant. When the rate at which the reactant is supplied by diffusion is just balanced by the rate at which it is consumed by reaction at the electrode, we can write

Rate of supply of reactant by diffusion = rate of reaction at electrode

$$\frac{ADC}{\delta} = \frac{i}{nF} \tag{7-64}$$

where

i = current
n = number of electrons in reaction
F = the Faraday constant

Then

$$i = \frac{nFAD\ C}{\delta} \qquad (7\text{-}65)$$

For a particular electrode system, the same temperature and physical conditions such as the stirring rate and the same electrode reaction, the term $nFAD/\delta$ is a constant and therefore

$$i = KC \qquad (7\text{-}66)$$

The measurement of the current in a cell with a polarized electrode can therefore be used to measure the concentration of the reactant-producing polarization.

Polarographic measurements can be conducted with either galvanic or electrolytic cells. With galvanic cells the potential required to drive the reaction at the polarizing electrode is generated by making the polarized electrode part of a galvanic cell. With an electrolytic cell the potential is supplied by an external voltage source such as a battery. The indicating or polarized electrode is usually an inert metal (such as platinum) or a mercury droplet; the other electrode is usually a reference electrode (e.g., a calomel electrode).

Polarographic methods using the dropping mercury electrode found wide acclaim for measuring metal concentrations in waters, but the emergence of atomic absorption spectrometry (AAS) has largely replaced polarography of this type for metals analysis. However, it should be realized that the two techniques are capable of measuring different things. Atomic absorption spectrometry measures the total metal content, that is, all species containing a specific metal, unless some effort is made to separate one species from another prior to AAS analysis. Polarography and other electrochemical methods are capable of measuring individual species, excluding complex forms, for example, if they are involved in unique electrode reactions.

1. Dissolved Oxygen Electrode.   One of the current wide applications of polarographic techniques in water chemistry is using solid inert electrodes, sometimes enclosed in a membrane that selectively passes the reactant of interest. An example of this type of cell is provided by the "dissolved oxygen probe" or "dissolved oxygen electrode." This electrode may be either of the electrolytic or galvanic type discussed above. The Makereth "oxygen electrode"[32] for example, is a galvanic cell employing a lead anode and a silver cathode in an aqueous $KHCO_3$ electrolyte, all isolated from solution by an oxygen-permeable polyethylene membrane (Fig. 7-40). Lead ions (Pb^{2+}) are produced at the anode. This oxidation drives the cathodic reaction in which oxygen is reduced to OH^-. Some

[32] F. J. H. Makereth. "An Improved Galvanic Cell for Determination of Oxygen Concentrations in Fluids," *J. Sci. Instrum.,* 41: 38 (1964).

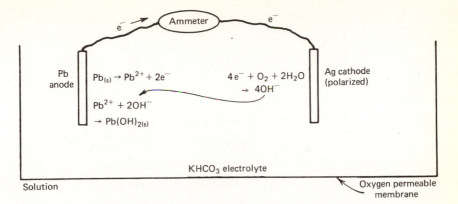

Fig. 7-40. Schematic of the Mackereth dissolved oxygen electrode.

OH^- ions migrate to the anode and react with Pb^{2+} to produce $Pb(OH)_{2(s)}$. The overall cell reaction is

$$O_{2(aq)} + 2Pb_{(s)} + 2H_2O \rightleftharpoons 2Pb(OH)_{2(s)}$$

and its rate is controlled by the rate of supply of oxygen to the silver cathode. Therefore, the current in the external circuit is controlled by the rate of the oxygen supply to the cathode. This supply depends on the diffusion rate of oxygen across the membrane. Since the oxygen concentration inside the membrane is kept low (virtually zero), the diffusion rate is a function of the oxygen concentration outside the membrane in the bulk solution. The current is, therefore, proportional to the bulk solution oxygen concentration. When using these electrodes, it is important that the bulk oxygen concentration close to the membrane does not become depleted. Therefore, it is essential to stir the bulk solution to ensure that the layer close to the membrane surface is constantly renewed with water of a composition representative of the bulk solution.

2. *Amperometric Titration of Chlorine.*   The second wide application of amperometric methods in water chemistry is the use of amperometric titration devices. In these instruments the polarization current is read out on a dial. A reagent is added that reacts with the polarizing reactant so that its bulk concentration is decreased. The dial reading decreases. Reagent is added until there no longer is a decrease in current upon the addition of further reagent (Fig. 7-41). This indicates that all of the reacting species has been consumed because the current flowing in the cell no longer is influenced by a reagent that reacts with it. Free chlorine (HOCl and OCl^-) can be determined by amperometric titration with the reductant phenylarsine oxide (C_6H_5AsO). The reaction is

$$HOCl + C_6H_5AsO + H_2O \rightarrow HCl + C_6H_5AsO(OH)_2$$

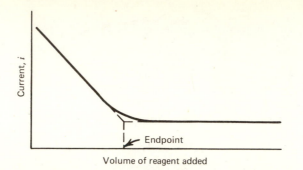

Fig. 7-41. Amperometric titration endpoint.

A typical aperometric titrator is shown in Fig. 7-42. Combined chlorine can also be measured by this technique if the proper solution conditions are used.[33]

7.11. PROBLEMS

1. Write balanced equations for oxidation of NH_3 with $HOCl$ assuming the following major end products.
    (a) $N_{2(g)}$ and Cl^-.
    (b) $NO_{(g)}$ and Cl^-.
    (c) NO_3^- and Cl^-.
       For the three cases, determine the moles of $HOCl$ required per mole of NH_3 and the $HOCl$ in mg/liter as Cl_2 required per mg NH_3–N/liter. If the stoichiometry observed in the laboratory is 3.6 moles $HOCl$/2 moles of NH_3, which of the three cases is likely to predominate?

2. What is the theoretical COD, TOC, and ultimate BOD of the following solutions?
    (a) 100 mg/liter of phenol, C_6H_5OH
    (b) 50 mg/liter of ethane, C_2H_6
    (c) 6 mg/liter of NH_3 (BOD only). Assume that NO_3^- is the nitrogen end product.
    (d) 10 mg/liter of Fe^{2+} (COD only). Assume that $Fe(OH)_{3(s)}$ is the end product.
        Note: TOC is the mass of carbon in the organic(s) under consideration. Theoretical COD and ultimate BOD are the amounts of oxygen required to oxidize the compounds in question to CO_2, H_2O, and other end products as stated.

3. A solution contains 10^{-8} M Cl_2, 10^{-5} M $HOCl$, 10^{-7} M OCl^-, 10^{-5} M NH_2Cl, and 10^{-5} M $NHCl_2$. What is the total concentration of chlorine residual in mg/liter as Cl_2?

[33] *Standard Methods of the Examination of Water and Wastewater,* Am. Publ. Health Assoc., 14th ed., 1976.

Fig. 7-42. Amperometric titrator. Courtesy of the Fischer &
Porter Co., Warminster, Pa.

4. Propionitrile has the general formula CH_3CH_2CN. Assume that it is completely
    oxidized by dichromate in the COD test.
    (a) Write the balanced reaction for CH_3CH_2CN with $Cr_2O_7^{2-}$. The nitrogen
        product of the reaction is NH_3.
    (b) What is the COD of a 50 mg/liter solution of CH_3CH_2CN?

(c) Add 10.0 ml of a 500 mg/liter solution of propionitrile to a COD flask. Also add 10.0 ml of 0.25 N $Cr_2O_7^{2-}$ solution plus the necessary reagents. Dilute to 50 ml total volume. What is the molarity of the $Cr_2O_7^{2-}$ in solution in the COD flask *after* the reaction is completed? (Assume 100 percent oxidation of CH_3CH_2CN.)

5. In the metal plating industry, cyanides are frequently eliminated from waste discharges by treatment with chlorine at high pH (alkaline chlorination process). Two steps are involved: (1) the oxidation of cyanide, CN^-, to cyanate, CNO^-, and (2) oxidation of CNO^- to N_2 and CO_2.

(a) Write balanced reactions for each of the two steps given that OCl^- is the oxidizing chlorine species and that Cl^- is formed as the reduced chlorine species in each step.

(b) How much chlorine, in mg/liter as Cl_2, is required per mg/liter of CN^- for each step?

6. Ozone (O_3) and chlorine dioxide (ClO_2) are both strong oxidizing agents. Given the half-cell reactions for these species as follows:

$$O_3 + 2H^+ + 2e^- \rightleftharpoons O_2 + H_2O; \qquad E° = 2.07 \text{ volts}$$
$$ClO_2 + e^- \rightleftharpoons ClO_2^-; \qquad E° = 1.15 \text{ volts}$$

(a) Write the stoichiometric equation for the reaction that occurs when O_3 is mixed with $NaClO_2$ in aqueous solution.

(b) What is the value of the equilibrium constant for this reaction? Is ozone a stronger or weaker oxidizing agent than chlorine dioxide?

(c) What is the standard free energy change for this reaction?

7. (a) Find the balanced reaction and the corresponding $E°$ for the oxidation of Fe^{2+} by dissolved oxygen, $O_{2(aq)}$.

(b) If the water is in equilibrium with the atmosphere with respect to O_2 (i.e., $P_{O_2} = 0.21$) and pH = 7, $[Fe^{3+}] = 10^{-4}$, and $[Fe^{2+}] = 10^{-7}$, is the reaction at equilibrium?

8. Given:

$$2e^- + Cl_2 \rightleftharpoons 2Cl^-; \qquad E° = +1.36 \text{ volts}$$
$$Cl_2 + H_2O \rightleftharpoons HOCl + H^+ + Cl^-; \qquad K_h = 3.8 \times 10^{-4} \qquad \text{and } K_w = 10^{-14}$$

calculate $E°$ for the reaction

$$2e^- + HOCl \rightleftharpoons Cl^- + OH^-$$

9. Given:

$$MnO_4^- + 8H^+ + 5e^- \rightleftharpoons Mn^{2+} + 4H_2O; \qquad E° = +1.51 \text{ volts}$$
$$Mn^{2+} + 2H_2O \rightleftharpoons 2e^- + 4H^+ + MnO_{2(s)}; \qquad E° = -1.23 \text{ volts}$$

What is $E°$ for the half-reaction representing the reduction of MnO_4^- to $MnO_{2(s)}$?

10. In the COD test, the $Cr_2O_7^{2-}$ remaining after oxidation of the organic matter is determined by titration with Fe^{2+}. When $C_{T,Cr_2O_7} = 10^{-4}$, $C_{T,Cr(III)} = 10^{-4}$,

$C_{T,Fe(II)} = 10^{-8}$, $C_{T,Fe(III)} = 10^{-5}$, and $[H^+] = 1\,M$, is the reaction for the reduction of the $Cr_2O_7^{2-}$ at equilibrium? ($E^{o\prime}_{Fe(III)/Fe(II)} = 0.68$ volt; $E^{o\prime}_{Cr_2O_7^{2-}/Cr^{3+}} = 1.33$ volts.)

11. 10^{-4} mole of Cl_2 is added to 1 liter of water at 25°. What is the pH of the solution and the HOCl concentration? (Neglect ionic strength effects.)

12. 10^{-4} mole of bromine, Br_2, is added to a liter of water. The resulting pH is 7.0. Given that the following reactions are at equilibrium,

$$Br_2 + H_2O \rightleftharpoons H^+ + Br^- + HOBr; \qquad K_h = 6 \times 10^{-9}$$
$$HOBr \rightleftharpoons H^+ + OBr^-; \qquad K_a = 3.7 \times 10^{-9}$$

and that $[Br^-] = 1.1 \times 10^{-4}$, *including the Br^- originally present in the sample*, what are the molar concentrations of Br_2, HOBr, and OBr^-?

13. The (O_{2_g}) in the atmosphere is in equilibrium with two aqueous solutions, one at pH 9 and the other at pH 6. What is the pϵ of each of these solutions?

14. (a) The pϵ of a solution can be controlled by controlling the solution composition. To show this, calculate
   (1) The pϵ of a pH 10 solution in equilibrium with the atmosphere at 25°C.
   (2) The pϵ of a pH 10 solution to which HS^- has been added until $[HS^-]/[SO_4^{2-}] = 10$ at equilibrium.
   (*Note:* This solution is *not* in equilibrium with the atmosphere.)
   (b) Also, assuming that iron species are present in solution, calculate the $[Fe^{2+}]/\{Fe(OH)_{3(s)}\}$ ratio in the solution in part (a,2).

15. (a) Draw the pC-pϵ diagram for the sulfur system at pH $= 10$, $C_{T,S} = 10^{-4}\,M$ and 25°C. Note that $S_{(s)}$ is not a stable species under these conditions. Thus the following reaction can be used,

$$SO_4^{2-} + 9H^+ + 8e^- = HS^- + 4H_2O$$

   (b) Assume that $O_{2(g)}$ is in equilibrium with an aqueous system in which $[HS^-] + [SO_4^{2-}] = 10^{-4}$, pH $= 10$, $[HS^-] = [SO_4^{2-}]$, and HS^- and SO_4^{2-} are in equilibrium with each other. What is the partial pressure of oxygen under these conditions? Can significant concentrations of HS^- exist in solution, at equilibrium, if measurable concentrations of dissolved oxygen (>0.05 mg/liter) are present?

16. As water is transported through an iron distribution main, it develops a reddish color. Analysis of the water shows that pH $= 7.5$, dissolved oxygen $= 5$ mg/liter, calcium hardness $= 100$ mg/liter as $CaCO_3$, $Na^+ = 30$ mg/liter, the total alkalinity $= 150$ mg/liter as $CaCO_3$, and the temperature $= 25°C$.
   (a) Diagram the "corrosion cell" that is probably causing the red water. Indicate the important elements of the cell and give the chemical reactions that are taking place as well as the point in the corrosion cell where they are occurring.
   (b) What is the value of the Langelier index?
   (c) Give three specific actions that could reasonably be expected to eliminate or significantly reduce the problem.

17. A secondary effluent contains 15 mg/liter of NH_3—N. Assume that the oxidation of NH_3 to $N_{2(g)}$ is the only reaction which takes place in the effluent.
    (a) What chlorine dose (in mg/liter as Cl_2) is required to insure a free residual of 1.0 mg/liter as Cl_2?
    (b) How much chloride ion (in mg/liter) will be added to the effluent, given the dose in part (a) and given that the reaction with NH_3 is complete? (Assume that $Cl_{2(g)}$ is the source of chlorine.)
    (c) Assuming that the ammonia was initially present as NH_4^+ and taking into account the hydrogen ion resulting from the addition of $Cl_{2(g)}$, how much hydrated lime $(Ca(OH)_2)$ in moles/liter will have to be added so that no pH change results from chlorination with the dose in part (a)?

7.12. ADDITIONAL READING

OXIDATION-REDUCTION REACTIONS

Butler, J. N., *Ionic Equilibrium—A Mathematical Approach*, Addison-Wesley, Reading, Mass., 1964.

Garrels, R. M., and C. L. Christ, *Solutions, Minerals, and Equilibria*, Harper, New York, 1965.

Stumm, W., and J. J. Morgan, *Aquatic Chemistry*, Wiley-Interscience, New York, 1970.

CORROSION

Evans, U. R., *The Corrosion and Oxidation of Metals*, St. Martins, New York, 1960.

Pourbaix, M., *Atlas of Electrochemical Equilibria in Aqueous Solution*, Pergamon, Elmsford, N.Y., 1966.

Scully, J. C., *The Fundamentals of Corrosion*, 2nd ed., Pergamon, Elmsford, N.Y., 1975.

CHLORINE CHEMISTRY

Jolley, R. L., ed., *Water Chlorination-Environmental Impact and Health Effects*, Vols. I and II, Ann Arbor Science, Ann Arbor, Mich., 1978. (Several chapters, including those by J. C. Morris, J. D. Johnson, A. A. Stevens et al., R. L. Jolley et al., W. H. Glaze et al., R. F. Christman et al., and others are of interest.)

ELECTROCHEMICAL MEASUREMENTS

Willard, H. H., L. L. Merritt, Jr., and J. A. Dean, *Instrumental Methods of Analysis*, 5th ed., Van Nostrand, New York, 1974.

APPENDIX
1
IONIZATION FRACTIONS

A1.1 IONIZATION FRACTIONS FOR A MONOPROTIC ACID

$p^cH = pK_a + \Delta p^cH$ (or $pH = pK_a + \Delta pH$ when $\mu \simeq 0$)	$\alpha_0 = [HA]/C_{T,A}$	$\alpha_1 = [A^-]/C_{T,A}$
$p^cK_a - 5$	1.0000	0.0000
$p^cK_a - 4$	0.9999	0.0001
$p^cK_a - 3$	0.9990	0.0010
$p^cK_a - 2$	0.9901	0.0099
$p^cK_a - 1.6$	0.9755	0.0245
$p^cK_a - 1.5$	0.9694	0.0306
$p^cK_a - 1.3$	0.9523	0.0477
$p^cK_a - 1.2$	0.9407	0.0593
$p^cK_a - 1.0$	0.9091	0.0909
$p^cK_a - 0.8$	0.8633	0.1367
$p^cK_a - 0.7$	0.8337	0.1663
$p^cK_a - 0.6$	0.7993	0.2007
$p^cK_a - 0.5$	0.7599	0.2401
$p^cK_a - 0.4$	0.7153	0.2847
$p^cK_a - 0.3$	0.6663	0.3337
$p^cK_a - 0.2$	0.6131	0.3869
$p^cK_a - 0.1$	0.5573	0.4427
p^cK_a	0.5000	0.5000
$p^cK_a + 0.1$	0.4427	0.5573
$p^cK_a + 0.2$	0.3869	0.6131
$p^cK_a + 0.3$	0.3337	0.6663
$p^cK_a + 0.4$	0.2847	0.7153
$p^cK_a + 0.5$	0.2401	0.7599
$p^cK_a + 0.6$	0.2007	0.7993
$p^cK_a + 0.7$	0.1663	0.8337
$p^cK_a + 0.8$	0.1367	0.8633
$p^cK_a + 1.0$	0.0909	0.9091
$p^cK_a + 1.2$	0.0593	0.9407
$p^cK_a + 1.3$	0.0477	0.9523
$p^cK_a + 1.5$	0.0306	0.9694
$p^cK_a + 1.6$	0.0245	0.9755
$p^cK_a + 2.0$	0.0099	0.9901
$p^cK_a + 3.0$	0.0010	0.9990
$p^cK_a + 4.0$	0.0001	0.9999
$p^cK_a + 5.0$	0.0000	1.0000

The values in the table are for a monoprotic acid-conjugate base pair and are based on the equations,

$$\alpha_0 = \frac{[H^+]}{[H^+] + {}^cK_a} = \frac{[HA]}{C_{T,A}} \tag{A1-1}$$

and

$$\alpha_1 = \frac{{}^cK_a}{[H^+] + {}^cK_a} = \frac{[A^-]}{C_{T,A}} \tag{A1-2}$$

When ionic strength effects are negligible, ${}^cK_a = K_a$.

A1.2.   IONIZATION FRACTIONS FOR MULTIPROTIC ACIDS

The exact equations for a diprotic acid, H_2A, are

$$\alpha_0 = \frac{[H^+]^2}{E} = \frac{[H_2A]}{C_{T,A}} \tag{A1-3}$$

$$\alpha_1 = \frac{[H^+]{}^cK_{a,1}}{E} = \frac{[HA^-]}{C_{T,A}} \tag{A1-4}$$

$$\alpha_2 = \frac{{}^cK_{a,1}{}^cK_{a,2}}{E} = \frac{[A^{2-}]}{C_{T,A}} \tag{A1-5}$$

where $E = [H^+]^2 + [H^+] \, {}^cK_{a,1} + {}^cK_{a,1}{}^cK_{a,2}$. When ionic strength effects are negligible, ${}^cK_{a,1} = K_{a,1}$ and ${}^cK_{a,2} = K_{a,2}$.

For the triprotic acid, H_3A, the equations are:

$$\alpha_0 = \frac{[H^+]^3}{D} = \frac{[H_3A]}{C_{T,A}} \tag{A1-6}$$

$$\alpha_1 = \frac{[H^+]^2 \, {}^cK_{a,1}}{D} = \frac{[H_2A^-]}{C_{T,A}} \tag{A1-7}$$

$$\alpha_2 = \frac{[H^+] \, {}^cK_{a,1}{}^cK_{a,2}}{D} = \frac{[HA^{2-}]}{C_{T,A}} \tag{A1-8}$$

$$\alpha_3 = \frac{{}^cK_{a,1}{}^cK_{a,2}{}^cK_{a,3}}{D} = \frac{[A^{3-}]}{C_{T,A}} \tag{A1-9}$$

where $D = [H^+]^3 + [H^+]^2 \, {}^cK_{a,1} + [H^+] \, {}^cK_{a,1} \, {}^cK_{a,2} + {}^cK_{a,1}{}^cK_{a,2}{}^cK_{a,3}$. Again, if ionic strength effects are negligible, ${}^cK_{a,1} = K_{a,1}$, ${}^cK_{a,2} = K_{a,2}$, and ${}^cK_{a,3} = K_{a,3}$.

The values given in the table can be used to determine α values for diprotic and triprotic acids, even though the table was developed for monoprotic acids. Assumptions are required, as described below, but these assumptions are often applicable and computation time is significantly reduced when these computations can be made.

The values given in the table apply to diprotic acids as follows:

1. For $p^cH = (p^cK_{a,1} \pm \Delta p^cH)$ in the region $p^cH \leq \dfrac{(p^cK_{a,1} + p^cK_{a,2})}{2}$

$$\alpha_0, \text{diprotic} = \alpha_0, \text{monoprotic}$$
$$\alpha_1, \text{diprotic} = \alpha_1, \text{monoprotic}$$
$$\alpha_2, \text{diprotic} \cong 0$$

2. for $p^cH = (p^cK_{a,2} \pm \Delta p^cH)$ in the range $pH \geq \dfrac{(p^cK_{a,1} + p^cK_{a,2})}{2}$

$$\alpha_0, \text{diprotic} \cong 0$$
$$\alpha_1, \text{diprotic} = \alpha_0, \text{monoprotic}$$
$$\alpha_2, \text{diprotic} = \alpha_1, \text{monoprotic}$$

This approach will give the non-zero values of α accurate to within 1% of the values given by the exact formulae (Eqs. A1-3 to A1-5), if the ^cK_a values are separated by 4 orders of magnitude or more. The maximum error is approximately 8 percent if the ^cK_a values differ by only 2 orders of magnitude. The error occurs only for p^cH between the two p^cK values and, if necessary, the exact equations can be used to give more accurate values in this region. Also, because $^cK_{a,1} = ([H^+]\alpha_1)/\alpha_0$ and $^cK_{a,2} = ([H^+]\alpha_2)/\alpha_1)$ the third value of α can be calculated with an intermediate degree of accuracy if two of the values can be obtained from the table.

Similarly, for a triprotic acid,

1. For $p^cH = (p^cK_{a,1} \pm \Delta p^cH)$ in the region $pH \leq \dfrac{(p^cK_{a,1} + p^cK_{a,2})}{2}$

$$\alpha_0, \text{triprotic} = \alpha_0, \text{monoprotic}$$
$$\alpha_1, \text{triprotic} = \alpha_1, \text{monoprotic}$$
$$\alpha_2 \cong 0$$
$$\alpha_3 \cong 0$$

2. For $p^cH = (p^cK_{a,2} \pm \Delta p^cH)$ in the region $\dfrac{(p^cK_{a,1} + p^cK_{a,2})}{2} \leq p^cH \leq \dfrac{(p^cK_{a,2} + p^cK_{a,3})}{2}$,

$$\alpha_0, \text{triprotic} \cong 0$$
$$\alpha_1, \text{triprotic} = \alpha_0, \text{monoprotic}$$
$$\alpha_2, \text{triprotic} = \alpha_1, \text{monoprotic}$$
$$\alpha_3, \text{triprotic} \cong 0$$

3. For $p^cH = (p^cK_{a,3} \pm \Delta p^cH)$ in the region $\dfrac{(p^cK_{a,1} + p^cK_{a,2})}{2} \leq p^cH,$

$$\alpha_0, \text{triprotic} \cong 0$$
$$\alpha_1, \text{triprotic} \cong 0$$
$$\alpha_2, \text{triprotic} = \alpha_0, \text{monoprotic}$$
$$\alpha_3, \text{triprotic} = \alpha_1, \text{monoprotic}$$

The accuracy of the non-zero α values for pH between the pK values is as given above for diprotic acids, and if more accuracy is required, Eqs. A1-6 to A1-9 can be used. An intermediate level of accuracy is often acceptable for the value that cannot be determined from the table, in which case the formula for K_a can be used, for example, $K_{a,1} = ([H^+]\alpha_1)/\alpha_0$, $K_{a,2} = ([H^+]\alpha_2)/\alpha_1$, and $K_{a,3} = ([H^+]\alpha_3)/\alpha_2$.

APPENDIX
2
ANSWERS TO PROBLEMS

CHAPTER 2

1. (a) 2.5×10^{-7} mole/liter-sec
   (b) 2.5×10^{-10} mole/cc-sec
   (c) 1.5×10^{-8} mole/cc-min

2. 1.70×10^{-4} sec^{-1}; $t = 157$ min

3. (a) $k = 1.25 \times 10^{-5}$ liter/mole-sec
   (b) $k_{30} = 3.89 \times 10^{-5}$ liter/mole-sec

4. (a) $k = \dfrac{\ln 2}{t_1} = \dfrac{0.693}{t_1}$
   (b) $t_{90} = 3.322\, t_1$

5. $[A]^{-1}$ versus t is linear whereas $\ln [A]$ versus t is curvelinear. Thus the rate is second order.

   $k = 0.192$ liter/mmole-sec

6. $t = 7.7 \times 10^{-8}$ sec

7. (a) $k_2 = k_1 \exp [(E_a/RT_1T_2)(T_2 - T_1)]$
   $\quad\quad = k_1\, \theta^{(T_2 - T_1)}$ where $\theta = \exp (E_a/RT_1T_2)$
   (b) $E_a = 7.8$ kcal/mole
   (c) $E_a = 7.6$ kcal/mole

8. $E_a = 10.7$ kcal/mole

9. (a) $k = 0.0279$ hr^{-1}
   $\quad\quad t = 165$ hr (or 6.9 days)
   (b) Total time required $= 3(12) + 1.1 = 37.1$ hr

10. (a) $t = 3.1 \times 10^{-4}$ sec
    (b) $E_a = 20.8$ kcal

11. Plot $(t/y)^{1/3}$ versus t (days) and fit this plot with a straight line. Analysis of the plot in accordance with Eq. 2-45 yields

    $k = 0.25$ (base e)
    $k = 0.11$ (base 10),   and $L_0 = 2.8$ mg/liter.

CHAPTER 3

1. -212.8 kcal/mole CH_4; -13.26 kcal/g CH_4

2. $\overline{\Delta H_f^\circ} = -26.4$ kcal/mole

3. Temperature rise $= 35.7°C$
    4.35 kg or 4.35 liter of H_2O at standard conditions will be vaporized.

4. $CH_3CHOHCOOH + \frac{1}{2}O_{2(g)} + CH_3COCOOH + H_2O_{(\ell)}$; $\Delta H° = -46.9$ kcal

5. $\Delta G° = -208.2$ kcal, the *maximum* amount of energy available.

6. (a) $K = 1.83 \times 10^{-5}$
    (b) $\Delta G < 0$, $\therefore$ Reaction goes to right.

7. Percent as $HNO_3 = 8.8\%$

8. $K = 10^{11.07}$
    $[Mg^{2+}] = 8.6 \times 10^{-4}\,M$

9. $\Delta G° = -37.6 - (-56.7) = +19.1$
    $\Delta H° = -54.96 - (-68.32) = +13.36$
    $\Delta G° = -RT \ln K$
    $K_{25°} = 10^{-14}$
    $\ln \dfrac{K_{40}}{K_{25}} = \dfrac{-\Delta H°}{R}\left(\dfrac{1}{T_{40}} - \dfrac{1}{T_{25}}\right)$
    (a) $K_{40} = 2.95 \times 10^{-14}$
    (b) $\Delta H°$ is positive, $\therefore$ reaction is endothermic.

10. (a) $K = 10^{-1}\,\dfrac{\text{moles}}{\text{liter-atm}}$

    (b) $P_{H_2S} = 10^{-3.5}$ atm

11. (a) $\Delta G = -15.7$ kcal
    $\therefore$ Mn^{2+} is still being oxidized after 10 days.
    (b) At equilibrium, assuming pH $= 8.5$ and $P_{O_2} = 0.21$ atm
    $[Mn^{2+}] = 2.15 \times 10^{-17}\,M$

12. (a) $\mu = 0.13$
    (b) $^cK = 10^{-6.68}$
    (c) $k_s = 0.17$

13. (a) $\Delta G < 0$, $\therefore$ Reaction goes to right.
    (b) pH $= 7.51$

14. $\Delta G > 0$, $\therefore$ Reaction goes to left.

15. $\Delta G° = +31.12$
Since $P_{CO_2} = K = 1.5 \times 10^{-23}$, whenever the sludge is exposed to an atmosphere with a P_{CO_2} of $10^{-3.5}$ atm the reaction tends to go to the left.

16. (a) $\Delta \overline{G}°_{f,O_{2(aq)}} = 3.95$ kcal/mole
(b) $K_2 = 7.74 \times 10^{-4}$
(c) At 25°C, $P_{O_2} = 0.20$
$[O_{2(aq)}] = 2.62 \times 10^{-4}$ M (8.4 mg/liter)
at 50°C, $P_{O_2} = 0.18$
$[O_{2(aq)}] = 1.39 \times 10^{-4}$ (4.5 mg/liter)
at 100°C, $P_{O_2} = 0$
$[O_{2(aq)}] = 0$
(d) Since 9.5 > 8.4 (from part c) the solution is not at equilibrium.

CHAPTER 4

1. (a) pH = 3.0
(b) pH = 6.98

2. (a) pH = 10
(b) pH = 8.0

3. pH 10.1

4. (a) pH = 7.95
(b) 99.8%

5. (a) pH = 9.25
(b) pH = 7.5
(c) $pK_b = pK_w - pK_a = 6.5$

6. (a) $C_{T,Na} = 2 \times 10^{-2}$ M = $[Na^+]$
$C_{T,PO_4} = 1 \times 10^{-2}$ M = $[H_3PO_4] + [H_2PO_4^-] + [HPO_4^{2-}] + [PO_4^{3-}]$
(b) $[H^+] + 2[H_3PO_4] + [H_2PO_4^-] = [PO_4^{3-}] + [OH^-]$
(c) $[Na^+] + [H^+] = [OH^-] + [H_2PO_4^-] + 2[HPO_4^{2-}] + 3[PO_4^{3-}]$

7. Approximate proton condition: $[H^+] = [HPO_4^{2-}]$
When $\mu = 0$, pH = 5.6
When $\mu = 10^{-2}$, $\gamma_{\pm 1} = 0.9$, $\gamma_{\pm 2} = 0.66$ (from Fig. 3-4)
and $^cK_{a,2} = 10^{-7.02}$
$p^cH = 5.50$ and pH = 5.45

8. Proton condition: $[H^+] + 2[H_2CO_3^*] + [HCO_3^-] = [OH^-]$
At 15°C, pH = 9.35
At 25°C, pH = 9.00

9. Proton condition: $[H^+] = [SO_4^{2-}] + [OH^-]$
pH = 3.0

10. (a) $\alpha_0 = 0.0909$; $\alpha_1 = 0.9091$ ($pK_a = 4.2$)

(b) $\alpha_0 = 0.3337$; $\alpha_1 = 0.6663$; $\alpha_2 = 0$ (approx.)
or $\alpha_2 = 0.00013$ (exact) ($pK_{a,1} = 6.3$, $pK_{a,2} = 10.3$)

(c) $\alpha_0 = 0$ (approx.), or $\alpha_0 = 1.7 \times 10^{-7}$ (exact); $\alpha_1 = 0.1367$;
$\alpha_2 = 0.8633$; $\alpha_3 = 0$ (approx.), or
$\alpha_3 = 4.3 \times 10^{-5}$ (exact) ($pK_{a,1} = 2.1$, $pK_{a,2} = 7.2$, $pK_{a,3} = 12.3$)

11. (a) $p^cH = 4.15$, $p^cK = 3.11$
$\alpha_0 = 0.0848$; $\alpha_1 = 0.9152$

(b) $p^cH = 6.55$, $p^cK_{a,1} = 6.21$, $p^cK_{a,2} = 10.12$
$\alpha_0 = 0.3141$; $\alpha_1 = 0.6859$; $\alpha_2 = 0$ (approx.),
$\alpha_2 = 1.8 \times 10^{-4}$ (exact)

(c) $p^cH = 7.95$, $p^cK_{a,1} = 2.01$, $p^cK_{a,2} = 7.01$, $p^cK_{a,3} = 12.01$
$\alpha_0 = 0$ (approx.), $\alpha_0 = 1.2 \times 10^{-7}$ (exact); $\alpha_1 = 0.1046$;
$\alpha_2 = 0.8954$; $\alpha_3 = 0$ (approx.), $\alpha_3 = 7.8 \times 10^{-5}$ (exact)

12. Charge balance: $[Na^+] + [H^+] + [NH_4^+] = [OH^-]$ where
$[Na^+] = [NH_{3(aq)}] + [NH_4^+]$
$pH = 7.7$

13. Charge balance: $[NH_4^+] + [H^+] + [Na^+] = [Ac^-] + [OH^-]$
where $[Na^+] = 10^{-2}$ and $[Ac^-] = 10^{-2} - [HAc]$
$pH = 10.6$

14. Proton condition: $[H_2CO_3^*] + [NH_4^+] + [H^+] = [CO_3^{2-}] + [OH^-]$
$p^cK_{a,1} = 6.21$, $p^cK_{a,2} = 10.12$, $p^cK_{a,NH_4^+} = 9.3$
$p^cH = 9.8$, $pH = 9.85$

15. (a) Moles Na_2CO_3 required = moles H^+/liter = $10^{-2.7}$ moles/liter
(b) $\beta = 9.6 \times 10^{-5}$

16. $2Na_2CO_3 + H_2SO_4 \rightarrow 2HCO_3^- + 4Na^+ + SO_4^{2-}$
2×10^{-1} moles Na_2CO_3/liter

17. $H_2PO_4^-$ is an acceptable buffer. 1.85×10^{-2} moles of phosphate salt added
to 1 liter of solution, followed by adjustment of the pH to 8.0, will give the
desired result.

$\beta = 5.12 \times 10^{-3}$ moles/liter

18. (a) At pH 8.3, $[HCO_3^-] = 2 \times 10^{-3} M$. At pH = 6.3, $[HCO_3^-] = [H_2CO_3^*] = 1 \times 10^{-3}$
$HCO_3^- + H^+ \rightarrow H_2CO_3^*$

Assuming complete reaction of the H^+ added with HCO_3^-, 1×10^{-3} moles
H^+ can be added/liter. Dilution ratio = 0.1 liter waste/1 liter stream.
(b) $\beta = 1.15 \times 10^{-3}$ moles/liter

19. (a) As the pH is lowered below 8.9, the solution is supersaturated if no CO_2
escapes.
(b) $pH_{CO_2} = 4.35$
(c) $pH_{CO_2} = 4.5$

20. | Total Alkalinity | Total Acidity |
| --- | --- |
| (a) Decrease | Increase |
| (b) Decrease | Increase |
| (c) No effect | No effect |
| (d) No effect | Increase |
| (e) Increase | No effect |

21. (a) Total alkalinity $= 3 \times 10^{-3}$ eq/liter

$\qquad = 3 \times 10^{-3}$ eq/liter $\times$ 50,000 mgCaCO$_3$/eq $= 150$ mg/liter

$\qquad$ as CaCO$_3$

$\qquad$ 0.015 liter of 0.02 N H$_2$SO$_4$ must be added.

(b) 0.04 liter can be added per liter of natural water.

22. (a) $[CO_3^{2-}] = 3.99 \times 10^{-7} M$

(b) Caustic and carbonate alkalinity $= 0$

$\qquad$ Total alkalinity $= 100$ mg/liter as CaCO$_3$

(c) Mineral acidity is zero

$\qquad$ CO$_2$ acidity $= 10^{-3}$ eq/liter

$\qquad$ Total acidity $= 4 \times 10^{-3}$ eq/liter

(d) pH $= 7.0$

23. (a) Carbonate alkalinity $= 2.4$ meq/liter

$\qquad$ Total alkalinity $= 5.6$ meq/liter

$\qquad$ Caustic alkalinity $= 0$

(b) $\dfrac{(V_{mo} - V_p)(0.02)}{50} = \dfrac{(8 - 6)(0.02)}{50} = 0.0008$ eq/liter

$\qquad$ Total acidity $= 8 \times 10^{-4}$ eq/liter

(c) $C_{T,CO_3} = 3.2 \times 10^{-3}$ moles/liter

(d) pH$_{CO_2} = 4.4$ pH$_{CO_3^{2-}} = 10.8$

(e) Approximate procedure:

$\qquad$ pH $= 10.8$, $[CO_3^{2-}] = 2.4 \times 10^{-3} M$, $[HCO_3^-] = 0.8 \times 10^{-3} M$

$\qquad$ Exact procedure:

$\qquad$ pH $= 10.55$, $[CO_3^{2-}] = 2.05 \times 10^{-3}$, M

$\qquad$ $[HCO_3^-] = 1.15 \times 10^{-3} M$, $[H_2CO_3^*] = 6.5 \times 10^{-8} M$

24. $C_{T,CO_3} =$ total alkalinity $-$ carbonate alkalinity $= 5 \times 10^{-3} M$

$\mu = 0$:

$\qquad$ $[H^+] = 10^{-9.55} M$, $[HCO_3^-] = 4.1 \times 10^{-3} M$, $[CO_3^{2-}] = 9.2 \times 10^{-4} M$

$\mu = 0.01$:

$\qquad$ $[H^+] = 10^{-9.4} M$, $[HCO_3^-] = 4.1 \times 10^{-3} M$, $[CO_3^{2-}] = 9.3 \times 10^{-4} M$

25. Predominant reaction: H$^+$ + HCO$_3^- \rightarrow$ H$_2$CO$_3^*$

$[H_2CO_3^*]_{pH=6.7} - [H_2CO_3^*]_{pH=7.0} = 0.86 \times 10^{-4} M =$ H$^+$ added per liter of lake water

$\qquad$ Thus 17.2×10^6 ft^3 of rain are required.

CHAPTER 5

2. 1.12×10^{-3} moles/liter; 112 mg/liter as $CaCO_3$

3. $[HgOH^+] = 1 \times 10^{-7} M \, (< [Hg(OH)_2^\circ] = 9.99 \times 10^{-6} M)$

4. (a) 0.40×10^{-3} moles $Ca(OH)_2$/liter
    (b) The amount in part (a) for coagulation plus 4.02×10^{-5} moles $Ca(OH)_2$/liter to raise the pH.

5. Using the equilibrium constants for the complexes as given in Section 5-6, 25% of $C_{T,Ca}$ and 28% of $C_{T,Mg}$ are in the form of sulfato complexes.

6. Assume $4[Ni(CN)_4^{2-}] \gg [HCN] + [CN^-]$
    Knowing $[HCN] + [CN^-] = 4[Ni^{2+}]$
    Calculate $[Ni^{2+}] = 1.4 \times 10^{-7}$

    $$\therefore \quad \frac{[Ni(CN)_4^{2-}]}{C_{T,Ni}} \cong 1.0$$

7. (a) $[MgEDTA^{2-}] = 1.1 \times 10^{-5} M$
    (b) $[MgEDTA^{2-}] = 7.4 \times 10^{-5} M$
    $[CaEDTA^{2-}] = 1.93 \times 10^{-3} M \; \therefore \; 5\%$ of $C_{T,Ca}$ is not complexed at the equivalence point for calcium. Note that in the standard procedure for calcium titration, the pH is such that OH^- will complex some of the Mg^{2+}. Thus, less EDTA than calculated above will be associated with the Mg^{2+} and more will be associated with the calcium.

8. pH = 5.6 (The concentrations of the $Cu(OH)_3^-$ and $Cu(OH)_4^{2-}$ complexes are negligible.)

9. From Fig. 5-2, the trimer and hexamer complexes are negligible.
    $C_{T,Pb} = 0.04 M = [Pb^{2+}] + 4[Pb_4(OH)_4^{4+}]$
    $[Pb_4(OH)_4^{4+}] = 0.004 M$

CHAPTER 6

1. (a) 8.4×10^{-5} moles of $SrSO_{4(s)}$ will precipitate/liter.
    (b) $[Sr^{2+}] = 1.6 \times 10^{-5} M$
    $[SO_4^{2-}] = 1.02 \times 10^{-3} M$

2. (a) From Fig. 6-3, pH = 10.65.
    (b) OH^- added: $7.36 \times 10^{-3} M$

3. 9.98×10^{-5} moles Mg^{2+}/liter will precipitate.

4. Using Fig. 6-4, when $-\log [Ca^{2+}] = 4.3$, $pCO_3 = 4$ (or $[CO_3^{2-}] = 10^{-4}$).
    When $-\log [Mg^{2+}] = 3.5$, $pCO_3 = 1.5$ (or $[CO_3^{2-}] = 10^{-1.5}$).

5. Using Fig. 6,3, when $[Mg^{2+}] = 3 \times 10^{-3} M$, pH $= 9.9$,
            when $[Mg^{2+}] = 4 \times 10^{-5}$, pH $= 10.8$.

6. (a) Neglecting $[HCrO_4^-]$, $S = 8.55 \times 10^{-5}$ and pH $= 7.0$
    (b) Taking $HCrO_4^-$ into account,
        $(K_{so}/4\alpha_1)^{1/3} (2 - 2\alpha_1 - \alpha_0) + [H^+] - K_w/[H^+] = 0$
        $S = 8.61 \times 10^{-5}$, pH $= 8.2$
        so solubility is increased by 6×10^{-7} moles/liter.

8. $P_s = 3.16 \times 10^{-7}$

9. Draw a diagram following the procedure given for Fig. 6-11. $FeCO_{3(s)}$ will precipitate at pH 7.3 whereas $Fe(OH)_{2(s)}$ will precipitate at 8.7.

10.

	Total Alkalinity	$[Ca^{2+}]$
(a) KOH	Increase	Decrease
(b) $Ca(NO_3)_2$	Decrease	Increase
(c) KCl	No effect	No effect
(d) Na_2CO_3	Increase	Decrease
(e) CO_2	Increase	Increase

Reason for (a):
    $OH^- + HCO_3^- \rightarrow CO_3^{2-}$; increase in $[CO_3^{2-}\pm b$ by this step $= X$, but the solution is now supersaturated. Then,

$$CO_3^{2-} + Ca^{2+} \rightarrow CaCO_{3(s)};$$ moles $CaCO_{3(s)}$/liter that ppt $= Y$.

$$([Ca^{2+}]_0 - Y)([CO_3^{2-}]_0 + X - Y) = K_{so}$$

Knowing $[Ca^{2+}]_0 X [CO_3^{2-}]_0 = K_{so}$, by expanding, simplifying and assuming both X and Y are much less than $[Ca^{2+}]_0$ and $[CO_3^{2-}]_0$, we find that

$$X - Y = \frac{[CO_3^{2-}]_0 Y}{[Ca^{2+}]_0}$$

and, thus, that $X > Y$. Then,

$[Ca^{2+}] = [Ca^{2+}]_0 - Y$; $\therefore [Ca^{2+}]$ decreases
$[CO_3^{2-}] = [CO_3^{2-}]_0 + X - Y$; $\therefore [CO_3^{2-}]$, and thus total alkalinity, increases.

Reason for (e),
    Addition of CO_2 will have no effect on total alkalinity if $CaCO_{3(s)}$ is not present. However, it will cause the solution to become undersaturated with $CaCO_{3(s)}$ because the solution pH is lowered. Since $CaCO_{3(s)}$ is present, some will dissolve, thus increasing $[Ca^{2+}]$ and total alkalinity.

13. $$\frac{2\alpha_0 K_{so}}{K_H P_{CO_2} \alpha_2} + [H^+] = \frac{K_w}{[H^+]} + \frac{\alpha_1 K_H P_{CO_2}}{\alpha_0} + \frac{2\alpha_2 K_H P_{CO_2}}{\alpha_0}$$

    (a) Using constants from Table 6-1, when $P_{CO_2} = 10^{-1.5}$ atm, using Appendix 1, we find that

    pH $= 6.98$, $[Ca^{2+}] = 2.25 \times 10^{-3} M$, total alkalinity $= 4.55 \times 10^{-3}$ eq/liter,

    $[HCO_3^-] = 4.55 \times 10^{-3} M$, total hardness $= 2.25 \times 10^{-3}$ moles/liter

(b) When $P_{CO_2} = 10^{-3.5}$ atm, solution yields (see Example 6-15):

pH = 8.3, total alkalinity (eq/liter) = $[HCO_3^-] = 10^{-3}$, $[Ca^{2+}] = 5 \times 10^{-4}$

Thus 1.75×10^{-3} moles $CaCO_3$/liter will precipitate. This precipitation will tend to occur as CO_2 is released during treatment or pumping; thus, CO_2 release should be controlled.

14. L.I. = 0.7

15. pCH = 8.18, pH = 8.21 $[Ca^{2+}] = 9.75 \times 10^{-4}$ M at equilibrium.
Precipitation potential = $1 \times 10^{-3} - 9.75 \times 10^{-4} = 2.5 \times 10^{-5} M$
$2.5 \times 10^{-5} \times 10^5 = 2.5$ mg/liter as $CaCO_3$.
If ionic strength effects are not accounted for, the precipitation potential is 5.2 mg/l as $CaCO_3$ (see Example 6-18).

16. Equations 3 to 7 in Example 6-18 plus

(Total alkalinity)$_{final}$ = (Total alkalinity)$_{initial}$ + 2 (Ca(OH)$_2$ moles/liter)$_{added}$
$(C_{T,CO_3})_{initial} = (C_{T,CO_3})_{final}$
$[Ca^{2+}]_{initial} + Ca(OH)_2$ added, moles/liter = $[Ca^{2+}]$

17. First determine (Ca(OH)$_2$, moles/liter)$_1$ to saturate the water as in Problem 6-16. Then, use equations 3 to 7 in Example 6-18 plus

(Total alkalinity)$_{final}$ = (Total alkalinity)$_{initial}$ + 2 (Ca(OH)$_2$ moles/liter)$_2$
$(C_{T,CO_3})_{final} = (C_{T,CO_3})_{initial}$
$[Ca^{2+}]_{final} = [Ca^{2+}]_{sat} + 5 \times 10^{-5} M$
$Ca(OH)_2$ added, moles/liter = (Ca(OH)$_2$)$_1$ + (Ca(OH)$_2$)$_2$

CHAPTER 7

1.

	mole ratio req'd $[HOCl]/[NH_3] = [Cl_2]/[NH_3]$	wt ratio (Cl_2/NH_3-N)
(a)	3:2	7.6:1
(b)	5:2	12.5:1
(c)	4:1	20:1

Since the stoichiometry is just greater than 3:2, reaction (a) likely predominates. Formation of a small amount of product such as NO_3^-, which requires a larger ratio, probably accounts for the amount of the ratio in excess of 3:2.

2. (a) COD = 238 mg/liter
TOC = 76.7 mg/liter
Ult. BOD = 238 mg/liter
(b) COD = 187 mg/liter
TOC = 40 mg/liter
Ult. BOD = 187 mg/liter
(c) Ult. BOD = 22.6 mg/liter
(d) COD = 1.4 mg/liter

3. 2.85 mg/liter as Cl_2

4. (a) $3CH_3CH_2CN + 7Cr_2O_7^{2-} + 56H^+ \rightarrow 14Cr^{3+} + 9CO_2 + 3NH_3 + 31H_2O$
   (b) $COD = 102$ mg/liter
   (c) Concentration of $Cr_2O_7^{2-}$ remaining $= 4.2$ mmoles/liter.

5. (a) $CN^- + OCl^- \rightarrow CNO^- + Cl^-$
   $2CNO^- + 3OCl^- + 2H^+ \rightarrow N_2 + 3Cl^- + 2CO_2 + H_2O$
   (b) For step 1: 2.73 mg/liter as Cl_2 per mg CN^-/liter.
   For step 2: 4.1 mg/liter as Cl_2 per mg CN^-/liter.

6. (a) $O_3 + 2ClO_2^- + 2H^+ \rightarrow O_2 + H_2O + 2ClO_2$
   $E° = 0.92$ volt
   (b) $K = 1.3 \times 10^{31}$
   Because O_3 will oxidize ClO_2^-, O_3 is the stronger oxidizing agent.
   (c) $\Delta G° = -nFE° = -42.4$ kcal

7. (a) $4Fe^{2+} + O_{2(aq)} + 4H^+ \rightleftharpoons 4Fe^{3+} + 2H_2O; E° = +0.50$
   (b) $4Fe^{2+} + O_{2(g)} + 4H^+ \rightleftharpoons 4Fe^{3+} + 2H_2O; E° = +0.46$
   From the Nernst equation, $E = -0.14$ volt
   Since $E < 0$, the reaction is tending to proceed to the left.

8. $E° = +1.05$ volt

9. $E° = +1.70$ volt

10. $E' = E°' - (0.059/n) \log Q$
    $E°' = 1.33 - 0.68 = 0.65$ volt

    $$Q = \frac{(C_{T,Cr(III)})^2 (C_{T,Fe(III)})^6}{[H^+]^{14} C_{T,Cr_2O_7} (C_{T,Fe(II)})^6} = 10^{14}$$

    $E' = 0.51$ volt
    Since $E' > 0$, the reduction of $Cr_2O_7^{2-}$ is continuing.

11. $pH = 4$, $[HOCl] = 10^{-4}$

12. $[HOBr] \cong 1 \times 10^{-4}$, $[OBr^-] = 3.7 \times 10^{-6} M$, $[Br_2] = 1.83 \times 10^{-7} M$

13. $pH = 9$, $p\epsilon = 11.6$
    If $pH = 6$, $p\epsilon = 14.6$

14. (a) (1) $p\epsilon = 10.6$
        (2) $p\epsilon = -7.2$
    (b) $[Fe^{2+}] = 1.6 \times 10^{-5} M$

15. (b) $p\epsilon = -7$, $P_{O_2} = 10^{-72}$ atm
    Since P_{O_2} decreases as $p\epsilon$ decreases (and as the ratio $[HS^-]/[SO_4^{2-}]$ increases), the stable form of S is SO_4^{2-} when measurable O_2 is present.

16. (b) L.I. $= 0$

17. (a) 115.1 mg/liter as Cl_2
    (b) Total $Cl^- = 114.6$ mg/liter
    (c) 2.14 mmoles $Ca(OH)_2$/liter

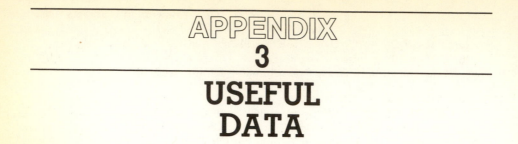

APPENDIX
3
USEFUL DATA

The tables and figure that follow are collected here for the convenience of students working on the problems in the text. The tables of constants should not be used for general reference, however, because a critical review of the research on which they are based has not been made.

TABLE 3-1 Thermodynamic Constants for Species of Importance in Water Chemistry[a]

Species	$\Delta \overline{H}_f^\circ$, kcal/mole	$\Delta \overline{G}_f^\circ$, kcal/mole
$Ca_{(aq)}^{2+}$	−129.77	−132.18
$CaCO_{3(s)}$, calcite	−288.45	−269.78
$CaO_{(s)}$	−151.9	−144.4
$C_{(s)}$, graphite	0	0
$CO_{2(g)}$	−94.05	−94.26
$CO_{2(aq)}$	−98.69	−92.31
$CH_{4(g)}$	−17.889	−12.140
$H_2CO_{3(aq)}^*$	−167.0	−149.00
$HCO_{3(aq)}^-$	−165.18	−140.31
$CO_{3(aq)}^{2-}$	−161.63	−126.22
CH_3COO^-, acetate	−116.84	−89.0
$H_{(aq)}^+$	0	0
$H_{2(g)}$	0	0
$Fe_{(aq)}^{2+}$	−21.0	−20.30
$Fe_{(aq)}^{3+}$	−11.4	−2.52
$Fe(OH)_{3(s)}$	−197.0	−166.0
$Mn_{(aq)}^{2+}$	−53.3	−54.4
$MnO_{2(s)}$	−124.2	−111.1

TABLE 3-1   continued

Species	$\Delta \overline{H}^\circ_f$, kcal/mole	$\Delta \overline{G}^\circ_f$, kcal/mole
$Mg^{2+}_{(aq)}$	-110.41	-108.99
$Mg(OH)_{2(s)}$	-221.00	-199.27
$NO^-_{3(aq)}$	-49.372	-26.43
$NH_{3(g)}$	-11.04	-3.976
$NH_{3(aq)}$	-19.32	-6.37
$NH^+_{4(aq)}$	-31.74	-19.00
$HNO_{3(aq)}$	-49.372	-26.41
$O_{2(aq)}$	-3.9	3.93
$O_{2(g)}$	0	0
$OH^-_{(aq)}$	-54.957	-37.595
$H_2O_{(g)}$	-57.7979	-54.6357
$H_2O_{(\ell)}$	-68.3174	-56.690
$SO^{2-}_{4(aq)}$	-216.90	-177.34
$HS^-_{(aq)}$	-4.22	3.01
$H_2S_{(g)}$	-4.815	-7.892
$H_2S_{(aq)}$	-9.4	-6.54

Source. R. M. Garrels and C. L. Christ, Solutions, Minerals, and Equilibria, Harper & Row, New York, 1965; and Handbook of Chemistry and Physics, Chemical Rubber Publishing Company, Cleveland, Ohio.

[a] For a hypothetical ideal state of unit molality, which is approximately equal to that of unit molarity.

TABLE 4·1   Acidity and Basicity Constants for Substances in Aqueous Solution at 25°C

Acid		$-\log K_a = pK_a$	Conjugate Base		$-\log K_b = pK_b$
$HClO_4$	Perchloric acid	-7	ClO_4^-	Perchlorate ion	21
HCl	Hydrochloric acid	~ -3	Cl^-	Chloride ion	17
H_2SO_4	Sulfuric acid	~ -3	HSO_4^-	Bisulfate ion	17
HNO_3	Nitric acid	0	NO_3^-	Nitrate ion	14
H_3O^+	Hydronium ion	0	H_2O	Water	14
HIO_3	Iodic acid	0.8	IO_3^-	Iodate ion	13.2
HSO_4^-	Bisulfate ion	2	SO_4^{2-}	Sulfate ion	12
H_3PO_4	Phosphoric acid	2.1	$H_2PO_4^-$	Dihydrogen phosphate ion	11.9
$Fe(H_2O)_6^{3+}$	Ferric ion	2.2	$Fe(H_2O)_5OH^{2+}$	Hydroxo iron (III) complex	11.8
HF	Hydrofluoric acid	3.2	F^-	Fluoride ion	10.8
HNO_2	Nitrous acid	4.5	NO_2^-	Nitrite ion	9.5
CH_3COOH	Acetic acid	4.7	CH_3COO^-	Acetate ion	9.3
$Al(H_2O)_6^{3+}$	Aluminum ion	4.9	$Al(H_2O)_5OH^{2+}$	Hydroxo aluminum (III) complex	9.1
$H_2CO_3^*$	Carbon dioxide and carbonic acid	6.3	HCO_3^-	Bicarbonate ion	7.7

Acid	Formula	pK	Conjugate base	Name	pK
Hydrogen sulfide	$H_2S_{(aq)}$	7.1	HS^-	Bisulfide ion	6.9
Dihydrogen phosphate	$H_2PO_4^-$	7.2	HPO_4^{2-}	Monohydrogen phosphate ion	6.8
Hypochlorous acid	$HOCl$	7.5	OCl^-	Hypochlorite ion	6.4
Hydrocyanic acid	$HCN_{(aq)}$	9.3	CN^-	Cyanide ion	4.7
Boric acid	H_3BO_3	9.3	$B(OH)_4^-$	Borate ion	4.7
Ammonium ion	NH_4^+	9.3	$NH_{3(aq)}$	Ammonia	4.7
Orthosilicic acid	H_4SiO_4	9.5	$H_3SiO_4^-$	Trihydrogen silicate ion	4.5
Phenol	C_6H_5OH	9.9	$C_6H_5O^-$	Phenolate ion	4.1
Bicarbonate ion	HCO_3^-	10.3	CO_3^{2-}	Carbonate ion	3.7
Monohydrogen phosphate	HPO_4^{2-}	12.3	PO_4^{3-}	Phosphate ion	1.7
Trihydrogen silicate	$H_3SiO_4^-$	12.6	$H_2SiO_4^{2-}$	Dihydrogen silicate ion	1.4
Bisulfide ion	HS^-	14	S^{2-}	Sulfide ion	0
Water	H_2O	14	OH^-	Hydroxide ion	0
Ammonia	$NH_{3(aq)}$	~23	NH_2^-	Amide ion	−9
Hydroxide ion	OH^-	~24	O^{2-}	Oxide ion	−10

TABLE 4-2   Ion Product of Water

Temperature,°C	K_w	pK_w	pH of a "neutral" solution ($\{H^+\} = \{OH^-\}$)
0	0.12×10^{-14}	14.93	7.47
15	0.45×10^{-14}	14.35	7.18
20	0.68×10^{-14}	14.17	7.08
25	1.01×10^{-14}	14.00	7.00
30	1.47×10^{-14}	13.83	6.92
40	2.95×10^{-14}	13.53	6.76

Source. H. S. Harned and B. B. Owen, *The Physical Chemistry of Electrolyte Solutions*, 3rd. Ed., Reinhold, New York, 1958.

TABLE 4-7   Temperature Dependence of Some Important Carbonate Equilibrium Constants

Reaction	Temperature, °C						
	5	10	15	20	25	40	60
1. $CO_{2(g)} + H_2O \rightleftharpoons CO_{2(aq)}$; pK_H	1.20	1.27	1.34	1.41	1.47	1.64	1.8
2. $H_2CO_3^* \rightleftharpoons HCO_3^- + H^+$; $pK_{a,1}$	6.52	6.46	6.42	6.38	6.35	6.30	6.30
3. $HCO_3^- \rightleftharpoons CO_3^{2-} + H^+$; $pK_{a,2}$	10.56	10.49	10.43	10.38	10.33	10.22	10.14
4. $CaCO_{3(s)} \rightleftharpoons Ca^{2+} + CO_3^{2-}$; pK_{so}	8.09	8.15	8.22	8.28	8.34	8.51	8.74
5. $CaCO_{3(s)} + H^+ \rightleftharpoons Ca^{2+} + HCO_3^-$; $p(K_{so}/K_{a,2})$	-2.47	-2.34	-2.21	-2.10	-1.99	-1.71	-1.40

Source. T. E. Larson and A. M. Buswell, "Calcium Carbonate Saturation Index and Alkalinity Interpretations," *J. Amer. Water Works Assoc.*, 34, 1664 (1942).

TABLE 5-1   Equilibrium Constants for Mononuclear Hydroxo Complexes

	$\log \beta_1{}^a$	$\log \beta_2$	$\log \beta_3$	$\log \beta_4$	$\log K_{so}{}^b$
Fe^{3+}	11.84	21.26		33.0	-38
Al^{3+}	9			34.3	-33
Cu^{2+}	8.0		15.2	16.1	-19.3
Fe^{2+}	5.7	$(9.1)^c$	10	9.6	-14.5
Mn^{2+}	3.4	(6.8)	7.8		-12.8
Zn^{2+}	4.15	(10.2)	(14.2)	(15.5)	-17.2
Cd^{2+}	4.16	8.4	(9.1)	(8.8)	-13.6

[a] β_i is the equilibrium constant for the reaction, $M^{n+} + i\ OH^- \rightarrow M(OH)^{(n-i)+}$.
[b] K_{so} is the equilibrium constant for the reaction $M(OH)_{n(s)} \rightleftharpoons M^{n+} + nOH^-$.
[c] ( ) indicates an estimated value.

TABLE 6-1   Solubility Product Constants at 25°C[a]

Solid	pK_{so}	Solid	pK_{so}
$Fe(OH)_3$ (amorph)	38	$BaSO_4$	10
$FePO_4$	17.9	$Cu(OH)_2$	19.3
$Fe_3(PO_4)_2$	33	$PbCl_2$	4.8
$Fe(OH)_2$	14.5	$Pb(OH)_2$	14.3
FeS	17.3	$PbSO_4$	7.8
Fe_2S_3	88	PbS	27.0
$Al(OH)_3$ (amorph)	33	$MgNH_4PO_4$	12.6
$AlPO_4$	21.0	$MgCO_3$	5.0
$CaCO_3$ (calcite)	8.34	$Mg(OH)_2$	10.74
$CaCO_3$ (aragonite)	8.22	$Mn(OH)_2$	12.8
$CaMg(CO_3)_2$ (dolomite)	16.7	$AgCl$	10.0
CaF_2	10.3	Ag_2CrO_4	11.6
$Ca(OH)_2$	5.3	Ag_2SO_4	4.8
$Ca_3(PO_4)_2$	26.0	$Zn(OH)_2$	17.2
$CaSO_4$	4.59	ZnS	21.5
SiO_2 (amorph)	2.7		

[a] Equilibrium constants for the reaction $A_zB_{y(s)} \rightleftharpoons zA^{y+} + yB^{z-}$.

TABLE 7-1   Standard Electrode Potentials at 25°C

Reaction	$E°$ Volt	$p\epsilon° \left(= \dfrac{1}{n}\log K\right)$
$H^+ + e^- \rightleftharpoons \frac{1}{2}H_{2(g)}$	0	0
$Na^+ + e^- \rightleftharpoons Na_{(s)}$	-2.72	-46.0
$Mg^{2+} + 2e^- \rightleftharpoons Mg_{(s)}$	-2.37	-40.0
$Cr_2O_7^{2-} + 14H^+ + 6e^- \rightleftharpoons 2Cr^{3+} + 7H_2O$	$+1.33$	$+22.5$
$Cr^{3+} + e^- \rightleftharpoons Cr^{2+}$	-0.41	-6.9
$MnO_4^- + 2H_2O + 3e^- \rightleftharpoons MnO_{2(s)} + 4OH^-$	$+0.59$	$+10.0$
$MnO_4^- + 8H^+ + 5e^- \rightleftharpoons Mn^{2+} + 4H_2O$	$+1.51$	$+25.5$
$Mn^{4+} + e^- \rightleftharpoons Mn^{3+}$	$+1.65$	$+27.9$
$MnO_{2(s)} + 4H^+ + 2e^- \rightleftharpoons Mn^{2+} + 2H_2O$	$+1.23$	$+20.8$
$Fe^{3+} + e^- \rightleftharpoons Fe^{2+}$	$+0.77$	$+13.0$
$Fe^{2+} + 2e^- \rightleftharpoons Fe_{(s)}$	-0.44	-7.4
$Fe(OH)_{3(s)} + 3H^+ + e^- \rightleftharpoons Fe^{2+} + 3H_2O$	$+1.06$	$+17.9$
$Cu^{2+} + e^- \rightleftharpoons Cu^+$	$+0.16$	$+2.7$
$Cu^{2+} + 2e^- \rightleftharpoons Cu_{(s)}$	$+0.34$	$+5.7$
$Ag^{2+} + e^- \rightleftharpoons Ag^+$	$+2.0$	$+33.8$
$Ag^+ + e^- \rightleftharpoons Ag_{(s)}$	$+0.8$	$+13.5$
$AgCl_{(s)} + e^- \rightleftharpoons Ag_{(s)} + Cl^-$	$+0.22$	$+3.72$
$Au^{3+} + 3e^- \rightleftharpoons Au_{(s)}$	$+1.5$	$+25.3$
$Zn^{2+} + 2e^- \rightleftharpoons Zn_{(s)}$	-0.76	-12.8
$Cd^{2+} + 2e^- \rightleftharpoons Cd_{(s)}$	-0.40	-6.8
$Hg_2Cl_{2(s)} + 2e^- \rightleftharpoons 2Hg_{(\ell)} + 2Cl^-$	$+0.27$	$+4.56$
$2Hg^{2+} + 2e^- \rightleftharpoons Hg_2^{2+}$	$+0.91$	$+15.4$
$Al^{3+} + 3e^- \rightleftharpoons Al_{(s)}$	-1.68	-28.4
$Sn^{2+} + 2e^- \rightleftharpoons Sn_{(s)}$	-0.14	-2.37
$PbO_{2(s)} + 4H^+ + SO_4^{2-} + 2e^- \rightleftharpoons PbSO_{4(s)} +$ $2H_2O$	$+1.68$	$+28.4$
$Pb^{2+} + 2e^- \rightleftharpoons Pb_{(s)}$	-0.13	-2.2
$NO_3^- + 2H^+ + 2e^- \rightleftharpoons NO_2^- + H_2O$	$+0.84$	$+14.2$
$NO_3^- + 10H^+ + 8e^- \rightleftharpoons NH_4^+ + 3H_2O$	$+0.88$	$+14.9$
$N_{2(g)} + 8H^+ + 6e^- \rightleftharpoons 2NH_4^+$	$+0.28$	$+4.68$
$NO_2^- + 8H^+ + 6e^- \rightleftharpoons NH_4^+ + 2H_2O$	$+0.89$	$+15.0$
$2NO_3^- + 12H^+ + 10e^- \rightleftharpoons N_{2(g)} + 6H_2O$	$+1.24$	$+21.0$
$O_{3(g)} + 2H^+ + 2e^- \rightleftharpoons O_{2(g)} + H_2O$	$+2.07$	$+35.0$
$O_{2(g)} + 4H^+ + 4e^- \rightleftharpoons 2H_2O$	$+1.23$	$+20.8$
$O_{2(aq)} + 4H^+ + 4e^- \rightleftharpoons 2H_2O$	$+1.27$	$+21.5$
$SO_4^{2-} + 2H^+ + 2e^- \rightleftharpoons SO_3^{2-} + H_2O$	-0.04	-0.68
$S_4O_6^{2-} + 2e^- \rightleftharpoons 2S_2O_3^{2-}$	$+0.18$	$+3.0$
$S_{(s)} + 2H^+ + 2e^- \rightleftharpoons H_2S_{(g)}$	$+0.17$	$+2.9$
$SO_4^{2-} + 8H^+ + 6e^- \rightleftharpoons S_{(s)} + 4H_2O$	$+0.35$	$+6.0$
$SO_4^{2-} + 10H^+ + 8e^- \rightleftharpoons H_2S_{(g)} + 4H_2O$	$+0.34$	$+5.75$
$SO_4^{2-} + 9H^+ + 8e^- \rightleftharpoons HS^- + 4H_2O$	$+0.24$	$+4.13$
$2HOCl + 2H^+ + 2e^- \rightleftharpoons Cl_{2(aq)} + 2H_2O$	$+1.60$	$+27.0$
$Cl_{2(g)} + 2e^- \rightleftharpoons 2Cl^-$	$+1.36$	$+23.0$
$Cl_{2(aq)} + 2e^- \rightleftharpoons 2Cl^-$	$+1.39$	$+23.5$
$2HOBr + 2H^+ + 2e^- \rightleftharpoons Br_{2(\ell)} + 2H_2O$	$+1.59$	$+26.9$

Table 7-1, continued

Reaction	$E°$ Volt	$p\epsilon° \left(= \frac{1}{n} \log K \right)$
$Br_2 + 2e^- \rightleftharpoons 2Br^-$	+1.09	+18.4
$2HOI + 2H^+ + 2e^- \rightleftharpoons I_{2(s)} + 2H_2O$	+1.45	+24.5
$I_{2(aq)} + 2e^- \rightleftharpoons 2I^-$	+0.62	+10.48
$I_3^- + 2e^- \rightleftharpoons 3I^-$	+0.54	+9.12
$ClO_2 + e^- \rightleftharpoons ClO_2^-$	+1.15	+19.44
$CO_{2(g)} + 8H^+ + 8e^- \rightleftharpoons CH_{4(g)} + 2H_2O$	+0.17	+2.87
$6CO_{2(g)} + 24H^+ + 24e^- \rightleftharpoons C_6H_{12}O_6(glucose) + 6H_2O$	-0.01	-0.20
$CO_{2(g)} + H^+ + 2e^- \rightleftharpoons HCOO^-(formate)$	-0.31	-5.23

Source. L. G. Sillen and A. E. Martell, *Stability Constants of Metal Ion Complexes.*
The Chemical Society, London, Special Publication No. 16, 1964.
W. Stumm and J. J. Morgan, *Aquatic Chemistry,* Wiley-Interscience, New York,
1970.

AUTHOR INDEX

SUBJECT INDEX